Macromolecular Symposia 212

Electronic Phenomena in Organic Solids

Prague, Czech Republic
July 14–18, 2002

Symposium Editor:
J. Kahovec, Prague, Czech Republic

pp. 1–580 · April 2004
ISBN 3-527-31045-2

Macromolecular Symposia publishes lectures given at international symposia and is issued irregularly, with normally 14 volumes published per year. For each symposium volume, an Editor is appointed. The articles are peer-reviewed. The journal is produced by photo-offset lithography directly from the authors' typescripts.

Further information for authors can be found at http://www.ms-journal.de

Suggestions or proposals for conferences or symposia to be covered in this series should also be sent to the Editorial office (E-mail: macro-symp@wiley-vch.de).

Macromolecular Symposia:
Annual subscription rates 2004
Macromolecular Full Package: including Macromolecular Chemistry & Physics (18 issues), Macromolecular Rapid Communications (24), Macromolecular Bioscience (12), Macromolecular Theory & Simulations (9), Macromolecular Materials and Engineering (12), Macromolecular Symposia (14):

Europe	Euro	6.424 / 7.067
Switzerland	Sfr	11.534 / 12.688
All other areas	US$	7.948 / 8.743

print only **or** electronic only / print **and** electronic

Postage and handling charges included. All Wiley-VCH prices are exclusive of VAT. Prices are subject to change.

Single issues and back copies are available. Please ask for details at: service@wiley-vch.de

Orders may be placed through your bookseller or directly at the publishers: WILEY-VCH Verlag GmbH & Co. KGaA, P. O. Box 10 11 61, 69451 Weinheim, Germany, Tel. +49 (0) 62 01/6 06-400, Fax +49 (0) 62 01/60 61 84. E-mail: service@wiley-vch.de

For USA and Canada: Macromolecular Symposia (ISSN 1022-1360) is published with 14 volumes per year by WILEY-VCH Verlag GmbH & Co. KGaA, Boschstr. 12, 69451 Weinheim, Germany. Air freight and mailing in the USA by Publications Expediting Inc., 200 Meacham Ave., Elmont, NY 11003, USA. Application to mail at Periodicals Postage rate is pending at Jamaica, NY 11431, USA. POSTMASTER please send address changes to: Macromolecular Symposia, c/o Wiley-VCH, III River Street, Hoboken, NJ 07030, USA.

Macromolecular Symposia

Articles published on the web will appear several weeks before the print edition. They are available through:

www.ms-journal.de

www.interscience.wiley.com

Electronic Phenomena in Organic Solids
9th International Conference Electrical and Related Properties of Organic Solids,
Prague (Czech Republic), 2002

Author Index

Preface

The 21st Discussion Conference "Electrical and Related Properties of Polymers and Other Organic Solids" was the 62nd meeting in the series of the Prague Meetings on Macromolecules. The meeting was organized under the auspices of the International Union of Pure and Applied Chemistry (IUPAC) in the Institute of Macromolecular Chemistry, Academy of Sciences of the Czech Republic, from July 14–18, 2002. Simultaneously, this meeting was the 9th International Conference on Electrical and Related Properties of Organic Solids (ERPOS).

The first ERPOS conference was held in 1974, in Karpacz (Poland), as the Summer School on Electrical Properties of Organic Solids. The success of this first meeting gave rise to regular ERPOS conferences which, over 28 years, have grown into a well established series. The conferences ERPOS-1 to ERPOS-5 were organized by physical chemists from the Institute of Organic and Physical Chemistry of the Technical University of Wroclaw.

At that time, several groups at universities and institutes of the Academy of Sciences of the former Eastern block were active in studying the chemistry and physical properties of molecular materials. Personal contact of scientists with their colleagues from abroad, however, was quite difficult due to economic and political reasons. Presentation of the results on an international forum was thus often impossible. Therefore, the idea of bringing together scientists from both parts of the divided world and creating a meeting point seemed timely.

Italian scientists from the Istituto di Fotochimica e Radiazioni d'Alta Energia, Bologna, organized the ERPOS-6 conference in beautiful Capri. The 7th conference returned to Poland, being again organized by the Institute of Physical and Theoretical Chemistry of the Technical University of Wroclaw. During that meeting, in view of political changes in Europe, it was decided to vary the venues of the ERPOS conferences and Prague (Czech Republic) was chosen as the next venue. In the course of the years, ERPOS conferences have gathered scientists from all over the world actively engaged in the study of various properties (electrical, spectroscopic, optical, structural and many others) of organic materials giving them a unique opportunity to exchange ideas with researchers of different backgrounds and interest. A high scientific level of the ERPOS conferences was possible due to active participation of world-recognized scientists who kindly agreed to serve in international advisory committees, delivered lectures reporting on the state of the art in their fields, but also due to scientists bringing "hot" subjects, and many young scientists and students attending these meetings with enthusiasm. No one is able to quantity how many fruitful cross-disciplinary collaborations have been initiated and how many scientific papers appeared as a result of these meetings.

The Prague meeting in 2002 was organized at a fascinating time characterized by the strong input of organic materials, both low-molecular-weight and polymers, into the application sphere. The Nobel Prize winners, Professors Alan J. Heeger,

Alan MacDiarmid and Hideki Shirakawa characterized the time in their Nobel Foundation lecture as "the era of the fourth generation of polymeric materials".

The Prague meeting aimed to provide a forum for scientists specializing in electronic properties of organic materials. The topics of the conference included various aspects of molecular materials, liquid crystals, organized structures of macromolecules and supramolecules, properties of nanostructures and materials such as charge carrier generation, transport, trapping and recombination, photoelectrical phenomena, electroluminescence, molecular electronics, nonlinear optical phenomena, photorefractivity, optical switching, data recording and storage. A total of 126 participants from 24 countries contributed to the scientific program of the conference. There were 12 main lectures, 25 special lectures and 93 poster presentations. An interesting panel discussion "From Molecular Crystals to Polymers and Single Molecules" was moderated by Professors R. W. Munn, J. Sworakowski and R. M. Metzger. The discussion was opened by the latter with his lecture "Unimolecular Rectifiers".

All the conference contributions and accompanying discussions were very helpful in getting a better understanding of electrical and optical properties of organic solids, and in particular charge carrier generation, transport and trapping, non-linear optics, photorefractivity, transistors, photodetectors, light-emitting diodes, rectifiers, and molecular devices. We believe that the papers presented as main and special lectures and posters, collected in this volume, will provide the same benefit to the readers.

The participants created not only an excellent professional forum but also a very agreeable company. We wish to express our gratitude to all participants and sponsors for supporting the meeting, to the organizing committee for their very good work and to the contributors for their carefully prepared papers.

S. Nešpůrek,
Conference Chairman

J. Pfleger,
Conference Co-chairman

J. Kahovec,
PMM Editor

Polarization Energies, Transport Gap and Charge Transfer States of Organic Molecular Crystals

Zoltán G. Soos,* Eugene V. Tsiper

Department of Chemistry, Princeton University, Princeton N.J. 08540, USA

Summary: A self-consistent calculation of electronic polarization in organic molecular crystals and thin films is presented in terms of charge redistribution in nonoverlapping molecules in a lattice. The polarization energies P_+ and P_- of a molecular cation and anion are found for anthracene and perelynetetracarboxylic dianhydride (PTCDA), together with binding energies of ion pairs and transport gaps of PTCDA films on metallic substrates. The 500 meV variation of $P_+ + P_-$ with film thickness agrees with experiment, as do calculated dielectric tensors. Comparisons are made to submolecular calculations in crystals.

Keywords: charge-transfer states; electronic polarization; organic molecular crystals; thin film devices; transport states

Electronic polarization and transport gap

Organic molecular crystals feature weak intermolecular forces, van der Waals contacts, and electronic excitations that correlate with gas-phase transitions. They are typically insulators with dielectric constant $\kappa \sim 3$. Since a charge in a cavity of radius $a \sim 5$Å in a dielectric medium has energy $e^2(1 - 1/\kappa)/2a \sim 1$ eV, electronic polarization has a central role for ionic states. Gutmann and Lyons[1] considered long ago a molecular ion or two embedded in a crystal lattice. Subsequent treatments are discussed by Pope and Swenberg[2] and by Silinsh and Capek,[3] who showed that electronic polarization substantially exceeds lattice (polaronic) relaxation in acenes. As sketched in Fig. 1, the ionization potential of a molecular crystal is reduced from the gas-phase value by the polarization energy, P_+, of a molecular cation, while the electron affinity is increased by P_-, the polarization energy of a molecular anion. The energy needed to create a well-separated pair of ions in the crystal is the transport gap,

$$E_t = I - A - P \tag{1}$$

DOI: 10.1002/masy.200450801

where $P = P_+ + P_-$ is primarily electronic polarization. The corresponding energy for creating an ion pair at finite separation $\mathbf{r}$ is associated with charge-transfer (CT) states,

$$E_{CT}(\vec{r}) = E_t - V(\vec{r}) \tag{2}$$

Here $\mathbf{r}$ is a lattice vector and (2) defines V(**r**), which is sometimes approximated by an effective dielectric constant.[4] Both expressions assume zero overlap between molecules, while recent studies of mixed Frenkel and CT excitons invoke finite overlap between neighbors.[5] Our discussion below is restricted to zero overlap and to electronic polarization in a fixed lattice.

Recent advances in preparation and characterization of ordered thin films have made possible organic electronic devices, such as light-emitting diodes and thin-film transistors, whose operation relies on charge transport.[6,7] Thin films mitigate the low mobility of organic molecular crystals or conjugated polymers. Although E_t or E_{CT} traditionally refers to bulk (infinite) crystals,[1-3] the transport gap of thin films requires the ionization potential and electron affinity at surfaces and photoelectron spectra yield surface properties. We present below the first evaluation of P for crystalline thin films on metallic substrates.[8] Electronic polarization directly affects E_t and thus the matching of energy levels for efficient charge injection. Moreover, in sharp contrast with inorganic semiconductors, the transport gap in several prototypical molecules used in organic devices exceeds[9] the optical gap, E_{opt} in Fig. 1, by about an eV. This reflects small overlap or organic narrow bands in organic crystals and is closely related to electronic polarization.

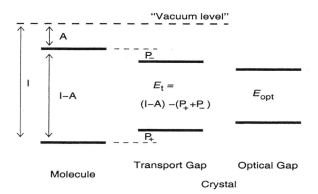

Fig. 1.Schematic energy diagram showing (a) the gas-phase adiabatic ionization potential and electron affinity of a molecule, (b) the transport levels of relaxed molecular ions with polarizations P_+ and P_- in the crystal and (c) the optical gap of the neutral molecule.

Molecular exciton theory describes the optical spectra and excitations of organic crystals. Intermolecular interactions are treated as perturbations to gas-phase (isolated) molecules that do not overlap. First-order corrections to excitation energies typically suffice. Even in this approximation, however, electronic polarization is more challenging because, by definition, first-order corrections to wave functions are required. Zero overlap nevertheless provides the major simplification of purely electrostatic interactions between molecules in the crystal lattice.[10] Such interactions redistribute the gas-phase charge density $\rho_G(\mathbf{r})$ and molecules can still be considered separately, albeit in a complicated crystalline potential.

Munn,[11] Silinsh[3] and coworkers have treated electronic polarization in infinite crystals, primarily the acenes, using the electrostatics of submolecules. The polarizability tensor α of an isolated molecule is taken from theory or experiment and partitioned equally among the rings or heavy atoms of a conjugated molecule. An ion in the crystal induces dipoles at submolecules and the resulting electrostatic problem is solved self consistently by a Fourier transform technique. As well recognized, the choice of submolecules is chemically motivated but arbitrary. Submolecules deal exclusively with induced dipoles and relate polarization to the molecular α. Instead, we incorporate α and induced atomic dipoles as corrections to charge redistribution.[10]

Self-consistent treatment of charge redistribution

Zero overlap reduces the problem to electrostatic interactions between molecules whose charge density $\rho(\mathbf{r})$ can be viewed as confined to volumes that, in one-component crystals, are given by the volume per molecule. The electrostatic energy is a functional of $\rho(\mathbf{r})$ and electronic polarization is the change, $\rho(\mathbf{r}) - \rho_G(\mathbf{r})$, from gas phase to crystal. The problem is formally similar to pairs of molecules[12] or changes in atomic charge densities in a molecule.[13] Direct quantum chemical treatments[12,13] do not require zero overlap but are restricted to small systems and contain other approximations. We are examining such studies and related work on atomic multipole expansions[14,15] as corrections to approximating $\rho_G(\mathbf{r})$ by discrete atomic charges. In addition to zero overlap, we take[10]

$$\rho_G(\vec{r}) \approx \rho^{(0)}(\vec{r}) \equiv \sum_{a,j} \rho_i^{a(0)} \delta(\vec{r} - \vec{r}_i^{\,a}) \tag{3}$$

Here $\mathbf{r}_i^a$ is the position of atom i in molecule a in the crystal lattice and $\rho_i^{a(0)}$ is its gas-phase atomic charge. To avoid the complicated sums-over-states encountered in perturbation theory, we consider solid-state or semiempirical models in which the electrostatic potential φ_i^a is a site energy ε_i^a. Site energies in Hückel theory represent heteroatoms and yield inductive effects.[16] As a practical alternative, we include all valence electrons and find the ground state energy E_0 and $\rho^{(0)}(\mathbf{r})$ of an isolated molecule or molecular ion using INDO/S.[17] We then construct orthogonal molecular orbitals to obtain Löwdin charges $\rho_i^{a(0)}$ in (3). Finally, we evaluate the atom-atom polarizability tensor Π and the polarizability α^C due to charge redistribution,[10]

$$
\begin{aligned}
\Pi_{ij} &\equiv \left(\partial^2 E_0 / \partial\varphi_i \partial\varphi_j\right)_0 \\
\alpha^C &\equiv \sum_{ij} \Pi_{ij} \vec{r}_i \vec{r}_j
\end{aligned}
\tag{4}
$$

Here i,j refer to atoms of one molecule or ion, the partial derivative is evaluated at $\varphi_i = \varphi_j = 0$ and dyadic notation is used for the tensor α^C.

We now have a discrete approximation for electronic polarization. The crystal structure and Löwdin charges $\rho_i^{a(0)}$ define an electrostatic potential $\varphi_i^{a(0)} = \varphi^{(0)}(\mathbf{r}_i^a)$ at each atom due to all other molecules. These charges are sources for molecules with inversion symmetry. Since α^C is in general different from α, we introduce atomic polarizabilities $\tilde{\alpha} = \alpha - \alpha^C = \sum_i \tilde{\alpha}_i$ that generate induced dipoles in the electric field $-\nabla\varphi_i^a$; the partitioning into $\tilde{\alpha}_i$ is based on the number of valence electrons at atom i. Self-consistent atomic charges and induced dipoles lead to a system of four linear equations per atom,[10]

$$
\begin{aligned}
\rho_i^a &= \rho_i^{a(0)} - \sum_j \Pi_{ij} \varphi_j^a \\
\vec{\mu}_i^a &= -\tilde{\alpha}_i \cdot \vec{\nabla}\varphi_i^a
\end{aligned}
\tag{5}
$$

These equations can be solved iteratively for systems of 10^5 atoms or more. Charges redistribute within molecules in the potential generated by atomic charges and induced dipoles, while dipoles are induced by the crystal's electric fields.

The polarization energy of the crystal is extensive and depends on the specified location of ions in the lattice. The general expression in terms of self-consistent potentials and gas-phase charges is[10]

$$
E_{\text{tot}} = \frac{1}{2} \sum_a \sum_i \rho_i^{a(0)} \varphi_i^a
\tag{6}
$$

The translational symmetry of a lattice of neutral molecules yields E_{tot} by Madelung techniques. Alternant hydrocarbons such as acenes have small atomic charges and negligible polarization, < 5 meV per molecule. The stabilization is 330 meV per molecule for perylenetetracarboxylic dianhydride (PTCDA) due mainly to atomic charges of CO groups. Charge redistribution in the neutral lattice is significant for molecules containing heteroatoms and differs qualitatively from submolecules, where E_{tot} vanishes in the absence of charges.

In the presence of an ion, the linear equations (5) are solved[10] relative to the self-consistent charges of the neutral lattice in spheres of radius $R \sim 100$ Å centered on the ion and containing up to $M \sim 2000$ molecules. The crystal structure relates R and M. Thus P_+ or P_- in Fig. 1 for a single ion is the difference between two extensive quantities, a lattice with a molecular ion and a neutral lattice. Convergence goes as $1/R$, or $M^{1/3}$, for a single charge and as $1/R^3$, or M^1, for a CT state. As seen in Fig. 2, $P = P_+ + P_-$ converges properly for anthracene and PTCDA crystals; "charges only" refers to $\tilde{\alpha} = 0$ in (5) while the other lines have α computed using B3LYP density functional theory[18] with a large basis (6-311++G(d,p)). The slopes in Fig. 2 are related[10] to the crystal's dielectric constants,[19] which can be found independently using (5) for a crystal in a uniform electric field.[20] The slopes agree within 3% in all cases. We obtain P = 1.82 eV for PTCDA and 2.20 eV for anthracene.

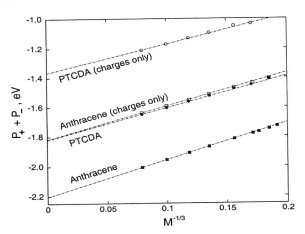

Fig. 2. Convergence of $P = P_+ + P_-$ for anthracene and PTCDA with $M^{1/3}$, the number of molecules in a sphere of radius R. Straight lines are linear fits and „charges only" means $\tilde{\alpha} = 0$.

The binding energy $V(\mathbf{r})$ in (2) between a molecular cation and anion is anisotropic in general and contains both a direct Coulomb interaction and electronic polarization.[4] We center a molecular cation at the origin, place the anion at the crystallographic position $\mathbf{r}$ = (na,mb,lc), and solve Eq. (5) in spheres that contain both ions.[10] The CT energies of anthracene and PTCDA in Fig. 3 are for zero overlap and self-consistent charges and induced dipoles in (5). $V(\mathbf{r})$ converges rapidly as 1/M. We have a practical approach to electronic polarization energies of molecular ions or ion pairs in infinite crystals using discrete atomic charges (3) and induced atomic dipoles based on the best available molecular polarizability α. We avoid partitioning into submolecules. Our numerical results[10] agree best with the largest number of submolecules, 14 at carbon atoms in anthracene[21] and 11 at the centers of rings and CO bonds in PTCDA.[22] Self-consistent treatment of charge redistribution rests on the well-defined limit of zero overlap and the atom-atom polarizability tensor Π in (4). The approximation of $\rho_G(\mathbf{r})$ by $\rho^{(0)}(\mathbf{r})$ is improved below to first order. Nearest-neighbor mixing of Frenkel and CT excitons raise separate spectroscopic issues.[5] Finite overlap is essential for charge transport, for example as hopping of a charge and its polarization cloud, and can be included as a perturbation.

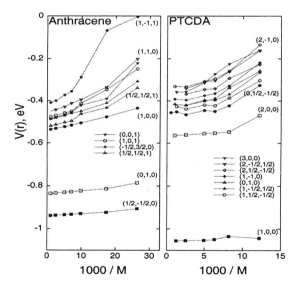

Fig. 3. Interaction energy V(**r**) in Eq. (2) for ion pairs at the origin and **r** = (na,mb,lc) in spheres of M molecules.

Electronic polarization in crystalline thin films

As noted above, organic electronic devices contain thin films on metallic substrates and photoelectron spectra (UPS) probe surfaces.[23] Inverse photoelectron spectra (IPES) is a recent tool for measuring electron affinities or P_-. Hence electronic polarization in thin films has direct implications for transport states. The structure of organic thin films is a separate topic, and multiple phases are possible. PTCDA films[6] are close to having a (102) plane at the organic-metal interface. Given the film's structure, a plane of the bulk crystal can be used to define an infinite slab of $N = 1, 2,...$ layers. This is an excellent approximation in view of the weak distance dependence of electrostatic interactions.

The simplest approximation for an inert metal is a constant-potential surface, as sketched in Fig. 4, which introduces a parameter h for the separation of the first layer and generates N layers of images charges.[8] The translational symmetry of the slab yields atomic charges and induced dipoles by solving Eq. (5) for neutral molecules. Instead of spheres, we compute P_+ and P_- of molecular ions using $2N$-layer thick pillboxes of radius R. We again solve Eq. (5) self-consistently relative to the neutral slab for M molecules and M images, now in a pillbox, and extrapolate as $1/R$ for the infinite slab.[8] UPS and IPES probe transport levels and P at the outermost layer in Fig. 4, while charge injection from the metal is related to P at the interface and P can be computed in any layer. We set $\phi = 0$ since P is independent of the magnitude of ϕ.

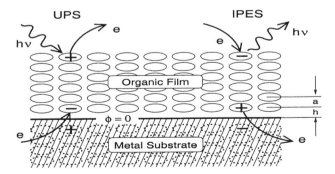

Fig. 4. Idealized model for electronic polarization in crystalline thin films on a metallic substrate at separation h. The N layers of the film appear as image charges. UPS and IPES generate a cation and anion at the surface, while charge injection generates ions in the interface layer.

The calculated $P = P_+ + P_-$ in Fig. 5 are for PTCDA at the surface of $N = 1, 2, 3, 5$ and 10 layer films at the indicated $h = xa$, where $a = 3.214$ Å is the crystal spacing between (102) planes. Since the metal-organic interface is N layers away, P cannot depend on h as $N \to \infty$ and all curves converge to $P_{surf} = 1.41$ eV. Electronic polarization at a (102) surface of an infinite PTCDA crystal is 410 meV less than in the bulk. For a free-standing film, we obtain 0.64 eV in Fig. 5 on extrapolating to infinite h.

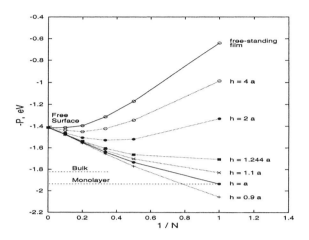

Figure 5. Calculated P at the outer surface of PTCDA films of $N = 1,2,3,5$ and 10 layers at several values of $h = xa$, where $a = 3.214$ Å is the spacing between (102) planes. The polarization at the free surface, a free-standing (102) monolayer and the bulk are indicated.

While h is not known, van der Waals radii suggest $h \sim a$. The monolayer ($N = 1$) then has $P_{mon} = 1.93$ eV, which exceeds the bulk value. Image charges represent the greater polarizability of metals compared to the crystal, while the vacuum at the outer surface in not polarizable. This is the physical basis for decreasing P at the surface with N at small h and increasing P with N for large h. We find $P_{mono} - P_{surf} \sim 500$ meV for PTCDA and striking agreement with experiment.[8] As an excellent film former as well as a good hole conductor, PTCDA is a prototypical molecule for organic devices and has been studied on Au, Ag and other substrates.[24] Photoelectron spectra on Ag show a 450 meV increase of E_t between a monolayer and a 64 Å ($N \sim 20$) film.[8] The corresponding increase of 500 meV on Au is based on scanning tunneling spectra[8] of the monolayer and photoelectron spectra of 50-100 Å films.[9] As expected from their Fermi energies,

Au and Ag form opposite surface dipoles with PTCDA,[24] but surface dipoles cancel in P.

Figure 6 shows the variation of E_t across a 10-layer PTCDA film, with $h = a$ and $n = 1$ next to the metal.[8] Surface effects extend several layers into the sample, as is natural for electrostatic interactions, and greater polarizability at $n = 1$ is also expected. These are direct applications for calculations of electronic polarization in thin films. Transport-level shifts of ~200 meV at the interface are large compared to kT. They are neglected in current modeling of charge injection, as is the long-range nature of polarization.

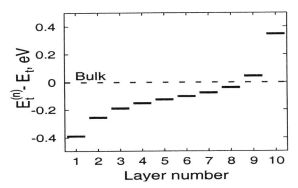

Fig. 6. Variation of the transport gap, $E_t^{(n)} - E_t$, across a 10-layer PTCDA film with $n = 1$ next to the metal and $n = 10$ at the outer surface. The dashed line is the bulk value, $P = 1.84$ eV.

UPS data for solid-state ionization potentials, $I - P_+$ in Fig. 1, are fairly common;[2,3] their detailed interpretation still has open questions. The frequent assumption of $P_+ = P_-$ is more convenient than correct. Molecular quadrupoles lead to divergent, shape-dependent contributions in crystals for P_+ or P_- that cancel in P. Measurements of P_+ in UPS and P_- in IPES are typically referenced to the substrate's Fermi energy. In terms of submolecules, Munn has evaluated corrections $W_{q\text{-}Q}$ due to fixed molecular quadrupoles.[25,11] They increase P_+ for anthracene by 0.18 eV and decrease P_- by the same amount. At least qualitatively, observed[26] shifts of $I - P_+$ in films, monolayers or even lower coverage can be understood in terms of electronic polarization.[8]

Discussion and Conclusions

We have presented a new, self-consistent calculation of electronic polarization in organic molecular crystals. The principal approximations are no overlap between molecules, which makes it possible to limit quantum theory to individual molecules, and semiempirical theory for electrostatic potentials, which yields the atom-atom polarizability tensor Π in (4) that governs charge redistribution. The discrete problem of self-consistent atomic charges and induced dipoles leads to the linear equations (5) whose solutions are discussed above. Experimental comparisons to date have been in terms of E_t for PTCDA films and electronic parts of the dielectric tensor[20] of crystalline anthracene and PTCDA. Theoretical comparisons are to submolecular results for P_+, P_- and $V(\mathbf{r})$ of CT states.[10] We anticipate wider comparisons, especially to thin films. The procedure is general and INDO/S is readily applicable to conjugated molecules used in organic devices. Molecular polarizabilities α are more demanding, but if necessary, such corrections can be found in a smaller basis. The crystal structure and α are inputs, as in the submolecular approach, but we explicitly compute atomic charges and charge redistribution. Induced atomic dipoles due to $\tilde{\alpha} = \alpha - \alpha^C$ are 10-20% corrections that are particularly important for polarization normal to the molecular plane.

The approximation (3) of discrete atomic charges $\rho^{(0)}(\mathbf{r})$ reduces the polarization problem to four linear equations per atom. Although atomic charges are not observables, the widespread usage of Mulliken, Löwdin and other charges speak to their appeal. But the potential $\varphi(\mathbf{r}')$ generated by any localized charge distribution $\rho_G(\mathbf{r})$ *is* unique at any $\mathbf{r}'$ outside the charge distribution. There is interest[14,15] in expanding $\varphi(\mathbf{r}')$ for molecules in terms of atomic multipoles (charges, dipoles, and perhaps quadrupoles) chosen by least-squares fitting procedures. In the same spirit, we associate atomic multipoles with $\Delta\rho(\mathbf{r}) = \rho_G(\mathbf{r}) - \rho^{(0)}(\mathbf{r})$, the difference between the actual charge density of an isolated molecule or ion and the INDO/S atomic charges in (3). For zero overlap, the potential $\Delta\Phi_a(\mathbf{r}_i^a)$ at atom i of molecule a due to $\Delta\rho(\mathbf{r})$ of all other molecules can readily be computed;[27] convergence is rapid because there are no monopole ($1/r$) contributions by construction.

The potential $\Delta\Phi_a(\mathbf{r})$ represents fixed sources that are neglected in (3) and hence in the linear equations (5). The polarization energy E_{tot} in (6) can also be expresses as self-consistent atomic charges ρ_i^a and induced dipoles μ_i^a coupled to $\varphi_i^{a(0)}$ and $\nabla\varphi_i^{a(0)}$, respectively, the potential due to

$\rho^{(0)}(\mathbf{r})$. The extra potential due to higher atomic multipoles increases E_{tot} by

$$E_{tot}^{(1)} = \sum_a \sum_i [\rho_i^a \Delta\Phi(r_i^a) + \mu_i^a \cdot \nabla\Delta\Phi(r_i^a)] \tag{7}$$

The correction is first order, since ρ_i^a and μ_i^a are not recalculated in the new potential. For acenes, PTCDA and other conjugated molecules with inversion symmetry, the π-system's quadrupole makes the largest contribution to $\Delta\Phi_a(\mathbf{r})$. The first term of (7) then gives the charge-quadrupole contribution to P_+ or P_-. This is qW_{q-Q} term[3,11] for fixed quadrupoles and submolecules. With $\Delta\Phi_a(\mathbf{r})$ based on a B3LYP-6311++G(d,p) calculation[18] of $\rho_G(\mathbf{r})$, we find -0.23 eV for anthracene, compared to $W_{q-Q} = -0.18$ eV for three fractional charges at ring centers .[27] The PTCDA result is larger and implies contributions beyond the perylene π-system. First-order corrections in $\rho_G(\mathbf{r}) - \rho^{(0)}(\mathbf{r})$ can readily be added to self-consistent electronic polarizabilities.

In PTCDA films, the molecular plane is nearly parallel to the substrate. Other film formers such as α-sexithiophene or pentacene, by contrast, have molecular planes nearly perpendicular to the metal. Image charges in Fig. 4 then generate fields along the long axis, the direction of greatest polarizability. Care must be taken with iterative solution Eq. (5) for pentacene films on metals,[28] whose transport gaps as a function of thickness are quite different than PTCDA. Even in terms of ideal inert surfaces, electronic polarization of thin films remains to be explored and applied to issues such as matching transport states for charge injection. Electronic polarization is a major effect in organic molecular crystals that can now be evaluated in a well-defined approximation. We anticipate many applications to thin films in organic electronic devices.

It is a pleasure to thank A. Kahn, R.A. Pascal, Jr., J.M. Sin, and M. Hoffmann for stimulating discussions and access to unpublished results; P. Petelenz and R.W. Munn for correspondence about submolecules and distributed polarizabilities; and the National Science Foundation for partial support through the MRSEC program under DMR-9400632.

12

[1] F. Guttmann, L.El Lyons, „*Organic Semiconductors*" , Wiley, New York, 1967. Chap. 6.

[2] M. Pope and C.E. Swenberg, „*Electronic Processes in Organic Crystals*", Clarendon, Oxford, 1982.

[3] E.A. Silinsh, V. Cápek, „*Organic Molecular Crystals*", AIP Press, New York, 1994.

[4] P.J. Bounds, W. Siebrand, Chem. Phys. Lett. 1980, *75*, 414.

[5] M.H. Hennessy, R.A. Pascal, Jr., Z.G. Soos, Mol. Cryst. Liqu. Cryst. **2001**, *335*, 41; M. Hoffmann, Z.G. Soos, Phys. Rev. B **2002**, in press.

[6] S.R. Forrest, Chem. Rev. **1997**, *97*, 1783.

[7] G. Horowitz, Adv. Mat. **1998**, *10*, 365; H.E. Katz, J. Mater. Chem. **1997**, *7*, 369.

[8] E.V. Tsiper, Z.G. Soos, W. Gao, A. Kahn, Chem. Phys. Lett. **2002**, (in press).

[9] I.G. Hill, A. Kahn, Z.G. Soos, R.A. Pascal, Jr., Chem. Phys. Lett. **2000**, *327*, 181.

[10] E.V. Tsiper and Z.G. Soos, Phys. Rev. B **2001**, *64*, 195124.

[11] J.W. Rohleder, R.W. Munn, Magnetism and Optics of Molecular Crystals, Wiley, New York, 1992

[12] A. Morita, S. Kato, J. Am. Chem. Soc. **1997**, *119*, 4021.

[13] A.J. Stone, Mol. Phys. **1985**, *56*, 1065.

[14] A.J. Stone, S.L. Price, J. Phys. Chem. **1988**, *92*, 3325.

[15] C. Chipot, J. Anqyan, G. Ferenczy, H. Scheraga, J. Phys. Chem. **1993**, *97*, 6628; M.M. Francl and L.E. Chirlian, Rev. Comp. Chem. **2000**, *14*, 1.

[16] L. Salem, The Molecular Orbital Theory of Conjugated Systems, Benjamin, New York, 1966, Ch. 1.

[17] M.C. Zerner, G.H. Loew, R.F. Kirchner, U.T. Mueller-Westerhoff, J. Am. Chem. Soc. **1980**, *102*, 589.

[18] M.J. Frisch et al, GAUSSIAN 98 (Gaussian Inc., Pittsburgh, 1999).

[19] P.J. Bounds, R.W. Munn, Chem. Phys. **1979**, *44*, 103.

[20] Z.G. Soos, E.V. Tsiper, R.A. Pascal, Jr., Chem. Phys. Lett. **2001**, *342*, 652.

[21] H. Reis, M.G. Papadopoulos, P. Calaminici, K. Jug, A.M. Köster, Chem. Phys. **2000**, *261*, 359.

[22] G. Mazur, P. Petelenz, Chem. Phys. Lett. **2000**, *324*, 161.

[23] W.R. Salaneck, K. Seki, A. Kahn, J.J. Pireaux, eds. *Conjugated Polymer and Molecular Interfaces*, Marcel Dekker, New York, 2001.

[24] I.G. Hill, A. Rajagopal, A. Kahn, Y. Hu, Appl. Phys. Lett. **1998**, *73*, 662.

[25] I. Eisenstein, R.W. Munn, Chem. Phys. **1983**, *77*, 43.

[26] I.G. Hill, A.J. Makinen, Z.H. Kafafi, J. Appl. Phys. **2000**, *88*, 889; W.R. Salaneck, Phys. Rev. Lett. **1978**, *40*, 60.

[27] J.M. Sin, E.V. Tsiper, Z.G. Soos, Europhys. Lett. (submitted)

[28] E.V. Tsiper, Z.G. Soos, to be published.

Macromol. Symp. **2004**, *212*, 13-24

Exciton Dissociation in Conjugated Polymers

Vladimir Arkhipov,[1,2] *Heinz Bässler,*[1] *Evgenia Emelyanova,*[3] *Dirk Hertel,*[1] *Vidmantas Gulbinas,*[4] *Lewis Rothberg*[5]

[1]Institute of Physical, Nuclear, and Macromolecular Chemistry, Philipps University of Marburg, Hans-Meerwein-Strasse, 35032 Marburg, Germany
[2]IMEC, Kapeldreef 75, B-3001 Heverlee-Leuven, Belgium
[3]Semiconductor Physics Laboratory, University of Leuven, Celestijnenlaan 200D, B-3001 Heverlee-Leuven, Belgium
[4]Institute of Physics, A. Gostauto 12, LT-2600 Vilnius, Lithuania
[5]Chemistry Department, University of Rochester, Rochester, NY 14627, USA

Summary: In conjugated polymers, a majority of photogenerated charges form metastable geminate pairs (GPs), of which only some fraction can dissociate completely. Both the yield of GP photogeneration and the probability of further dissociation of GPs into free charges depend upon an external electric field. In the present article we discuss several experimental methods to detect the existence of geminate pairs such as delayed field collection of charges, field quenching of fluorescence, and field-assisted photoinduced optical absorption. It is shown that the field dependences of the exciton dissociation into GPs and of the free carrier photogeneration yield are rather similar. This is in contrast with the traditional Onsager theory, which assumes field-independent yield of primary photoionization and disregards the field dependence of the initial separations between carriers in GPs.

Keywords: charge photogeneration; conjugated polymers; geminate pairs; photoluminescence; photophysics

Introduction

It is a generally adopted notion that, in a clean molecular crystal, photogeneration of charge carriers proceeds as described by Onsager's model of radiation-induced conductivity.[1,2] One assumes that a highly excited electronic state autoionizes and, after thermalization of the electron, forms a coulombically bound electron-hole pair at an initial separation r_0. Alternatively, that pair can be generated by a direct charge-transfer (CT) transition at a CT-energy above the S_1 exciton energy. The pair can either fully dissociate in the course of field- and temperature-dependent diffusive (Brownian) motion or recombine to form a singlet whose energy is too low for further thermally assisted dissociation.

In the approaches pertaining to the dissociation of an optical excitation, it has implicitly been assumed that generation of the parent state is an instantaneous, field- and temperature-independent process. Based upon a 3D-continuum treatment, Onsager derived a formula

DOI: 10.1002/masy.200450802

describing the field and temperature dependences of the yield, $\eta(F,T)$, in a factorized form, as[1]

$$\eta(F,T) = \varphi_0 \varphi_{esc}(F,T) \tag{1}$$

where φ_0 is the initial yield of pair formation. The mathematical treatment is well documented in the literature, suffice to mention that

$$\lim_{F,T \to \infty} \eta(F,T) = \varphi_0 \tag{2}$$

i.e. that both field and temperature dependences of the yield must converge to a common asymptotic value. The fact that, in conjugated polymers, the field and temperature dependences of the photogeneration quantum efficiency do not extrapolate to a common asymptotic value, casts doubts on the applicability of the Onsager formalism to those systems. Although one can fit $\varphi(F)$ data by 3D Onsager's theory for several conjugated polymers[3], the significance of such an analysis is questionable because a super-linear field dependence is not at all unique and cannot be considered as firm evidence for a particular mechanism. For instance, intrinsic photogeneration at higher photon energies has been explained successfully in terms of hot exciton dissociation[4,5] whose underlying concept is different from the 1D Onsager approach.

There is compelling evidence that, within the spectral range of the $S_1 \leftarrow S_0$ transition, photogeneration in conjugated polymers must be a two-step process, the first one being the dissociation of geminately bound electron-hole pairs from a precursor exciton and the second one is further escape of charge carriers from the Coulomb potential well. One of the essential questions to be asked is how can one distinguish both processes experimentally and, if so, what are the respective field dependences. Relevant experimental techniques are quenching of photoluminescence by an electric field, delayed collection of optically created charge carriers and time-resolved transient absorption. This question is also related to the lifetime of coulombically bound intermediates between excitons and pairs of coulombically unbound free charges. Metastability can only occur if charges, comprising the geminate pair or charge transfer state, are localized. Photoemission from tetracene crystallites, studied by the Millikan technique[2,6], revealed this phenomenon.[7] The lifetime of such pairs can increase by many orders of magnitude[8] and it is straightforward to conjecture that this increase is due to the inherent disorder.

Photoluminescence quenching by electric fields

If an exciton is dissociated by an electric field, it can no longer contribute to luminescence. Dissociation of an excited singlet state of a conjugated polymer requires field-assisted transfer of one of the constituent charges to a neighboring chain or chain segment. To first-order approximation, this would occur if the gain of electrostatic energy, $eF\Delta x$, were compensated for the energy expense for the charge transfer in zero field. For $\Delta x = 1$ nm and $F = 2 \times 10^6$ V/cm, $eF\Delta x = 0.2$ eV is estimated. If the energy difference between an exciton and a metastable geminate pair was of that order of magnitude, one would expect photoluminescence (PL) quenching to occur in electric fields in excess of 1 MV/cm.

PL quenching experiments were performed with films of poly(phenyl-p-phenylenevinylene) (PPPV) doped into polycarbonate (PC)[9] as well as with an oligomeric model compound, tris(stilbene)amine (TSA) doped into PC and various polystyrene derivatives.[10] TSA was chosen because it contains stilbene substituents resembling the PPV repeat unit and it absorbs in the same spectral region as PPPV does. The samples were prepared in sandwich configuration between ITO and Al contacts as commonly used for electroluminescence studies. The quenching efficiency, or relative fluorescence reduction Q, was experimentally determined as,

$$Q(F) = \frac{I(0) - I(F)}{I(0)} \quad , \tag{3}$$

where $I(0)$ and $I(F)$ are the fluorescence intensities normalized to the intensity of the incident light intensity and measured at zero bias and with a field F applied to the sample, respectively. Within the wavelength region from 490 nm ($h\nu = 2.5$ eV) to 380 nm (3.2 eV), the quenching efficiency is independent of photon energy and increases superlinearly with electric field (Figure 1). A decrease in the quenching efficiency is seen only at the very absorption tail (425 nm, i.e. 2.92 eV for TSA/PC and 480 nm, i.e. 2.58 eV, for PPPV/PC). This indicates that the actual quenching step is not sensitive to the occurrence of any vibrational or electronic excess

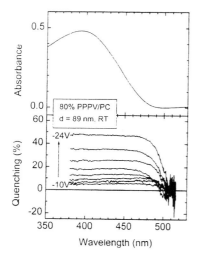

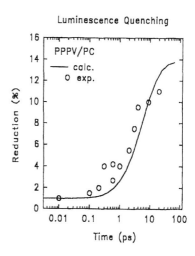

Figure 1. Absorption spectrum (a) and spectral dependence of the fluorescence quenching (b) in a PPPV:PC film. The parameter is the applied (reverse) voltage.

Figure 2. Field-induced reduction of transient PL intensity. Circles are experimental data, the solid line is a fit based on the model of nearest-neighbor dissociating jumps.[14]

energy. From PL studies with sub-ps time resolution[11,12], it is known that vibrational cooling in PPPV occurs within 100 fs. Subsequent electronic relaxation of the vibrationally cold singlet exciton within the manifold of localized states is a dispersive rate process, the fastest steps occurring on a 1 ps time scale. Any spontaneous exciton dissociation occurring during those relaxation processes would not show up in the field quenching studies since only field-induced processes are monitored.

The decrease in relative PL reduction at the absorption tail can be rationalized as a natural consequence of energetic disorder. The energy of localized excited states is subject to a distribution that is often approximated as Gaussian. The number of sites that lie energetically lower than an originally excited chromophore and that are available for energy relaxation decreases with decreasing excitation energy. Hence, the probability of finding a charge-accepting neighbor that is both spatially and energetically accessible drops with decreasing excitation energy. For this reason, exciton dissociation is less efficient for excitation into the low-energy tail of the DOS distribution.

The temporal evolution of field-assisted PL quenching was studied on a sample, composed of 20 % PPPV and 80 % PC.[13] Applying a field of $F = 2 \times 10^6$ V/cm results in a quenching of the luminescence intensity up to 30 % independent of photon energy. The luminescence spectra shift with applied electric field in quantitative accord with the prediction of the second-order Stark effect. Figure 2 shows the time evolution of the luminescence quenching Q.[14] The prompt 1% reduction suggests that only a small fraction of the overall quenching is due to transfer of oscillator strength to charge transfer states with lower radiative yield.

A model of field-assisted off-chain dissociation of optical excitations

In a conjugated polymer consisting of segments of typically 5 nm in length, an optical excitation can dissociate into an on-chain geminate pair if the gain of the electrostatic energy within the segment is sufficient to stabilize the carriers on the opposite ends of a segment and to prevent them from recombination within the Coulomb potential well.[15] The external field also assists in further dissociation of on-chain geminate pairs by stimulating carrier jumps to other segments. Therefore, most of the on-chain geminate pairs must eventually dissociate into free carriers and contribute to the photoconductivity. However, it is known that the observed quantum yield of free carrier photogeneration is normally much less than the field-induced quenching of the photoluminescence, implying that only a small fraction of geminate pairs can subsequently dissociate into free carriers.[3] This argues against on-chain carrier separation as the main mechanism of the non-radiative relaxation of optical excitations in conjugated polymers.

Dissociation of an optical excitation into a metastable coulombically bound $e...h$ pair is energetically feasible if the binding energy of the latter, $E_b^{(GP)}$, is larger than that of the former, E_b. Both these energies are eigenvalues of the corresponding Hamiltonians and they can be represented as sums of the potential, $E_p^{(GP)}$, E_p, and kinetic energies, $E_k^{(GP)}$, E_k, respectively. Since very little is known about relative contribution of those energies to the total binding energy of a geminate pair, in the following treatment we consider the kinetic energy $E_k^{(GP)}$ as a parameter. The dissociation is energetically possible if the distance between carriers in a geminate pair is such that the following inequality is valid,

$$E_b^{(eff)} \leq \frac{e^2}{4\pi\varepsilon_0\varepsilon r} + eFrz \quad,$$

(4)

where r is the distance between carriers in a geminate pair, $z = \cos\vartheta$ with ϑ being the angle between the external field F and the direction of the carrier jump over the distance r, and the effective binding energy of the exciton $E_b^{(eff)}$ is defined as,

$$E_b^{(eff)} = E_b + E_k^{(GP)} \quad . \tag{5}$$

Equation (4) determines the region in space within which the sites are located available for dissociation of excitons into geminate pairs.

Another limiting factor is the rate of exciton dissociation. The rate v of energetically downward tunneling carrier jumps over the distance r is given by,

$$v(r) = v_0 \exp(-2\gamma r) \quad , \tag{6}$$

where γ is the inverse localization radius and v_0 the attempt-to-jump frequency. The probability w that, during its lifetime τ on a given segment, an exciton will dissociate into a geminate pair of radius r is given by the Poisson distribution of probability as,

$$w(r) = 1 - \exp[-v_0\tau \exp(-2\gamma r)] \quad . \tag{7}$$

The average number of dissociation sites, $<n_{diss}>$, can be evaluated by integrating the density of neighboring sites, $N_n(r)$, over the dissociation volume. The result reads,

$$\langle n_{diss} \rangle (F) = \int_{V_{diss}} d\mathbf{r} \; w(r) \; N_n(r) \quad . \tag{8}$$

The coordinate dependence of the function $N_n(r)$ accounts for possible correlations between positions of neighboring sites. The probability W that no sites are located within the dissociation volume is also determined by the Poisson distribution as

$$W = \exp\left[-\langle n_{diss} \rangle (F)\right] \quad . \tag{9}$$

The dissociation yield η_{diss} , i.e., the probability to find a dissociation site around at least one of n_v sites, visited by an exciton during its lifetime, is then given by

$$\eta_{\text{diss}} = 1 - \exp\left[-n_v \langle n_{\text{diss}} \rangle (F)\right] \quad . \tag{10}$$

Since the quenched PL intensity is proportional to the dissociation probability η_{diss}, the use of Eq. (3) leads to the following expression for the quenching parameter Q,

$$Q(F) = 1 - \exp\left\{-n_v \left[\langle n_{\text{diss}} \rangle (F) - \langle n_{\text{diss}} \rangle (0)\right]\right\} \quad . \tag{11}$$

Field dependences of the quenching parameter are plotted in Figure 3 parametric in the effective exciton binding energy for a completely disordered system without any correlation between positions of nearest sites. At weak and moderate fields, all these dependences remarkably well feature a square field dependence of Q. At strong fields the functions $Q(F)$ approach unity and saturate.

Photoinduced transient optical absorption

Another way to delineate exciton dissociation is to monitor the evolution of the transient optical absorption due to the generated charge. Upon adding or removing a charge carrier to or from a molecule, a radical anion or cation is generated whose absorption spectrum is red-shifted relative to the absorption of the neutral parent molecule. In conjugated polymers associated optical transitions are typically in the vicinity of 2 eV and 0.6 eV. The former are easily detectable by pump-probe spectroscopy involving a white light continuum generated by the pump laser. Such experiments were performed by Graupner et al.[16] on MeLPPP upon applying an electric field. A transient absorption feature was indeed observed at 1.9 eV where charge carriers are normally identified as polarons although that optical transition gives no information on the relaxation of the polymer chain upon ionization. It has been concluded that (i) charges are generated directly from excitons without any additional intermediate states and (ii) the process occurs on the time scale of 10 ps, i.e. less than the lifetime of singlet excitons, suggesting that they are reactive before thermalization towards the bottom states of the excitonic DOS is complete.

Analogous experiments were performed recently by Gulbinas et al.[17], the only difference being the photon dose applied upon excitation. The experiments were done using 100 fs laser pulses at $h\nu = 3.1$ eV with the dose of 14 μJ/(cm^2×pulse), i.e. almost by a factor of 100 less than that used in Ref. [16]. It turned out that a high intensity of bimolecular exciton

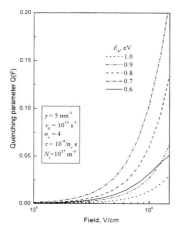
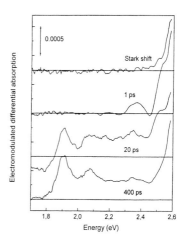

Figure 3. Field dependence of the PL quenching parameter calculated for different values of the exciton binding energy in a material with uncorrelated positions of neighboring localized states.

Figure 4. EDA spectra at different delay times after excitation. The upper curve shows the electric field-induced differential absorption, measured without excitation.

annihilation has limited their lifetime. At an incident intensities of around 14 $\mu J/(cm^2 \times pulse)$, this effect is absent as indicated by the fact that the fluorescence decay is governed by the intrinsic exciton lifetime of 300 ps.

Figure 4 shows the electromodulated differential absorption (EDA) spectra at different delay times after excitation at a field of 2.2×10^6 V/cm. Positive signals correspond to quenching of stimulated emission and to electrostimulated induced absorption by polarons at 1.9 eV and 2.1 eV. The top EDA spectrum, measured without optical excitation, is due to a Stark shift of the excitonic absorption edge. In the following analysis, the Stark shift contribution was subtracted from the spectra measured at positive delays. The EDA spectrum at 1 ps delay displays a sizeable positive contribution in the stimulated emission region but there is no yet signal at energies corresponding to absorption from excitons and polarons. At 20 ps delay, induced absorption with pronounced bands at 1.9 eV and 2.1 eV and a negative signal at the low-energy part are observed. At 400 ps the new absorption bands are even more pronounced, whereas the signal below 1.85 eV has disappeared. The induced absorption has a spectrum similar to that reported in Ref. [16] and is unambiguously assigned to the charged species, i.e.

polarons. Most remarkably, their concentration grows within the entire exciton lifetime.

From the analysis of the transient absorption, the rate at which singlet excitons dissociate into charged species, i.e. the exciton breaking rate γ_b, can be inferred. It is time-dependent and can be fitted by an algebraic law, $\gamma_b(t) \propto (t_0/t)^{0.4}$, with a cut-off value of 2.2×10^{10} s^{-1} at short times. Such time dependences of rate constants are often found in random systems[5] and, in this case, can be explained by a distribution of charge transfer rates in a disordered solid MeLPPP polymer film. The dissociation rate has its maximum at short times, when the nearest-neighbor geminate pair formation dominates, but at longer times jumps to more distant sites may prevail depending on the electric field and the density of states distribution.

In summary, the EDA experiments on MeLPPP showed that (i) the photodissociation of singlet excitons into geminately bound electron-hole pairs is facilitated by external electric field and (ii) at an excess photon energy of 0.4 eV above the $S_1 \leftarrow S_0$ (0 - 0) transition, the dissociation predominantly proceeds via vibrationally relaxed S_1 excitons during their entire lifetime with a rate decreasing as $t^{-0.4}$.

Delayed collection of optically generated charge carriers

A unique way to distinguish between primary exciton dissociation and subsequent dissociation of an intermediate geminate pair is the time-delayed collection field (TDCF) technique.[18] It allows to measure electric-field-induced fluorescence quenching and charge carrier generation yield simultaneously. The π-conjugated polymer under investigation was MeLPPP which has an extraordinary low intrachain disorder. Sandwich-type devices with configuration ITO/SiO(150 nm)/MelPPP/SiO(50 nm)/Al were used. In the TDCF experiment, the photoresponse R of the device at a bias field F_{appl} is measured by applying a collection field F_{coll} with a time delay t_d of about 100 ns after pulsed laser excitation. Simultaneously, the fluorescence emitted from the device and the intensity of the inducing light are recorded. The fluorescence intensity I and photoresponse are measured either without electric field or as functions of the field by varying F_{exc}. Due to the SiO blocking layers, no charge injection occurred, which is proven by the fact that both R and I are independent of the bias polarity. Typical excitation intensities were 5...30 μJ/(cm$^2 \times$pulse).

Fluorescence quenching $Q(F)$ was calculated via Eq. (3) and the relative $R=\Delta V/I_{exc}$, where ΔV is the photoinduced voltage drop across the device and I_{exc} is the incident light intensity. The quantum yield of charge generation η can be calculated via $\eta(F) = q/n$ where n is the number of absorbed photons and q the number of photogenerated charges, $q = \Delta VC/e$ with C being the device capacitance and e the elementary charge.

Upon application of an external electric field during optical excitation, the fluorescence of MeLPPP was drastically reduced (Figure 5). For electric fields in excess of 6×10^5 V/cm, the fluorescence quenching Q rises superlinearly as a function of electric field. The magnitude of Q reaches values of 20 % – 50 % at $F_{exc} = 2\times10^6$ V/cm depending on excitation photon energy. The absolute value of fluorescence quenching compares favorably with experimental data obtained on PPV derivatives[19] and with the results of previous measurements on MeLPPP.[20,21] One can see from Figure 5 that the excess energy of absorbed photons relative to the energy of a relaxed singlet exciton facilitates dissociation of excitons into geminate pairs.

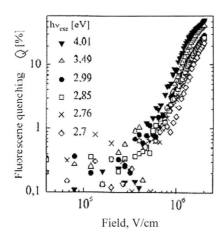

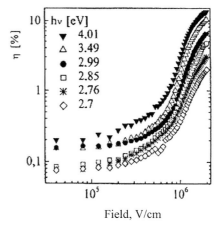

Figure 5. Field dependence of the PL quenching in MeLPPP parametric in the photon energy of photoexcitation.

Figure 6. Field dependence of the charge carrier photogeneration yield in MeLPPP for different photon energies of photoexcitation.

In Figure 6 the quantum yield of charge carrier photogeneration is shown as a function of applied electric field for different excitation energies. At $F_{exc} = 2\times10^6$ V/cm, the yield reaches values from 2 % to 13 % depending on photon energy. The dotted lines illustrate the 3D Onsager field dependences of the yield calculated for $T = 295$ K, $\varepsilon = 3.5$, and for two values of

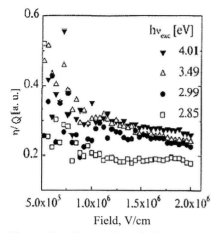

Figure 7. The ratio of the carrier photogeneration yield and the PL quenching parameter as a function of the electric field during photoexcitation.

the initial distance between carriers in geminate pairs, $r_0 = 1.1$ and 1.3 nm. It is apparent from Figure 6 that the experimental field dependence might be reasonably well described by the Onsager model at strong fields. However, this is an accidental agreement. At zero applied field during excitation, the charge collection efficiency saturates at $F_{coll} > 4 \times 10^5$ V/cm. This proves that the superlinear increase in charge generation efficiency at applied fields above 5×10^5 V/cm must reflect the increase in field-assisted generation of geminate pairs from singlet excitations and not the secondary step of dissociation of GPs into free carriers as described by the Onsager theory. This shows that the basic assumption of the Onsager theory of field-independent geminate pair formation is not fulfilled. Intuitively this is obvious. Since it costs energy to overcome the Coulombic potential barrier to generate free carriers from geminate pairs, it is difficult to assume that the compensation of the exciton binding energy is not assisted by an applied electric field. According to the above reasoning, the ratio of η/Q should not reveal a noticeable field dependence that is depicted in Figure 7. Although the data scatter at weak applied fields, η/Q remains practically constant at $F_{exc} > 8 \times 10^5$ V/cm.

Conclusion

The basic message of the article is that in conjugated polymers excitons require additional energy in order to dissociate into charge carriers. This energy can be supplied by a sufficiently strong electric field. It can promote an electron from the excited chain segment to an adjacent chain and form an intermediate charge transfer state, i.e. geminately bound electron-hole pair. The field dependence of the fluorescence quenching is a measure of the primary dissociation rate. Since this initial step is field-sensitive, it is obvious that an Onsager-like description for the photoconduction in conjugated polymers is inappropriate.

Acknowledgments

Financial support by NSF, Deutsche Forschungsgemeinschaft, Stiftung Volkswagenwerk, and the EU-TMR program HPRI-CT is gratefully acknowledged.

[1] L. Onsager, *Phys. Rev.* **1938**, *54*, 554.

[2] H. Pope, C. E. Swenberg, *Electronic Processes in Organic Crystals and Polymers,* 2nd ed., Oxford, Univ. [3] S. Barth, H. Bässler, H. Rost, H. H. Hörhold, *Phys. Rev. B* **1997**, *56*, 3844.

[4] V. I. Arkhipov, E. V. Emelianova, H. Bässler, *Phys. Rev. Lett.* **1999**, *82*, 1321.

[5] V. I. Arkhipov, E. V. Emelianova, S. Barth, H. Bässler, *Phys. Rev. B* **2000**, *61*, 8207.

[6] M. Pope, J. Burgos, J. Giachino, *J. Chem. Phys.* **1965**, *43*, 3367.

[7] M. Pope, C. E. Swenberg, A. Dourandin, *Mol. Cryst. Liq. Cryst.* **2001**, *355*, 77.

[8] J. Mort, M. Morgan, S. Grammatica, J. Noolandi, K. M. Hong, *Phys. Rev. Lett.* **1982**, *48*, 1411.

[9] M. Deussen, M. Scheidler, H. Bässler, *Synth. Met.* **1995**, *73*, 123.

[10] M. Deussen, P. H. Bolivar, G. Wegmann, H. Kurz, H. Bässler, *Chem. Phys.* **1996**, *207*, 147.

[11] R. Kersting, U. Lemmer, R. F. Mahrt, K. Leo, H. Kurz, H. Bässler, E. O. Göbel, *Phys. Rev. Lett.* **1994**, *70*, 3820.

[12] B. Mollay, U. Lemmer, R. Kersting, R. F. Mahrt, H. Kurz, H. F. Kauffmann, H. Bässler, *Phys. Rev. B* **1994**, 50, 10769.

[13] R. Kersting, U. Lemmer, M. Deussen, H. J. Bakker, R. F. Mahrt, H. Kurz, V. I. Arkhipov, H. Bässler, *Phys. Rev. Lett.* **1994**, *73*, 1440.

[14] V. I. Arkhipov, H. Bässler, M. Deussen, E. O. Göbel, R. Kersting, H. Kurz, U. Lemmer, R. F. Mahrt, *Phys. Rev. B* **1995**, *52*, 4932.

[15] M. C. J. M. Vissenberg, M. J. M. de Jong, *Phys. Rev. Lett.* **1996**, *77*, 4820.

[16] W. Graupner, G. Cerullo, G. Lanzani, M. Nisoli, E. J. W. List, G. Leising, S. De Silvestri, *Phys. Rev. Lett.* **1998**, 81, 3259.

[17] V. Gulbinas, Y. Zaushitsin, V. Sundström, D. Hertel, H. Bässler, A. Yartsev, *Phys. Rev. Lett.*, in press.

[18] D. Hertel, E. V. Soh, Z. D. Popovic, H. Bässler, L. J. Rothberg, *Chem. Phys. Lett.*, in press.

[19] M. Estaghamatian, Z. D. Popovic, G. Xu, *J. Phys. Chem.* **1996**, *100*, 13716.

[20] S. Tasch, G. Kranzelbinder, G. Leising, U. Scherf, *Phys. Rev. B* **1997**, *55*, 5079.

[21] M. Wohlgenannt, W. Graupner, G. Leising, Z. V. Vardeny, *Phys. Rev. B* **1999**, *60*, 5321.

Recombination Radiation from Organic Solids

Jan Kalinowski

Department of Molecular Physics, Technical University of Gdańsk,
ul. G. Narutowicza 11/12, 80-952 Gdańsk, Poland
E-mail: kalinovski@polnet.cc

Summary: The recombination radiation from organic solids, defined as the light emission following the fusion of oppositely charged carriers into an electrically neutral state, is discussed as a phenomenon underlying the function of organic light-emitting diodes (LEDs). Its intensity and spectral range depend on the population and nature of the emissive states, which differ, in general, from those created using light. These differences are pointed out and shown to be a result of the reverse pathways of the mutual transformation of localized molecular excitons and coulombically-correlated charge-pair excited states formed either by photoexcitation or electron-hole recombination. Spectral features of the radiation produced by the recombination of statistically independent charge carriers are discussed in terms of two molecules-based excited states like *excimers* or *electromers* in single-component materials and *exciplexes* or *electroplexes* in multicomponent materials. Consequences for optical and electrical characteristics of organic LEDs are discussed and illustrated by examples. Progress in the fundamental and applied research may be expected based on properties of recombination-produced electronic excited states.

Keywords: charge recombination processes; electronically excited states; organic electroluminescence; organic light-emitting diodes; recombination radiation

Introduction

The oppositely charged carriers (e.g. holes and electrons) recombining (or combining) in an organic solid release a considerable amount of energy which can be emitted in the form of electromagnetic radiation. This is called *recombination radiation*. If electrons and holes are injected from electrodes applied to a layer or a system of layers of organic materials, we deal with *organic recombination electroluminescence* (EL) and the devices are called *organic light-emitting diodes* (LEDs).[1] Due to their wide-spectrum display performance, improved stability and relatively simple device architecture, organic LEDs show advantages compared with conventional inorganic semiconductor materials.[2,3] Worldwide efforts in their commercialization

© 2004 International Union of Pure and Applied Chemistry   DOI: 10.1002/masy.200450803

are one of the most important reasons making recombination radiation in organic solids a hot scientific topic. Analyses of fundamental processes of charge injection and transport, and properties of emissive states underlying organic LED operation mechanisms enable better understanding of the recombination processes and form a firm basis for tailoring organic EL devices.

In the present paper, the processes governing the recombination radiation in organic solids are discussed with particular attention paid to recombination mechanisms and the nature of emissive states produced in the electron-hole recombination process.

Types of Charge Recombination

The charge recombination process can be defined as fusion of a positive (e.g. hole) and a negative (e.g. electron) charge carrier into an electrically neutral entity or, following its evolution, successive excited states. The *initial* (or geminate) (IR) and *volume-controlled* (VR) recombination can be distinguished on the basis of the charge-carrier origin (Fig. 1).[4]

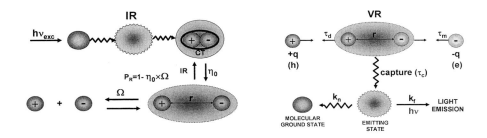

Figure 1. Initial (IR) and volume-controlled (VR) recombination (for explanation, see text).

The IR is the recombination process following the initial carrier separation, from an unstable, locally excited state, forming a nearest-neighbor charge-transfer (CT) state. It typically occurs as a part of intrinsic photoconduction in organic solids due to generation of charge from light-excited molecular states. The probability (P_{IR}) of the IR can be expressed by the primary (often assumed to be electric field-independent) quantum yield in carrier pairs for the absorbed photon, η_0, and the (e…h) pair dissociation probability, Ω, $P_{IR} = 1 - \eta_0\Omega$. The electric field effect on the effective charge separation can be observed in generation-controlled photoconduction or electric field-induced quenching of luminescence since the external electric field-enhanced dissociation decreases the number of emitting states.[5]

If the oppositely charged carriers are generated independently far away from each other (for example, injected from electrodes), the VR takes place, and the carriers are statistically independent of each other, the recombination process is kinetically bimolecular. It naturally proceeds through a coulombically-correlated electron-hole pair (e...h) leading to various emitting states in the ultimate recombination step (mutual carrier capture). The capture probability is defined by $P_R^{(2)} = (1+\tau_c/\tau_d)^{-1}$, where τ_c and τ_d is the capture and dissociation time, respectively. The classic treatment of carrier recombination can be related to the notion of the recombination time. The recombination time is a combination of the carrier motion time (τ_m), i.e. the time to get the carriers within the capture radius (it is often identified with the Coulombic radius $r_C = e^2/4\pi\varepsilon_0\varepsilon kT$), and the elementary capture time (τ_c), $\tau_{rec}^{-1} = \tau_m^{-1} + \tau_c^{-1}$. Following the traditional description of recombination processes in ionized gases, a Langevin-like and Thomson-like recombination[4,6] can be defined if $\tau_c \ll \tau_m$ and $\tau_c \gg \tau_m$, respectively (cf. Fig. 2).

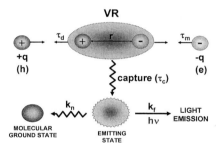

Figure 2. VC recombination scheme allowing to distinguish between Langevin and Thomson recombination (for explanation, see text).

In solid state physics, these two cases have been distinguished by comparison of the mean free path for optical phonon emission with the average distance across a sphere of critical radius r_c.[7] A signature of the Langevin recombination mechanism is a field and temperature independence of the bimolecular recombination rate constant (γ_{eh}) of the effective mobility (μ_m) ratio. This is often the case (especially for low electric fields). However, as we will see later on, it becomes a field-dependent process at higher electric fields, suggesting the Thomson-like recombination to set in.

The Nature of Excited States

A variety of emissive states can be created either by photoexcitation or a charge carrier recombination process. Let us first consider a simpler case of single-component organic solids (Fig. 3). A molecular (well-localized) excited state is a typical product of photoexcitation. Light absorption creates excited molecular singlet (M_S^*) which can relax by the intersystem-crossing to an excited molecular triplet (M_T^*). Their radiative decay produces fluorescence and phosphorescence. On the other hand, the interaction of an excited molecular state with a ground-state molecule leads to the formation of locally-excited *excimer* or *charge-transfer excimer* if the distance between molecules is shorter than 0.4 nm. If it is larger than 0.4 nm, one can deal with direct transition of a LUMO-located electron of one molecule to HOMO-located hole of another molecule (cross-transition); such an "emitting state" has been called *electromer*. Electromer emission occurs when, due to a defect or disorder, the electron transfer from LUMO to LUMO is impeded. The level shift makes the emission red-shifted with respect to the molecular emission as well as to the excimer emission. It is important to note that "electromer" emission requires the charge carriers to be separated, for example by the autoionization process.

In the volume recombination process, such separated charge pairs (e...h) are formed when statistically independent carriers approach each other. Again, molecular and bimolecular excited states can be created. The difference between these two ways of creation of emissive states is clearly apparent. The charge pair states, as a rule, precede final emissive states under recombination and follow the localized states under photoexcitation. This is a good reason for expecting the bimolecular excited states to be produced more efficiently in the charge recombination process. Thus, the emission spectrum is expected to be different or at least more complex than that under photoexcitation. There are few striking examples confirming this expectation. In Fig. 4 we see the photoluminescence spectrum (PL) of an amine derivative (TAPC) in polycarbonate with its recombination radiation spectrum (EL).

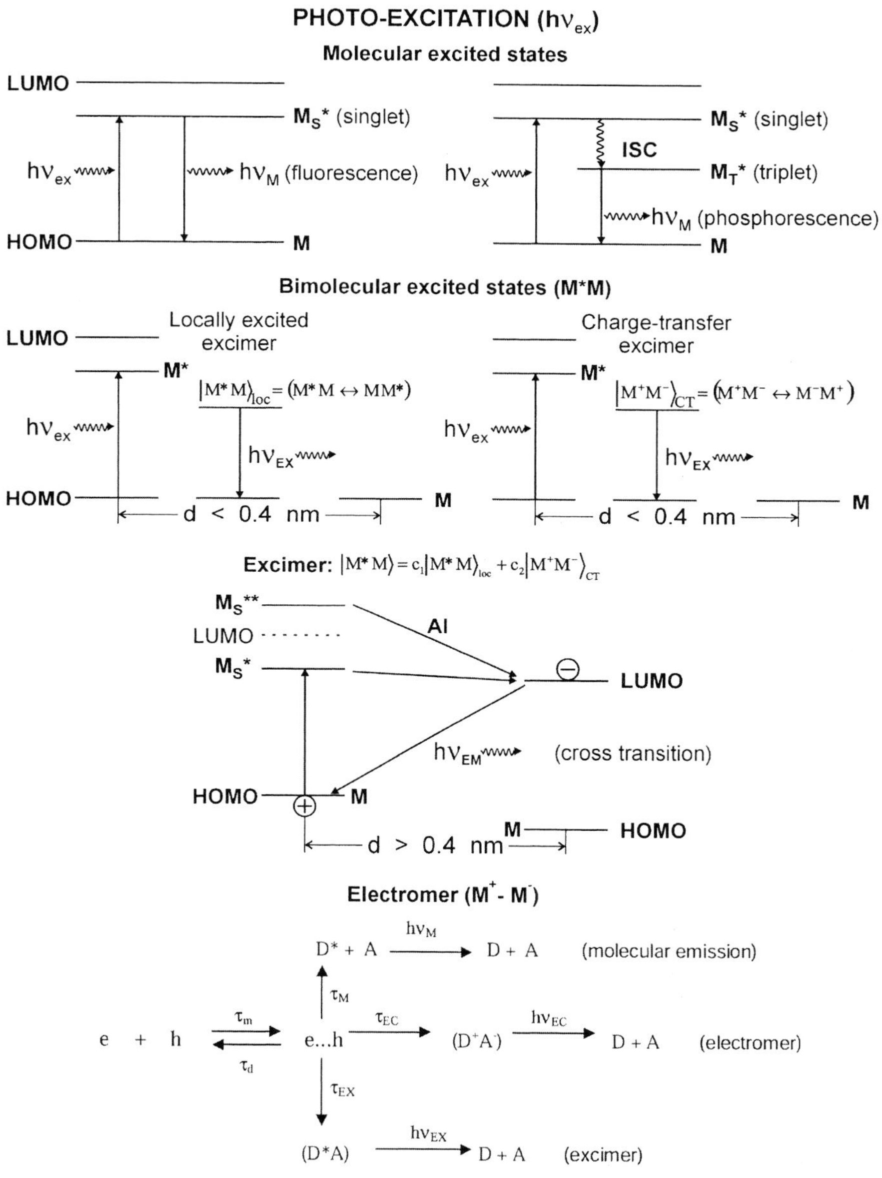

PHOTO-EXCITATION ($h\nu_{ex}$)

Molecular excited states

Bimolecular excited states (M*M)

Excimer: $|M^*M\rangle = c_1|M^*M\rangle_{loc} + c_2|M^+M^-\rangle_{CT}$

Electromer (M^+- M^-)

Figure 3. Photoexcited and electron-hole recombination-produced excited states in single-component organic solids.

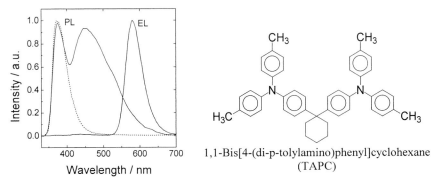

1,1-Bis[4-(di-p-tolylamino)phenyl]cyclohexane
(TAPC)

Figure 4. Optically excited (PL) and recombination radiation (EL) spectra of TAPC thin films. The dashed curve is the PL spectrum of TAPC in a 10^{-5}M dichloromethane solution.[8]

While in the photoexcited spectrum the molecular emission combined with red-shifted band of the excimer emission is apparent, in the recombination radiation only a single strongly red-shifted emission band could be detected and ascribed to the electromer states. Another striking example of the difference between photoexcited and charge recombination-generated emission spectra is due to anthracene (Fig. 5).

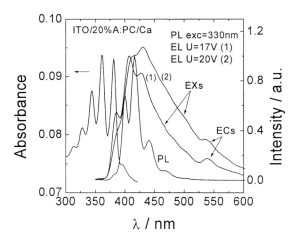

Figure 5. Optically excited (PL) and recombination (EL) emission from anthracene (A) dispersed in polycarbonate (PC). EXs denotes emission from excimers, ECs emission from electromers. Absorbance of the 20%A:PC film is shown for comparison.[9]

Again, the PL spectrum is dominated by the emission from the molecular states, the recombination radiation (EL) reveals the broad band of a series of excimers (EXs) and electromers (ECs). The latter are completely absent in the PL spectrum. One could argue that the longest-wavelength maximum in the EL spectra is due to an impurity (e.g. an oxidation product of the compound). Such a possibility, considered earlier to explain a weak emission at 540-550 nm in the thermoluminescence of γ-irradiated anthracene in squalane,[10] has been, however, questioned on the basis of time-resolved experiments. Nevertheless, this point as well as the recombination of dimer cations with the anion[11] still requires further studies.

In the case of two- or multicomponent materials with different ionization potentials and electronic affinities, the situation becomes more complex. Let us take a look at the two-component system displayed in Fig. 6.

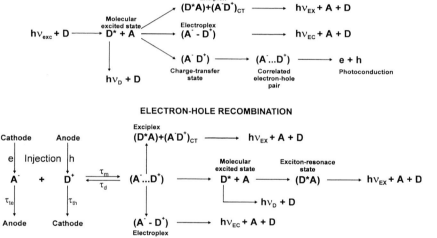

Figure 6. Excited states in two-component [electron donor (D) – electron acceptor (A)] system formed under photoexcitation and in the electron-hole recombination process.

Light absorbed by, say, electron donor molecules produces excited donor singlets (D*). Their radiative relaxation shows up as donor fluorescence ($h\nu_D$). Interacting with the ground-state electron acceptors, they can form exciplexes and *electroplexes*, the latter being two different molecule, analogue of electromers in single-component solids.

Organic Electroluminescence

If the recombining carriers [e.g. (A^-, D^+) in an organic acceptor-donor system] are formed by electrons and holes injected from electrodes as shown in Fig. 6, we deal with recombination electroluminescence (EL). In Fig. 7 an example of one of the most efficient double-layer device allowing observation of organic EL is presented.

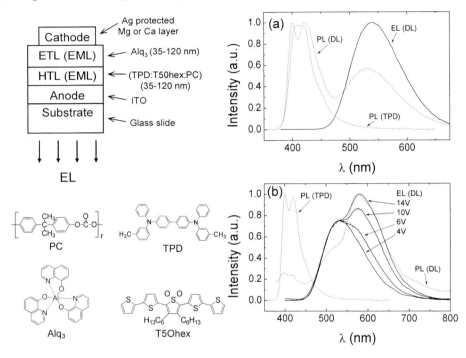

Figure 7. A double-layer (DL) organic LED and its EL spectra as compared with the PL spectra of the component materials.[12]

Electrons injected from cathode are transported through an electron-transporting layer (ETL), holes injected from anode are transported through a hole-transporting layer (HTL). Their recombination can occur either in the ETL, HTL or both. Then, different excited states are produced and the emission spectrum is expected to be a combination of their emission characteristics. The system consists of the Alq_3 electron-transporting layer and a hole-transporting layer composed of TPD and a thiophene derivative (T5Ohex) dispersed in the polycarbonate matrix. The PL spectrum contains three bands belonging to three different molecules (TPD – blue; Alq$_3$ – green; and T5Ohex – red). Typically, in the EL spectrum the TPD emission disappear and the contribution of two remaining components depends on the electric field: the dominant red emission at high fields disappears at low fields, passing through decreasing proportions at intermediate fields. This is an example showing how the EL emission color can be changed varying only the voltage applied to the device.

Another example is shown in Fig. 8. A mixture of the electron-donor (TPD) and electron-acceptor (PBD) molecules can form a single-layer (SL) system (Fig. 8a) or these molecules can be brought into an intimate contact on the interface of two independent layers; TPD in polycarbonate and 100% evaporated PBD (Fig. 8b). Their emission spectra are different (Figs. 8a, b, bottom).

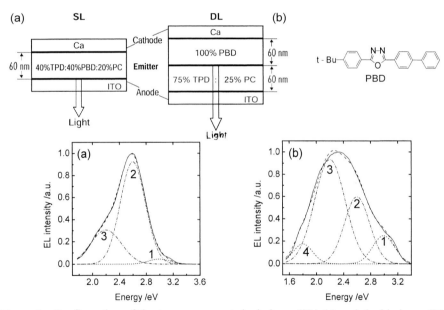

Figure 8. Configuration of the two-component single-layer (SL) (a) and double-layer (DL) (b) LEDs and corresponding EL spectra with their Gaussian profile analysis showing emission from different excited states.[13]

In the SL system a narrower spectrum is dominated by band 2, which in the broader spectrum of the DL system decreases; a slightly pronounced shoulder 3 in the previous spectrum becoming now a dominating feature. This is an excellent example of coexistence of exciplex and electroplex emissions. In the energy level scheme of the active molecules, we can see how they are formed (Fig. 9). The locally excited exciplex is created by a LUMO PBD → LUMO TPD electron transfer, the molecular TPD singlet forming an exciplex with the nearest neighbor molecule of PBD (process 2'). Due to an energy barrier for slightly more distant molecules, a cross transition takes place (process 3). As a result, a red-shifted emission 3 appears in the spectrum. A strong attractive electric field at the TPD/PBD interface impedes the external electric field-assisted charge-pair dissociation, the electroplex component becoming the major feature of the DL system emission spectrum.

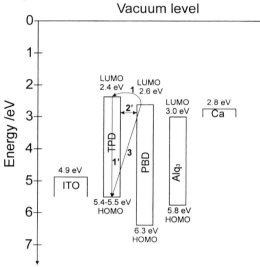

Figure 9. The energy level scheme of the material components used for manufacturing the cells described in Fig. 8. Selected electronic transitions are indicated by lines with arrows.[12]

EL Quantum Efficiency

From the above it has been clear that single-, double- and even more-layer thin film systems form devices which are called organic light-emitting diodes (LEDs). One of their most important characteristics is the EL quantum efficiency defined as the quantum flux [$\Phi/<h\nu>$] divided by the carrier flux [j/e] resulting from their driving current, j:

$$\varphi_{\text{EL}}^{(\text{ext})} = \frac{\Phi / \langle h\nu \rangle}{j/e}$$

(1)

where $\langle h\nu \rangle$ is the averaged photon energy.

The common feature of the EL quantum efficiency of organic LEDs is its nonmonotonic evolution with driving current (electric field applied to the device). Typical examples of electrofluorescence efficiencies for a series of organic LEDs are presented in Fig. 10. As a rule, the quantum efficiency increases in the low-field region and decreases above a certain field (usually close to 1 MV/cm); the current-voltage characteristics do not follow the space-charge-limited current pattern. The emitter layer composed of neat or diphenylpentacene (DPP)-doped Alq_3 is here placed between the Alq_3 electron-transporting layer and TPD hole-transporting layer. DPP forms holes as well as electron traps in Alq_3. Their EL quantum efficiency (QE) shows a maximum for a certain field, but its position and width depend on the emitter composition.

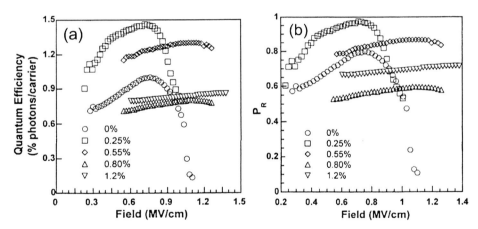

Figure 10. The quantum efficiency (a) and recombination probability, P_R, (b) as a function of electric field for the three-layer EL devices ITO/TPD/Alq_3:%DPP/Alq_3/Mg:Ag with varying concentration of DPP (mol % given in the figures).[14]

The well pronounced maximum for neat and low-doped emitters shifts remarkably and broadens with increasing concentration of DPP.

By definition, the measured QE of EL is proportional to the radiative decay efficiency of the emissive states (φ_r) with a series of proportionality coefficients ξ, P_S, P_R.

$$\varphi_{EL}^{(ext)} = \xi P_S P_R \varphi_r \qquad (2)$$

ξ is the light output coupling factor responsible for all light losses in the LED structure, P_S-probability of formation of a singlet emitting state, and P_R – recombination probability which is the product of the recombination probability due to the carrier motion [$P_R^{(1)}$] and final carrier capture [$P_R^{(2)}$],

$$P_R = P_R^{(1)} P_R^{(2)} = \left(1 + \tau_m / \tau_t\right)^{-1} \left(1 + \tau_c / \tau_d\right)^{-1} \qquad (3)$$

Having ξ, P_S and φ_r, the recombination probability can be extracted from the experimental data of the EL quantum efficiency. The results indicate that the maxima in the recombination probability underlie the maxima in the efficiency. From definitions of the recombination and transit times $\tau_{rec}/\tau_t=(8e\mu_h/9\gamma\varepsilon_0\varepsilon)(j_{SCL}/j)$. Assuming for granted the Langevin recombination mechanism, i.e. the mobility μ_h recombination rate ratio being independent of electric field, we find that the decrease in the τ_{rec}/τ_t ratio is due to the fact that the actual current approaches its space-charge-limited behavior: as field increases, j_{SCL}/j decreases, and P_R increases approaching unity. However, exceeding the field ca. 0.8 MV/cm, P_R becomes a decreasing function of electric field. This means either the Langevin formalism breaks down, μ_h/γ increasing as F increases, or $P_R \cong P_R^{(2)}$, i.e. the capture-controlled (Thomson-like) recombination sets in. Increasing electric field reduces the dissociation time τ_d, the τ_c/τ_d ratio increases leading to decreasing probability of recombination and, consequently, to a drop in the EL quantum efficiency. We believe the second possibility is correct since electric field-induced quenching of fluorescence[15] and phosphorescence[16] of various luminescent materials have been observed experimentally.

The importance of selection of proper device structure to maximize the EL quantum efficiency from a given type of the emissive states is worth mentioning. Emission from exciplexes is of considerable interest since their spectra can fall in the middle of the visible region while component molecules emit in the violet or even UV region. We have recently shown that the EL quantum efficiency of exciplex emitting diodes can reach its typical value of ca. 1 %

photon/electron for the efficient emitting singlet molecular states, fabricating a DL structure with properly selected electron- and hole-transporting layers (Fig. 11). The exciplex-forming molecules of m-MTDATA and PBD are brought into an intimate contact on the interface of these materials evaporated successively one on the other. The exciplex appears in the green region in contrast to m-MTDATA and PBD which emit in the blue and violet region.

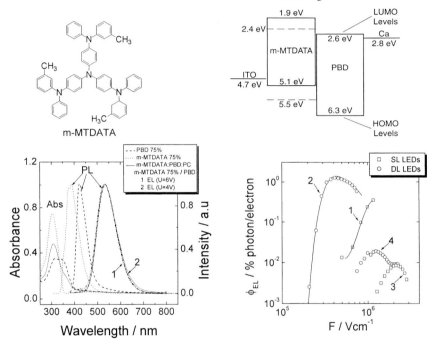

Figure 11. Efficient exciplex emitting diode ITO/m-MTDATA/PBD/Ca. Its composition and energy level scheme of the materials used are shown in the upper part of the figure; absorption, PL and EL spectra together with the field-dependent EL quantum efficiency are shown in the bottom part. Comparison with a LED with a TPD HTL is given (dashed levels in the energy scheme, and curves 3 and 4 for the efficiency).[17]

At a certain field, the QE of EL reaches and even slightly exceeds 1 % photon/electron which is almost two orders of magnitude higher than the (TPD-PBD) exciplex emission efficiency. This is due to a high energy barrier for holes at the m-MTDATA interface as compared with that at the

TPD/PBD interface. We note that this is not the case with a SL structure formed by a mixture of these components. This example shows how, by manipulating the diode structure and selecting proper materials, the desired spectral and performance requirements can be fulfilled.

Concluding Remarks

Physics of recombination radiation from organic solids reveals some particular spectral features as compared with photoexcited emission. They are associated with a diversity of two-molecular excited states, opening prospects for new ideas in the fundamental research and utilization in practical light-emitting devices.

Acknowledgments

The author acknowledges the financial support by the Research Grant of the US Navy, Office of Naval Research (ONR). A useful correspondence with Dr. Brian Brocklehurst of the Sheffield University is gratefully acknowledged.

[1] J. Kalinowski, *J. Phys. D: Appl. Phys.* **1999**, 32, R179.
[2] J. W. Allen, *J. Luminescence* **1994**, 61/62, 912.
[3] H. Sixl, H. Schenk, N. Yu, *Phys.Bull.* **1998**, 54(3), 225.
[4] J. Kalinowski, *Mol. Cryst. Liq. Cryst.* **2001**, 355, 231.
[5] J. Szmytkowski, W. Stampor, J. Kalinowski, Z. H. Kafafi, *Appl. Phys. Lett.* **2002**, 80, 1465.
[6] J. Kalinowski, M. Cocchi, V. Fattori, P. Di Marco, G. Giro, *Jpn. J. Appl. Phys.* **2001**, 40, L282.
[7] M. Lax, *Phys. Rev.* **1960**, 119, 1502.
[8] J. Kalinowski, G. Giro, M. Cocchi, V. Fattori, P. Di Marco, *Appl. Phys. Lett.* **2000**, 76, 2352.
[9] J. Kalinowski, G, Giro, M. Cocchi, V. Fattori, R. Zamboni, *Chem. Phys.* **2002**, 277, 387.
[10] M. Al-Jarrah, B. Brocklehurst, M. Evans, *J. Chem. Soc. Faraday Trans. 2* **1976**, 72, 1921.
[11] B. Brocklehurst, *Int. J. Radiat. Phys. Chem.* **1974**, 6, 483.
[12] J. Kalinowski, M. Cocchi, G. Giro, V. Fattori, P. Di Marco, *J. Phys. D: Appl. Phys.* **2001**, 34, 2282.
[13] G. Giro, M. Cocchi, J. Kalinowski, P. Di Marco, V. Fattori, *Chem. Phys. Lett.* **2000**, 318, 137.
[14] J. Kalinowski, L. C. Picciolo, H. Murata, Z. H. Kafafi, *J. Appl. Phys.* **2001**, 89, 1866.
[15] W. Stampor, J. Kalinowski, P. Di Marco, V. Fattori, *Appl. Phys. Lett.* **1997**, 70, 1935.
[16] J. Kalinowski, W. Stampor, J. Mężyk, M. Cocchi, D. Virgili, V. Fattori, P. Di Marco, *Phys. Rev. B* **2002**, 66, 235321.
[17] M. Cocchi, D. Virgili, G. Giro, P. Di Marco, J. Kalinowski, Y. Shirota, *Appl. Phys. Lett.* **2002**, 80, 2401.

Macromol. Symp. **2004**, *212*, 39-50

Interfacial Phenomena in Polymer Films and Generation of Maxwell Displacement Current

Mitsumasa Iwamoto

Department of Physical Electronics, Tokyo Institute of Technology,
2-12-1 O-okayama, Meguro-ku, Tokyo 152-8552, Japan
E-mail: iwamoto@pe.titech.ac.jp

Summary: Interfacial electrostatic phenomena in ultrathin polyimide films have been examined, and the space charge distribution and electronic density of states have been determined. The presence of excess negative charges at the film-metal interface of nanometer thickness has been revealed and the alignment of the surface Fermi level of polymer films and Fermi level of metals have been elucidated. Taking into account the interfacial space charge, a step structure observed in the I-V characteristic of metal-polyimide-rhodamine-polyimide-metal junction, very similar to Coulomb staircase, is well explained. Furthermore, the electrical breakdown mechanism of a nanometer-thick polyimide film is found quite different from that of micrometer-thick films, owing to the presence of this interfacial nanometric space charge. Finally, for a profound understanding of the behaviour of surface monolayer, the Maxwell displacement current measurement coupled with optical second harmonic generation measurement has been employed.

Keywords: Maxwell-displacement current; polyimide; SHG; single electron tunnelling; surface potential

Introduction

Many organic materials that are interesting in electronics have been synthesized and discovered during last several decades.[1,2] We can see one of the most remarkable achievements of Heeger, MacDiarmid and Shirakawa awarded by the Nobel Prize in Chemistry 2000, due to their contribution to the discovery and development of conductive polymers.[3-5] In the hope of observing novel and useful electric and optical properties, many investigations have been carried out to build organic devices, taking into account specific physical properties of organic films such as flexibility, lightness and others. Plastic solar cells, flexible-type field-effect transistors (FETs), electroluminescent (EL) devices and others have been developed along with the development of new organic materials.[6] Insightful ideas have also been proposed to open new methods and device technologies in molecular electronics.[7] However, these are no longer sufficient. Profound understanding of nanointerfacial phenomena is essential. It is needless to say

  DOI: 10.1002/masy.200450804

that one of the most primitive but important applications of organic thin films is to use them as tunneling barriers. Thus there has been a growing interest in the preparation of high-quality organic ultrathin films during the last several decades. As a result, the preparation technique has greatly advanced, and many important organic films have been successfully prepared.[8] Among them are polyimide (PI) and polyethylene Langmuir-Blodgett (LB) films, which function as tunneling spaces.[9] Briefly, for example, the monolayer thickness of the prepared PI LB film is 0.4 nm. The PI LB films are almost pinhole-free, and they are thermally and chemically stable up to a temperature of 400 °C. They exhibit excellent electric-insulating properties, and function as tunneling barriers in tunnel junctions such as Josephson junctions. The success of the preparation of excellent insulating ultrathin films and other semiconductor films motivated the authors to study the interfacial electronic phenomena in electric-insulating films, semiconductor films, etc,[10] because the I-V characteristic of thin films is controlled by electrostatic interfacial phenomena. The surface potential method, Maxwell displacement current measurement and optical second harmonic generation (SHG) measurement have been employed in elucidating nanometric interfacial phenomena. In this paper, the study of polymer films to clarify the interfacial phenomena has been reviewed. Then a step structure, similar to Coulomb staircase, observed in current-voltage (I-V) characteristics of junctions using rhodamine-dendrimer molecules [11,12] is discussed in association with the interfacial nanometric phenomena.[13,14] Furthermore, the electrical breakdown process of ultrathin film is discussed in association with the space charge phenomena. Finally, Maxwell displacement current measurement coupled with optical second harmonic generation measurement used to clarify the surface molecular motion is briefly introduced.

Interfacial Electrostatic Phenomena in Ultrathin Films

For a better understanding of the interfacial electrostatic phenomena, it is helpful to use ultrathin films whose thickness is less than the electrostatic double layer formed at the film/metal interface, and then to gain information on the distribution of the electronic density of states as well as on the space charge distribution of excess charges in the films prepared on metal electrodes. LB films are suitable because they can be prepared on solid substrates by the layer-by-layer deposition with monolayer thickness. In combination with the LB deposition technique and the surface potential method, one can examine space charge distribution as well as the density of states in ultrathin films.[13,14] PI LB films deposited on Au, Cr, and Al base electrodes were examined after the heat treatment for more than one hour at a temperature of 150 °C in vacuum. The surface potential of PI LB films was measured with reference to the potential of the clean base metal electrodes. The surface potentials gradually decreased as the number of deposited layers increased, and then reached a constant saturated potential at the number of 20-50 layers. These results indicated that PI LB films acquired electrons from metal, and an electrostatic layer was thus formed at the metal/PI LB film interface within a region of nanometer thickness. Interestingly, the surface potential value in PI LB films shifted negatively as the temperature increased, indicating that the tendency of PI to accept electrons becomes stronger as the temperature increases. The surface potential V_S built across PI LB films on metal is given by

$$V_S = \int_0^D \frac{x\rho(x)}{\varepsilon_0 \varepsilon_r} \, dx = \frac{\overline{x}Q}{\varepsilon_0 \varepsilon_r}, \tag{1}$$

$$\text{with } \overline{x} = \frac{\int_0^D x\rho(x)\,dx}{Q} \quad \text{and } Q = \int_0^D \rho(x)\,dx.$$

Here, ε_0 is dielectric permittivity of vacuum, ε_r (= 3.2) is the relative permittivity of PI, D is the film thickness, x is the distance from the metal electrode, $\rho(x)$ is the space charge density at location x, $\overline{x}$ is the mean location of the excess charges displaced from the electrodes, and Q is the total charge displaced from the metal electrodes. Differentiation of surface potential V_S with respect to the film thickness D gives a quantity proportional to the space distribution of charges ρ (D). Using the surface potential measurement, the relationship between V_S and the number of deposited layers n, proportional to the film thickness D, is calculated. Thus ρ (D) can be estimated. Figure 1 shows an example of the space charge density ρ (D) in PI LB film at a temperature of 25 °C. The space charge density decreases steeply as the number of layers increases. Most of the excess charges exist in PI LB films within the distance of 4 nm from electrodes. Ca. 1 - 10 % of monomer units of PI accept electrons from metal electrodes in this region, in which the density of PI molecule unit is about 3×10^{27} m^{-3}. A linear relationship with a slope of unity was observed between the work function of metals and the saturated surface potential of PI. This relationship indicates that a thermodynamic equilibrium was established at the PI LB film/metal interface. In other words, electronic charges are transferred at the interface between PI films and metals until the surface Fermi level of PI and the Fermi level of metal electrodes are in coincidence.[14] Similar results were obtained for other LB films.[15] As described above, we may have the prediction that excess electronic charges exist at the film-metal interface.

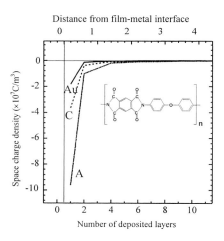

Figure 1. Interfacial space charge distribution in polyimide LB film at film-metal interface at room temperature.

MIM junction with rhodamine molecule

As mentioned in the previous section, excessive electronic charges transferred from metals to PI LB films exist at the metal/PI interface of an LB film . This interfacial space charge forms a very high-electric field of the order of $10^8 - 10^9$ V/m at the interface and directly affects the electron transport through PI LB films as well as the electrical breakdown in the films. The Schottky-type behavior in the current-voltage (*I-V*) and capacitance-voltage (*C-V*) characteristics of polyimide LB films can be successfully explained by taking into account the space charge field.[16] Similar discussion with consideration of interfacial space charge is also valid to explain the I-V characteristic of electron tunneling junctions in a structure of Au/ PI/ rhodamine dendrimer / PI/ Al, where a step *I-V* structure, very similar to Coulomb staircase behavior, is observed.[12,13] In more detail, the junctions with structures Au/PI (25)/PI:Rh-G2 (molar ratio 1:1, and 500:1)/PI (30)/Al junctions (junction area 0.25 mm^2) were prepared using Au and Al electrodes. Rh-G2 is rhodamine dendrimer whose chemical structure is shown in Fig. 2. The number in the parenthesis represents the number of deposited layers. It was supposed that the prepared Au/PI (25)/PI:Rh-G2 (1)/PI (30)/ Al junction (electrode area : 0.25 mm^2) consisted of about 5.4×10^8 small elements with the same structure. Here the number of small elements was calculated using the limiting molecular area estimated from the surface pressure area isotherm of Rh-G2 monolayer.

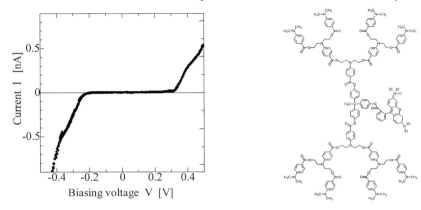

Figure 2. The *I-V* characteristics of Au/PI (25)/PI:Rh-G2 (1)/PI (30)/Al junction. The rhodamine dendrimer (Rh-G2) structure is given.

Figure 2 shows a typical *I-V* characteristic of the junction at a temperature of 30 K, where a dc step voltage was applied to the top Al electrode with reference to the base Au electrode with a step of 0.003-0.005 V. In the figure, the step structure is clearly observed; the *I-V* steps can be seen at voltages of 0.3 V and -0.25 V. Similar experimental results were obtained for junctions with on top Au electrode. On the other hand, no step structure was seen in junctions without rhodamines. These results suggest that Rh-G2 molecules make a main contribution to the creation of step structure. The *I-V* characteristic is controlled by the single-electron tunneling via rhodamine molecule. However, to explain the behavior of the *I-V* step structure, it is necessary to take into account the interfacial space charge and residual charge. In what follows, we show the analysis, where the rhodamine molecule is assumed to function as a quantum dot (central electrode).

According to the theory of Coulomb blockade,[17] the *I-V* step structure appears at a voltage of

$$V = \frac{1}{C_f}(n + \frac{1}{2})e \quad (n = 0, \pm 1, \pm 2, \pm 3, \ldots), \tag{2}$$

if the contribution of space charge, i.e., the so-called "back ground charge", and residual charge are discarded. Here C_f is the capacitance formed between the quantum dot and metal electrodes, and e is electron charge. Using the image charge method in the electromagnetic field theory (see Fig. 3a, the capacitance C_0 formed between a sphere metal (quantum dot) and plane metal electrode is calculated as[11,18]:

$$C_0 = 4\pi \varepsilon_0 r F(y) \quad \text{with} \quad F(y) = \sum_{i=0}^{\infty} \frac{\sinh y}{\sinh (i+1) y} \tag{3}$$

assuming r is the radius of the sphere metal; y is given by $y = \ln(s + \sqrt{s^2 - 1})$ with $s = z_h/r + 1$. z_h is the distance between the sphere and plane electrode (see Fig. 3a). It is instructive to note that the capacitance of the sphere with radius r in a free space is given by $4\pi\varepsilon_0 r$. Therefore, in Eq.(3), it expresses the effect of the sphere approaching the plane electrode. C_0 is nothing but the geometrical capacitance and it is replaced by $C_M = \varepsilon\, C_0$ when the sphere-plane electrode system is placed in a medium with permittivity ε.

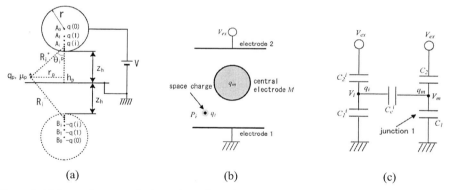

(a) (b) (c)

Figure 3. (a) Image charges and sphere-plane electrode system. (b) Model structure of our junction. (c) Equivalent circuit.

Figure 3b shows the model structure of our junction, where central electrode M is sandwiched between two electrodes separated by tunneling barriers.[19] In this case, the tunneling barrier is PI LB film. The central electrode M is a spherical electrode with a radius r, and this is the rhodamine molecule (Rh-G2). This modeling is very simple, but does not lose the underlying physics for the understanding of the *I-V* step structure shown in Fig. 2. Between the central electrode M and outer two electrodes 1 and 2, capacitance C_1 and C_2 are formed. C_1 and C_2 are the capacitance C_M mentioned above, and the capacitance C_f is given by $C_1 + C_2$. Thus the *I-V* step structure appears at a voltage given by Eq. (2) when the contribution of space charge and residual charge is minor. However, this is not true when the interfacial space charge is present as shown in Fig. 1.

First, in order to clarify the space charge behavior, charge q_i is assumed to site in position P_i. This charge couples to the central electrode and outer two electrodes with coupling capacitance $C_C{}^i$, $C_1{}^i$ and $C_2{}^i$, respectively, and then the equivalent circuit model is obtained as shown in Fig. 3c. Charge q_M is induced on central M due to the presence of charge q_i. Using this equivalent circuit and then to calculate the energy $\varDelta W$ due to the increase in charge q_m from $-ne$ to $-ne+1e$ (n is the number of electrons) by electron tunneling across the capacitance C_1, we obtain the external voltage V_{ex} for the electron tunneling as[18,19]:

$$V_{ex} \geq \frac{1}{C_2 g_2}\left\{\left(n+\frac{1}{2}\right)e - \frac{C_c^i}{C_\Sigma^i}q_i\right\} \tag{4}$$

with $g_2 = 1 + \dfrac{C_c^i}{C_\Sigma^i}\cdot\dfrac{C_2^i}{C_2}$ and $C_\Sigma^i = C_1^i + C_2^i + C_2^i$. Here g_2 represents the effect of coupling capacitance and $q_i C_c^i / C_\Sigma^i$ represents the space charge effect. However, if we take into account the difference in the work function between central metal electrode and two outer electrodes, Eq. (4) is replaced by

$$V_{ex} \geq \frac{1}{C_2 g_2}\left\{\left(n'+\frac{1}{2}\right)e - V_D C_f - \frac{C_c^i}{C_\Sigma^i}q_i\right\}, \quad (n'=\cdots-2,-1,0,1,2,\cdots), \tag{5}$$

with $n' = n + k$, $C = C_1 + C_2$ and $f = 1 + f_i$ with $\dfrac{C_c^i}{C_\Sigma^i}\left(\dfrac{C_1^i + C_2^i}{C_1 + C_2}\right)$. Here k is an integer chosen to

satisfy the relation $\left|V_D C_f + \dfrac{C_c^i}{C_\Sigma^i}q_i - ke\right| \leq \dfrac{e}{2}$. The reason is that the accurate alignment in the Fermi levels of central electrode and outer electrodes never happens, and the displacement of one electron from outer electrodes into central electrode builds up the additional potential $e/(C_1+C_2)$. V_D in Eq.(5) is the residual voltage and expresses the difference in the Fermi energies between the central metal electrode and outer electrodes under a short circuit condition. Furthermore, ke represents the charge that should be redistributed due to the additional potential established in the short circuit condition. The induced charge $q_i C_c^i / C_\Sigma^i$ on electrode M can be calculated if the potential V_i in position P_i in Laplace field is known. Similarly, in the presence of excessive charges q_i (i = 1, 2, 3, …) surrounding the central metal M, the following relation (6) is obtained in replacing Eq.(5). That is

$$V_{ex} \geq \frac{1}{C_2 G_2}\left\{\left(j+\frac{1}{2}\right)e - V_D CF - \Delta Q\right\}, \ (j = \cdots -2,-1,0,1,2,\cdots).\qquad(6)$$

Here $F = 1 + \sum_i f_i$, $G_2 = 1 + \sum_i (C_c^i / C_\Sigma^i) \cdot (C_2^i / C_2)$, and $\Delta Q = \sum_i (C_c^i / C_\Sigma^i) q_i$.

From Eq.(6) we find that the space charge field and residual charge make a significant contribution to the I-V characteristic.

On the basis of the above analysis, the I-V step structure shown in Fig. 2 can be argued as follows: The induced charge ΔQ is approximately given by

$$\Delta Q = C_1 G_1 V_{s1} - C_2 G_2 V_{s2} \quad \text{with} \quad G_1 = 1 + \sum_i (C_c^i / C_\Sigma^i) \cdot (C_1^i / C_1).\qquad(7)$$

Here V_{s1} and V_{s2} are surface potentials built at the film-metal interface, given by Eq. (1). The surface potentials of about -200 mV and -1000 mV are built at PI/Au and Al/PI interface at room temperature. If the space charge density is high at the interface, the relation $C_1 G_1 = C_2 G_2$ is approximately satisfied. Using Eq. (7), the shift of step voltage $|\Delta V|$ is estimated to be less than 1200 mV. However, this shift must satisfy the relation given by $|\Delta V| \leq e/2C_2 G_2 \, (= e/2C_1 G_1)$ (see Eqs. (5) and (6)). For our junctions of Au/PI/Rh-G2/PI/Al, the capacitance formed between Rh-G2 and metal electrode is estimated to be of the order of 10^{-19} F. Thus, $|\Delta V| \leq 800$ mV is expected. This speculation well supports our experimental result shown in Fig. 2. Figure 4 shows the theoretical and experimental dI/dV-V characteristics of Au /PI /Rh-G2 /PI /Al junction at a temperature of 33 K. The theoretical line was calculated in a manner similar to that in the analysis of *I-V* characteristics of double barrier tunneling junction.[20] For $C_2 = 2.39 \times 10^{-19}$ F, $C_1 = 2.76 \times 10^{-19}$ F and $Q - Q_0 + ke = 8.37 \times 10^{-21}$ C, the calculated line shows an agreement with the experimental plots. As mentioned above, the *I-V* step structure is explained by taking into account the presence of space charge and residual charge.

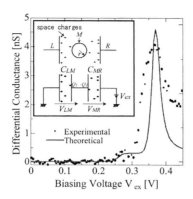

Figure 4. The theoretical and experimental d*I*/d*V*-*V* characteristic of Au-/PI/Rh-G2/PI/Al junction.

The *I-V* measurement was carried out under the scanning tunneling microscope (STM) operation and the step-structure was obtained. Furthermore, we recently succeeded in the detection of a similar *I-V* step structure using metal PI-metal junctions with including C60 molecules prepared by the spin-coating technique.

Electric breakdown of ultrathin films

The study of electrostatic interfacial phenomena at organic film/metal interfaces is obviously important in electronics and electrical insulation, and they have been a continuous research subject since the discovery of contact electrification.[21,22] As mentioned earlier, the nanometric interfacial electrostatic charges make a significant contribution to electric conduction. Similarly, the electrical breakdown will be controlled by the formation of space charge, though to show this is not an easy task, owing to the difficulty in preparation of sophisticated organic ultrathin films. The information on interfacial space charge distribution obtained by the surface potential study will help in elucidating the electrical breakdown process. To examine the electrical breakdown mechanism, we measured the surface potential of PI LB films on Al, Ag and Au electrodes, which were charged by a dc voltage in a needle plane electrode system. We then studied the electrical breakdown process in ultrathin PI LB films in association with the interfacial electrostatic phenomena.[23] Briefly, a needle electrode (point angle $30°$ and curvature radius 10μ m) was placed above the sample in a vacuum vessel, at a distance d of 1 mm from the sample. The samples were charged by applying a dc voltage between -2.5 kV and $+2.5$ kV to the needle electrode for 20 min, and then the surface potential of the charged samples was examined. Whilst the samples were charged, some of them were electrically broken. When 50 % of the samples were electrically broken by the application of biasing voltage, we defined this voltage as the breakdown voltage. The surface potential of charged samples decreased as the dc biasing voltage increased, as shown in Fig. 5. Usually, electrically insulating polymer films are charged with electronic homo-charges when the polymers are non-polar and do not possess extrinsic ionic impurities. That is, the surfaces of films are negatively charged by the application of negative voltages to the needle electrode with respect to the plane electrode, whereas these are positively charged by the application of positive voltages. However, the experimental results for PI LB films were opposite to this prediction. Furthermore, the breakdown voltage increases as the work function of metals increases (work-function, Au > Cu > Al), and it depends on the polarity of biasing voltages. As mentioned earlier, the surface potential V_s is built across the PI LB films. At equilibrium, the surface Fermi level of PI LB films and the Fermi level of metals are brought into coincidence at the interface: the electronic states of PI, whose electronic energy is higher than the Fermi level of the metal, can donate electrons to the metal if the states are filled with electrons before electrification. By contrast, the electronic states of PI whose electronic energy is lower than the Fermi level of the metal, can accept electrons from the metal if the states are vacant. On the basis of this model, we may argue the result shown in Fig. 5 as follows. PI LB films are negatively charged even when biasing voltage is zero (see Fig. 1). The PI LB films are more negatively charged by applying a positive biasing voltage to the needle electrode, whereas they are gradually positively charged as the biasing voltage becomes more negative. A possible explanation is that the mean location of electronic charges $\bar{x}$ and the total charge Q given by Eq. (1) increase by application of positive biasing voltage, whereas they decrease by the application of negative biasing voltage. Thus we may expect that electrons are injected from base electrodes to PI LB films via interfacial states at the film/metal interface when the needle electrode is positively biased, whereas holes are injected into PI LB films when the needle is negatively biased. In other words, the main contribution of excessive charge is electrons injected from metal electrodes, and not electron charges deposited on film surface from the side of needle electrode. The experimental result shown in Fig. 5 can be explained in this way. Similarly, the dependence on the number of deposited layers, the nature of metal electrode and others can be explained in the same way.

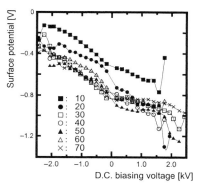

Figure 5. Relationships between the surface potentials and the number of deposited layers for PI LB films on Al electrodes. The abrupt change of the surface potential around dc biasing voltage of +1.5 and –2.0 kV is due to the breakdown.

MDC and SHG study of surface monolayers

As mentioned earlier, the electrical transport property of monolayers is strongly dependent on the nanometric dielectric phenomena observed at the interface. In this sense, it is also very important to clarify the behavior of polar molecules at the interface in association with the dielectric phenomena, though the dipolar polarization was not so important for PI LB films. Generally, the orientational distribution of polar molecules gives rise to the spontaneous polarization at the interface, and Maxwell displacement current and SHG are generated from non-centrosymmetric structured monolayers[10,24] Monolayers of liquid crystals, polymers and other materials have been examined using this technique. For example, the azobenzene dendrimer monolayer on water surface has been examined using the MDC measurement, and the dipole moment was determined in association with the molecular motion of constituent molecules. The calculated dipole moment was in good agreement with the calculation based on the semiempirical molecular orbital (MO) calculation.[25] Further, MDC and SHG measurements have been employed for liquid crystal monolayers, and the orientational order parameters of constituent LC molecules have been determined.[26] This information will be important to clarify the electrical transport in ultrathin films as well as for the control of electron motion such as light-driven molecular current, [27] and others.

Conclusion

The surface potential of PI LB films in a needle plane electrode system was examined. It was concluded that the main contribution to the creation of the surface potential was electron excess charges injected from the metal electrodes via interfacial states in PI LB films at the film/metal interface. Furthermore, the interfacial space charge also makes a significant contribution to the I-V characteristic of organic junctions and the electrical breakdown of ultrathin films.

[1] *Special issue on Functional Organic Materials for Devices*, J. Mater. Chem. **1999**, 9, 1853.
[2] *Organic Thin Films*, Materials Chemistry Discussion No. 2, J. Mater. Chem. **2000**, 10, 1.
[3] A. J. Heeger, Rev. Mod. Phys. **2001**, 73, 681 (Nobel Lectures).
[4] A. G. MacDiamid, Rev. Mod. Phys. **2001**, 73, 701 (Nobel Lectures).
[5] H. Shirakawa, Rev. Mod. Phys. **2001**, 73, 713 (Nobel Lectures).
[6] C. J. Brabec, N. S. Sariiftci and J. C. Hummelen, Adv. Funct. Mater. **2001**, 11, 15.
[7] C. Joachim, J. K. Gimzewski and A. Aviram, Nature **2000**, 408, 541.
[8] H. S. Nalwa, *Supramolecular Photosensitive and Electractive Materials*, Academic Press, San Diego 2001.
[9] M. Iwamoto and M. Kakimoto, in: *Polyimides, Fundamentals and Application,* M.K. Ghosh and K.L. Mittal, Eds., Marcel Dekker, Inc., New York, 1996, p. 815/884ff.
[10] M. Iwamoto and C. X. Wu, *The Physical Properties of Organic Monolayers*, World Scientific, Singapore 2001.
[11] Y. Noguchi, Y. Majima and M. Iwamoto, J. Appl. Phys. **2001**, 90, 1368.
[12] Y. Noguchi, M. Iwamoto, T. Kubota and S. Mashiko, J. Appl. Phys. **2002**, 94, in press.
[13] M. Iwamoto, A. Fukuda and E. Itoh, J. Appl. Phys. **1994**, 75, 1607.
[14] E. Itoh and M. Iwamoto, J. Appl. Phys. **1997**, 81 1790.
[15] M. Iwamoto, J. Mater. Chem. **2000**, 10, 99.
[16] C. Q. Li, Y. Noguchi, H. C. Wu and M. Iwamoto, Jpn. J. Appl. Phys. **2001**, 40, 4575.
[17] S. Carraral, in *Nanoparticles and Nanostructured Films*, F.H. Fendl, Ed., Wiley-VCH, Weinheim, 1998, Chap.15.
[18] M. Iwamoto, Y. Noguchi, T. Kubota and S. Mashiko, Curr. Appl . Phys. **2002**, in press.
[19] Y . Noguchi, M. Iwamoto, T. Kubota and M. Mashiko, Jpn. J. Appl. Phys. **2002**, 41, 2749.
[20] A . E. Hanna and M .Tinkham, Phys. Rev. B **1991**, 44, 5919.
[21] J. Lowell and A. C. Rose-Innes, Adv. Phys. **1980**, 29 947, and references therein.
[22] L. H. Lee, J. Electrostatics **1994**, 32, 1, and references therein.
[23] M. Fukuzawa and M. Iwamoto, IEEE Trans. Dielec. Electri. Insul. **2001**, 8, 832.
[24] Y. R. Shen, *The Principles of Nonlinear Optics*, Wiley, New York, 1984.
[25] T. Manaka, D. Shimura and M. Iwamoto, Chem. Phys. Lett. **2002**,355, 164.
[26] A. Tojima, T. Manaka and M. Iwamoto, J. Chem. Phys. **2001**, 115, 9010.
[27] S. Nespurek and J. Sworakowski, Thin Solid Films **2001**, 393, 168.

Macromol. Symp. **2004**, *212*,51-61   51

Luminescence Properties of PPV Derivatives Bearing Substituted Carbazole Pendants

Jong Hyun Park,[1] *Kyungkon Kim,*[1] *Young-Rae Hong,*[1] *Jung-Il Jin,**[1] *Byung-Hee Sohn*[2]

[1] Division of Chemistry and Molecular Engineering, Center for Photo- and Electro-Responsive Molecules, Korea University, 5-1, Anam-Dong, Seoul 136-701, Korea
[2] Polymer Lab, Samsung Advanced Institute of Technology, Kiheung-eup,Yongin-shi, Kyungki-do 449-901, Korea

Summary: A series of PPV derivatives bearing substituted and unsubstituted carbazole and 2-ethylhexyloxy pendants were prepared and their photo- (PL) and electroluminescence (EL) properties were studied. Substituted carbazole structures were *N*-phenylcarbazole and 3,6-dimethoxycarbazole. The substituents on the carbazole pendants caused little change in UV-vis absorption, PL, and EL when compared with the polymer bearing the unsubstituted carbazole pendants. The presence of the benzene ring between the main chain and the carbazole pendant increased the threshold electric field in EL. We could obtain maximum brightness of ca. 17,000 – 30,000 cd/m^2 for the polymers carrying the unsubstituted and dimethoxy substituted carbazole pendants.

Keywords: conjugated polymers; luminescence

Introduction

In recent years, there have been published many reports concerning the photo- and electroluminescence properties of poly(*p*-phenylenevinylene) (PPV) derivatives[1-8] and other polyconjugated polymers.[9-12] The Cambridge group's[1] pioneering work on the light-emitting diodes (LED) fabricated with PPV generated a strong impetus to researches on organic polymer LED devices, some of which are expected to become commercialized in the near future. Previously, we reported the EL characteristics of LED devices prepared with PPV derivative bearing diphenyloxadiazole,[13,14] dialkylfluorene[15], and carbazole pendants[14,16] directly bonded to the PPV backbone. All of the PPV derivatives are green-light emitting polymers exhibiting excellent device performance. In particular, the PPV derivative (CzEh-PPV) having 2-ethylhexyloxy (Eh) and carbazole (Cz) pendants directly attached to the benzene rings in the backbone revealed a very high external quantum efficiency and a low threshold electric field. In addition, CzEh-PPV showed much more balanced mobility of the two carriers, i.e., positive hole and negative electron than PPV and MEh-PPV.[17,18] The low threshold electric field for this

                      DOI: 10.1002/masy.200450805

polymer could be ascribed to the formation of new intragap states when the polymer is contacted with the cathode metal such as calcium, which was supported by the near-edge X-ray absorption fine structure (NEXAFS) spectroscopy. [17,18]

In this report we would like to compare the PL and EL properties of two new PPV derivatives bearing substituted carbazole pendants, MCzEh-PPV and PCzEh-PPV:

CzEh-PPV   MCzEh-PPV   PCzEh-PPV

MCzEh-PPV carries two methoxy groups on the benzene rings in the carbazole moiety while in PCzEh-PPV exists an additional benzene ring bridge between the backbone and the carbazole moiety. These polymers were prepared by the Gilch polymerization[19] of the corresponding bis(chloromethyl) monomers.

Results and Discussion

Synthesis of MCzEh-PPV and PCzEh-PPV

MCzEh-PPV and PCzEh-PPV were polymerized by the Gilch polymerization[19] of the two monomers, 1,4-bis(chloromethyl)-2-(3,6-dimethoxycarbazol-9-yl)-5-[(2-ethylhexyl)oxy]benzene (6) and 1,4-bis(chloromethyl)-2-(4-carbazol-9-ylphenyl)-5-[(2-ethylhexyl)oxy]benzene (13). Synthetic routes to the two monomers are shown in Schemes 1 and 2. The polymers were purified by Soxhlet extraction with methanol. Both polymers were readily soluble in common organic solvents such as tetrahydrofuran, chloroform, and 1,1,2,2-tetrachloroethane. The molecular weights of the two polymers measured by gel permeation chromatography (GPC) were $\overline{M_n}$ = 280,000 and 42,000, respectively. Their polydispersity indices were 3.0 and 2.1.

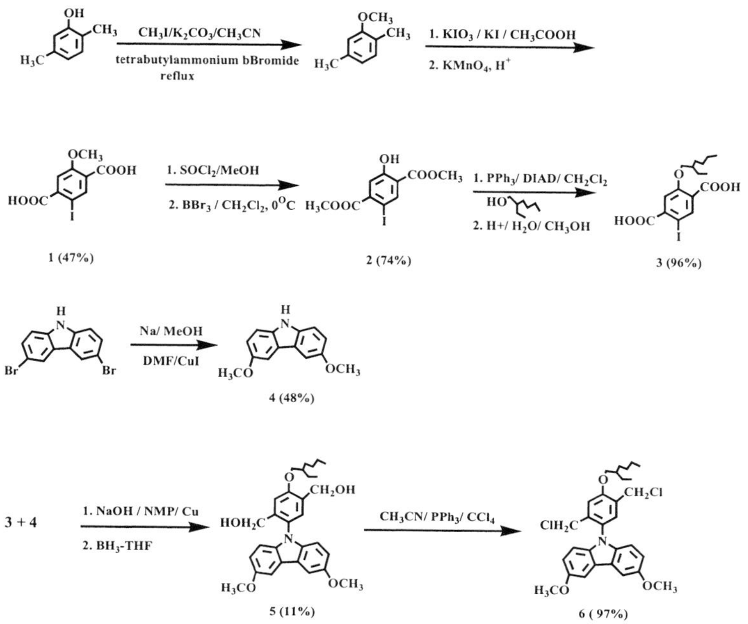

CH$_3$I/K$_2$CO$_3$/CH$_3$CN

tetrabutylammonium bBromide

reflux

1. KIO$_3$ / KI / CH$_3$COOH

2. KMnO$_4$, H$^+$

1. SOCl$_2$/MeOH

2. BBr$_3$ / CH$_2$Cl$_2$, 0°C

1. PPh$_3$/ DIAD/ CH$_2$Cl$_2$

2. H+/ H$_2$O/ CH$_3$OH

1 (47%)

2 (74%)

3 (96%)

Na/ MeOH

DMF/CuI

4 (48%)

3 + 4

1. NaOH / NMP/ Cu

2. BH$_3$-THF

CH$_3$CN/ PPh$_3$/ CCl$_4$

5 (11%)

6 (97%)

Scheme 1. Synthetic route to MCzEh-PPV monomer

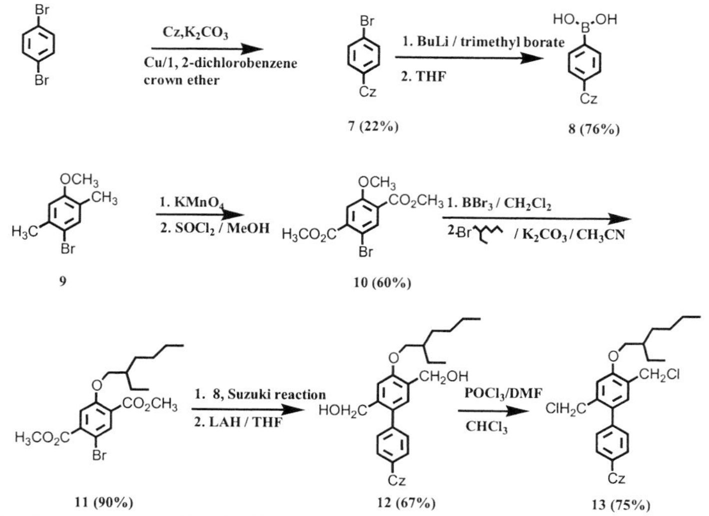

Cz,K$_2$CO$_3$

Cu/1, 2-dichlorobenzene

crown ether

1. BuLi / trimethyl borate

2. THF

7 (22%)

8 (76%)

1. KMnO$_4$

2. SOCl$_2$ / MeOH

1. BBr$_3$ / CH$_2$Cl$_2$

2. Br / K$_2$CO$_3$/ CH$_3$CN

9

10 (60%)

1. 8, Suzuki reaction

2. LAH / THF

POCl$_3$/DMF

CHCl$_3$

11 (90%)

12 (67%)

13 (75%)

Scheme 2. Synthetic route to PCzEh-PPV monomer

Uv-vis Absorption and PL Spectra, Electronic Energy Levels

Figure 1 compares the UV-vis absorption spectra of films of CzEh-PPV, MCzEh-PPV, and PCzEh-PPV. The absorption maxima (λ_{max}) responsible for the π - π^* transitions of the backbone's π-electron systems are located at 461, 461, and 475 nm, respectively, for CzEh-PPV, MCzEh-PPV, and PCzEh-PPV. Insertion of the benzene ring between the main chain and the carbazole moiety causes a red shift, but attachment of methoxy substituents on the benzene rings of the carbazole moiety influenced little the λ_{max} position although absorption by the carbazole moiety is significantly red-shifted and appears at 313 and 373 nm from 295 and 341 nm. The electron-donor property of the methoxy group is well reflected by this red shift. In addition, the methoxy groups broadened the absorption peak for the π - π^* transition.

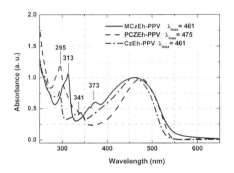

Figure 1. UV-vis absorption spectra of polymers.

The optical band gaps estimated from the absorption edges are 2.35 eV (528 nm), 2.28 eV (544 nm) and 2.30 eV (540 nm), respectively, for CzEh-PPV, MCzEh-PPV, and PCzEh-PPV. The presence of the methoxy and phenyl substituents slightly reduced the band gap energy.

The PL spectra of the polymers obtained for the excitation wavelength of 430 nm are given in Figure 2. All the three spectra reveal a similar feature that shows some vibronic details. The intensive major peak (at 530, 538 and 544 nm for CzEh-PPV, PCzEh-PPV and MCzEh-PPV, respectively) in the shorter wavelength region carries overlapped shoulder peaks on the longer wavelength side. Here again, we observe that the spectrum of MCzEh-PPV is broadest extending to 700 nm. All the three polymers emit green light, but MCzEh-PPV with an orange shade.

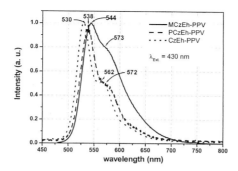

Figure 2. PL spectra of polymer thin films (thickness 80 nm).

We tried to obtain information on the electronic energy levels, i.e., HOMO and LUMO levels, of the polymers by cyclovoltammetry[20,21] and the results are summarized in Table 1. The optical bandgaps estimated from the absorption edges in the UV-vis spectra were utilized in the evaluation of the LUMO levels. We note that the HOMO and LUMO levels have not been altered much by either methoxy groups in MCzEh-PPV or the benzene ring in PCzEh-PPV when compared with HOMO and LUMO levels of CzEh-PPV, although the HOMO levels are elevated slightly more than LUMO levels.

Table 1. HOMO-LUMO values of polymers determined by CV.

	HOMO (eV)	LUMO (eV)
MczEh-PPV	5.40	3.14
PCzEh-PPV	5.44	3.15
CzEh-PPV	5.53	3.22

EL Spectra and Characteristics of LED Devices

We constructed LED devices having the configuration of ITO/ PEDOT:PSS (20 nm)/ polymer (80 nm)/ Li:Al. Poly[3,4-(ethylenedioxy)thiophene] (PEDOT) doped with poly(styrenesulfonic

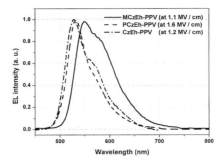

Figure 3. EL spectra of polymers (device structure: ITO/ PEDOT : PSS (20nm) / polymer (80 nm)/ Li:Al).

acid) (PSS) supplied by Bayer (σ = 10 Scm^{-1}) was spin-coated on the ITO-coated glass anode. Then the emitting polymer layer was spin-coated onto it using a solution (1 wt.%) in chlorobenzene. Finally, the cathode layer was vacuum-deposited using an alloy of aluminium and lithium (0.12 % Li). Figure 3 compares the EL spectra of the three devices obtained at the electric field indicated in the figure. The general profile of the EL spectra is very similar to those of the corresponding PL spectra shown in Figure 2.

Figure 4 compares the current density-electric field (Figure 4a) and electric field - emitted light intensity curves (Figure 4b) for the three devices. The threshold electric fields taken at the electric field for the current density of 0.1 mA/mm^2 are 0.82, 0.70, and 1.3 MV/cm, respectively, for CzEh-PPV, MCzEh-PPV, and PCzEh-PPV. The methoxy group reduced the threshold electric field whereas the phenylene group increased it. The maximum brightness observed was the highest (6,080 cd/m^2, at 1.4 MV/cm) for MCzEh-PPV and the lowest (280 cd/m^2 at 1.8 MV/cm) for PCzEh-PPV. The electric field – light output curve of Figure 4b tells us that PCzEh-PPV utilizes the injected carriers least efficiently among the three polymers.

Since the HOMO and LUMO levels of the present polymers are not much different from each other, it is conjectured that mobilities of carriers are least balanced in PCzEh-PPV resulting in poor device performance. Earlier, we[17,18] reported that CzEh-PPV exhibits a much greater balance in the mobility of carrier ($\mu_{h+}/\mu_{e-} \approx 5$) than PPV ($\mu_{h+}/\mu_{e-} \approx 200$) and MEh-PPV ($\mu_{h+}/\mu_{e-} \approx 100$).

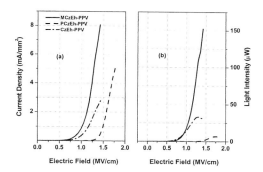

Figure 4. (a) Electric field vs. current density and (b) light intensity curves of polymers (device structure: ITO/ PEDOT:PSS (20 nm)/ polymer (80 nm)/ Li:Al).

Probably, the presence of the benzene ring between the backbone and the carbazole moiety in PCzEh-PPV deprives the carbazole structure from its ability to stabilize the positive carrier which would bring about the improvement in the balance of carriers mobilities as in CzEh-PPV. Our recent optically detected magnetic resonance (ODMR) study[22] clearly demonstrates that the carbazole unit directly attached to the PPV backbone is involved in the stabilization of positive polarons formed along the backbone.

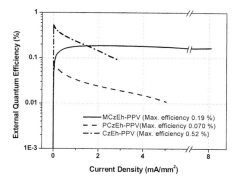

Figure 5. Comparison of external quantum efficiency of polymers (device structure: ITO/ PEDOT:PSS (20 nm)/ polymer (80 nm)/ Li:Al).

The device constructed with MCzEh-PPV not only showed the highest maximum brightness, but also excellent device stability as shown in Figure 5. This figure compares the external quantum efficiency of the three devices. Although the initial quantum efficiency is the greatest for CzEh-PPV, at current density higher than 2 mA/mm^2 it becomes greater for MCzEh-PPV than for CzEh-PPV. Moreover, the device efficiency of MCzEh-PPV remained practically constant at about 0.2 % until it failed. On the other hand, the device efficiency of the other two polymer devices decreased rapidly as the current density is increased. The device stability observed for MCzEh-PPV is rather remarkable, although its electronic structure is not much different from the other two.

Highly Luminescent Devices

After observing that MCzEh-PPV performs the best among the present polymers, we decided to construct another LED device of ITO/ PEDOT:PSS(20nm)/ polymer (80 nm)/ Ca/ Al and to study its performance. Earlier, we[16] learned that the ITO/ PEDOT:PSS (20nm)/ polymer (80 nm)/ Ca/ Al device produced higher brightness than the ITO/ PEDOT:PSS (20nm)/ polymer (80 nm)/ Li:Al device, especially when we used PPV derivatives for the fabrication of the devices. In order to protect the calcium cathode, aluminium was vacuum-deposited on the cathode.

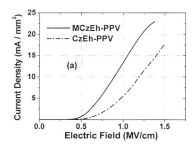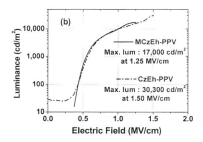

Figure 6. Comparison of (a) electric field vs. current density and (b) electric field vs. luminance properties of MCz-PPV and CzEH-PPV (device structure: ITO/ PEDOT:PSS (20 nm)/ polymer (80 nm)/ Ca/ Al).

Figure 6a and 6b compare characteristics of the devices fabricated with MCzEh-PPV and CzEh-PPV. We can see that the two devices exhibit very similar electrical properties. The threshold electric fields are 0.40 and 0.50 MV/cm for 0.1 mA/mm^2, respectively, for MCzEh-PPV and CzEh-PPV. The dependence of luminescence on the applied electric field (Figure 6b) of the two devices is very much the same up to 1.25 MV/cm, where the device of MCzEh-PPV failed after reaching the maximum brightness of 17,000 cd/m^2 at the applied electric field of 1.25 MV/cm. The CzEh-PPV device's maximum brightness was 30,300 cd/m^2 at the applied field of 1.50 MV/cm. The two devices exhibited the brightness of 200 cd/m^2 commonly at the applied field of 0.46 MV/cm. The major difference between the two devices, however, can be noted in their luminance efficiency as shown in Figure 7. The luminance efficiency in lm/W of the device of MCzEh-PPV decreases from 0.65 lm/W to 0.26 lm/W, as the applied field was increased to 1.25 MV/cm. This decrease in efficiency is much less than the decrease observed for the CzEh-PPV device, which revealed the initial luminance efficiency higher than 3 lm/W, decreasing rapidly to and stabilized at 0.45 lm/W. The device efficiency at 200 cd/m^2 was 0.60 lm/W for MCzEh-PPV and 2.95 lm/W for CzEh-PPV.

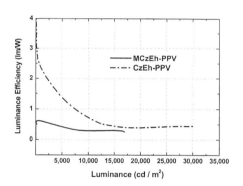

Figure 7. Comparison of luminance efficiency of MCz-PPV and CzEH-PPV (device structure: ITO/ PEDOT:PSS (20 nm)/ polymer (80 nm)/ Ca/ Al).

This observation again proves that MCzEh-PPV shows much more consistent device performance than CzEh-PPV. The scientific reason for this is not yet clear. One major difference between the polymers is in their average molecular weights: the MCzEh-PPV sample utilized in this study is of a very high molecular weight, $\overline{M_n}$ = 280,000 and $\overline{M_w}$ = 850,000, whereas $\overline{M_n}$ and $\overline{M_w}$ values of the CzEh-PPV sample are much lower (49,800 and 72,000, respectively). It is well known that, in general, low-molecular-weight polymers exhibit poorer thermal stability and experience easier molecular migration. These two factors are expected to lower the LED device performance as Joule heat is evolved when the device is in operation. Therefore, the particular stability of the MCzEh-PPV device is ascribed to its very high molecular weight.

Conclusion

We have prepared two new PPV derivatives, MCzEh-PPV and PCzEh-PPV, bearing substituted carbazole moieties as pendants and their EL properties were compared with CzEh-PPV. Among the three polymers, MCzEh-PPV imparted to the LED device the highest stability most probably owing to its high molecular weight. CzEh-PPV and MCzEh-PPV performed significantly better than PCzEh-PPV in LEDs. The maximum brightness of LED devices fabricated with MCzEh-PPV and CzEh-PPV easily reached $17,000 - 30,000$ cd/m^2 when device configuration was ITO/ PEDOT:PSS/ polymer/ Ca/ Al. All the three polymers emit green light.

Experimental

Synthesis of polymers: Synthetic details will be published elsewhere.[23]

Device fabrication and characterization: The patterned ITO-coated glass slides were cleaned by successive sonication in acetone, propan-2-ol and distilled water each for 13 min, and finally cleaned by UV - ozone for 13 min. The conducting polymer solution of poly(3,4-ethylenedioxy-2,5-thiophene) doped with poly(styrenesulfonate) (PEDOT:PSS) (Bayer) was spin-coated (Laurell, U.S.A) at 3300 rpm for 1 min and dried at 150 °C for 30 min. The electric conductivity of this film measured by the four-line probe method was 10 Scm^{-1}. The solution (1wt %) of polymer in purified 1,1,2,2-tetrachloroethane was spin-coated at 1000 rpm for 1 min and the film was subjected to thermal treatment at 150 °C for 1 h under vacuum. The Ca cathode 2000 Å thick was vacuum-deposited from the tungsten boat at a deposition rate of 2 Å/s under the pressure of 3.0×10^{-7} torr. An Al capping layer was then evaporated to protect the Ca cathode at a deposition rate of 4 Å/s under the same pressure. The active area of the EL devices was 4 mm^2. Photoluminescence (PL) and electroluminescence (EL) spectra were recorded on a PC1 photon counting spectrofluorometer (ISS Inc., USA). Current-voltage (I-V) characteristics and the intensities of EL emission were simultaneously measured with a Keithley 238 SMU electrometer and a BM7 luminance meter (Topcon Technologies, Inc., USA). Device fabrication and all the measurements were performed in a dry argon filled glove box without exposing to air.

Cyclovoltammetry:[20,21] The polymer thin films were spin-coated on the ITO-coated glasses. Chlorobenzene was used as the solvent. The electrolyte solution employed was 0.1 M tetrabutylammonium perchlorate in dimethylformamide. The Ag/AgCl and Pt wire (600 μm in diameter) electrodes were utilized as reference and counter-electrodes, respectively.

Acknowledgment

This work was supported by the Korea Science and Engineering Foundation through the Center for Electro- and Photoresponsive Molecules, Korea University. Kyungkon Kim and Jong Hyun Park were the recipients of Brain Korea 21 assistantship supported by the Ministry of Education.

[1] J. H. Burroughes, D. D. C. Bradley, A. R. Brown, R. N. Marks, K. Mackay, R. H. Friend, P. L. Burn, A. B. Holmes, *Nature* **1990**, *347*, 539.
[2] D. Braun, A. J. Heeger, *Appl. Phys. Lett.* **1991**, *58*, 1982.
[3] T. Zyung, D.-H. Hwang, I.-N. Kang, H.-K. Shim, W.-Y. Hwang, J.-J. Kim, *Chem. Mater.* **1995**, *7*, 1499.
[4] S.-J. Chung, J.-I. Jin, K. K. Kim, *Adv. Mater.* **1997**, *9*, 551.
[5] A. Kraft, A. C. Grimsdale, A. B. Holmes, *Angew. Chem., Int. Ed.* **1998**, *37*, 402.
[6] T. Ahn, S. Y. Song, H.-K. Shim, *Macromolecules* **2000**, *33*, 6764.
[7] S.-H. Jin, M.-S. Jang, H.-S. Suh, H.-N. Cho, J.-H. Lee, Y.-S. Gal, *Chem. Mater.* **2002**, *14*, 643.
[8] H.-K. Shim, J.-I. Jin, *Adv. Polym. Sci.* **2002**, *158*, 193.
[9] M. Gross, D. C. Muller, H.-G. Nothofer, U. Scherf, D. Neher, C. Brauchle, K. Meerholz, *Nature* **2000**, *1*, 207.
[10] D. Neher, *Macromol. Rapid Commun.* **2001**, *22*, 1366.
[11] S. Kim, D. W. Chang, S. Y. Park, K. Kim, J.-I. Jin, *Bull. Korean Chem. Soc.* **2001**, *22*, 1407.
[12] N. S. Cho, D.-H. Hwang, J.-I. Lee, B.-J. Jung, H.-K. Shim, *Macromolecules* **2002**, *35*, 1224.
[13] D. W. Lee, K. Y. Kwon, J.-I. Jin, Y. Park, Y. R. Kim, I. W. Hwang, *Chem. Mater.* **2001**, *13*, 565.
[14] S.-J. Chung, K.-Y. Kwon, S.-W. Lee, J.-I. Jin, C. H. Lee, C. E. Lee, Y. Park, *Adv. Mater.* **1998**, *10*, 1112.
[15] B.-H. Sohn, K. Kim, D. S. Choi, Y. K. Kim, S. C. Jeoung, J.-I. Jin, *Macromolecules* **2002**, *35*, 2876.
[16] K. Kim, Y. R. Hong, S. W. Lee, J.-I. Jin, Y. Park, B. H. Sohn, W. H. Kim, J. K. Park, *J. Mater. Chem.* **2001**, *11*, 3023.
[17] K. Kim, Y. Park, C. E. Lee, J. W. Jang, J.-I. Jin, *Macromol. Symp.*, in press.
[18] J.-I. Jin, C. E. Lee, J.-S. Joo, Y. Park, in *Nanotechnology Toward the Organic Photonics,* H. Sasabe, Ed., Gootech Press, Chitose (Japan) 2002, Chapter 28.
[19] H. G. Gilch, W. L. Weelwright, *J. Polym. Sci., Part A: Polym. Chem.* **1966**, *4*, 1337.
[20] M. Helbig, H.-H. Horhold, *Makromol. Chem.* **1993**, *194*, 1607.
[21] R. Cervini, X.-C. Li, G. W. C. Spencer, A. B. Holmes, S. C. Moratti, R. H. Friend, *Synth. Met.* **1997**, *84*, 359.
[22] C.-H. Kim, J. Shinar, D.-W. Lee, Y.-R. Hong, J.-I. Jin, in *Annual APS March Meeting 2002,* Indiana, 2002, Session B11.
[23] J. H. Park, MS thesis, Korea Univ. Seoul, **2002**.

Electrical Rectification by Monolayers of Three Molecules

Robert M. Metzger

Laboratory for Molecular Electronics, Department of Chemistry, Box 870336, The University of Alabama, Tuscaloosa, AL 35487-0336, USA
E-mail: rmetzger @ bama.ua.edu

Summary: Rectification, i.e. asymmetrical electrical conduction by a Langmuir-Blodgett monolayer of dicyano{4-[1-cyano-2-(1-hexadecylquinolin-1-ium-4-yl)vinyl]phenyl} methanide (**1**), occurs between both Al and Au electrodes: this rectification arises from the asymmetry of the molecule, and the interplay between the zwitterionic ground state D^+-π-A^- and the less dissociated first electronic excited state D^0-π-A^0. Two more monolayer rectifiers have been found: 1-butyl-2,6-bis{2-[4-(dibutylamino)phenyl] vinyl}pyridin-1-ium iodide (**2**) is an interionic rectifier with back charge transfer between the iodide ion and the pyridinium ring. 1a-[4-(dimethylamino)phenyl]-1aH-1a-aza-1(2)a-homo(C_{60}-I_h)[5,6]fullerene (**3**) is a moderate rectifier, with a rectification ratio of 2.

Keywords: Aviram-Ratner proposal; electrical conduction through molecules; Langmuir-Blodgett monolayers; unimolecular rectification

Introduction

"Molecular electronics" *(sensu stricto)*, or "molecular-scale electronics" is the study of electrical and electronic processes measured or controlled on a molecular scale.[1] A second, much wider, definition of molecular electronics *(sensu lato)*, or "molecule-based electronics" encompasses electronic processes by molecular assemblies of any scale, including crystals and conducting polymers.[1] This review falls within the former, narrower definition of the field, and focuses on our own work on electrical rectification, or asymmetric conduction, through a monolayer of organic molecules.

We have "reached out and touched" individual molecules with two metal electrodes, and exploited their structure, to control the flow of electrical signals from them and to them. We have measured electrical rectification, or asymmetrical conductivity, in Langmuir-Blodgett monolayers of three molecular species: first, dicyano{4-[1-cyano-2-(1-hexadecylquinolin-1-ium-4-yl)vinyl]phenyl} methanide, ($C_{16}H_{33}Q$-3CNQ, **1**; Fig. 1), a zwitterionic D^+-π-A^- molecule,[2-4] second, 1-butyl-2,6-bis{2-[4-(dibutylamino)phenyl]vinyl}pyridin-1-ium iodide (($Bu_2N\varphi V)_2BuPy^+I^-$, **2**; Fig. 1), a charge-transfer salt,[5] and third, 1a-[4-

(dimethylamino)phenyl]-1aH-1a-aza-1(2)a-homo(C_{60}-I_h)[5,6]fullerene (DMAn-NC$_{60}$, **3**; Fig. 1), a D-σ-A molecule.[6]

These results are an experimental realization of a molecular rectifier proposed in 1974 by Aviram and Ratner (AR).[7] The interest in such systems is due to the hope that molecular-scale electronics, in one embodiment or other, will become practical for integrated circuits. The four-decade-old progressive miniaturization (and concomitant increase in speed) of integrated circuits uses silicon and inorganic semiconductors, and is well described by Moore's "law", which has seen a remarkable doubling in circuit speeds, at first every 2 years, then every 18 months or so.[8] The increased processor and memory speeds have been driven by market forces, but may fail in the future at some scale of miniaturization (50 nm or 10 nm between adjacent circuit elements?), because of cost, or inherent limitations on inorganic materials, or limitations in electron-beam lithography.

The AR Ansatz

The first concrete suggestion for unimolecular electronics was the 1974 AR proposal that a one-molecule rectifier could be achieved with a D-σ-A molecule **4**, where D is a good one-electron donor, with relatively low first ionization potential, where σ is some saturated covalent "sigma" bridge, and where A is a good one-electron acceptor, with relatively high electron affinity, when this molecule is placed between two appropriate metal contacts M_1 and M_2 [7]. The σ bridge should partially decouple the molecular orbitals mostly localized on the donor moiety D from the molecular orbitals mostly localized on the acceptor moiety A, yet allow for intramolecular charge-transfer band. The molecular ground state of D-σ-A has a low dipole moment (written as D^0-σ-A^0), while the first excited state is much more polar, and is written as the zwitterion D^+-σ-A^-.

The suggested mechanism consists of two, not necessarily simultaneous, resonant electron transfers across metal-organic interfaces:

$$M_1 + D^0\text{-}\sigma\text{-}A^0 + M_2 \rightarrow M_1^- + D^+\text{-}\sigma\text{-}A^0 + M_2 \tag{1}$$
$$M_1^- + D^0\text{-}\sigma\text{-}A^0 + M_2 \rightarrow M_1^- + D^+\text{-}\sigma\text{-}A^- + M_2^+ \tag{2}$$

This creates the excited state D^+-σ-A^-. To restore the molecule to the initial state, there is an inelastic down-hill intramolecular electron transfer:

$$M_1^- + D^+\text{-}\sigma\text{-}A^- + M_2^+ \rightarrow M_1^- + D^0\text{-}\sigma\text{-}A^0 + M_2^+ \tag{3}$$

Overall, one electron has migrated from M_2 to M_1. This intramolecular electron transfer requires that the molecule has reasonable oscillator strength in the optical intervalence transfer (IVT) band linking D^0-σ-A^0 to D^+-σ-A^-.

In their proposed molecule **4,** Aviram and Ratner suggested D = tetrathiafulvalene, A = tetracyanoquinodimethane, because these were, respectively, a good organic donor D, and one of the best organic acceptors A. The AR mechanism uses an IVT electronic transition, which is fast (ps to ns), compared with translations, conformational transitions, or molecular rearrangements.

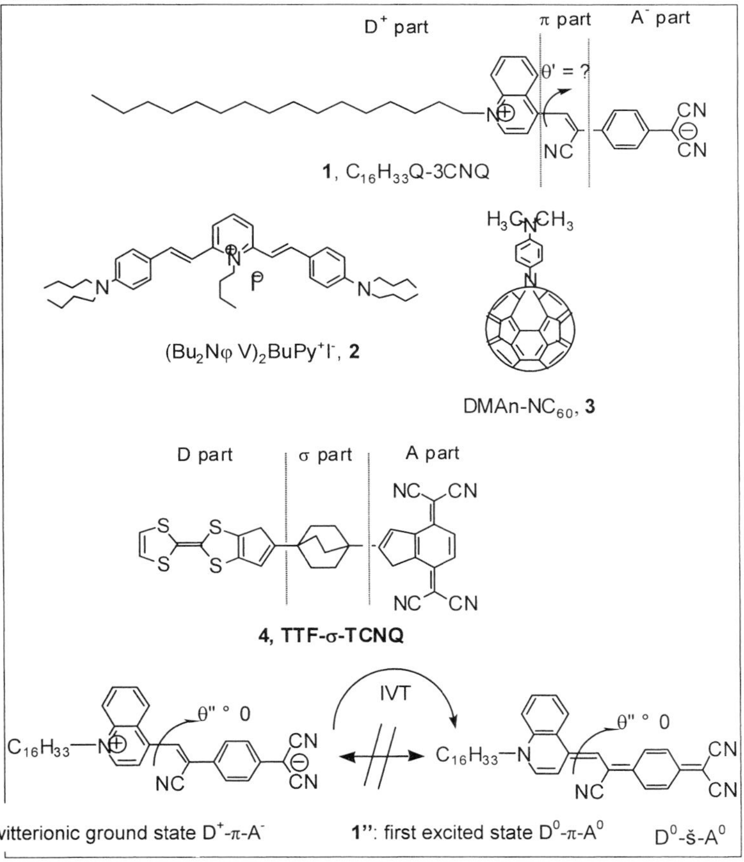

Figure 1. Relevant molecular structures.

Assembly

To design unimolecular rectifiers or, some day, unimolecular transistors, one must choose how the designed molecules will be assembled and measured. The molecules can be placed on an electrode surface by either physisorption or chemisorption. Physisorption includes the random deposition from a vapor onto a solid substrate ('chemical vapor deposition'), or the transfer of an ordered monolayer (Langmuir film or Pockels-Langmuir[9] monolayer) from the air-water interface to a solid substrate, forming a Langmuir-Blodgett (LB) monolayer or, if the transfer is repeated, an LB multilayer. The unimolecular rectifiers described below were assembled as LB films. Chemisorption forms a covalent bond between molecule and electrode, (e.g. thiols and similar compounds bonded to gold, or chlorosilanes bonded to hydroxyl-covered silicon surfaces ("self-assembled monolayers").

The Top Electrode

It is easy enough to transfer an organic monolayer or multilayer atop a sufficiently flat metal layer either by the LB method,[10] or by chemisorption.[11] It is much more difficult to then put a second metal electrode atop the organic layer without damaging the organic layer. J. Roy Sambles made dramatic improvements, first by using Mg as the top electrode,[12-16] then by using the "cold gold" method.[17] Samples showed that **1** rectified when sandwiched between Pt and Mg.[11,12]

Rectification of $C_{16}H_{33}Q$-3CNQ

We confirmed unimolecular rectification by **1**; we avoided asymmetries in the current-voltage plots due to electrodes of dissimilar metals: we used the same metallic electrode on both sides of an LB monolayer or multilayer of **1** (at first Al [2,18,19], now Au [3-6]). We also characterized what in **1** is responsible for rectification, both chemically and spectroscopically.[6,20] Compound **1** is slightly soluble in polar solvents and forms microcrystallites; a single-crystal structure determination was not possible.[2] Its cyclic voltammogram showed a reversible reduction at $E_{1/2}$ = -0.54 V vs SCE (this potential resembles that of the weak acceptor 1,4-benzoquinone, **8**) [2]. The ground-state static electric dipole moment of **1** is μ_{GS} = 43 ± 8 D at infinite dilution in CH_2Cl_2 [2]. There is an intense, hypsochromic absorption band; this is the intervalence transfer (IVT) band [2,20]. The near-infrared fluorescence lifetime of the first excited state **1"** in solution is as short as 1.4 ps [21]. Using the Stokes shift or theory, the excited-state dipole moment is between 3 and 9 D [20]. **1** is clearly zwitterionic in the ground state (D^+-π-A^-, **1'**), and less dissociated (D^0-π-A^0, **1"**) in the first excited state. The twist angle θ' in structure **1** is non-zero for steric reasons, and prevents the π electron bridge from allowing complete mixing of the quinolinium electrons with the electrons on the 3CNQ part, i.e. states **1'** and **1"** are non-degenerate. Molecule **1** forms amphiphilic Pockels-Langmuir monolayers at the air-water interface, with a collapse pressure of 34 mN m^{-1} and a collapse area of 50 Å^2 at 20 °C [2]. It transfers well on the upstroke, with transfer ratios around 100 % onto hydrophilic glass, quartz, or aluminum [2] or fresh hydrophilic Au [3,4]. It transfers poorly on the downstroke onto HOPG graphite (transfer ratio ≈ 50 %) [2]. Successive layers transfer onto HOPG on the upstroke, with 100% transfer ratios, forming Z-type LB multilayers [2].

The LB monolayer thickness of **1** is 23 Å [2] and 29 Å [4] by X-ray diffraction, 23 Å by spectroscopic ellipsometry, 22 Å by surface plasmon resonance [2,22], and 25 Å by X-ray photoelectron spectrometry (XPS) [22], yielding an average monolayer thickness of 23 Å. Using the calculated maximum molecular length of 33 Å, one estimates a tilt angle of $\cos^{-1}$ (23 / 33) = 46° from the surface normal [2]. The XPS spectrum of one monolayer of **1** on Au shows two N(1s) peaks, one attributable to the quinolinium N, and one attributable to the three CN species [22]. Molecules of **1**, adhering by the two terminal CN groups onto a hydrophilic substrate, are tilted about 45° from the surface normal, and present alkyl chains to the air. This is confirmed by a grazing-angle FTIR study of **1** on Al [2] or on Au [22]. LB multilayers of **1** exhibit an intense IVT absorption band at 575 nm [2].

To perform rectification measurements, LB monolayers and multilayers of **1** were sandwiched between macroscopic Al electrodes [2], or between Au electrodes [3,4]. In both cases, the sample holder was cryo-cooled to 77 K. For Au deposition, 10^{-3} Torr of Ar gas was added to the evaporation chamber [17], to cool the Au atom vapor, and the substrate was shielded from direct thermal radiance from the heated Au source [3,4].

As expected, a monolayer or multilayer of arachidic acid, $C_{19}H_{39}COOH$, sandwiched between Al or Au electrodes, has a sigmoidal and almost symmetrical curve under both positive bias and negative bias [2,4]. When a monolayer of **1** is placed between Al electrodes (with their inevitable patchy and defect-ridden covering of oxide), then a dramatically asymmetric current is seen (Fig. 2). The rectification ratio, RR, is defined as the current at a positive bias V divided by the absolute value of the current at the corresponding negative bias $-V$:

$$RR(V) = I(V) / | I(-V) | \qquad (4)$$

For **1**, RR = 26 at 1.5 V [2]; this corresponds to a current I = 0.33 electrons molecule^{-1} s^{-1} [2]. This same asymmetry is seen also for multilayers of **1**, for a sample covered by Mg pads topped by Al pads [2] (as before [12]), for monolayers and multilayers of **1** on graphite studied by scanning tunneling spectroscopy [2], and even for a solution of **1** in dimethyl sulfoxide placed in the scanning tunneling microscope [2]. The RR varies somewhat from pad to pad, as does the total current, in part because these are two-probe measurements, with all electric resistances (Al, Ga/In or Ag paste, wires, etc.) in series. A thorough review of all data suggested that any molecule which exhibits RR < 2 at maximum bias V should not be considered a rectifier [18]. Some samples of **1** between Al electrodes have enhanced currents, albeit smaller, under negative bias, instead of positive bias [18]. As high potentials are scanned repeatedly, the *I-V* curves show progressively less asymmetry; the rectification ratios decrease gradually with measurement, i.e. with repeated cycling of the bias across the monolayer. Putting 1.5 V across a monolayer of thickness 2.3 nm creates an electrical field of 0.65 GV m^{-1}; under such large fields, many zwitterionic molecules in the monolayer may turn around, end over end, to minimize the total energy. Measurements of the temperature dependence of rectification of **1** between Al electrodes in the range 105 K $< T <$ 390 K established that the asymmetry is not temperature-dependent [19].

Although it seemed clear that with symmetrical Al electrodes we had measured the rectification of several molecules in parallel (unimolecular rectification) [2], some asked whether the oxide covering of the Al electrodes could somehow be responsible for the *I-V* asymmetry. Using the cold-gold evaporation technique [17], the current rose dramatically, and asymmetric conduction was seen again (Figs 3 and 4) [3,4]. Figure 3 shows the best rectification ratio at 2.2 V (RR = 27.53), obtained with Au electrodes [4]. The maximum current registered was 90,400 electrons molecule^{-1} s^{-1} [4]. Some pads exhibit, as in Fig. 4, a saturation in the forward current [4] predicted by physical models for conduction through a molecule or monolayer of molecules (e.g. the AR model [7]). Rectification by a one-molecule thick layer of $C_{16}H_{33}Q$-3CNQ **(1)** is an established fact.

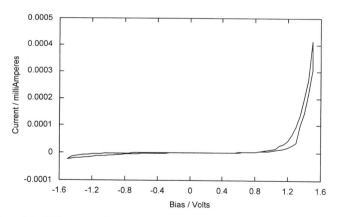

Figure 2. Plot of the DC current *I* versus the DC applied voltage *V* (*I-V* plot) through a single monolayer of $C_{16}H_{33}Q$-3CNQ **(1)** sandwiched between Al electrodes (top Al pad area 4.5 mm^2, thickness 100 nm), using Ga/In eutectic and Au wires. The DC voltage is swept at 10 mV s^{-1} [2].

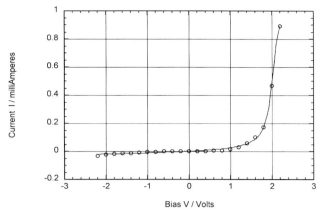

Bias V / Volts

Figure 3. Current-voltage (*I-V*) plot for a cell Au | $C_{16}H_{33}Q$-3CNQ (**1**) monolayer | Au. The resistance at 2.2 V is R = 2.47 kΩ; the current at 2.2 V is I = 9.83 × 10^3 electrons molecule^{-1} s^{-1}. The rectification ratio at 2.2 V is RR = 27.53 in the first cycle (shown), but decreases to 10.1, 4.76, 2.44, and 1.86 in cycles 2, 3, 4, and 5, respectively (not shown). [4].

A plausible mechanism for rectification by **1** is a minor change in the AR proposal, so that Eqs. (1), (2), and (3) are replaced by:

$$M_1 + D^+\text{-}\pi\text{-}A^- + M_2 \rightarrow M_1 + D^0\text{-}\pi\text{-}A^0 + M_2 \qquad (5)$$
$$M_1 + D^0\text{-}\pi\text{-}A^0 + M_2 \rightarrow M_1^- + D^+\text{-}\pi\text{-}A^- + M_2^+ \qquad (6)$$

where the first step is the electric field-driven excitation from ground to excited state, followed by electron transfer across the two molecule | metal interfaces [2].

Two New Rectifiers

Recently, two more molecules were found: 1-butyl-2,6-bis{2-[4-(dibutylamino)phenyl]vinyl}pyridin-1-ium iodide $(Bu_2N\varphi V)_2BuPy^+I^-$, **2** [6], and an azafullerene, 1a-[4-(dimethylamino)phenyl]-1aH-1a-aza-1(2)a-homo(C_{60}-I_h)[5,6]fullerene, DMAn-NC$_{60}$, **3** [5].

$(Bu_2N\varphi V)_2BuPy^+I^-$ (**2**) forms a PL film at the air-water interface, and transfers to hydrophilic substrates as a Z-type multilayer. The monolayer thickness was 0.7 nm by spectroscopic ellipsometry, and 1.15 nm by surface plasmon resonance (λ = 532 nm), and 1.3 nm by X-ray diffraction [6]. The films exhibit an absorption maximum at 490 nm (which is slightly hypsochromic in solution), attributable to iodide-to-pyridinium back charge transfer, and a second harmonic signal $\chi^{(2)}$ = 50 pm V^{-1} at normal incidence (λ =1064 nm) and 150 pm V^{-1} at 45° [6]. XPS of a multilayer of **2** on Au finds only 30 % of the expected signal from the iodide; the iodide anion may be partially replaced by a more abundant hydroxide anion during LB transfer [6]. The rectification (Fig. 5) shows a decrease of rectification upon successive cycles. Some cells have initial rectification ratios as high as 60. The favored direction of electron flow is from the gegenion to the pyridinium ion, i.e. in the direction of back charge transfer; the rectification in **2** is attributed to an interionic process, rather than to an intramolecular process [5].

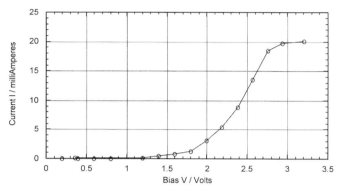

Figure 4. *I-V* plot for a cell Au | $C_{16}H_{33}Q$-3CNQ (**1**) monolayer | Au that shows saturation in the forward current $I = 20$ mA at 3.2 V (this cell broke down at 3.4 V) [4].

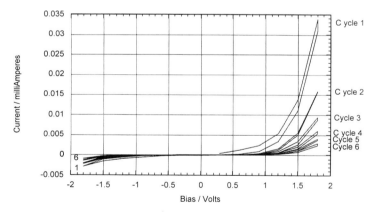

Figure 5. *I-V* plots for $(Bu_2N\phi V)_2BuPy^+I^-$ (**2**) measured in an Au | LB monolayer of **2** | Au cell, for six successive cycles of measurement. The rectification ratios are RR = 12, 7, 5, 4, 3, 3, for cycles 1 through 6, respectively [6].

The azafullerene (DMAn-NC$_{60}$, **3**) consists of a weak electron donor (*N,N*-dimethylaniline) bonded to a moderate electron acceptor (N-capped C$_{60}$). It is a blue compound, with a significant IVT peak at 720 nm [5]. Molecules **3** probably transferred onto Au in a staggered mode, since the measured molecular area at the transfer pressure (22 mN m^{-1}) is only 70 Å^2, whereas the area of C_{60} is about 100 Å^2 [5]. The XPS film thickness is 2.2 nm [5]. Angle-resolved N(1s) XPS spectra confirm that the two N atoms are closer to the bottom Au electrode than is the C_{60} cage [5]. The monolayer was covered with 17 nm thick Au pads deposited by the cold-gold technique. The current-voltage plot shown in Fig. 6 shows a "marginally" rectifying current [18] in the forward direction, with RR = 2 [5].

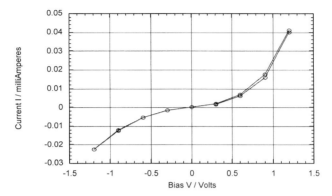

Figure 6. Current-voltage plots for DMAn-NC$_{60}$ (**3**) measured in a Au | LB monolayer of **3** | Au cell, in which the current stays small: this is probably the current due to the molecules **3** and not to the Au filaments [5].

Caution

Not all asymmetric currents seen in metal | molecule | metal cells are due to asymmetries in the molecular energy levels, however. At any molecule | metal interface a set of dipoles may form, that become Schottky barriers. Also, if a chromophore is placed asymmetrically within a metal | molecule | metal sandwich, a current asymmetry will be produced [23].

Conclusion

Unimolecular rectification, a quarter century after it was proposed, is now a reality.

Acknowledgments

This work was generously supported by the United States National Science Foundation (grants NSF-DMR-FRG-00-95215 and DMR-00-99674), and was made possible by the help, the ideas, the patience and the friendship of many colleagues, students, and post-doctoral fellows.

72

[1]   R. M. Metzger, in *Lower-Dimensional Systems and Molecular Electronics*, R. M. Metzger, P. Day, and G.
       C. Papavassiliou, Eds, *NATO ASI Ser. B* **248** (Plenum Press, New York, NY, 1991), p. 659.
[2]   R. M. Metzger, B. Chen, U. Höpfner, M. V. Lakshmikantham, D. Vuillaume, T. Kawai, X. Wu, H.
       Tachibana, T. V. Hughes, H. Sakurai, J. W. Baldwin, C. Hosch, M. P. Cava, L. Brehmer, and G. J.
       Ashwell, *J. Am. Chem. Soc.* **119**, 10455 (1997).
[3]   T. Xu, I. R. Peterson, M. V. Lakshmikantham, and R. M. Metzger, *Angew. Chem. Int. Ed.* **40**, 2119 (2001).
[4]   R. M. Metzger, T. Xu, and I. R. Peterson, *J. Phys. Chem. B* **105**, 7280 (2001).
[5]   R. M. Metzger, J. W. Baldwin, W. J. Shumate, I. R. Peterson, P. Mani, G. J. Mankey, T. Morris, G.
       Szulczewski, S. Bosi, M. Prato, A. Comito, and Y. Rubin, *J. Phys. Chem. B*, submitted.
[6]   J. W. Baldwin, R. R. Amaresh, I. R. Peterson, W. J. Shumate, M. P. Cava, M. A. Amiri, R. Hamilton, G. J.
       Ashwell, and R. M. Metzger, *J. Phys. Chem. B*, in press.
[7]   A. Aviram and M. A. Ratner, *Chem. Phys. Lett.* **29**, 277 (1974).
[8]   G. E. Moore, *Electronics* **1965** (19 April), 114.
[9]   E. Torres, C. A. Panetta, and R. M. Metzger, *J. Org. Chem.* **52**, 2944 (1987).
[10]   K. B. Blodgett, *J. Am. Chem. Soc.* **57**, 1007 (1935).
[11]   W. C. Bigelow, D. L. Pickett, and W. A. Zisman, *J. Colloid Sci.* **1**, 513 (1946).
[12]   G. J. Ashwell, J. R. Sambles, A. S. Martin, W. G. Parker, and M. Szablewski, *J. Chem. Soc., Chem.
       Commun.* 1374 (1990).
[13]   A. S. Martin, J. R. Sambles, and G. J. Ashwell, *Phys. Rev. Lett.* **70**, 218 (1993).
[14]   R. M. Metzger, C. A. Panetta, N. E. Heimer, A. M. Bhatti, E. Torres, G. F. Blackburn, S. K. Tripathy, and
       L. A. Samuelson, *J. Mol. Electron.* **2**, 119 (1986).
[15]   N. J. Geddes, J. R. Sambles, D. J. Jarvis, W. G. Parker, and D. J. Sandman, *Appl. Phys. Lett.* **56**, 1916
       (1990).
[16]   N. J. Geddes, J. R. Sambles, D. J. Jarvis, W. G. Parker, and D. J. Sandman, *J. Appl. Phys.* **71**, 756 (1992).
[17]   N. Okazaki and J. R. Sambles, in *Extended Abstracts of the International Symposium on Organic
       Molecular Electronics, Nagoya, Japan,* 66 (2000).
[18]   D. Vuillaume, B. Chen, and R. M. Metzger, *Langmuir* **15**, 4011 (1999).
[19]   B. Chen and R. M. Metzger, *J. Phys. Chem. B* **103**, 4447 (1999).
[20]   J. W. Baldwin, B. Chen, S. C. Street, V. V. Konovalov, H. Sakurai, T. V. Hughes, C. S. Simpson, M. V.
       Lakshmikantham, M. P. Cava, L. D. Kispert, and R. M. Metzger, *J. Phys. Chem B* **103**, 4269 (1999).
[21]   M. Fujitsuka, O. Ito, and R. M. Metzger, unpublished results.
[22]   T. Xu, G. J. Szulczewski, T. A. Morris, R. R. Amaresh, Y. Gao, S. C. Street, L. D. Kispert, R. M. Metzger,
       and F. Terenziani, *J. Phys. Chem. B,* in press.
[23]   C. Krzeminski, C. Delerue, G. Allan, D. Vuillaume, and R. M. Metzger, *Phys. Rev. B* **64**, #085405 (2001).

Charge-Transfer Exciton Trapping at Vacancies in Molecular Crystals: Photoconductivity Quenching, Optical Damage and Detonation

R. W. Munn, * D. Tsiaousis*

Department of Chemistry, UMIST, Manchester M60 1QD, United Kingdom

Summary: Calculations are presented that show that vacancies can trap charge-transfer (CT) states in anthracene, acetanilide and hexahydro-1,3,5-trinintro-1,3,5-triazine (RDX). Such trapping provides a mechanism for photoconductivity quenching by geminate recombination, and for optical damage and detonation by concentrating optical or mechanical energy stored in CT states.

Keywords: charge transfer; molecular crystals; theory; trapping; vacancies

1. Introduction

Charge-transfer (CT) states in molecular crystals consist of an anion–cation pair (or an electron–hole pair on different molecules). Such states are well known as obvious precursors to photoconductivity,[1] they may play a role in optical damage, where multiphoton ionization is a possible mechanism,[2] and they have been postulated as initial stages in detonation.[3]

Information on the mechanism of optical damage in molecular crystals is sparse, but it appears not to occur homogeneously and there are questions as to how uniform illumination yields the necessary concentration of energy. The initial mechanism of detonation remains a subject of investigation, but it is known that detonation proceeds at "hot spots" where energy is presumably trapped,[4] and is more facile in less dense and less perfect crystals.[5, 6] These observations suggest a possible role for vacancies, which are necessarily present in some thermodynamic concentration in real crystals. If vacancies trap CT states, they could serve as sites to concentrate the energy that these states store from the optical or mechanical mechanism that generates them. (Recombination would also at once reduce the efficiency of photoconduction.) The energy thereby released could also disrupt the crystal lattice near the vacancy: as Pope and Swenberg remark,[7] "The recombination of a hole and electron releases considerable energy, and

DOI: 10.1002/masy.200450807

one of the problems that must be considered is how this energy is disposed of." The larger defective region so produced could presumably trap CT states even more efficiently, and optical damage could thereby ensue. Alternatively, energy concentrated and released in this way could initiate the chemical decomposition involved in detonation, and as Ramaswamy and Field remark,[8]" It is always of prime interest to understand how energy is concentrated in an explosive to form hot spots."

The possibility that vacancies could trap CT states is plausible, given that vacancies can trap single charge carriers, as indicated by theoretical calculations for the aromatic hydrocarbon crystals[9] and some experimental evidence.[10] The energy of an ion in a crystal is lower than that of the isolated ion by the polarization energy P. This is reduced near a vacancy by an amount ΔP because of the loss of the polarizability of the missing molecule by this mechanism, and so by the same mechanism a vacancy scatters charges. However, the loss of the permanent multipole moment of the missing molecule leads to trapping or scattering in different directions, and trapping by this mechanism can readily outweigh the scattering due to ΔP.

Similar arguments can be developed for vacancies and CT states,[11] but need modification because the relevant polarization energy of the CT states is E_p, which is the polarization relative to the total polarization energy for the separated charges in the crystal, and there is in addition the stabilizing Coulomb energy E_C between the electron and hole. Partial cancellation between the effects of the electron and the hole in the CT state makes E_p positive, which serves to reduce the magnitude of the bare Coulomb energy E_C to the screened Coulomb energy E_{scr}. Thus polarization is responsible for the screening that weakens the attraction between the separated changes. A vacancy reduces the polarization because of the loss of the polarizability of the missing molecule, and thereby reduces the screening. As a result, the charges attract more strongly and the vacancy acts as a trap for CT states. The loss of the permanent multipole moment at the vacancy should also trap some CT states in some directions, and in this case the trapping can be reinforced by the polarization energy change (rather than offset as for a single carrier).

We have therefore calculated the energetics of charge-transfer states near vacancies in three crystals: anthracene (a benchmark crystal where photoconductivity has been extensively studied), acetanilide (one of the few crystals where optical damage has been studied in any detail), and 1,3,5-trinitrohexahydro-1,3,5-triazin or RDX (a well-known energetic material where detonation has been extensively studied). We first summarize our method, and then report results for each crystal, followed by overall conclusions.

2. Method

Our method[11] combines previous treatments for single charges and CT states in perfect crystals and for single charges near vacancies, which use a Fourier-transform method to obtain explicit solutions for polarization energies. These treatments entail the calculation of wave-vector modulated lattice monopole and dipole sums over a mesh of wave-vectors in the first Brillouin zone. Polarization energies are then calculated by summation over wave-vectors, making the mesh finer until convergence is achieved (a process facilitated by using the dependence on mesh size to extrapolate to the continuum limit).

For the calculations on anthracene, these treatments suffice, because they also treat charge–quadrupole energies W_Q. However, acetanilide and RDX are both polar molecules, so that a treatment of charge–dipole energies is also required. This is complicated because it is also necessary to take account of the dipoles induced in the crystal environment by the other molecules, including the effect of the vacancy. Details of the method are given elsewhere.[12] The principal results are that the charge–dipole energy is zero in the perfect crystal if it does not carry a macroscopic dipole moment, and that the change in charge–dipole energy ΔW_D can be obtained by a modification of the method used to calculate the change in polarization energy ΔP, with the permanent dipole field in the perfect crystal $\mathbf{F}^P$ and the loss of the permanent dipole moment at the vacancy $\boldsymbol{\mu}$ as new source terms. The algebraic expression for ΔP is

$$\Delta P = \tfrac{1}{2}\tilde{\mathbf{F}}^0 \cdot \boldsymbol{\alpha} \cdot \mathbf{Q} \cdot \mathbf{F}^0 / \varepsilon_0 v, \tag{1}$$

where $\mathbf{F}^0$ is the field due to the charges, $\boldsymbol{\alpha}$ is the polarizability of the missing molecule at the vacancy, v is the unit-cell volume, and the quantity $\mathbf{Q}$ corrects for the missing polarizability in the dielectric response of the crystal. The expression for ΔW_D is

$$\Delta W_D = \left(\tilde{\mathbf{F}}^0 \cdot \boldsymbol{\alpha} \cdot \mathbf{Q} \cdot \mathbf{F}^P - \boldsymbol{\mu} \cdot \mathbf{Q} \cdot \mathbf{F}^0\right)/\varepsilon_0 v. \tag{2}$$

All the calculations treat each molecule as a set of point-polarizable submolecules, one at each heavy (non-hydrogen) atom. Effective molecular polarizabilities relevant to the crystal environment are derived by analysing crystal dielectric or optical data with published crystal structure data. The quadrupole moment for anthracene is that used previously.[13] For acetanilide and RDX, the dipole moments are calculated by standard high-level quantum-chemical techniques, using the in-crystal geometry. The change in molecular geometry from the free state to the crystal is the most immediately obvious and tractable environmental effect on the molecules and their properties.[14] In RDX it is particularly significant because the conformation of the nitro groups attached to the ring (which is in the chair conformation) changes from all-axial in the gas phase[15] to two axial and one pseudo-equatorial in the crystal.[16]

3. Results

3.1 Anthracene

Anthracene forms monoclinic crystals with two equivalent molecules in the unit cell,[17] as shown in Figure 1, right. The total energy (polarization energy plus charge–quadrupole energy) for an electron in the perfect crystal is -1.43 eV and for a hole -0.93 eV. The lowest CT states with the hole fixed at the origin have the electron at $(\frac{1}{2},\pm\frac{1}{2},0)$ and $(0,1,0)$ with total energies E_p relative to the separate electron and hole of -0.86 eV and -0.73 eV, respectively. There are no other CT states below -0.5 eV. A vacancy at the origin changes the energy of an electron at $(\frac{1}{2},-\frac{1}{2},0)$ by -0.15 eV and at $(\frac{1}{2},\frac{1}{2},0)$ by 0.12 eV; for a hole at these sites, the energy changes by 0.26 eV and 0.01 eV. These changes are dominated by those in the charge–quadrupole energy.

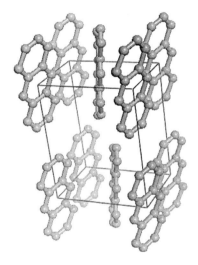

Figure 1. Unit cell of anthracene (hydrogens omitted), looking at the ac face.

For the CT state with the hole at the origin and the electron at $(\frac{1}{2},\frac{1}{2},0)$, a vacancy at a neighbouring site changes the magnitude of the polarization energy change ΔE_p relative to the separate charges by $0.03 - 0.05$ eV. The corresponding change in the charge–quadrupole energy reaches -0.16 eV for the vacancy at $(1,0,0)$ and -0.24 eV for the vacancy at $(0,1,0)$. Vacancies at these two sites (both nearest neighbours of the CT electron) therefore yield traps at -0.19 eV and

–0.29 eV, respectively. Hence the postulated existence of trapping of CT states by vacancies is confirmed.

Similar results are found for the CT state with the hole at the origin and the electron at (0,1,0). A vacancy at (½,½,0) yields a trap at –0.29 eV, and a vacancy at (½,1½,0) yields a trap at –0.21 eV; again, both sites are nearest neighbours of the CT electron. Because the lowest CT states are only 0.13 eV apart, it follows that the trap manifolds based on them overlap. However, they remain distinct from those based on higher CT states.

In these cases, ΔE_p is negative, consistent with the idea that a vacancy reduces the dielectric screening. However, for some CT states a vacancy on or near the charge-transfer axis can actually increase the screening by up to 0.04 eV. This may be associated with the fact that in an anisotropic dielectric continuum, screening depends not on the component of the dielectric tensor along the CT axis but rather on the components in the plane perpendicular to this axis.[18]

However, overall it is clear that vacancies constitute traps for the lowest CT states, with a depth of several times kT at room temperature. Vacancies could therefore have a significant adverse effect on the yield of photogenerated change by facilitating geminate recombination. A more detailed account of these results is given elsewhere.[11]

We have also investigated the effect of divacancies on CT states. These obviously entail rather complicated calculations, involving two vacant sites and two charged molecules. Overall, we find that the energy changes for divacancies near the lowest two CT states differ very little from the sum of the energy changes for the two separate vacancies. For example, the magnitude of the polarization energy change ΔE_p relative to the separate charges lies in the range $0.05 - 0.08$ eV for several divacancy arrangements, values that are close to the sum of those for the separate vacancies. The change in charge–quadrupole energy ΔW_Q^{CT} for the divacancy is already simply the sum of those for the separate vacancies, and dominates the total energy. Then for the CT state with the hole at the origin and the electron at (½,½,0), a divacancy at (1,0,0) and (0,1,0) constitutes a trap of depth 0.48 eV, while for the CT state with the hole at the origin and the electron at (0,1,0), a divacancy at (½,½,0) and (–½,1½,0) constitutes a trap of depth 0.578 eV.

3.2 Acetanilide

Acetanilide forms orthorhombic crystals with eight equivalent molecules in the unit cell,[19] as shown in Figure 2, below. This complicates identification and classification of the CT states. We divide the molecules into two sets, one labelled 1 – 4 and the other labelled 5 – 8, related to 1 – 4 respectively by inversion, and use (k; l_1,l_2,l_3) to denote molecule k in unit cell (l_1,l_2,l_3). Previous work treating the molecule as three submolecules found one surprisingly small local-field component,[20] but the present treatment with all heavy atoms as submolecules removes this anomaly. Details of the analysis that gives the effective molecular polarizability and the local field will be given elsewhere.[21] The dipole moment of an isolated molecule with the crystal structure is calculated using density functional theory as 3.62 D; in the crystal environment the molecules polarize one another through fields of their dipoles[22] and we find the effective moment is 5.41 D.

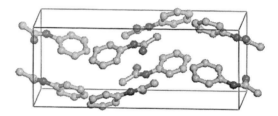

Figure 2. Unit cell of acetanilide (hydrogens omitted), looking at the *ac* face.

The polarization energy of an electron or hole in the perfect crystal is –1.16 eV; the charge–dipole energy is zero assuming that the crystal does not carry a net dipole moment. The lowest CT states with the hole fixed on molecule (1;0,0,0) have the electron on molecule (5;0,–1, –1) with an energy of –1.30 eV, on molecule (7;0,0,0) with an energy of –1.11 eV, and on molecule (3;0,0,0) in the origin cell with an energy of –0.93 eV. All other CT states lie at –0.75 eV or above.

For a charge at molecule (1;0,0,0), a vacancy at these sites changes the polarization energy by 0.09, 0.06 and 0.04 eV respectively. The corresponding changes in the charge–dipole energy for a hole at molecule (1;0,0,0) are –0.10, 0.24 and –0.23 eV (opposite sign for an electron). Thus the total energy changes are –0.01, 0.30 and –0.19 eV for a hole and 0.19, –0.18 and 0.27 eV for an electron. These results show that a vacancy can trap carriers of either sign, but at different sites, through the change in charge–dipole energy. This confirms that vacancies can act as traps in crystals of polar molecules, as well as in crystals of quadrupolar molecules like the linear condensed polyacenes.

Because of the eight molecules in the unit cell, the number of arrangements of a vacancy relative to each CT state is very high. However, the results are rather uniform. For most positions of a vacancy in a cell adjacent to either charge of the CT state, there is a trap with a depth of 0.01 – 0.02 eV. In a few cases the trap depth reaches 0.05 eV, for example for the lowest CT state with the electron at molecule (5;0,–1,–1) and a vacancy at molecule (7;0,0,–1) and for the next lowest CT state with the electron at molecule (7;0,0,0) and a vacancy at molecule (5;0,–1,–1). These values are dominated by the polarization energy change ΔE_p. The changes in charge–dipole energy cancel almost exactly, in contrast to the changes in charge–quadrupole energy in anthracene. This difference may reflect the weaker angular dependence of the dipole potential in acetanilide and the larger number of molecules in the unit cell, but further analysis is required.

Calculations have also been performed for a few divacancy arrangements near the lowest CT state. As for anthracene, these confirm that a divacancy constitutes a deeper trap than either vacancy alone, and that to a good approximation the divacancy trap depth is the sum of those for the separate vacancies it comprises. This confirms the feasibility of a mechanism of optical damage whereby a vacancy traps CT states that recombine and release energy to form a larger defective region that acts as a yet deeper trap, so that the process is cumulative and autocatalytic in the sense that the products favour further reaction. This mechanism is also plausible since experimental evidence indicates that the mechanisms of defect formation and crystal destruction are the same.[23]

3.3 RDX

RDX presents a case that is formally very similar to acetanilide. There are again eight equivalent molecules in the unit cell,[16] as shown in Figure 3, right; we label the molecules as before. Detailed dielectric analysis will be reported elsewhere,[24] together with calculations of the dipole

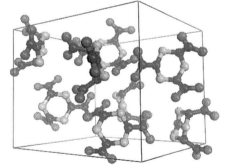

Figure 3. Unit cell of RDX (hydrogens omitted), with the *ac* face on the left.

moment, which is 7.40 D for the isolated molecule and 9.40 D in the crystal.

The polarization energy of an electron or hole in the perfect crystal is -1.14 eV, with zero charge–dipole energy. The lowest CT state with the hole fixed on molecule $(1;0,0,0)$ has the electron on molecule $(5;-1,0,0)$ with an energy of -1.27 eV. After that, there are three states with the electron on molecule $(7;0,0,0)$, on molecule $(3;-1,0\ 0)$ and on molecule $(6;0,0,0)$ with energies of -0.88, -0.83 and -0.81 eV. Thus the lowest CT state is well separated from the manifold of higher CT states.

For a charge at molecule $(1;0,0,0)$, a vacancy at molecule $(5;-1,0,0)$ changes the polarization energy by 0.10 eV, with the corresponding change in the charge–dipole energy for an electron as -0.52 eV. Thus the total energy change is -0.42 eV, and this vacancy can trap electrons. Other electron traps are found for vacancies at molecule $(3;-1,-1,0)$, depth 0.28 eV, and at molecule $(6;0,0,0)$, depth 0.19 eV. Hole traps are found for vacancies at molecule $(8;0,0,0)$, depth 0.28 eV, and for molecules $(3;-1,0,0)$ and $(6;-1,0,0)$, depth 0.22 eV. Vacancies in RDX thus form rather deeper traps for single carriers than in acetanilide, because the dipole moment is larger.

As for acetanilide, the effect of vacancies is dominated by the polarization energy changes ΔE_p. The deepest traps lie at -0.04 eV, for vacancies at molecule $(3;-1,0,0)$ and molecule $(7;0,0,0)$, which are symmetrically related to the CT state. Thus vacancies can act as rather shallow traps for CT states in RDX.

4. Conclusions

We have presented calculations of the energies of charge-transfer states near vacancies in anthracene, acetanilide and RDX. The results confirm qualitative arguments that vacancies can trap CT states. The polarization energy change due to the vacancy is almost always negative, i.e. trapping. The charge–multipole energy change due to the vacancy is positive or negative, depending on the direction, and dominates the polarization energy change for anthracene in most directions but tends to cancel for acetanilide and RDX. Trap depths for a single vacancy are as large as 0.3 eV for anthracene, but for acetanilide and RDX are shallower at no more than 0.05 eV, or only twice kT at room temperature. For acetanilide, defects that are microscopic precursors to macroscopic optical damage are found to accumulate at low temperatures but not at room temperature,[2] which is consistent with calculated trap depths comparable to kT at room temperature.

These results show that vacancies can have a significant influence in processes where CT states are important participants. This could be explored experimentally given crystals with differing vacancy concentrations. In anthracene, the efficiency of photogeneration of charge carriers via CT states will be reduced by geminate recombination of CT states trapped at vacancies. In acetanilide, the energy released by this process would be available to cause further disruption of the lattice. Our calculations show that divacancies produce traps for CT states of depths roughly equal to the sum of the depths for the two separate vacancies, and this provides a mechanism for optical damage to accumulate. Similarly, in RDX vacancies can trap CT states produced by mechanical excitation and thereby concentrate the energy so that it can initiate the chemical process of detonation. This mechanism is consistent with the identification of "hot spots" for detonation as vacancies (or similar lattice imperfections).[4, 25, 26, 27] We find that the trapping is dominated by the polarization energy change at the vacancy, and hence is rather isotropic, which is consistent with the weakness of directional effects on detonation RDX but not in other crystals.[28] Our suggestion that the mechanisms for optical damage and initiating detonation may have features in common (specifically the trapping of CT states by vacancies) is consistent with the observation of laser-induced detonation,[8] often used to explore the mechanism of detonation.

We have concentrated here on our overall results, which show how different processes typical of three different crystals can share an underlying mechanism, namely trapping of CT states by vacancies. Detailed results for the individual crystals are given elsewhere,[11, 21, 24] as is the derivation of the change in the charge–dipole energy.[12] All our results use exact algebraic solutions, including the change in the dielectric response of a crystal with a defect.[29] Such rigorous treatments of electrical interactions in perfect and imperfect crystals can yield quantitative insights into a wide range of problems, as the present work illustrates.

Acknowledgements

We are grateful to C. J. Eckhardt and T. Luty for drawing our attention to the possible role of CT states in the initiation of detonation, and to P. J. Smith for assistance with quantum-chemical calculations. This work was initiated with funding from EU TMR Network "DELOS", contract FMRX-CT96-0047

[1]   L. Sebastian, G. Weiser, G. Peter and H. Bässler, *Chem. Phys.* **1983**, *75*, 103.
[2]   T. J. Kosic, J. R. Hill and D. D. Dlott, *Chem. Phys.* **1986**, *104*, 169.
[3]   T. Luty, P. Ordon and C. J. Eckhardt, *J. Chem. Phys.* **2002**, *117*, 1775.
[4]   A. B. Kunz and M. M. Kuklja, *Thermochim. Acta* **2002**, *384*, 279.
[5]   F. P. Bowden and Y. D. Yoffe, *"Initiation and Growth of Explosion in Liquids and Solids"*, Cambridge University Press, London 1952, p. 64.
[6]   A. W. Campbell, W. C. Davis, J. B. Ramsay and J. R. Travis, *Phys. Fluids* **1961**, *4*, 511.
[7]   M. Pope and C. E. Swenberg, *"Electronic Processes in Organic Crystals and Polymers"*, 2nd ed., Oxford, New York 1999, p. 501.
[8]   A.L. Ramaswamy and J. E. Field, *J. Appl. Phys.* **1996**, *79*, 3842.
[9]   I. Eisenstein and R. W. Munn, *Chem. Phys.* **1983**, *77*, 47.
[10] E. A. Silinsh, I. J. Muzikante, A. J. Rampans and L. F. Taure, *Chem. Phys Letters.* **1984**, *105*, 617.
[11] D. Tsiaousis and R. W. Munn, *J. Chem. Phys.* **2002**, *117*, 1833.
[12] D. Tsiaousis and R. W. Munn, unpublished results.
[13] P. J. Bounds and R. W. Munn, *Chem. Phys.* **1981**, *59*, 41.
[14] R. W. Munn, M. Malagoli and M. in het Panhuis, *Synthetic Metals* **2000**, *109*, 29.
[15] I. F. Shishkov, L. V. Vilkov, M. Kolonits and B. Roszondai, *Struct. Chem.* **1991**, *2*, 57.
[16] C. S. Choi and E. Prince, *Acta Cryst. B* **1972**, *28*, 2857.
[17] C. P. Brock and J. D. Dunitz, *Acta Cryst. B* **1990**, *46*, 795.
[18] P. J. Bounds, W. Siebrand, I. Eisenstein, R. W. Munn and P. Petelenz, *Chem. Phys.* **1985**, *95*, 197.
[19] S. W. Johnson, J. Eckert, M. Barthes, R. K. McMullan and M. Muller, *J. Phys. Chem.* **1995**, *99*, 16253.
[20] R. W. Munn and M. Hurst, *Chem. Phys.* **1990**, *147*, 35.
[21] D. Tsiaousis and R. W. Munn, unpublished results.
[22] R. W. Munn and M. Hurst, *Chem. Phys.* **1990**, *147*, 35.
[23] J. R. Hill, T. J. Kosic. E. L. Chronister and D. D. Dlott, in: *Springer Proceedings in Physics, Vol. 4. Time-resolved vibrational spectroscopy"*, A. Laubereau and M. Stockburger, Eds, Springer, Berlin 1985, p. 107.
[24] D. Tsiaousis and R. W. Munn, unpublished results.
[25] L. Phillips, *J. Phys:. Condens. Matter* **1995**, *7*, 7813.
[26] M. M. Kuklja, E. V.Stefanovich and A. B. Kunz, *J. Chem. Phys.* **2000**, *112*, 2215.
[27] M. M. Kuklja, *J. Phys. Chem. B* **2001**, *105*, 10161.
[28] P. Maffre and M. Peyrard, *Phys. Rev. B* **1992**, *45*, 9551.
[29] I. Eisenstein, R. W. Munn and P. J. Bounds, *Chem. Phys.* **1983**, *74*, 307.

Macromol. Symp. **2004,** *212,* 83-91

Materials for Polymer Electronics Applications –
Semiconducting Polymer Thin Films and Nanoparticles

U. Asawapirom,[1] *F. Bulut,*[2] *T. Farrell,*[1] *C. Gadermaier,*[5] *S. Gamerith,*[5] *R. Güntner,*[1]
T. Kietzke,[3] *S. Patil,*[1] *T. Piok,*[5] *R. Montenegro,*[4] *B. Stiller,*[3] *B. Tiersch,*[1]
K. Landfester,[4] *E. J. W. List,*[5] *D. Neher,*[3] *C. Sotomayor Torres,*[2] *U. Scherf**[1]

[1] Universität Potsdam, Institut für Chemie, Karl-Liebknecht-Str.24/25, Haus 25,
  14476 Golm, Germany
  E-mail: scherf@rz.uni-potsdam.de
[2] BUGH Wuppertal, FB Elektrotechnik, Gaußstr.20, 42097 Wuppertal, Germany
[3] Universität Potsdam, Institut für Physik, Am Neuen Palais 10, D-14469 Potsdam,
  Germany
[4] Max-Planck-Institut für Kolloid- und Grenzflächenforschung, 14424 Golm,
  Germany
[5] TU Graz, Institut für Festkörperphysik, Petersgasse 16, 8010 Graz, Austria

Summary: The paper presents two different approaches to nanostructured semiconducting polymer materials: (i) the generation of aqueous semiconducting polymer dispersions (semiconducting polymer nanospheres SPNs) and their processing into dense films and layers, and (ii) the synthesis of novel semiconducting polyfluorene-*block*-polyaniline (PF-*b*-PANI) block copolymers composed of conjugated blocks of different redox potentials which form nanosized morphologies in the solid state.

Keywords: block copolymer; nanoparticle; nanostructure; polyaniline; polyfluorene; semiconducting polymer

Introduction

Solid layers of conjugated polymers have been successfully included as active layers into various electrical and electrooptical devices such as light-emitting diodes (LEDs),[1] solar cells[2] and field-effect transistors (FETs).[3] In most cases, these layers have been deposited from solutions of the polymers in organic solvents. However, deposition from those solvents brings about several problems, in particular when dealing with large-area or multilayer devices; the deposition from aqueous systems would be most desirable. Recently, polymer conductors, such as poly[(ethylenedioxy)thiophene] (PEDOT) doped with poly(styrenesulfonate) (PSS) have been deposited from water-based dispersion,[4] but this approach is focused on the deposition of

DOI: 10.1002/masy.200450808

electrically conducting polyelectrolytes (PEDOT, or polyaniline PANI). Very recently, we could demonstrate for the first time the film formation from semiconducting, preferably fluorescent polymer nanospheres (SPN), deposited onto solid substrates from aqueous dispersions.[12]

For LED and FET applications, the active semiconducting polymers are mostly used as single-component materials, such as isotropic films or layers of uniform morphology. For other applications, in particular the use of conjugated materials in photovoltaic devices (solar cells, photodetectors), more complex materials have become more and more attractive. [5,6] The efficiency of such devices is often limited by the short exciton diffusion range in conjugated polymers relative to their optical absorption depth. Structures in which a heterojunction is distributed throughout the film have become increasingly important, e.g. conjugated polymers doped with fullerene acceptors or fullerene/polymer multilayer systems.[7,8] Since only the light absorbed close to the heterojunction between the components results in charge generation, the width of this active region is limited by the exciton diffusion range of photogenerated charge carriers (in conjugated polymers typically 10-20 nm [9,10]). Novel systems, which are able to undergo internal nano- and microphase separation, came in the focus of interest, since such micro- or nanoscopically structured systems allow for an effective charge separation (and transport) of optically generated electron/hole pairs.

Several approaches to the generation of such systems have been used: (i) Mixtures of two immiscible polymers can segregate (demix) to complex, phase-separated structures.[11] The length scale of phase separation is often in the "meso" length regime (10-100 nm), and is convenient but often less controlled. (ii) Semiconducting polymer nanospheres (SPNs) can be processed with a second, bulk component into nanostructured heterojunction composites. (iii) In conjugated-conjugated block copolymers the scale length of phase separation and the electronic nature of the phases is expected to be directly related to the chemical structure and the length of both blocks. It will be possible to fine-tune the scale length of phase separation to be in the range of the diffusion lengths of the (optically formed) excitons. That would allow for a sufficient charge carrier transport of holes and electrons.

Such conjugated-conjugated block copolymers represent a so far unknown class of electronically active polymers. In particular those composed of two conjugated blocks of different oxidation and reduction potentials are very attractive. If they form suited nanophase- and microphase-separated structures, efficient charge separation can occur, favourably followed by an efficient transport of the charge carriers to the corresponding electrodes.

Semiconducting Polymer Nanospheres

We could recently demonstrate for the first time the film formation from semiconducting, preferably fluorescent polymer nanospheres (SPN), deposited onto solid substrates from aqueous dispersions. [12]

Nanoparticles from semiconducting polymers in aqueous phase have been produced via the miniemulsion process. Miniemulsions are understood as stable emulsions consisting of stable droplets with a size of 50 - 500 nm by shearing a system containing oil, water, a surfactant, and a highly water-insoluble compound, the so-called hydrophobe.[13,14] The surfactant stabilizes the droplets against collisions; mass exchange (Ostwald ripening) between the droplets is suppressed by the use of the hydrophobe. For our SPN dispersions, we have started from artificial latexes consisting of a solution of the preformed semiconducting polymer. After evaporation of the solvent, the polymer dispersion is obtained.

Solutions of different conjugated semiconducting polymers in chloroform (with concentrations between 1.5 and 5.4 %) were successfully miniemulsified in water by using sodium dodecyl sulfate (SDS) as surfactant. Me-LPPP is a solution processable poly(p-phenylene)-type ladder polymer which has been widely used as active semiconducting material in electronic devices (light emitting diodes LEDs, solid state lasers, photodiodes).[15,16] Polyfluorene (PF) derivatives are characterized by a unique combination of semiconducting and liquid-crystalline (LC) properties, and have been applied as high-performance blue emitters in LEDs based on organic semiconducting polymers.[17-19] Poly(cyclopentadithiophene)s (PCPDT) as heterocyclic PF analogues are characterized by a reduced band gap (HOMO/LUMO) energy relative to PF, they are promising materials for potential use in organic materials-based field-effect transistors (FETs) and solar cells.[20] After evaporation of the solvent (chloroform), stable polymer dispersions with solid contents of 5-10 % and a polymer particle size of 50-250 nm were obtained.

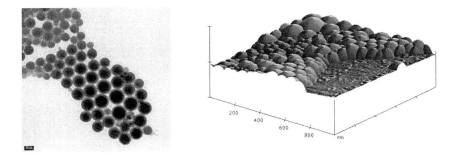

Me-LPPP PF PCPDT

Chemical structure of Me-LPPP (R_1: hexyl, R_2: methyl, R_3: 4-decylphenyl), PF (R_1: 2-ethylhexyl - PF2/6, and 3,7,11-trimethyldodecyl - PF11112, respectively), and PCPDT (R_1: 3,7-dimethyloctyl).

In Figure 1a, a TEM picture of a dispersion with Me-LPPP particles shows spherical and, due to their high T_g, hard particles. It has to be mentioned that for this measurement, the dispersion is highly diluted leading to isolated clusters of particles.

Figure 1: a) (left) TEM picture of Me-LPPP particles; b) (right) 3D AFM picture of semiconducting nanosphere monolayer, prepared by spin-coating a Me-LPPP dispersion on glass (part of the material has been removed by scratching)

Homogeneous layers of SPNs could be prepared by spin-coating the dispersion with the negatively charged particles onto a glass substrate. The resulting films consist of a monolayer of closely packed particles, as seen by AFM; the micrographs do not reveal any cracks within an area of $5 \times 5 \ \mu m^2$ (Figure 1b). Because Me-LPPP has its glass transition temperature well above the onset of decomposition (300-350 °C), the particulate structure also does not change during annealing at 200 °C for 2 h. The absorption and photoluminescence (PL) emission spectra recorded on these films (Figure 2) are identical to those reported for layers of Me-LPPP from organic solvents.[16]

For homogeneous layers of polyfluorene (PF) particles, in contrast to Me-LPPP, annealing of these layers above T_g at 200 °C for 2 h (transition to the birefringent fluid (nematic) LC phase at 100-170 °C [21,22]) results in coalescence of particles and formation of larger structures.

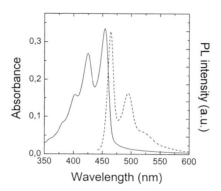

Figure 2: Absorption and photoluminescence spectrum of the SPN layer (Me-LPPP) of Figure 1b

In conclusion, we have shown that layers of conjugated semiconducting polymers can be deposited from aqueous dispersion prepared by the miniemulsion process. Dispersions of particles of different conjugated semiconducting polymers such as a ladder-type poly(p-phenylene) and several soluble derivatives of polyfluorene could be prepared with well controllable particle sizes ranging between 50 - 250 nm. Layers of these particles formed by spincoating exhibit a particulate structure, revealing the shape of individual polymer nanoparticles. Annealing above the glass transition temperature of the polymer results in coalescence of particles, larger domains of continuous structure being formed. We, therefore, propose that the concept of semiconducting polymer nanoparticles allows to form multilayer structures by, e.g., depositing a first layer from a solution of a polymer and overcoating it by semiconducting polymer nanospheres of the second polymer from an aqueous phase, followed by annealing and film formation. Most important, this will allow the formation of a multilayer structure from polymers, which are highly soluble in the same solvents, without introducing any additional chemical conversion steps.

Polyfluorene/Polyaniline Block Copolymers

The synthesis of conjugated 9,9-dialkylfluorene/2-alkylaniline block copolymers was described by us in two earlier publications.[23,24] The first step of the today favoured synthetic route towards PF/PANI block copolymers involves the aryl-aryl coupling of 2,7-dibromo-9,9-dialkylfluorenes in the presence of unprotected 4-bromoaniline according to Yamamoto. The coupling reaction leading to prepolymer **1** surprisingly tolerates unprotected 4-aminophenyl functions. In this way, the one-step synthesis of 4-aminophenyl end-functionalized polyfluorenes **1** was carried out. The molecular weight of the polyfluorene prepolymer **1** can be controlled by the feed ratio of bifunctional dibromofluorene/4-bromoaniline. The end-cappers are indeed chemically attached to the prepolymer chains, which can be proved by elemental analysis and ^{1}H NMR spectroscopy. The NMR analysis indicates functionalization with ca. 2.0 4-aminophenyl end groups per prepolymer **1** chain, the PF chains thus being completely terminated with 4-aminophenyl functions.

Synthetic route to polyaniline-*block*-polyfluorene-*block*-polyaniline (PF-*b*-PANI) copolymers.

The second reaction step of our improved PF-*b*-PANI synthesis then involves the oxidative coupling (reagent: $(NH_4)_2S_2O_8/H^+$) of NH_2-terminated prepolymer **1** with 2-undecylaniline. The AB-type character of the aniline monomer guarantees the exclusive formation of PF/PANI block copolymers **2** during this reaction step without crosslinking, branching or multiblock formation. The undecyl-substituted aniline monomer was chosen because of the increased solubility of the corresponding, alkylated PANI building block.[25] The poly(2-undecylaniline) homopolymer is, in contrast to the insoluble unsubstituted polyaniline, slightly soluble in several polar organic solvents (e.g. THF), but insoluble in toluene.[25] Based on this property, the homocoupling product poly(2-undecylaniline), which is formed as (unwanted) side-product remains as insoluble residue during an extraction of the reaction product with toluene.

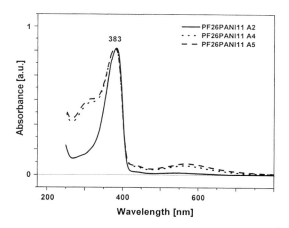

Figure 3: UV-Vis spectra of PF-*b*-PANI block copolymers **2** with different PANI block lengths (R_1: 2-ethylhexyl, R_2: undecyl)

An independent proof that the PANI blocks are really covalently attached to the PF blocks came from GPC analysis with a parallel UV-Vis detection at the distinctly different absorption maxima of the PF (380 nm) and the PANI block (540-560 nm), and RI detection (control experiment). In particular, the identical peak molecular weights M_p indicate, that the absorbing species at ca. 380 and 540-560 nm correspond to molecules with (nearly) identical molecular weights. Also considering the low molecular weight of the poly(2-undecylaniline) homopolymer which is formed as side-product (M_n < 5000), these results are a strong proof of the presence of the expected PF/PANI block copolymer **2**.

The UV-VIS spectrum of **2** (Figure 3) indicates the presence of the poly(2-undecylaniline) blocks, for which a red-shifted, low-intensity absorption band centred at ca. 540-560 nm is characteristic (λ_{max} for poly[9,9-bis(2-ethylhexyl))fluorene] homopolymer 385 nm, for poly(2-undecylaniline) homopolymer 550 nm). The relatively weak intensity of the PANI absorption at 540-560 nm is a result of the relatively low absorbance of poly(2-undecylaniline) at this wavelength.

The next step was the characterization of the morphology of **2** by AFM and TEM (Figure 4). First, tapping (contact) mode AFM, and TEM images of block copolymer **2** are depicted in Figure 2: (a) an AFM image of a spin-coated film of a PF/PANI block copolymer **2** (film thickness ca. 100 nm), (b) a TEM image of a spin-coated film of the same block copolymer **2** (film thickness ca. 100 nm).

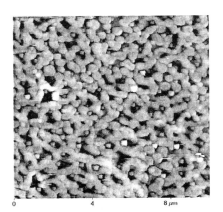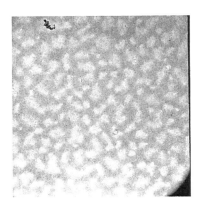

Figure 4: a) (left) AFM image (tapping mode) of a spin-coated film of the PF-*b*-PANI block copolymer **2** (film thickness ca. 100 nm, distance between the arrows: 0.6 μm); b) (right) TEM image of a spin-coated film of the same PF-*b*-PANI block copolymer **2** (film thickness ca. 100 nm)

Clearly, the copolymers self-assemble into branched cylindrical morphologies already during spin-coating, with a diameter of the nanocylinders of ca. 50-300 nm. Annealing (2 h / 200 °C) of the films leads to formation of denser packed films but under conservation of the nanostructered morphology.

In summary, we have developed a simple synthesis of novel PANI-PF-PANI triblock copolymers. The next step will be a more detailed investigation of the formed nanostructures, with respect to the relative and absolute block lengths, as well as the processing conditions.

[1]   R. H. Friend, R. W. Gymer, A. B. Holmes, J. H. Burroughes, R. N. Marks, C. Taliani, D. D. C. Bradley, D. A. Dos Santos, J. L. Bredas, M. Logdlund, W. R. Salaneck, *Nature* **1999**, *397*, 121.
[2]   C. J. Brabec, N. S. Sariciftci, J. C. Hummelen, *Adv. Funct. Mater.* **2001**, *11*, 15.
[3]   A. R. Brown, C. P. Jarrett, D. M. Deleeuw, M. Matters, *Synth. Met.* **1997**, *88*, 37
[4]   B. L Groenendaal, F. Jonas, D. Freitag, H. Pielartzik, J. R. Reynolds, *Adv. Mater.* **2000**, *12*, 481.
[5]   N. S. Sariciftci, A. J. Heeger, *Int. J. Mod. Phys.* **1994**, *B 8*, 237.
[6]   N. S. Sariciftci, *Prog. Quantum Electron.* **1995**, *19*, 131.
[7]   S. Morita, A. Zakhidov, K. Yoshino, *Solid State Commun.* **1992**, *82*, 249.
[8]   B. Kraabel, J.C. Hummelen, D. Vacar, D. Moses, N.S. Sariciftci, A.J. Heeger, F. Wudl, *J. Chem. Phys.* **1996**, *104*, 4267.
[9]   A. Haugeneder, C. Kallinger, W. Spirkl, U. Lemmer, J. Feldmann, U. Scherf, E. Harth, A. Gügel, K. Müllen, in: Z.H. Kafafi (Ed.), *Proc. SPIE* **1997**, *3142*, 140.
[10] A. Haugeneder, M. Neges, C. Kallinger, W. Spirkl, U. Lemmer, J. Feldmann, U. Scherf, E. Harth, A. Gügel, K. Müllen, *Phys. Rev. B* **1999**, *59*, 15346.
[11] D. Vacar, E. Maniloff, D. McBranch, A.J. Heeger, *Phys. Rev. B* **1997**, *56*, 4573.
[12] K. Landfester, R. Montenegro, U. Scherf, R. Güntner, U. Asawapirom, S. Patil, D. Neher, T. Kietzke, *Adv. Mater.* **2002**, *14*, 651.
[13] K. Landfester, *Macromol. Rapid Commun.* **2001**, *22*, 896.
[14] K. Landfester, *Adv. Mater.* **2001**, *13*, 765.
[15] U. Scherf, *J. Mater. Chem.* **1999**, *9*, 1853.
[16] G. Lanzani, S. De Silvestri, G. Cerullo, S. Stagira, M. Nisoli, W. Graupner, G. Leising, U. Scherf, K. Müllen, "Photophysics of Methyl Substituted Poly(para-Phenylene)-Type Ladder Polymers", in: *Semiconducting Polymers*, G. Hadziioannou and P. F. van Hutten, Eds, Wiley-VCH, Heidelberg, **1999**, Chap. 9, p. 235.
[17] Gross, D. C. Müller, H.-G. Nothofer, U. Scherf, D. Neher, C. Bräuchle, K. Meerholz, *Nature* **2000**, *405*, 661.
[18] T. Miteva, A. Meisel, W. Knoll, H.-G. Nothofer, U. Scherf, K. Müller, K. Meerholz, A. Yasuda, D. Neher, *Adv. Mater.* **2001**, *13*, 565.
[19] D. C. Müller, T. Braig, H.-G. Nothofer, M. Arnoldi, M. Gross, U. Scherf, O. Nuyken, K. Meerholz, *ChemPhysChem* **2000**, *1*, 207.
[20] U. Asawapirom, U. Scherf, *Macromol. Rapid Commun.* **2001**, *22*, 746.
[21] M. Grell, W. Knoll, D. Lupo, A. Meisel, T. Miteva, D. Neher, H.-G. Nothofer, U. Scherf, A. Yasuda, *Adv. Mater.* **1999**, *11*, 671.
[22] U. Scherf, E. J. W. List, *Adv. Mater.* **2002**, *14*, 477.
[23] C. Schmitt, H.-G. Nothofer, A. Falcou, U. Scherf, *Macromol. Rapid Commun.* **2001**, *22*, 624.
[24] R. Güntner, U. Asawapirom, M. Forster, C. Schmitt, B. Stiller, B. Tiersch, A. Falcou, H.-G. Nothofer, U. Scherf, *Thin Solid Films* **2002**, in press.
[25] A. Falcou, D. Marsacq, P. Hourquebie, A. Duchêne, *Tetrahedron* **2000**, *56*, 225.

Optical Grating Recording in Highly Organized Thin Films of Disperse Red 1

A. Miniewicz,[*1] *M. Solyga,*[1] *H. Taunaumang,*[2] *M.O. Tija*[2]

[1] Institute of Physical and Theoretical Chemistry, Wroclaw University of Technology, 50-370 Wroclaw, Poland
[2] Department of Physics, Institut Teknologi Bandung, Jalan Ganesa no.10 Bandung 40132, Indonesia

Summary: Optical grating recording with submicrometer spatial resolution, which can handle grey-level patterns, has been investigated in photochromic material made of Disperse Red 1 (DR1) molecules vacuum-deposited on a glass substrate. Holographic gratings of periods Λ within the range of 0.6 µm - 12 µm were recorded by 514.5 nm light from cw Ar^+ laser using a degenerate two-wave mixing technique. Despite the very small DR1 layer thickness ($\sim$ 0.1 µm), the diffraction efficiency measured in a Raman-Nath scattering regime reached 2 %. The obtained amplitude gratings were analysed with an optical microscope and Fourier transforms. Grating profiles were analysed in relation to exposure conditions and in correlation with molecular organisation. Polarising microscopy studies revealed the presence of light-induced optical anisotropy. Following that, we have checked the possibility of polarisation-sensitive recording in this medium.

Keywords: Disperse Red 1; grating recording; holographic materials; photochromism

Introduction

Properties of polymers containing photochromic azo dyes have received enormous attention. These materials could be suitable for applications in photonics, e.g. as holographic optical elements or 3-D data storage media [1-5]. Except polymers, a thin highly organised films of organic compounds can be envisaged for dense 2-D optical holographic storage of information providing their properties could be permanently and/or reversibly modified by laser light. Thin molecular layers of thickness of the order of 0.1 µm provide high spatial resolution suitable for holographic recording in a small-size region ($\sim$ 100 µm). One of the already known methods of

© 2004 International Union of Pure and Applied Chemistry

DOI: 10.1002/masy.200450809

producing such films is the physical vacuum deposition (PVD) technique. Using PVD with an option of heated substrate, the highly oriented thin layers of low-molecular-weight organic materials like DMANS ((dimethylamino)nitrostilbene) and p-NA (4-nitroaniline) were deposited [6-8]. Taunaumang et.al. [9] confirmed by x-ray, FTIR and other spectroscopic techniques that 4-[ethyl(2-hydroxyethyl)amino]-4-nitroazobenzene known as Disperse Red 1 (DR1), when evaporated on glass, Si or ITO-covered glass, can form highly organised molecular layers with molecules standing almost perpendicular to the surface. It is well known that molecules of DR1 embedded in polymer matrices can be easily switched by light between their two stable forms, *trans* and *cis*. This phenomenon should not occur in densely packed media due to the lack of free space around the molecule to permit the transition of *trans*-form into the bent *cis*-form of DR1. In our preceding paper, we investigated whether optical grating recording with 514.5 nm laser light is possible in highly organised layers of pure DR1 [10] and we have found that absorption-type gratings can be recorded. In this report, we would like to focus our attention on spatial resolution achievable in such a medium and, by discovering the possibility of polarisation grating recording, to shed more light on molecular mechanism responsible for the observed effects.

Experimental

DR1 with molecular mass 314.3 and melting point at 153 °C was obtained from Sigma. The DR1 thin films were evaporated onto glass and ITO-covered glass substrates under vacuum (2.7-4) x 10^{-3} Pa with crucible temperature kept at 173 °C. The temperature of the substrate was kept at 26 °C. After evaporation, no further annealing was performed. The layer thickness was determined with a Dektak IIA depth profilometer giving values of $0.07 - 0.1$ μm. Crystallinity of the obtained layers was examined with an X-ray Philips diffractometer ($\lambda = 1.5406$ Å). The most pronounced diffraction peak corresponded to an interlayer distance of 9.8 Å, which is close to the length of DR1 molecule in its elongated *trans* form. These and other investigations (FTIR, RAS FTIR) of molecular ordering [9] are consistent with the assumption that the PVD of DR1 on glass leads to near perpendicular long-molecular-axis alignment with respect to the substrate plane. It was also demonstrated that when during the PVD process substrates are kept at higher temperatures (55 and 75 °C), both the crystallinity and ordering increase [9]. Refraction index of the as-grown films measured by the standard prism coupling technique amounted to $n = 1.508$.

Results

Absorption spectra in the visible spectral region, shown in Fig. 1, exhibit a sharp peak at around 400 nm assigned to the formation of J-type or H-type molecular aggregates. Linear absorption coefficient estimated from the spectrum and direct light attenuation measurements at 514.5 nm, with the assumption of layer thickness $d = 0.1$ μm, am ounts to $\alpha \cong 40000$ cm^{-1}.

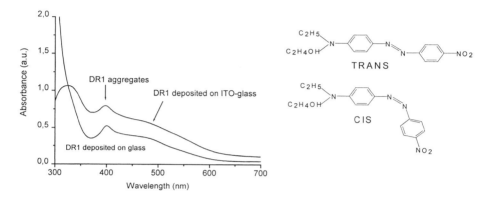

Figure 1. Absorption spectra of 0.1 μm thick layers of DR1 deposited by the PVD method on glass plate and ITO-covered glass plate kept at 26 °C during evaporation. On the right: *trans* and *cis* forms of the DR1 photoisomers are shown.

In the prepared DR1 layers, we recorded gratings using the standard holographic technique of two-wave mixing with an Ar$^+$-ion laser (Innova 90, Coherent) working at 514.5 nm wavelength. The technique was modified for the purpose of the present studies by introduction of a lens giving small spots of 294 or 167 μm in diameter measured at $1/e^2$ ($e = 2.718$) of the maximum intensity of a Gaussian beam at the center. The experimental set-up is shown in Fig. 2a. We used TEM$_{00}$ Gaussian beam with rotationally symmetric amplitude distribution given by

$A(\rho) = A_o \exp(-\rho^2 / w^2)$ and $I(r) = \dfrac{1}{2}\varepsilon_o cn |A(r)|^2$ where ρ is the cylindrical coordinate perpendicular to the direction of propagation z and w is the spot size. In the grating recording experiment, the intensity distribution $I(x)$ is periodically modulated and described by the relation:

$$I(x) = I_1 + 2\Delta I \cos(Kx) + I_2 \tag{1}$$

where $K = 2\pi/\Lambda$ is the grating wave vector and $\Delta I = \frac{n}{2}\varepsilon_0 c A_1 A_2^*$. In order to account for the cases

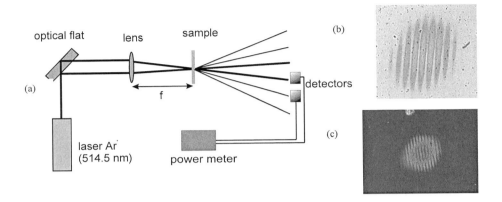

Figure 2. Experimental set-up for optical recording of gratings in DR1 layers deposited on glass. Measurements of 0th- and 1st-order diffraction intensities were made with a Labmaster Ultima (Coherent Inc.), two-channel power meter (a). Photographs of gratings under microscope (b) and under crossed-polarisers condition (c).

of grating recording in the media with anisotropic interaction, one defines $\Delta I = \left| tr\{\Delta M_{ij}\} \right|$, where the ΔM_{ij} is the interference tensor. In the s:s (VV) polarisation experiment, $\Delta I = \sqrt{I_1 I_2}$ while for the s:p (VH) configuration, $\Delta I = 0$ and there is no light intensity modulation along the x-direction (grating wavevector). In a typical experiment with $P_1 = 4.72$ mW, $P_2 = 3.28$ mW, modulation factor $m = 0.98$ and average light intensities at the fringe centres were 23 W/cm^2 and 72 W/cm^2 for $f_1 = 450$ mm and $f_2 = 255$ mm lenses, respectively. So recorded gratings in DR1 layers were permanent. Experimental diffraction efficiencies η were determined as the ratio of the power diffracted into first-order direction (P_{diff}) and the transmitted beam power (P_t), $\eta = \frac{P_{diff}}{P_t}$. From the value of Klein's parameter $Q = 0.0007$ ($Q = \dfrac{2\pi\lambda d}{n\Lambda^2}$, $\lambda = 0.5145$ μm, $n = 1.508$, $\Lambda = 17$ μm,

and $d = 0.1$ μm) it follows that we are in the Raman-Nath scattering regime of thin gratings. The transmittance of a thin grating can be expressed by the relationship $t(x) = \exp(\alpha(x)d)\exp(-ikdn(x))$ where $\alpha(x)$ and $n(x)$ are the x distributions of the index of absorption and index of refraction, respectively. An example of a microscopic image of a grating recorded in the DR1 film vacuum-deposited on glass is given in Fig. 2b ($\Lambda = 17$ μm). In the bright regions of the interference fringes, the absorption is increased compared with unilluminated areas. The grating profile in its central part can roughly be described by $\alpha(x) = \alpha_{av} + \Delta\alpha(1 + \sin(Kx))$ where $\alpha_{av} = 40000$ cm^{-1} is the average absorption coefficient of the layer outside the illuminated region and $\Delta\alpha$ is the amplitude of the absorption grating to be determined. From the Kramers-Kroning relations [11], it is known that any modulation of absorption coefficient in a material is accompanied by the respective modulation of its refractive index $n(x) = n_{av} + \Delta n(1 + \sin(Kx))$. This is the reason why we assumed the coexistence of the amplitude $\Delta\alpha$ and the phase Δn gratings in the layer. The observation of the grating area under the condition of crossed polarisers (cf. Fig. 2c) evidences that the illuminated region becomes, as a whole, optically anisotropic. In that region the light polarization is rotated. This experiment tells us that the virgin DR1 layer is optically isotropic (molecules are either completely disordered or stand almost perpendicularly to the glass surface). However, an increase in absorption upon illumination supports the assumption that initially the molecules were more or less perpendicular to the layer. The coloration indicates that the illuminated regions become more absorptive, i.e., the average molecular angle measured with respect to the surface decreases from 90° to lower values. The reason can be heat-induced oriented crystallite growth (of sizes $s \ll \lambda$) or concerted tilt and reorientation of the molecules similar to those observed in polymer matrices [1-5]. The prolonged illumination with an intense focused light ($\lambda = 514.5$ nm) eventually leads to the film degradation at the centre (sublimation of DR1) with microscopic crystallites appearing around the centre of the beam spot. This observation supports the heat-induced oriented-crystallite growth mechanism as the one responsible for the absorption grating formation.

Figure 3. Photograph of the amplitude grating recorded with a period of 0.59 μm made with an optical microscope and its Fourier transform showing a single period (cf. two spots around the centre of the Fourier spectrum).

In Fig. 3 we show that recording a very dense grating with period $\Lambda = 0.59$ μm is possible. The observation of such a high-frequency grating is close to the resolution limit of our optical microscope (Olympus BX 60), but the Fourier transform of the grating shows the well defined single period (cf. spots in the Fourier transform spectrum). Depending on the exposure, the profiles of absorption gratings change dramatically. In Fig. 4, the Fourier transform of the properly exposed grating shows only one Fourier component whereas overexposed grating shows up to three Fourier components indicating a severely distorted sinusoidal profiles.

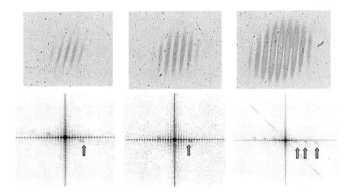

Figure 4. Gratings formed at different exposures and their Fourier transforms. Overexposure causes that the grating shows more than one Fourier component.

A closer inspection of the grating profiles (various cross-sections) obtained for a long-exposure grating is shown in Fig. 5. A bleaching in the central part of the grating and a nonsinusoidal hape of the fringes (cf. Fig. 5a) are noticeable while well sinusoidal fringes and no bleaching is observed for a section outside the central part (cf. Fig. 5b).

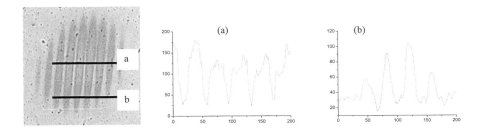

Figure 5. The sections along the grating showing a change in absorption grating profiles as a function of place, i.e. intensity level during recording process.

Kinetics of grating recording in DR1 layers

The kinetics of grating recording has been measured by monitoring the first-order diffraction power with respect to the exposure. The time evolution of the first-order diffraction efficiency (cf. Fig. 6) behaves totally differently when compared with grating recording experiments measured with chromophore-doped polymers where exponential growth with saturation is usually observed. In the case of thin DR1 layers, one observes the well-defined maximum of diffraction efficiency η followed at higher exposures by a minimum and subsequent maximum. Tentative explanation of such a behaviour must take into account the Gaussian distribution of intensity, which causes non-uniformity of the recorded grating amplitude across the grating area and the very limited number of fringes.

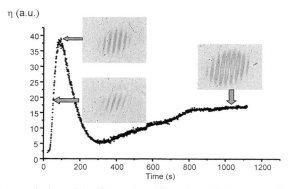

Figure 6. Time evolution of the first-order diffraction efficiency together with photos of gratings created at the marked points [10].

Thus, initially at the beam centre ($\rho = 0$), the grating amplitude grows faster due to higher incoming light intensity. The molecules are tilted and the amplitude of absorption grating reaches its maximum at the moment when the grating is perfectly sinusoidal with the maximum attainable amplitude, $\Delta\alpha_{max}$. This effect is well documented in Fig. 6 where the photographs of the gratings recorded in the DR1 film are shown. The decrease in diffraction efficiency might be associated with bleaching of the absorption in the vicinity of each fringe and in the whole central part of the grating. Grating evidently becomes less sinusoidal, so the diffraction in the first-order direction competes now with the diffraction into higher diffraction orders. The mechanism which is responsible for the subsequent rise in diffraction efficiency may originate from (i) slow rise in the first-order diffraction coming from side parts of the beam centre where the grating is still sinusoidal and (ii) slow but progressive mechanism associated with sublimation of the molecules. Due to this process, the material becomes again more transparent and more diffracted light can be collected with a detector placed in the first-order diffraction direction. All these processes are difficult to model. However, the indirect evidence that the above mentioned scheme may be realistic is given by the following experiment of grating recording without lenses and with a very short grating period $\Lambda = 0.6$ μm. In this case, due to lower total light intensity, sublimation, bleaching and side effects are limited and the kinetics of the first-order diffraction power resemble those measured in typical photochromic materials (cf. Fig. 7, VV grating recording).

Grating translation studies [12,13] as well as microscopic observations allowed to determine the ratio of amplitude ($\Delta\alpha$) and phase (Δn) grating strengths at the maximum of diffraction efficiency $\Delta\alpha / \Delta n \cong 7.2 \times 10^{-5}$ cm^{-1}. From these data one can estimate the contribution to the overall diffraction efficiency, $\eta_{tot} = 1.57$ %, coming from the refractive index grating, $\eta_{\Delta n} = 6.5 \times 10^{-2}$ %. Thus one can state that the contribution of pure phase grating to the observed diffraction is about 25 times lower than the contribution of pure absorption grating at 514.5 nm.

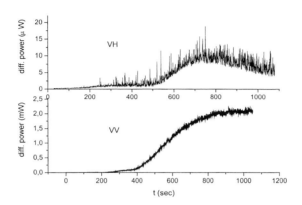

Figure 7. Examples of polarisation (VH) and intensity (VV) grating recording in DR1 layer without focusing the writing beams ($\Lambda = 0.6$ μm).

In Fig. 7 we show for the first time, to our knowledge, the formation of the polarisation grating in a highly ordered film of DR1 molecules on glass. Two recording beams were carefully orthogonally polarised with respect to each other. Polarisation grating (VH) written under the same laser power conditions as the intensity grating (VV) gives a much smaller diffraction power, 10 μW versus 2 mW. Note that under a lower light intensity than that used previously, the "incubation" time of grating formation is much longer and the characteristics of grating dynamics are different.

Conclusions

We investigated the process of holographic absorption grating recording in thin layers of Disperse Red 1 vacuum-deposited on glass substrate. We observed a relatively high self-diffraction efficiency at 514.5 nm reaching at the best 2 % for a 100-nm thick DR1 layer. The spatial resolution reaches 2000 lines/mm. From the presented results, we postulated the mechanism of photoinduced grating formation in highly organized thin films of DR1. Molecules during grating recording are addressed with strongly absorbed light and, due to local heating and/or *cis-trans* multiple excitations, change their initial orientation producing stronger absorption in the illuminated regions. The molecular reorientation is not completely random as it produces optical birefringence. The process is well controlled by experimental conditions, i.e., the growth rate depends on the light illumination level in a highly predictable manner.

Acknowledgements

A part of this work was supported by the Polish State Committee for Scientific Research under grant 4T 08A 03523.

[1] M. S. Ho, A. Natansohn, and P. Rochon, *Macromolecules* **1995**, *28*, 6124.
[2] Z. Sekkat, J. Wood and W. Knoll, *J. Phys. Chem.* **1995**, *99*, 17226.
[3] T.G. Pedersen, P.M. Johansen, N.C.R. Holme, P. Ramanujam, S. Hvilsted, *J. Opt. Soc. Am. B* **1998**, *15*, 1120.
[4] P.-A. Blanche, P.C. Lemaire, C. Maertens, P. Dubois, R. Jerome, *Opt. Commun.* **2000**, *185*, 1.
[5] P. Lefin, C. Fiorini, J-M. Nunzi, *Opt. Mater.* **1998**, *9*, 323.
[6] T. Kaino, A. Yokoo, M. Asabe, S. Tomaru, T. Kurihara, *Nonlinear Opt.*, **1995**, *14*, 135.
[7] T. Ehara, H. Hirose, H. Kobayashi, M. Kotani, *Synth. Met.*, **2000**, *109*, 43.
[8] N. Okamoto, H. Unoh, O. Sugihara, R. Matsushima, *Nonlinear Opt.*, **1995**, *14*, 245.
[9] H. Taunaumang, Herman, M.O. Tija, *Opt. Mater.* **2001**, *18*, 343.
[10] H. Taunaumang, M. Solyga, M.O. Tija, A. Miniewicz, *Opt. Mater.*, submitted
[11] H.J. Eichler, P. Guenter, D.W. Pohl, *Laser-Induced Dynamic Gratings*, Springer, Berlin **1986**, pp. 98-99.
[12] K. Sutter, P. Gunter, *J. Opt. Soc. Am. B*. **1990**, *7*, 2274.
[13] C.H. Kwak, A.J. Lee, *Opt. Commun.*, **2000**, *183*, 547.

Macromol. Symp. **2004**, *212,* 103-111

Manifestation of Electrode Surface States in Molecular Conduction

Giorgos Fagas,[*1] *Rafael Gutierrez,*[2] *Klaus Richter,*[1] *Frank Grossmann,*[2] *Rüdiger Schmidt*[2]

[1] Institut für Theoretische Physik, Universität Regensburg, 93040 Regensburg, Germany

[2] Institut für Theoretische Physik, TU-Dresden, 01062 Dresden, Germany

Summary: We present a conduction mechanism across molecular junctions which derives from conductance resonances that are not associated with particular molecular orbitals. Instead, the resonances are induced by states localized at the surface of the electrodes. To this end, we studied the conductance of a C_{60} molecule bridging two carbon nanotubes. A simple tight-binding model is employed to investigate analytically the basic features of the effect.

Keywords: charge transport; fullerenes; Landauer theory; modeling; nanotechnology

Introduction: Molecules as Building Blocks for Electronic Circuits

Triggered by recent advances in chemical synthesis, scanning probe microscopy and break junction techniques, the seminal idea of using molecular-scale conductors as active components of electronic devices has received a new impulse.[1] Rectification, nanomechanical oscillators as well as negative differential resistance have already been demonstrated at the nanoscale.[2]

The understanding of the basic physical mechanisms that determine electron transport at the nanoscale is essential for achieving conductance control and, thus, for opening the possibility of device applications. Detailed studies at either the semiempirical or first-principles level have related the conduction mechanisms to the molecular electronic structure, the topology of the molecule/electrodes interface, charging effects, the band lineup and inelastic effects in long molecular wires.

Recently, we have studied molecular junctions which include mesoscopic electrodes such as carbon nanotubes (CNTs).[3-5] We have suggested a setup where a C_{60} molecule is contacted to two CNTs as an effective all-carbon molecular switch by molecular rotation.[4] An essential feature of such devices is the existence of localized electronic states at the electrode surface.[5]

DOI: 10.1002/masy.200450810

Such states have important consequences in the electronic response of the junction and may in general be manifested in similar systems. An example includes localized states of anchor groups in a electrode/molecular wire/electrode bridge. Our approach is based on a combination of the Landauer theory of electronic transport with a tractable density-functional (DF) parametrized tight-binding Hamiltonian for calculating the electronic structure.

In the next section we present the theoretical method. Then the influence of surface states of the electrodes on the conductance of a CNT/C_{60}/CNT device is discussed. To reveal the effect, we compare with a simple tight-binding model followed by some concluding remarks.

Methodology: Landauer approach and Electronic Structure

We have recently developed an efficient method to calculate the conductance of two-terminal devices on the nanoscale.[5] The formulation is based on supplementing the Landauer picture of electronic transport with a density-functional parameterized tight-binding (DF-TB) approach to the electronic structure of the nanodevice. As shown in Figure 1, the device can be partitioned into three regions: two ideal (left and right) electrodes and a scattering region which includes the physical object of interest (e.g. molecule, cluster, nanowire) and, in most cases, some group of atoms belonging to the electrodes. The latter is particularly important when the electrodes' surface undergoes a structural relaxation which would introduce additional scattering. In this section the term 'molecule' is used to denote the scattering region.

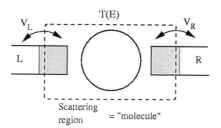

Figure 1. Schematic representation of a two-terminal device. The scattering region (enclosed in the dashed-line frame) with transmission probability $T(E)$ is connected to semi-infinite left (L) and right (R) electrodes.

The Landauer approach[6] then reduces the calculation of the conductance G to the determination of the transmission function $T(E)$ within an elastic scattering problem of independent electrons with energy E. The relation in the linear regime at zero temperature is

$$G = \frac{e^2}{\pi\hbar}T(E_F) \tag{1}$$

where E_F is the equilibrium Fermi energy. Using Green functions techniques $T(E_F)$ is expressed as[7]

$$T(E_F) = Tr\left[G_M(E_F - i\eta)\Gamma_L(E_F)G_M(E_F + i\eta)\Gamma_R(E_F)\right] \tag{2}$$

In Equation 2, G_M is the Green function of the scattering region while $\Gamma_{L(R)}$ describe its effective coupling to the electrodes as described below.

From the Green function of the complete system it is possible to extract G_M using projector operator techniques.[8] The above mentioned partition of the setup leads to the following block representation of the full Hamiltonian matrix (in some suitable basis representation)

$$H = \begin{pmatrix} H_L & V_{L-M} & 0 \\ V_{L-M}^+ & H_M & V_{R-M} \\ 0 & V_{R-M}^+ & H_R \end{pmatrix} \tag{3}$$

The matrices $V_{L-M(R-M)}$ couple atoms belonging to the left (right) electrode to the molecule, and no direct coupling between the electrodes is assumed. Note that $H_{L(R)}$ are infinite dimensional submatrices. By projecting onto the molecular subspace via a projector P_M, it can be shown that $G_M = P_M\,G\,P_M$ satisfies the M-dimensional matrix equation

$$(zS_M - H_M - \Sigma_L(z) - \Sigma_R(z))G_M(z) = 1; \qquad z = E + i\eta;\ \eta \to 0^+ \tag{4}$$

where S_M is the overlap matrix for the general case of a non-orthogonal basis set. The self-energies $\Sigma_{L(R)}$ include the coupling to the electrodes as well as information on their electronic structure. They are given by

$$\Sigma_{L(R)}(z) = (zS_{L(R)-M}^+ - V_{L(R)-M}^+)g_{L(R)}(z)(zS_{L(R)-M} - V_{L(R)-M}) \tag{5}$$

$S_{L(R)}$ are overlap matrices between orbitals centered on molecule and electrode atoms, and $g_{L(R)}(z)$ are electrode Green functions which are calculated using recursive techniques. Since the coupling matrices are in general short-range, they eliminate all contributions coming from atoms other than those nearest to the molecule. Hence, only surface Green functions are needed. Finally, the quantities $\Gamma_{L(R)}$ in Eq. 2 are given by $i[\Sigma_{L(R)}(E+i\eta)-\Sigma_{L(R)}(E-i\eta)]$.

Then, the issue of characterizing the electronic structure of the molecule as well as that of the electrodes when treating real systems has to be considered. The DF-parameterized TB approach[9] relies on a representation of the single-particle electronic Kohn-Sham eigenstates ψ_i of the system within a non-orthogonal basis set $\varphi_\alpha(\mathbf{r}\text{-}\mathbf{R}_\alpha)$ taken as a valence basis localized at the ionic positions $\mathbf{R}_\alpha$, namely

$$\psi_i(\mathbf{r}) = \sum_\alpha c_\alpha^i \varphi_\alpha(\mathbf{r} - \mathbf{R}_\alpha) \tag{6}$$

With this *Ansatz* the Kohn-Sham equations for ψ_i are transformed into a set of algebraic equations

$$\sum_\beta (H_{\alpha\beta}^{\mathrm{TB}} - S_{\alpha\beta}E_i)c_\beta^i = 0 \tag{7}$$

In this expression the many-body Hamiltonian has been additionally approximated by a two-center (tight-binding) Hamiltonian H^{TB}, and $S_{\alpha\beta}$ is the corresponding overlap matrix. Note that the formal structure of Eq. 7 is that of the Hückel Hamiltonian. In contrast to the latter, however, all necessary matrix elements are calculated numerically using the φ_α-basis, which allows us to avoid the introduction of any empirical parameters. [9]

Results: electrode surface states and molecular conduction

In this section we demonstrate an unconventional way of metallization of a molecular bridge via intrinsic electrode surface states, which manifests itself as conductance resonances within the molecular HOMO-LUMO gap. As an example we utilize the molecular junction shown in Figure 2. We consider a C_{60} molecule between two semi-infinite *capped* metallic (5,5) CNTs. The diameters of the CNTs and the fullerene are comparable (~ 0.7 nm). The center of the fullerene is placed at the mid-point between the two pentagons at the cap-edges.

Because the molecular junction under discussion is a non-periodic system, we simulate it as a supercell with open boundary conditions. In addition to the caps, the CNT part consists of eight unit cells. The CNT back-bonds were saturated with hydrogens in order to avoid nonphysical charge transfer towards the nanotube caps. The scattering region includes the C_{60} molecule, the caps and the nearest-lying nanotube unit cells. A minimal basis set consisting of the 2s2p valence orbitals of carbon and the 1s orbital of hydrogen was used in the calculations. Due to the proximity of both subsystems (the cap-molecule separation is about 0.2 nm) the C_{60} cage is expected to be slightly distorted. To take into account these modifications, we perform a structural optimization prior to each conductance calculation. Hereby we keep fixed all other unit cells that do not belong to the scattering region to simulate the bulk of the electrodes. It turns out that C_{60} adopts an ellipsoidal shape with its longer axis perpendicular to the transport direction.

A very important point is the determination of the equilibrium Fermi energy E_F of the whole system. As a first approximation, we have used the HOMO level obtained by diagonalizing the Hamiltonian matrix of the supercell after structural optimization of the molecular junction. With increasing supercell size, this level converges to the equilibrium Fermi level. We have tested that, with the cell size used in these calculations, the error in the Fermi energy lies within the precision of the DFTB method.

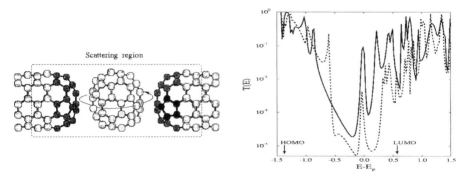

Figure 2: Geometric configuration of an all-carbon molecular junction (left). A C_{60} molecule bridges two (5,5) CNTs. Transmission spectra for two different molecular orientations (right). The HOMO and LUMO levels of the isolated C_{60} are indicated by arrows. Note the resonance at the Fermi energy.

Three features of this molecular junction should be highlighted:

(i)   It is an all-carbon device, which means that large charge-transfer effects at equilibrium between the electrodes and the molecule can be ruled out. This is also supported by a Mulliken population analysis showing a very small charge accumulation on the molecule.

(ii)   The lateral dimensions of the electrodes are of the same order as of the molecule. Thus, modifications of the interfacial atomic structure (produced, e.g., by some extrinsic mechanism like rotation or compression of the molecule) should strongly influence the electronic conductance.[3-5]

(iii)   The presence of topological defects (pentagons) on the electrode contact surfaces (caps) induces localized states around the equilibrium Fermi energy.[11] It is just this aspect that is related to the unconventional metallization of the molecule.

A typical transmission spectrum for two different orientations of C_{60} is shown in Figure 2 (lower panel). The rather strong variation of the peak intensities around the HOMO-LUMO gap is a direct consequence of point (ii) above and is thus related to modifications of the molecule/electrode interfacial coupling upon rotation.

Metallization by the electrode surface states is manifested by the conductance resonance near the Fermi energy. Its position and width remain nearly unaffected by molecular rotations, although its height strongly fluctuates. Since charge transfer is excluded, this resonance cannot be associated with a molecular orbital. Therefore, it is attributed to the states localized at the tips (caps) of the CNTs; see point (iii) above.

The basic features of the conductance around the Fermi energy described above can be nicely illustrated by considering a simple model Hamiltonian, schematically depicted in Figure 3. It consists of a dimer with hopping matrix element Γ coupled to two semi-infinite chains, which represent the electrodes. All atoms have a common onsite energy ε_0. A central feature of the model is the ability to trigger the presence of a localized state at the chain ends ('surface' state) which couples to the dimer and to the semi-infinite chains via parameters γ and $\gamma_{L(R)}$, respectively. The Hamiltonian reads

$$
\begin{aligned}
H &= H_L + H_R + H_{dimer} + H_{surface} \\
&= \varepsilon_0 \sum_{L,R} c_i^+ c_i + \beta \sum_{i=-\infty}^{0} (c_{i-1}^+ c_i + h.c.) + \beta \sum_{i=5}^{\infty} (c_{i-1}^+ c_i + h.c.) + \\
&\quad \varepsilon_0 \sum_{dimer} c_i^+ c_i + \Gamma(c_2^+ c_3 + h.c.) + \gamma(c_1^+ c_2 + c_3^+ c_4 + h.c.) + \\
&\quad \varepsilon_0 \sum_{surface} c_i^+ c_i + \gamma_L(c_0^+ c_1 + h.c.) + \gamma_R(c_4^+ c_5 + h.c.)
\end{aligned}
\tag{8}
$$

where the subscripts 2,3 and 1,4 correspond to the dimer and the surface atoms, respectively.
For this model Eq. 2 simplifies to

$$
T(\eta) = 4\kappa^2 \Delta_L(\eta) \Delta_R(\eta) |G_{14}(\eta)|^2
\tag{9}
$$

where the Green function matrix element is given by

$$G_{14}(\eta) = \frac{\left(\lambda/\kappa\right)^2}{(2\kappa\eta - 1)(2\kappa\eta + 1)(2\cdot\eta - \Sigma_R(\eta))(- 2\cdot\eta + \Sigma_L(\eta))}$$ (10)

and

$$\Sigma_{L(R)}(\eta) = \Lambda_{L(R)}(\eta) - i\Delta_{L(R)}(\eta) = \mu_{L(R)}^2\left(\eta - i\sqrt{1-\eta^2}\right); \qquad |\eta| < 1$$ (11).

In deriving Eqs 9-11 we have introduced the following parametrization: $\beta/\Gamma=\kappa$, $\gamma/\Gamma=\lambda$, $\gamma_{L(R)}$ /$\beta=\mu_{L(R)}$ and $\eta=-(E-\varepsilon_0)/2\beta$; the condition $|\eta|<1$ fixes the energy within the electrode band. Additionally, we have assumed that the coupling between the surface state and the dimer is weak ($\lambda<<1$), so that in the denominator of the G_{14}, all terms higher than first-order in λ have been neglected. This condition naturally arises if the molecular states are to be only slightly perturbed by the coupling to the electrodes. One clearly sees the pole structure of the Green function. Note that the dimer interacts only indirectly with the electrodes via the surface states and, hence, its HOMO ($\kappa\eta=-0.5$) and LUMO ($\kappa\eta=0.5$) are broadened by the electrodes only to (the neglected) second order in λ. To simulate a surface state, $\mu_{L(R)}$ should also be much smaller than unity.

To test our approximation, we compare in the left panel of Figure 3 the exact transmission function to that obtained from the above approximated expressions Eqs 9-11 in the limit of weak coupling. As can be seen, both curves are almost identical showing that Eq. 10 is a valuable approximation in this regime. Evidently, for $\mu_{L(R)}<<1$, the surface states correspond to the degenerate resonance at $\eta=0$ whose height is less than unity due to the asymmetric coupling to the electrodes. For weak coupling the latter is determined by $\lambda/\mu(=\mu_{L(R)})$.

By varying the surface state/molecule coupling, the width of the resonance is rather insensitive but the height is strongly modified in a similar fashion to our results on the CNT/C_{60}/CNT molecular junction. This is shown in the right panel of Figure 3. This follows from Eq. 10 which implies a λ-independent broadening and a height proportional to $(\lambda/\mu)^4$.

110

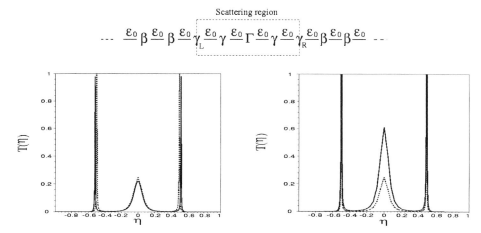

Figure 3.Upper panel: tight-binding model used to illustrate the conductance resonance induced by electrode surface states as found in the carbon-based molecular junction of Fig. 2. Lower panel: typical transmission function in the weak coupling regime as defined in the text. The resonance at zero arises from the electrodes. Left figure shows the comparison between the exact result and the weak coupling formula, Eqs 9-11. Here, $\kappa=1$, $\mu_{L(R)}=0.4$ and $\lambda=0.2$. Right figure illustrates the evolution of the resonance with respect to the coupling between the dimer and the electrode (λ equals 0.2 and 0.275 for the dashed and solid curve, respectively).

Concluding Remarks

By employing a novel molecular device, we pointed out the manifestation of surface resonant states of the electrodes in the conductance of molecular junctions. This presents an alternative way to metallization of a molecule and contrasts to the adopted view that electron transport across molecular junctions is simply dominated by hybridization of the molecular orbitals. The resulting conductance profile is generic to the appearance of surface resonant states within the HOMO-LUMO gap with characteristic additional resonances in this spectral window. To illustrate the effect, we analyzed a simple tight-binding model which reproduces all basic features. Moreover, in many setups anchor groups are used to bind the molecular bridge to the electrodes. These may also possess localized states with similar properties as the ones discussed in this paper. As pointed out,[12] sulfur atoms in thiol end groups offer such an example.

Acknowledgments

We would like to thank the organizers of the 2002 Prague Meeting on Macromolecules 'Electrical and Related Properties of Polymers and Other Organic Solids' for giving us the opportunity to present this work. Rafael Gutierrez acknowledges support by the Deutsche Forschungsgemeinschaft through the Forschergruppe 'Nanostrukturierte Funktionselemente in makroskopischen Systemen'. Giorgos Fagas is grateful to the Alexander von Humboldt Stiftung.

[1]   A. Aviram and M. Ratner, *Chem. Phys. Lett.* **1974**, *29*, 277
[2]   A. Aviram, M. Ratner, and V. Mujica (Eds), *Molecular Electronics II, Ann. N.Y. Acad. Sci.* **2002**, *960*; C. Joachim, J. K. Gimzewski, and A. Aviram, *Nature (London)* **2000**, *408*, 541
[3]   G. Fagas, G. Cuniberti, and K. Richter, *Phys. Rev. B* **2001**, *63*, 045216; G. Cuniberti, G. Fagas, and K. Richter, *Chem. Phys.* **2002**, *281*, 465
[4]   R. Gutierrez, G. Fagas, G. Cuniberti, F. Grossmann, R. Schmidt, and K. Richter, *Phys. Rev. B* **2002**, *65*, 113410
[5]   R. Gutierrez, G. Fagas, K. Richter, F. Grossmann, and R. Schmidt, to be published in *Europhys. Lett.*
[6]   R. Gutierrez, F. Grossmann, O. Knospe, and R. Schmidt, *Phys. Rev. A* **2001**, *64*, 0132
[7]   S. Datta, *Electronic transport in mesoscopic systems*, Cambridge University Press, **1995**
[8]   Y. Meir and N. S. Wingreen, *Phys. Rev. Lett.* **1992**, *68*, 2512
[9]   P. O. Löwdin, *J. Chem. Phys.* **1951**, *19*, 1396
[10]   O. Knospe, R. Schmidt, and G. Seifert, *Advances in Classical Trajectory Methods* **1999**, *4*, 153; Th. Frauenheim, G. Seifert, M. Elstner, Z. Hajnal, G. Jungnickel, D. Porezag, S. Suhai, R. Scholz, *Phys. Status Solidi (b)* **2000**, *217*, 41
[11]   R. Tamura and M. Tsukada, *Phys. Rev. B* **1995**, *52*, 6015; D. L. Carroll *et al*, *Phys. Rev. Lett.* **1997**, *78*, 281
[12]   J.K. Tomfohr and O. F. Sankey, *Phys. Rev. B* **2002**, *65*, 245105

Traps for Charge Carriers in Molecular Materials Formed by Dipolar Species: Towards Light-Driven Molecular Switch

Juliusz Sworakowski,[*1] *Stanislav Nešpůrek*[2]

[1] Institute of Physical and Theoretical Chemistry, Wrocław University of Technology, Wyb. Wyspiańskiego 27, Wrocław, Poland
E-mail: sworakowski@pwr.wroc.pl
[2] Institute of Macromolecular Chemistry, Academy of Sciences of the Czech Republic, Heyrovsky Sq. 2, 162 06 Prague 6, and Faculty of Chemistry, Technical University of Brno, Purkyňova 118, 612 00 Brno, Czech Republic
E-mail: nespurek@imc.cas.cz

Summary: The influence of polar species on the transport and trapping of charge carriers is discussed. Calculations performed on a model molecular lattice demonstrate that polar dopants locally modify the polarization energy thus creating traps for charge carriers in the vicinity of the dipole. The presence of polar dopants in disordered solids gives rise to a broadening of the density-of-states function. A scheme of a molecular switch has been put forward, based on electrostatic interactions between photochromic moieties and charge carriers travelling on a molecular wire (conjugated polymer chain).

Keywords: charge transport; dipolar traps; molecular switch; photophysics; polarization energy

Introduction

Owing to numerous current and emerging applications of molecular materials, there has been a continued interest in studies of electrical properties of low-molecular-weight single-crystal and polycrystalline materials, molecular glasses and polymers.[1-9] Many concepts, useful in explaining electrical properties of molecular solids stem from ideas developed at early stages of investigations.[6-9] In particular, the origin of local centres capturing charge carriers (carrier traps) in molecular materials has been explained[10-12] employing the electrostatic model put forward by Lyons[6,13] almost half a century ago. The traps are formed on sites where, locally, the ionization energy is lower than that in a perfect lattice (traps for holes) or the electron affinity is higher than the respective value in the perfect crystal (traps for electrons). Local modifications of these parameters can be achieved in two ways: *(i)* substitution of a host molecule with a chemically different guest, or *(ii)* a suitable local change of the polarization energy.

The polarization energy *(P)* is the energy of electrostatic interactions between a charge carrier momentarily localized at a given site of the lattice with a polarizable molecular solid.[14] If the molecular crystal is built of non-polar molecules, then, within the approach put forward by Lyons, [6,13] the latter parameter can be written as a sum of terms representing electrostatic interactions of increasing order (and decreasing importance)

$$P = P_{id} + P_{dd} + ... , \qquad (1)$$

where the successive terms stand for energies of interactions between the localized charge (ion) and induced dipoles, interactions between induced dipoles, etc. P_{dd} and higher-order terms usually account for ca. 20 % of the total effect[6] and may be neglected in semiquantitative (zeroth-order) calculations. In this case, only the first right-hand term is left in Eq. (1).

Important for the matter of this paper is the case when a non-polar crystal is doped with polar guest molecules. Irrespective of whether the guest molecules themselves act as chemical traps, their presence should locally modify electric field acting on neighbouring molecules. Thus one should expect local changes of the polarization energy.

Calculations on a model lattice

Details of the calculations have been described in earlier papers of the present authors.[15-17] The calculations reported in the present paper have been performed on a model 'anthracene-like' crystal: a primitive regular lattice with numerical values of appropriate parameters chosen so as to mimick properties of the anthracene crystal: the lattice constant was assumed equal to 0.62 nm (i.e., was adopted to yield the molecular volume equal to that of an anthracene molecule in a real crystal),[18] and the (isotropic) polarizability was chosen to amount to 2.78×10^{-39} m^{2}F $(25 \times 10^{-24}$ cm$^3)$.[19] The calculation of the polarization energy in the vicinity of a polar impurity requires, even within the zeroth-order approximation, that at least three terms be retained

$$P^{loc} \approx P_{id} + P_{im} + P_{md} , \qquad (2)$$

with P_{id} describing energy of interactions between induced dipoles and the charge, P_{im} - between the permanent dipole(s) and the charge, and P_{md} - between the permanent and induced dipoles. Within the point-dipole approximation, these energies can be calculated from the equations[14]

$$P_{id} = -\frac{\alpha}{2}\sum_j F_j^2 , \tag{3}$$

$$P_{md} = -\frac{e}{4\pi\varepsilon_o r_{km}^3}(\vec{m}\cdot\vec{r}_{km}), \tag{4}$$

$$P_{im} = \frac{\alpha}{4\pi\varepsilon_o}\sum_j\left[-\frac{3}{r_{mj}^5}(\vec{m}\cdot\vec{r}_{mj})(\vec{F}_j\cdot\vec{r}_{mj}) + \frac{1}{r_{mj}^3}(\vec{m}\cdot\vec{F}_j)\right], \tag{5}$$

$$\vec{F}_j = \frac{1}{4\pi\varepsilon_o}\left[\frac{1}{r_{mj}^5}\left[(\vec{m}\cdot\vec{r}_{mj})\vec{r}_{mj} - r_{mj}^2\vec{m}\right] + \frac{e}{r_{kj}^3}\vec{r}_{kj}\right]. \tag{6}$$

In the above equations, $\vec{m}$ is the permanent dipole moment of the polar dopant (a guest molecule situated at an m-th site), $\vec{F}_j$ is the effective field acting on a j-th molecule, and r's are the distances between the charge situated on a k-th molecule, the permanent dipole, and a neutral polarizable j-th molecule, the subscripts indexing the appropriate distances.

The calculations of local values of the polarization energy in the model crystal containing polar impurities, reported in this paper, were carried out for several tens of thousands of molecules adjacent to the polar dopant: in the case of an isolated dopant molecule, the values of P^{loc} were calculated for 3374 molecules (i.e., for all molecules located within the ±7 lattice constants from the dipole), with the summation going over 6858 induced dipoles adjacent to the charge localized successively on each molecule. Several values of the electrical moment of the permanent dipole (up to 10 D, i.e. 3.33×10^{-29} Cm) were employed in the calculations. Typical results are shown in Fig. 1.

The main results can be summarized as follows: *(i)* the presence of a polar impurity results in local perturbations of the polarization energies on adjacent molecules, the number of affected molecules depending on the dipole moment of the guest; *(ii)* the depth of traps E_t formed on molecules adjacent to the dopant is a nearly linear function of its dipole moment (cf. Fig. 2); *(iii)* the density-of-states (DOS) function is broadened, the broadening parameter increasing with increasing dipole moment (cf. Fig. 1).

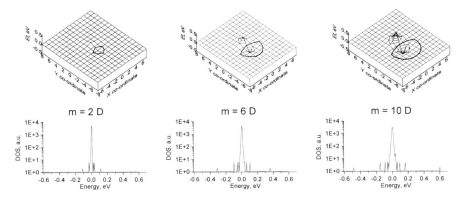

Figure 1. Upper row: Traps for holes created in a regular lattice in the vicinity of polar centres of various polarities. The dipoles are located at the origins, their positive poles pointing in the positive directions of the y axes. The nodes of the grids represent molecules, the figures showing two-dimensional cross-sections through three-dimensional lattices taken at $z = 0$ (trap depth axis). The thick contours mark the extent of the perturbation $\Delta P = E_t = -0.025$ eV. Lower row: Densities of local states associated with the presence of isolated dipoles. DOS in a perfect crystal is a δ function centred at zero.

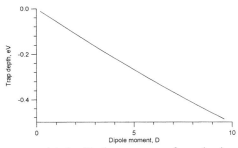

Figure 2. Depths of traps associated with the presence of a polar impurity, as function of its dipole moment. The line is a result of calculations performed on the model molecular lattice described in the text.

Due to severe simplifications introduced in the calculation procedure, the results obtained should be treated as semiquantitatve indications of trends. Nevertheless, they demonstrate that the presence of polar dopants results in creation of traps for both electrons and holes even though the dopant molecules themselves do not act as chemical traps.

Similar calculations have been performed for arrays of dipolar molecules,[17,20] mimicking the situation in polymers containing polar side groups. The results obtained demonstrate that, for moderately high dipole moments of guest molecules, the dipoles can be treated as isolated (non-interacting) ones if the distance between them exceeds ca. 5-6 lattice constants, i.e., if their concentration does not exceed ca. 0.1 %. Thus the effect of the dipole-dipole interaction on the charge transport cannot be neglected in molecular glasses or polymers in which the concentrations of polar species often exceed 1 %, being usually of the order of 10 %.[2]

Charge carrier mobilities influenced by polar guest molecules

As has been shown in the preceding section, the presence of dipolar species results in creation of traps, their depths and cross-sections depending on the dipole moment of the dopant. Simultaneously, high densities of spatially connected shallow local states are created which may act as transport states. It would be instructive to assess the effect of traps associated with the presence of polar impurities. In nearly-perfect solids, the trap-controlled mobilities follow the equation[21]

$$\mu = \mu_0 \left[1 + x_{\text{dip}} \exp\left(\frac{-E_t}{kT} \right) \right]^{-1} \tag{7}$$

where μ_0 is the drift mobility of the carriers in a perfect solid, x_{dip} is the mole fraction of the polar species, and E_t is the depth of trap created due to the presence of the dopant ($E_t < 0$). As follows from the results shown in Fig. 3, even for moderate dipole moments of the impurity, one should expect a significant decrease in mobility.

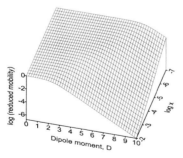

Figure 3. The reduced mobility (μ/μ_0) at 300 K, expressed in function of the mole fraction of dipolar dopants and their dipole moment. The relation between the dipole moment and the trap depth has been taken from the curve shown in Figure 2.

In disordered solids, the changes of the photochemically controlled mobilities are more difficult to assess since, apart from modification of the trap depths (cf., e.g., Ref. [22]), one should also expect a modification of the shape of the density-of-states functions. The importance of the latter effect can be estimated employing the model put forward by Bässler.[23,24] According to the model, DOS in disordered solids may be approximated by a Gaussian function

$$N(E) = \frac{N_{tot}}{(2\pi\sigma^2)^{1/2}} \exp\left(-\frac{(E-E_m)^2}{2\sigma^2}\right),$$ (8)

N_{tot} standing for the total density of states, E_m for the energy of the distribution maximum, and the variance σ being a measure of the disorder; typically, $\sigma \leq 0.1$ eV. At sufficiently high voltages, the mobility is field-dependent

$$\mu(\sigma, T, F) = \mu_0(\sigma, T)\exp\left(\vartheta F^{1/2}\right),$$ (9)

where ϑ is a field-independent coefficient and μ_0 stands for the zero-field mobility. The latter parameter follows a non-Arrhenius temperature dependence

$$\mu_0(\sigma, T) = \mu_{0,0} \exp\left[-\left(\frac{2\sigma}{3kT}\right)^2\right],$$ (10)

$\mu_{0,0}$ being usually identified with the mobility in a parent perfect material (on-chain mobility in the case of polymers). The effect of dipolar species has been rationalized invoking a dipolar contribution in the DOS function: the width of DOS is a superposition of van der Waals and dipolar components (σ_{vdW} and σ_{dip}, respectively)

$$\sigma^2 = \sigma^2_{vdW} + \sigma^2_{dip} \ ,$$ (11)

with the latter component related to the permanent dipole moment of the polar species and its mole fraction[25-27]

$$\sigma_{dip} = \frac{\kappa x^b_{dip} m}{\varepsilon a^2_0} .$$ (12)

Here, κ and b are constants, ε is the relative electric permittivity of the medium, and a_0 is the intersite distance. A combination of Eqs. (10-12) yields

$$\frac{\mu_0(\sigma,T)}{\mu_0(\sigma_{vdW},T)} = \exp\left[-\theta\left(\frac{x_{dip}^b m}{T}\right)^2\right], \tag{13}$$

where θ is a constant. Thus the contribution of the dipolar term manifests itself in the modification of the mobility: the mobility should decrease with increasing dipole moment of the dopant and with its concentration.

The model calculations described in the preceding section do not allow to extract any quantitative data concerning the broadening of the DOS function, hence no quantitative results can be obtained directly from the model. Qualitatively, however, the trend is correct: the presence of polar species results in broadening od DOS (cf. Fig. 1), hence one should expect a decrease in the carrier mobility as has indeed been found in earlier experiments[28,29] performed on polysilanes. Taking the values reported in Ref. [29], one may approximate the concentration dependence of the dispersion parameter in dilute systems ($x_{dip}<0.02$) by the following equation

$$\sigma^2 = 8.65\times10^{-3} + 0.23m^2 x_{dip}^2, \tag{14}$$

(σ in eV, m in D). Thus the carrier mobility measured in polysilane containing 2 mol % of polar additives (or polar side groups) with the permanent dipole moment equal to ca. 10 D should be almost three orders of magnitude lower than the respective value in the neat polymer.

The estimates presented above demonstrate that in all cases the presence of polar species should result in a decrease in the charge carrier mobility in macroscopic molecular samples though the magnitude of the effect may vary depending on the degree of perfection of the systems.

Modification of charge carrier mobilities by reversible photochemical reactions

The dependence of the depths and cross-sections of dipolar traps on the dipole moment of the polar dopants makes it possible to design a photoactive molecular system in which carrier mobilities are controlled by reversible changes of the dipole moment of a dopant. Suitably chosen photochromic systems come as an obvious choice: several molecular photochromic systems have been known to be weakly polar in their stable forms and zwitterionic (i.e., highly polar) in their metastable forms. For example, the dipole moments of certain spiropyrans and spirooxazines amount to ca. 1-2 D and over 10 D in their stable and metastable forms, respectively.

Taking the results obtained from the calculations carried out on the model lattice (presented in the preceding section – cf. Fig. 2) as a 'calibration curve', one can estimate depths of dipolar traps associated with the presence of photoactive dopants in the model molecular system. Assuming the dipole moments of the stable and metastable forms being respectively equal to 2 D and 10 D, one may estimate the depths of the traps to amount to ca. 0.1 eV and 0.5 eV, respectively. Photochemically driven reactions in the photochromic system should, in principle, allow for switching in a controlled way between these two states. It would be instructive to assess the effect of the switching between the 'low-moment' and the 'high-moment' states on the drift mobility. Making use of Eq. (7) and assuming $x_{dip} \approx 10^{-3}$ (corresponding to an average distance between the dipolar traps equal to 10 lattice constants), one arrives at $\mu^{(high\ m)}/\mu^{(low\ m)} \approx 4 \times 10^{-6}$ at ambient temperature. Similarly, the modulation of carrier mobilities in a highly disordered molecular system can be estimated taking the values obtained for polysilane[28,29] and using Eqs. (13) and (14). Taking $x_{dip} \approx 0.02$ and assuming the 'low' and 'high' dipole moments equal to 2 and 10 D, respectively, one obtains $\mu^{(high\ m)}/\mu^{(low\ m)} \approx 3 \times 10^{-3}$ at ambient temperature.

Towards a molecular switch

The estimates given in the preceding section point to the possibility of a controlled modulation of charge carrier mobilities in ordered and disordered molecular materials, by a light-induced reversible photochemical reaction. Moreover, one may envisage the construction of a molecular-scale switch acting on the same principle. The idea has been put forward in earlier papers of the

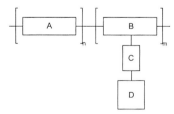

Figure 4. A scheme of a molecular switch based on the principle discussed in the paper. A, B – segments of a conjugated polymer chain; C – spacer; D – photochromic group.

present authors.[30,31] The switch, schematically depicted in Fig. 4, would consist of a molecular wire (a π- or σ-conjugated polymer chain) with suitable photochromic moieties, either placed in its vicinity or chemically attached as side groups, embedded in a neutral (electrically inactive) dielectric medium. In the vicinity of the polar centres formed on the photochromic units, new local electron states arise as a consequence of the electrostatic interaction of the charge carrier on the polymer wire with the dipoles of the side groups. Thus the electrons travel through the polymer wire in the system of potential wells whose parameters can be controlled by light. This feature leads to a modulation of the on-chain charge carrier mobility.

Discussing the architecture of the molecular switch, one should consider the role of the spacer. In particular, two limiting cases are possible: (*i*) a neutral spacer fixing the geometry of the photoactive side group and its distance from the main chain, and (*ii*) a spacer electronically connecting the side group with the main chain, i.e., allowing for the trapping of a charge carrier on the side group. In the former case, one deals with a dipolar trap created *on the chain* due to electrostatic interactions of the carrier with the dipole of the side group, whereas the latter case is equivalent to creation of a chemical trap *on the side group*, whose behaviour cannot be described within the electrostatic model employed throughout this work.

Acknowledgements

The research was supported by the Polish State Committee for Scientific Research (grant No 4 T09A 132 22), by the Ministry of Education, Youth and Sports of the Czech Republic (grant No OC D14.30) and by the Grant Agency of the Czech Republic (grant No 202/01/0518).

122

[1] P.N. Prasad, D.R. Ulrich, Eds., *Nonlinear Optical and Electroactive Polymers*, Plenum Press, New York 1988.

[2] P.M. Borsenberger, D.S. Weiss, *Organic Photoreceptors for Imaging Systems*, M. Dekker, New York 1993.

[3] V. Shibaev, Ed., *Polymers as Electrooptical and Photooptical Active Media*, Springer Verlag, Berlin 1996.

[4] S. Miyata, H.S. Nalwa, Eds., *Organic Electroluminescent Materials and Devices*, Gordon and Breach, London 1996.

[5] R. Farchioni, G. Grosso, Eds., *Organic Electronic Materials. Conjugated Polymers and Low Molecular Weight Organic Solids*, Springer Verlag, Berlin 2001.

[6] F. Gutmann, L.E. Lyons, *Organic Semiconductors*, J. Wiley, New York 1967.

[7] E.A. Silinsh, *Organic Molecular Crystals. Their Electronic States*, Springer Verlag, Berlin 1980.

[8] Proc. Oji Int. Seminar on Organic Semiconductors – 40 Years, Okazaki, *Mol. Cryst. Liq. Cryst.* **1988**, *171*, 1-355.

[9] M. Pope, C.E. Swenberg, *Electronic Processes in Organic Crystals and Polymers*, Oxford Univ. Press, New York 1999.

[10] J. Sworakowski, *Mol. Cryst. Liq. Cryst.* **1970**, *11*, 1.

[11] E.A. Silinsh, *Phys. Status Solidi A* **1970**, *3*, 817

[12] N. Karl, *Adv. Solid State Phys. (Festkörperprobleme)* **1974**, *14*, 261.

[13] L.E. Lyons, *J. Chem. Soc.* **1957**, 5001.

[14] C.J.F. Böttcher, *Theory of Electric Polarization*, 2nd ed., Vol. 1, Elsevier, Amsterdam 1973.

[15] J. Sworakowski, S. Nešpůrek, *Pol. J. Chem.* **1998**, *72*, 163.

[16] J. Sworakowski, *Proc. SPIE* **1999**, *37DP*, 83.

[17] J. Sworakowski, *IEEE Trans. Dielectr. Electr. Insul.* **2000**, *7*, 531.

[18] R. Mason, *Acta Crystallogr.* **1964**, *17*, 547.

[19] R.J.W. Lefevre, K.M.S. Sundaram, *J. Chem. Soc.* **1963**, 4442.

[20] J. Sworakowski, S. Nešpůrek, in: *Molecular Low Dimensional and Nanostructured Materials for Advanced Applications*, A. Graja, B.R. Bulka, F. Kajzar, Eds., Kluwer, Dordrecht 2002, p. 25 ff.

[21] D.C. Hoesterey, G.M. Letson, *J. Phys. Chem. Solids* **1963**, *24*, 1609.

[22] U. Wolf, H. Bässler, P.M. Borsenberger, W.T. Gruenbaum, *Chem. Phys.* **1997**, *222*, 259.

[23] H. Bässler, *Phys. Status Solidi B* **1981**, *107*, 9.

[24] H. Bässler, *Phys. Status Solidi B* **1993**, *175*, 15.

[25] A. Dieckmann, H. Bässler, P.M. Borsenberger, *J. Chem. Phys.* **1994**, *99*, 8136.

[26] R.H. Young, J.J. Fitzgerald, *J. Chem. Phys.* **1995**, *102*, 9380.

[27] A. Hirao, H. Nishizawa, *Phys. Rev. B* **1997**, *56*, R2904.

[28] S. Nešpůrek, H. Valerián, A. Eckhardt, V. Herden, W. Schnabel, *Polym. Adv. Technol.* **2001**, *12*, 306.

[29] S. Nešpůrek, J. Sworakowski, A. Kadashchuk, *IEEE Trans. Dielectr. Electr. Insul.* **2001**, *8*, 432.

[30] S. Nešpůrek, J. Sworakowski, *Thin Solid Films* **2001**, *393*, 168.

[31] S. Nešpůrek, P. Toman, J. Sworakowski, J. Lipiński, in: *Molecular Low Dimensional and Nanostructured Materials for Advanced Applications*, A. Graja, B.R. Bulka, F. Kajzar, Eds., Kluwer, Dordrecht 2002, p. 37 ff.

Charge Transfer Processes and Environmental Degrees of Freedom: Cooperativity and Non-Linearity

Anna Painelli,* Francesca Terenziani

Dip. Chimica GIAF, Università di Parma, INSTM-UdR Parma, 43100 Parma, Italy
E-mail: anna.painelli@unipr.it

Summary: To investigate supramolecular effects in samples with high concentration of push-pull chromophores, we propose a model for interacting polar and polarizable molecules. Each molecule is described in terms of the same two-state picture successfully adopted to model solvated chromophores and electrostatic interactions among different chromophores are introduced. Important supramolecular effects are observed even at the lowest mean-field level, showing up the possibility of tuning molecular polarity from the neutral to the zwitterionic regime or vice versa. Supramolecular effects in excitation spectra are more complex. Here we demonstrate large supramolecular effects beyond mean-field in static optical responses.

Keywords: charge transfer; cooperative effects; dyes; modeling; NLO

Introduction

π-Conjugated chromophores and, specifically, molecules where an electron-rich (donor, D) group is linked by a π-conjugated bridge to an electron-poor (acceptor, A) group attract a lot of interest both in view of their practical applications[1,2] and due to the possibility they offer in investigation of fundamental physical phenomena in terms of basic models.[3] DA-conjugated chromophores are used as laser dyes as well as active luminescent materials in organic light-emitting diodes, are common solvation probes, show large non-linear optical (NLO) responses and are good two-photon absorbers. One of these molecules is the first molecular rectifier.[2] All these properties are related to the presence of a low-lying excited state with a different electronic distribution with respect to the ground state (GS). The low-energy physics of these molecules is governed by this state and by its interplay with slow (vibrational and environmental) degrees of freedom.

Following the original suggestion of Oudar and Chemla,[4] we adopted a two-state electronic model for these molecules and extended it to account for molecular vibrations. The resulting model, the Mulliken DA-dimer with Holstein coupling,[5] can also account for solvation, provided one of the Holstein coordinates represents an effective solvation coordinate.[6] The model is simple enough to allow for exact non-adiabatic solution[7] or for the construction of exact adiabatic potential energy surfaces (PES) for the ground and excited states.[7,8] The non-adiabatic solution proved useful to understand the physics of vibrational amplification of static NLO responses,[7] whereas the adiabatic PES opened the way to detailed spectroscopic analysis. Specifically, steady-state absorption and fluorescence spectra and their evolution with the solvent polarity are well reproduced based on few microscopic parameters.[8] Time-resolved fluorescence, pump-probe and, more generally, time–resolved experiments involving the evolution of the system in either the ground- or excited-state PES can also be interpreted based on the same model, describing the relaxation of the system in the relevant PES within the Fokker-Planck approach.[9] Vibrational spectra can also be modeled, with vibrational solvatochromism and inhomogeneous broadening in polar solvents concurring, to originate impressive effects in resonance Raman spectra.[10,11]

The detailed analysis of spectral properties of solvated DA chromophores allowed us to show the non-linearity of the electronic system as the most characteristic feature of these molecules.[12] The simple two-state model we adopt for the electronic system in fact accounts for the molecular polarity and polarizability at all orders. Several anomalous spectroscopic features of DA chromophores that were not understood within standard (perturbative) approaches are a natural consequence of the large molecular (hyper)polarizability. Of particular relevance in this respect is the evolution with the solvent polarity of steady-state absorption and fluorescence band shapes as well as their non-specularity,[8,13] the temporal evolution of time-resolved fluorescence bands, and the appearance of temporary isosbestic points in pump-probe and femtosecond hole-burning spectra.[9] The vibrational solvatochromism itself and the related inhomogeneous broadening induced by the solvent polarity[10,11] are a signature of the large non-linearity that characterizes the electronic responses to vibrational and solvation perturbations. Large non-Condon effects in resonant and non-resonant non-linear responses are an additional consequence of the non-linearity, with effects that show up most clearly in two-photon absorption spectra,[14] in the frequency-dependence of the second-harmonic generation signal[15] and, more generally, in the large vibrational amplification of static NLO responses.[5,7]

This same non-linearity is expected to show up in samples with large concentrations of chromophores, including crystalline samples, aggregates, Langmuir-Blodgett films, and polymers doped and/or functionalized with DA chromophores. Electrostatic interactions, which for sure are important in a system of interacting polar chromophores, can lead to really impressive effects if the molecular (hyper)polarizability is accounted for, leading to amplification of the non-linearity that shows up by spectroscopic anomalies as well as by impressive collective effects in the GS like, e.g., phase transitions. In the following we will introduce a simple model for interacting polar and (hyper)polarizable chromophores and investigate supramolecular effects in the corresponding GS properties.

A model for interacting polar-polarizable chromophores

An array of N chromophores can be described by the following general Hamiltonian:

$$\hat{H} = \sum_i \left(2z_0\hat{\rho}_i - \hat{\sigma}_{xi}\right) + \frac{1}{2}\sum_{i,j\neq i} V_{ij}\hat{\rho}_i\hat{\rho}_j \tag{1}$$

where the first term describes each chromophore in terms of the same two-state model used for the isolated chromophore.[5] Specifically, the two basis states, $|DA>$ and $|D^+A^->$, are separated by an energy gap $2z_0$ and are mixed by a matrix element that, without loss of generality, we set to 1. The ionicity operator $\hat{\rho}_i = \left(1-\hat{\sigma}_{zi}\right)/2$ measures the polarity of the chromophore on site i, where $\hat{\sigma}_{xi}$, $\hat{\sigma}_{zi}$ represent the Pauli spin matrices. The last term in the above Hamiltonian describes inter-chromophore interactions, with V_{ij} measuring the electrostatic interaction between fully ionized chromophores on sites i and j. The Hamiltonian in Eq. (1) applies to molecular clusters of any size and dimension; here, however, we only discuss linear arrays of chromophores with the three different molecular arrangements shown in Fig. 1.

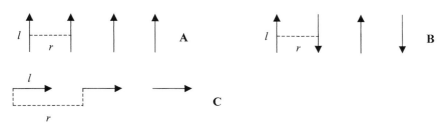

Fig 1. The three different geometrical arrangements discussed in the text. The arrow schematically represents a polar molecule.

As far as electrostatic interactions are concerned, we assign to each chromophore a fixed dipole length, l, so that:

$$V_{ij} = (\pm 1)^{|i-j|} 2vw \left[\frac{1}{d_{ij}} - \frac{1}{\sqrt{d_{ij}^2 + w^2}} \right]$$ (2)

where + and − sign refer to A and B clusters, respectively, and

$$V_{ij} = 2vw \left[\frac{1}{d_{ij}} - \frac{d_{ij}}{d_{ij}^2 - w^2} \right]$$ (3)

for C cluster. In all cases, d_{ij} is the distance between sites i and j in units of r, the distance between adjacent sites, and v measures the magnitude of the intrachromophore electrostatic interactions as relevant to a fully ionized chromophore ($v = e^2/l$). Finally, $w = l/r$ so that $w = 0$ describes non-interacting chromophores, and $w < 1$ C clusters. Interchromophore electron hopping is forbidden in the proposed Hamiltonian, and the direct product of the two basis functions centered on each site forms a complete basis set. The resulting $2^N \times 2^N$ Hamiltonian matrix is diagonalized exactly for clusters with up to 8 chromophores with periodic boundary conditions (PBC).

Results

Figure 2 reports the evolution of the GS chromophore polarity, $\rho = \langle \hat{\rho}_i \rangle$, independent of i for PBC, with w. The three columns refer to molecular arrangements A, B, C in Fig. 1; the top row shows results for $z_0 = 1$, corresponding to an isolated chromophore with a neutral (N) GS ($\rho = 0.15$ at $w = 0$), the bottom row shows results for $z_0 = -1$, corresponding to a zwitterionic (I) chromophore ($\rho = 0.85$ at $w = 0$). Interchromophore interactions disfavor charge separation for the A geometry, and ρ decreases with w in the leftmost panels in Fig. 2, whereas just the opposite occurs for geometry B and C (Fig. 2, middle and left panels). The behavior of an A cluster of I chromophores (Fig. 2b) and of B and C clusters of N molecules (Fig. 2c and e, respectively) are then particularly interesting. In the first case the isolated chromophore is zwitterionic, but, with increasing w (i.e. by decreasing the interchromophore distance), the molecular polarity decreases to the cyanine limit ($\rho = 0.5$) reaching the N regime for $r < \sim 0.7\, l$. Similarly, a N isolated chromophore can be driven to the I regime for large enough interactions in either B and C geometries at $r < 0.5\ l$ and $r < 1.4\ l$, respectively. The corresponding I to N and N to I crossovers are different in shape, as can be understood in a mean-field (mf) picture.

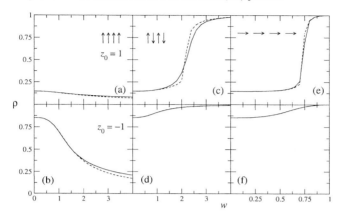

Fig 2. Molecular GS polarity, ρ, as a function of the strength of the interchromophore interaction, w, calculated for clusters of 8 molecules, with $v = 1$. Left, middle and right columns refer to geometries A, B and C, respectively; top and bottom rows correspond to $z_0 = 1$ and -1, respectively. Continuous and dashed lines refer to exact and mean-field results, respectively.

The mf treatment of the Hamiltonian has been described many years ago for C geometry[16] and the same treatment applies to other geometries as well. The mf Hamiltonian describes a collection of non-interacting chromophores, i.e., it only retains the first term in Eq. (1), but each chromophore feels the electric field generated by the surrounding chromophores in their GS configuration. The mf problem then reduces to a self-consistent two-state problem with the bare z_0 parameter in Eq. (1) substituted by $z = z_0 + M\rho$, with $M = 1/2\sum_{j \neq i} V_{ij}$. In this picture ρ only depends on z, and hence an explicit expression $z_0(M,z)$ is easily written. Specifically

$$\frac{\partial \rho}{\partial z_0} = \frac{d\rho}{dz}\left(1 - M\frac{d\rho}{dz}\right)^{-1} \tag{4}$$

so that for positive M (repulsive interchromophore interaction, A geometry), the slope of ρ vs z_0 decreases with M, justifying the smooth evolution of $\rho(w)$ in Fig. 2b. The opposite occurs for negative M (attractive interaction, B and C geometries). As a matter of fact, for large negative M a divergent $\partial \rho / \partial z_0$ is obtained, marking the occurrence of a discontinuous phase transition from the

N to the I regime.[16] Correspondingly, S-shaped mf $\rho(w)$ curves are calculated for $z_0 > 1$.

The large variation of ρ with w, and particularly the possibility of tuning the molecular polarity across the N-I boundary by simply modifying the relative orientation and/or the separation of the chromophores, demonstrate the importance of supramolecular interactions in clusters of polar-polarizable chromophores. Each polar molecule in fact originates a local electric field, which affects the GS polarity of surrounding molecules in a self-consistent loop that easily creates cooperativity as demonstrated most impressively by the possible occurrence of discontinuous N-I transitions. It is well known that molecular properties of DA chromophores strongly vary with molecular polarity and many efforts have been made to tune the molecular polarity to improve specific properties, including, e.g., static NLO responses.[1] Our results demonstrate large environmental effects in clusters, suggesting, on one hand, a large amount of tunability of molecular properties via supramolecular interactions and, on the other, the need for a careful modeling of the interactions themselves.

The problem of supramolecular interactions in clusters of polar-polarizable chromophores is more complex than recognized within a mf treatment. In fact the mf picture describes reasonably well GS properties like, e.g., the molecular polarity (cf. Fig. 2). However, mf approaches are inadequate for describing excited states and, more generally, molecular properties, including static NLO responses, which strongly depend on the excitation spectrum. The Hamiltonian in Eq. (1) actually describes the excitonic coupling and, indeed, it can be exactly mapped into an Hamiltonian where standard excitonic coupling terms can be explicitly observed. Excitonic (and ultra-excitonic) effects on the excited states of clusters of DA chromophores are particularly large and impressive and their analysis will be the subject of a separate publication.[17] Here we limit ourselves to investigation of (ultra)excitonic effects on static NLO responses of clusters of polar-polarizable chromophores. Figure 3 reports the static linear polarizability, α, and the first and second hyperpolarizabilities, β and γ, calculated, respectively, for geometry A and $z_0 = -1$ (cf. Fig. 2b), and geometry B and C for $z_0 = 1$ (cf. Fig. 2c and e, respectively), with $v = 1$ in all cases. Exact results (continuous lines) are compared with mf data (dashed lines), calculated by modeling the cluster as an oriented gas of mf chromophores. Whereas the mf estimate of the molecular polarity is a reasonable approximation to the exact results, the mf approach badly fails in the calculation of static (hyper)polarizabilities. Specifically, data in the first column of Fig. 3, relevant to an A cluster of I chromophores, demonstrate a large depression of static (hyper)polarizabilities, with effects that become appreciable when the interchromophore distance is about three times larger than the dipole length, i.e., well before the I-N crossover is reached (cf. Fig. 2b). A similar depression of static responses is observed in the case of a B cluster of N chromophores even if, due to the opposite sign of M, a qualitatively different behavior is observed for $\rho(w)$ with respect to the previous case (cf. Figs. 2c and 2b). Again sizeable deviations from the mf results are already observed at $r \sim 3\ l$, with effects that become prominent at the N-I crossover. The C cluster of N molecules shows a similar $\rho(w)$ evolution as the B cluster of N molecules (cf. Figs. 2c and 2e), but, as follows from the rightmost panels in Fig. 3, a *qualitatively different* behavior of optical susceptibilities. In this case, in fact, an amplification of static NLO responses is calculated with respect to the mf result, with the effects that are really huge at the N-I transition and increase fast with the order of the response.

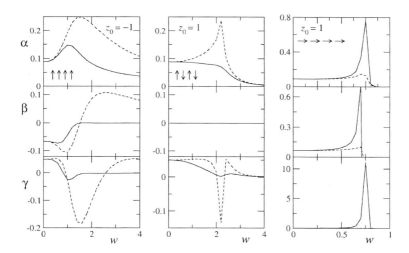

Fig 3. Static susceptibilities (α, β, γ) calculated as a function of the strength of the inter-chromophore interaction, w, for clusters of 8 molecules, with $v = 1$. Left panels: geometry A, $z_0 = -1$; middle panels: geometry B, $z_0 = 1$; right panels: geometry C, $z_0 = 1$. Continuous and dashed lines refer to exact and mean-field results, respectively. The β response vanishes exactly for the centrosymmetric B-type cluster.

Conclusions

We have demonstrated large supramolecular effects on ground-state properties, including static optical susceptibilities, in clusters of polar-polarizable molecules. The adopted model is relevant to samples of interacting DA chromophores, systems that are currently under active investigation, particularly for their possible use in advanced applications.[1,18] We underline the similarity between the N-I transition observed in C clusters of N molecules with the N-I transition observed in mixed-stack charge transfer (CT) crystals.[16,19] The amplification of α observed in C clusters (Fig. 3, upper-right panel) implies amplification of the static dielectric response, which is reminiscent of recent experimental[20] and theoretical[21] results on CT crystals. The model describes rigid clusters of rigid chromophores, but, similarly to what happens for isolated chromophores[5,7] and for CT crystals as well,[21] we believe that molecular vibrations will lead to a further amplification of non-linearity. Orientational disorder is not accounted for. Disorder is for sure present in aggregates as well as in samples where DA chromophores are dispersed in (or attached to) a polymeric matrix, and possibly has important effects there.[18] However, no reliable modeling of the properties of these intriguing materials can be achieved without fully accounting for the molecular (hyper)polarizability.

[1] M. Albota et al., *Science* **1998**, *281*, 1653.
[2] R. M. Metzger et al., *J. Am. Chem. Soc.* **1997**, *119*, 10455.
[3] A. Painelli, L. Del Freo, F. Terenziani, *Molecular Low Dimensional Materials for Advanced Applications*, A. Graja, V. M. Agranovich, F. Kajsar Eds., Kluver Academic Publisher, Netherlands 2002, p. 113.
[4] J. L. Oudar, D. S. Chemla, *J. Chem. Phys.* **1977**, *66*, 2664.
[5] A. Painelli, *Chem. Phys. Lett.* **1998**, *285*, 352.
[6] A. Painelli, *Chem. Phys.* **1999**, *245*, 183.
[7] L. Del Freo, F. Terenziani, A. Painelli, *J. Chem. Phys.* **2002**, *116*, 755.
[8] B. Boldrini, E. Cavalli, A. Painelli, F. Terenziani, *J. Phys. Chem. A* **2002**, *106*, 6286.
[9] F. Terenziani, A. Painelli, *Proceedings of the International School of Physics "Enrico Fermi", Course CXLIX*, V. M. Agranovich and G. C. La Rocca Eds., IOS Press, Netherlands 2002, p. 569.
[10] A. Painelli, F. Terenziani, *J. Phys. Chem. A* **2000**, *104*, 11041.
[11] A. Painelli, F. Terenziani, *J. Phys. Chem. A* **2000**, *104*, 11048.
[12] A. Painelli, F. Terenziani, *Synth. Met.* **2001**, *121*, 1465.
[13] A. Painelli, F. Terenziani, *Chem. Phys. Lett.* **1999**, *312*, 211.
[14] A. Painelli, L. Del Freo, F. Terenziani, *Chem. Phys. Lett.* **2001**, *346*, 470.
[15] A. Painelli, L. Del Freo, F. Terenziani, in *On the Non-linear Optical Responses of Molecules, Solids and Liquids: Methods and Applications*, M. Papadopoulos Ed., Research Signpost, India, in press.
[16] Z. G. Soos et al., *Ann. N.Y. Acad. Sci.* **1978**, *313*, 442.
[17] A. Painelli, F. Terenziani, in preparation.
[18] H. Ma et al., *J. Am. Chem. Soc.* **2001**, *123*, 986.
[19] Y. Anusooya-Pati, Z. G. Soos, A. Painelli, *Phys. Rev. B* **2001**, *63*, 205118.
[20] S. Horiuchi et al., *J. Am. Chem. Soc.* **2001**, *123*, 665.
[21] L. Del Freo, A. Painelli, Z. G. Soos, *Phys. Rev. Lett.* **2002**, *89*, 27402.

Macromol. Symp. **2004**, *212*, 131-139

Correlation between Large Polarons in Molecular Chains

Larissa Brizhik, Alexander Eremko*

Bogolyubov Institute for Theoretical Physics of the Ukrainian National Academy of Sciences, Metrologichna Str. 14-b, 03150 Kyiv, Ukraine

Summary: We studied few extra electrons in a molecular chain with the account of electron-phonon coupling in the adiabatic approximation. It is shown that the lowest state of two extra electrons in a chain corresponds to the singlet bisoliton state with one deformational potential well. Two electrons with parallel spins form a localised triplet state, which corresponds to the two-hump charge distribution function. Three extra electrons form an almost independent nonlinear superposition of a soliton and bisoliton states. In the case of four electrons, the two almost independent bisolitons are formed. These two states tend to separate in the chain at the maximal distance due to the Fermi repulsion, accounted for in the zero-order adiabatic approximation. This repulsion is partly compensated by the attraction between the solitons due to their exchange with virtual phonons, described by the non-adiabatic part of the Hamiltonian. The formation of solitons is characterised by the appearance of the bound soliton and bisoliton levels in the forbidden energy band. This constitutes the qualitative difference of the large polaron (soliton) states from the almost free electron states and small polaron states.

Keywords: collective state; conducting polymers; electron-phonon interaction; polaron; soliton

Introduction

The concept of a 'soliton' [1,2] or large polaron [3] in low-dimensional molecular systems has been intensively used to explain various phenomena in biological physics and condensed matter physics [2-4]. A large polaron state is formed due to the electron-phonon interaction at moderate enough values of interaction constant and theoretically is described in adiabatic approximation. There are clear experimental evidences for the existence of large polarons or solitons in conducting polymers [3,5]. In this respect, there arises a question about the interaction between large polarons. It was shown in [6] that due to the interaction with local deformation of the chain, two electrosolitons with opposite spins bind in a localized bound singlet spin state called 'bisoliton'. Here we consider the localized states of few extra electrons in the conducting band of a molecular chain in the adiabatic approximation.

DOI: 10.1002/masy.200450813

Model Hamiltonian

The state of electrons in an isolated conduction band accounting for the electron-phonon interaction and neglecting direct electron-electron interaction, is described by Fröhlich Hamiltonian $\hat{H} = \hat{H}_q + \hat{H}_{e-ph} + \hat{H}_{ph}$. This Hamiltonian can be represented in the site or quasimomentum representation:

$$\hat{H} = \sum_{n,\sigma} E_0 a^+_{n,\sigma} a_{n,\sigma} + \sum_{n \neq m,\sigma} J_{n,m} a^+_{n,\sigma} a_{m,\sigma} + \frac{1}{\sqrt{N}} \sum_{n,\sigma,q} \chi(q) e^{iqan} a^+_{n,\sigma} a_{n,\sigma} (b_q + b^+_{-q}) + \sum_q \hbar\omega_q b^+_q b_q , \quad (1a)$$

$$\hat{H} = \sum_{k,\sigma} E(k) \ a^+_{k,\sigma} a_{k,\sigma} + \frac{1}{\sqrt{N}} \sum_{k,\sigma,q} \chi(q) a^+_{k,\sigma} a_{k-q,\sigma} (b_q + b^+_{-q}) + \sum_q \hbar\omega_q b^+_q b_q , \quad (1b)$$

where creation and annihilation operators of an electron of the site n with the spin σ are connected with the creation and annihilation operators of an electron with the wavevector k by the unitary transformation $a^+_{k,\sigma} = \frac{1}{\sqrt{N}} \sum_n e^{ikan} a^+_{n,\sigma}$. The electron energy dispersion in the conduction band, $E(k)$, is related to the matrix elements of the exchange interaction, $J_{n,m}$, by the Fourier transformation, and the function $\chi(q)$ determines the short-range interaction of electrons with phonons of the frequency ω_q.

The ground electron state in the conduction band is determined from the Schrödinger equation:

$$\hat{H}|\Psi\rangle = E|\Psi\rangle , \quad (2)$$

Depending on the value of the electron-phonon coupling, the three different types of the ground electron states can be realised in the system: (1) almost free electrons, (2) large polaron states, (3) small polaron states. Physical properties and mathematical tools of describing these three principal states of charge carriers are different. The realisation of one of the three regimes is determined by the relations between the three main parameters of the system [7]: (1) the electron band width, which is determined by matrix elements, (2) characteristic (or maximal) phonon frequency, ω_0, (3) electron-phonon interaction energy, which can be represented in the form:

$$E_b = \frac{1}{N} \sum_q \frac{|\chi(q)|^2}{\hbar\omega_q} .$$ (3)

Here we discuss the state of few extra electrons in a molecular chain under adiabatic condition, which is necessary for the existence of large polarons. In this case the electron correlation is qualitatively different from the one in the two other cases. This is clear from the following analysis of the three limiting approximations.

Since the Hamiltonian (1) conserves the number of electrons, the state-vector of N_e electrons depends on the N_e creation operators. The solution of the Schrödinger equation can be written as $|\Psi\rangle = \hat{U}|\Psi_0\rangle$, where $\hat{U}(\{a^+\}, \{b^+\}) = \exp(\hat{S})$ is a unitary operator. Then Eq. (2) can be transformed into the following one:

$$e^{-\hat{S}} \hat{H} e^{\hat{S}} |\Psi_0\rangle \equiv \widetilde{H} |\Psi_0\rangle = E |\Psi_0\rangle .$$ (4)

1. In the **weak interaction limit**, the electron-phonon interaction is assumed to be small, i.e., it is proportional to a small parameter ε. In this case the operator $\hat{S}$ in (4) is also proportional to the small parameter. The operator expansion can be used: $\widetilde{H} = \hat{H} + [\hat{H}, \hat{S}] + \frac{1}{2}[[\hat{H}, \hat{S}], \hat{S}] + ...$ In the first order of the perturbation theory the operator $\hat{S}$ can be found from the condition $\hat{H}_{e-ph} + [\hat{H}_e + \hat{H}_{ph}, \hat{S}] = 0$, which gives

$$\hat{S} = \sum_{k,q,\sigma} f(k,q) a_{k,\sigma}^+ a_{k-q,\sigma} b_q^+ - \text{h..c.}, \quad f(k,q) = \frac{1}{\sqrt{N}} \frac{\chi^*(q)}{\hbar\omega_q + E(k+q) - E(k)} .$$ (5)

This transforms Hamiltonian (4) into the next one

$$\widetilde{H} = \sum_{k,\sigma} [E(k) + \Delta E(k)] a_{k,\sigma}^+ a_{k,\sigma} + \sum_{k,k',q,\sigma,\sigma'} V(k,k',q) a_{k+q,\sigma}^+ a_{k'-q,\sigma'}^+ a_{k',\sigma'} a_{k,\sigma} + O(\varepsilon^3) .$$ (6)

Here $\Delta E(k)$ determines the standard renormalisation of the electron energy and effective mass in the second order of the perturbation theory [8]. The new term in (6) accounts for the direct electron-electron interaction induced by phonons, $V(k,k',q) \sim |\chi(q)|^2$. This interaction is of the attraction type, it does not influence the state of an extra isolated electron, but it leads to the binding of the two isolated electrons into a Cooper pair [9]. In the case of a large number of electrons at the finite density, the ground electron state at the zero temperature is the superconducting state.

2. In the **small polaron approximation** the matrix elements of the exchange interaction, $J_{n,m}$, are assumed to be small in the Hamiltonian. The unitary transformation is chosen in the form [10]:

$$\hat{S} = \sum_{n,q,\sigma} f(n,q) a_{n,\sigma}^+ a_{n,\sigma} b_q^+ - h.c., \qquad f(k,q) = \frac{1}{\sqrt{N}} \frac{\chi^*(q)}{\hbar\omega_q}. \qquad (7)$$

In this case the transformed Hamiltonian in the nearest-neighbour approximation reads

$$\widetilde{\widetilde{H}} = \sum_{n,\sigma} [(E_n - E_b) a_{n,\sigma}^+ a_{n,\sigma} - Je^{-G} (a_{n,\sigma}^+ a_{n+1,\sigma} e^{-B_n^+} e^{B_n} + h.c.)] -$$

$$- \sum_{n,m,\sigma,\sigma'} V(n-m) a_{n,\sigma}^+ a_{m,\sigma'}^+ a_{m,\sigma'} a_{n,\sigma} + \sum_q \hbar\omega_q b_q^+ b_q . \qquad (8)$$

Here the term proportional to the exponential operators describes jumps of a polaron from a site to site with the creation of a certain number of phonons: $B_n^+ \sim b_q^+$. In view of the small value of this term at strong coupling, $G \sim |\chi(q)|^2 >> 1$, it can be considered as a perturbation, so that the zero-order term of the Hamiltonian accounts for the processes with the conservation of the total number of phonons only. Then, in the wavevector presentation, one extra electron in a chain is described by a small polaron state in a narrow conduction band, $\tilde{J} = J\exp(-G)$. Moreover, the direct attraction type polaron-polaron interaction via phonons appears in Hamiltonian (8), which is given by the term $V(n-m)$. Therefore, the two extra electrons in a chain form a bipolaron [11]. The adiabatic limit will be considered in more details in the next section.

Adiabatic limit

In this case the kinetic energy of atom vibrations, which is a part of phonon Hamiltonian, H_{ph}, is considered as a small term of the total Hamiltonian. Due to this, the vector state can be represented as the product of electron and phonon wavefunctions, which corresponds to the Born-Oppenheimer approximation. Accordingly, let us represent the unitary transformation, $\hat{U}$, as a product of $\hat{U}_e$ and $\hat{U}_{ph} = \exp(\hat{S})$.

$$\hat{S} = \frac{1}{\sqrt{N}} \sum_q \beta_q b_q^+ - h.c., \qquad a_{k,\sigma} = \sum_\lambda \psi_\lambda(k) A_{\lambda,\sigma} , \qquad (9)$$

where the coefficients of the unitary operators are chosen in such a form that the electron part of the Hamiltonian $\widetilde{\widetilde{H}}$ is diagonal, and, thus, they satisfy the equations:

$$E(k)\psi_\lambda(k) + \frac{1}{N}\sum_q \chi(q)(\beta_q + \beta^*_{-q})\psi_\lambda(k-q) = E_\lambda\psi_\lambda(k). \tag{10}$$

These two transformations are not independent, electron state is determined by the lattice configuration, which, in turn, depends on the electron state. The transformed Hamiltonian now takes the form:

$$\tilde{\tilde{H}} = W + \sum_{\lambda,\sigma} E_\lambda A^+_{\lambda,\sigma} A_{\lambda,\sigma} + \sum_q \hbar\omega_q b^+_q b_q + \sum_q \{[\beta_q + \frac{1}{\sqrt{N}}\sum_{\lambda,\sigma}\varphi_{\lambda,\lambda}(q)A^+_{\lambda,\sigma}A_{\lambda,\sigma}]b^+_q + \text{h.c.}\} +$$

$$+ \sum_{\lambda\neq\lambda',\sigma}\varphi_{\lambda,\lambda'}(q)A^+_{\lambda,\sigma}A_{\lambda',\sigma}(b_q + b^+_{-q}). \tag{11}$$

Here $\varphi_{\lambda,\lambda'}(q) = \chi(q)\sum_k \Psi^*_\lambda(k)\Psi_{\lambda'}(k-q)$. The last term in (11) describes the transitions between the adiabatic terms with the absorption and radiation of phonons and is the nonadiabaticity operator. Provided the condition of the adiabatic approximation is fulfilled, it can be neglected and the problem can be studied in the zero-order adiabatic approximation. Unlike in the considered cases of almost free electrons and small polarons, the direct electron-electron interaction is absent in the zero-order Hamiltonian. Nevertheless, as mentioned above, electrons are not independent and move in the common self-consistent potential. In this case the ground electron state can be represented as the product of N_e electron creation operators and of the vacuum state

$$|\Psi_0\rangle = \prod_{\lambda,\sigma} A^+_{\lambda,\sigma}|0\rangle \tag{12}$$

which is the eigenstate with the energy $E = W + \sum_\lambda n_\lambda E_\lambda$. Here $n_\lambda = 0,1,2$ are the occupation numbers of the adiabatic level and $\sum_\lambda n_\lambda = N_e$. The coefficients β_q ought to satisfy the relation

$$\beta_q + \frac{1}{\sqrt{N}}\sum_\lambda n_\lambda \varphi^*_{\lambda\lambda}(q) = 0. \tag{13}$$

Using the longwave (continuum) approximation and switching to spatial representation, we can give Eq. (10) in the form:

$$-\frac{d^2\Psi_\lambda}{dx^2} + U(x)\Psi_\lambda = \frac{2ma^2}{\hbar^2}E_\lambda\Psi_\lambda, \tag{14}$$

where the deformational potential, accounting for the expression (13), has the form

$$U(x) = -2g \sum_{\lambda} n_{\lambda} |\Psi_{\lambda}|^2, \tag{15}$$

where m is the effective electron mass and g is the electron-phonon interaction constant. Because the deformational potential (15) is determined by the occupied adiabatic levels, the problem reduces to solving the system of many-component nonlinear Schrödinger equations for the occupied states [12]. Non-occupied excited states are determined by Eq. (15) with the given potential.

One extra electron is described by the conventional nonlinear Schrödinger equation for a single adiabatic level, $n = 1$, whose solution corresponds to the soliton state with the energy $E_1 = -\hbar^2 \mu^2/(2ma^2)$, where $\mu = g/2$. The ground state of the two extra electrons in a chain also corresponds to one adiabatic level occupied by the two electrons with opposite spins, $n = 2$. The corresponding energy is $E_1 = -\hbar^2 \mu^2/(2ma^2)$, where $\mu = g$.

In the case of three extra electrons the ground state corresponds to the two occupied adiabatic levels with the occupation numbers $n_1 = 2$, $n_2 = 1$, which are determined by the two-component nonlinear Schrödinger equation which admits the solution [12]

$$\Psi_1 = \sqrt{\frac{\mu_1^2 - \mu_2^2}{g}} \frac{\mu_1 \cosh(\phi_2)}{D}, \quad \Psi_2 = \sqrt{\frac{\mu_1^2 - \mu_2^2}{g}} \frac{\mu_2 \sinh(\phi_1)}{D}, \tag{16}$$

where

$$D = \mu_1 \cosh(\phi_1)\cosh(\phi_2) - \mu_2 \sinh(\phi_1)\sinh(\phi_2), \quad \phi_1 = \mu_1(x+l), \quad \phi_2 = \mu_2(x-l). \tag{17}$$

The constant l in (17) can take arbitrary values. These two states correspond to energies $E_j = -\hbar^2 \mu_j^2/(2ma^2)$, with $\mu_1 = g$, $\mu_2 = g/2$. Therefore, the three electrons form a bisoliton and a soliton, which are described by essentially changed wavefunctions as compared with an isolated soliton function, and which have the total energy in the form of the sum of soliton and bisoliton energies. The constant l in this case determines the distance between soliton and bisoliton center of mass coordinates. This gives us an example of a nonlinear superposition of quasiparticles, when the energy is independent of the distance between them, although the total wavefunction is not a superposition of the functions. The corresponding wavefunctions (16) are shown in Fig. 1.

In the case of four extra electrons, the two adiabatic levels are occupied, with the occupation numbers $n_1 = n_2 = 2$. Similar situation takes place for the two electrons with polarised spins, so that $n_1 = n_2 = 1$. In these latter cases $\mu_1, \mu_2 \rightarrow \mu_0$, where $\mu_0 = g$ for two bisolitons, and $\mu_0 = g/2$ for the triplet state. The wavefunctions $\Psi_{1,2}$ describe the two bisolitons (two solitons) separated by the large distance:

$$L = \frac{1}{2\mu_0} \ln \frac{2\mu_0 \cosh(2\mu_0 l)}{\Delta} \to \infty \quad \text{at} \quad \Delta = \mu_1 - \mu_2 \to 0. \tag{18}$$

The total energy

$$E = -\frac{n^2}{6} Jg^2 + 2Jg^2 \cosh^2(gl) \exp(-2ngL), \tag{19}$$

where $n = 1$ for the triplet state, and $n = 2$ for the two bisolitons. The corresponding wavefunctions of four electrons are shown in Fig. 2. In this case the constant l characterises the level of collectivisation of the two separated potential wells.

The energy expression includes the term which depends on the distance between the center of mass coordinates, L, and describes the repulsion due to the Pauli principle.

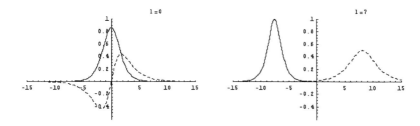

Fig.1. The wavefunctions of three extra electrons in a chain: (a) $l = 0$, (b) $l = 7$.

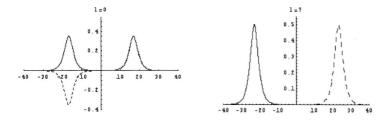

Fig.2. The wavefunctions of four extra electrons in a chain: (a) $l = 0$, (b) $l = 7$.

Non-adiabatic corrections

The account of the nonadiabatic term in the Hamiltonian (11) results in the additional direct interaction between the two solitons or bisolitons via the phonon field. This additional interaction partly compensates the Fermi repulsion and stabilizes the two-bisoliton and triplet state solutions. Considering the nonadiabatic term of the Hamiltonian as a perturbation, one can write the first-order energy correction in the form

$$\Delta E = \left\langle \Psi_0 \left| \hat{H}_{na} \frac{1}{E - H_0} \hat{H}_{na} \right| \Psi_0 \right\rangle. \tag{20}$$

Here the wavefunctions correspond to the zero-order adiabatic approximation $\hat{H}_0$, and $\hat{H}_{na}$ is the non-adiabatic part of the Hamiltonian, i.e., the last term in (11). To calculate the energy of (20), is necessary to find the total energy spectrum of the system. This includes the localised (bound) levels and delocalised (continuum) spectrum. The localised levels are always occupied, and unoccupied levels belong to the continuum spectrum. The corresponding calculations show that the energy correction term (20) contains two terms. One of these terms is independent of the distance between centre of mass coordinates and determines the renormalisation of their energy, while the second term depends on the distance. The latter term accounts for the additional attraction, and results in he stabilisation of the state. The total energy with the account of the correction term (20) has the form [13]:

$$E = -\frac{n^2}{6} Jg^2 \left(1 + \frac{3\gamma}{8\pi^6 g}\right) + 2Jg^2 \cosh^2(gl)\exp(-2ngL) - \frac{3Jg\gamma}{8\pi^3 L^2 \cosh(\lg)}. \tag{21}$$

Here γ is the so-called non-adiabaticity parameter determined as $\gamma = \hbar\omega_{max}/J$.

Because of the competition between the repulsion force due to the Pauli principle and attraction force due to the phonon field, the two bisolitons or two solitons with parallel spins are separated by the equilibrium distance, which is determined from the condition $dE/dL = 0$, where E is determined in Eq. (21).

Conclusion

We have shown that the lowest state of two extra electrons in a chain corresponds to the singlet bisoliton. Three extra electrons form almost independent nonlinear superposition of a soliton and bisoliton states. In the case of four electrons or two electrons with parallel spins, the two independent bisolitons (solitons) are formed. These two states tend to separate in the chain at the maximal distance due to the Fermi repulsion. This repulsion is partly compensated by the attraction between the solitons due to their exchange by the virtual phonons. The formation of large polarons is characterised by the appearance of the bound soliton and bisoliton levels in the forbidden energy band. This constitutes a qualitative difference of the large polaron (soliton) states from the almost free electron states and small polaron states. In the case of a finite density of extra electrons in a chain at the zero temperature, the charge density wave is formed in a system [14]. The repulsion between bisolitons leads to the formation of a periodic lattice of bisolitons with the distribution period $l = N/N_{bs}$, $N_{bs} = N_e/2$ that corresponds to the many-electron solution of the Peierls-Froehlich problem at zero temperature [15].

It is worth noting that the self-consistent deformation potential of the lattice is reflectionless. In the case of one extra electron or a singlet state of two electrons that occupy the same level (bisoliton), this potential has a single bound state. In the case of a triplet state of two electrons (two bisolitons), when Fermi statistics forbids one-level occupation, lattice deformation forms a reflectionless potential with two bound levels. In the case of an arbitrary number of electrons, the corresponding self-consistent deformation potential is a reflectionless single-band potential with one gap, which separates the occupied sublevels from the vacant ones [15].

Acknowledgement: This work was partly supported by a grant of the Programme of Fundamental Research of the Ukrainian National Academy of Sciences.

[1] A.S. Davydov, N.I. Kislukha, *Phys. Status Solidi,* (b) **1973**, *59*, 465.
[2] A.S. Davydov, *Solitons in Molecular Systems*, Reidel, Dordrecht, 1985.
[3] A.J. Heeger, et al., *Rev. Mod. Phys.,* **1988**, *60*, 781.
[4] A.C. Scott, *Phys. Rep.*, **1992**, *217*, 1.
[5] I. Gontia, et al., *Phys. Rev. Lett.,* **1999**, *82*, 4058.
[6] L.S. Brizhik, A.S. Davydov, *Fiz. Nizk. Temp.*, **1984**, *10*, 748.

[7] L.S. Brizhik, A.A. Eremko, *Synth. Met.,* **2000**, *109*, 117.
[8] H. Haken, *Quantenfeldtheorie des Festkörpers*, B.G.Teubner, Stuttgart, 1973.
[9] L.N. Cooper, *Phys. Rev.*, **1956**, *104*, 1189.
[10] Yu.A. Firsov (Ed.), *Polarons*, Nauka, Moscow, 1975.
[11] A.S. Alexandrov, J. Ranninger, *Phys. Rev. B,* **1981**, *23*, 1794; *24*, 1164.
[12] L.S. Brizhik, A.A. Eremko, *Physica D,* **1995**, *81*, 295.
[13] L.S. Brizhik, A.A. Eremko, *Ukr. J. Phys.*, **1999**, *44*, 1022.
[14] G. Gruner, *Rev. Mod. Phys.*, **1988**, *60*, 1129.
[15] A.A. Eremko, *Phys. Rev. B,* **1992**, *46*, 3721.

Macromol. Symp. **2004**, *212*, 141-157

Design of Molecular Magnets

J. V. Yakhmi

Technical Physics and Prototype Engineering Division, Bhabha Atomic Research Centre, Mumbai (Bombay) – 400 085, India

Summary: The conventional magnetic materials used in present-day technology, such as Fe, Fe_2O_3, Cr_2O_3, $SmCo_5$, $Nd_2Fe_{14}B$, etc. are all atom-based, whose synthesis requires high-temperature routes. Employing ambient-temperature synthetic organic chemistry, it has become possible to engineer a bulk molecular material with long-range magnetic order, primarily due to the weak nature of intermolecular interactions in it. Typical synthetic approach to design molecule-based magnets consists of choosing molecular precursors, each bearing an unpaired spin, and assembling them in such a way that there is no compensation of spins at the scale of the crystal lattice. Magnetism being a co-operative effect, the spin-spin interaction must extend to all the three dimensions, either through space or through bonds. Specific occurrence of 'spin delocalisation' and 'spin polarisation' in molecular lattices is helpful in bringing about ferromagnetic interaction by facilitating necessary intermolecular exchange interactions. Since the first successful synthesis of molecular magnets in 1986, a large variety of them have been synthesized, which can be classified on the basis of the chemical nature of the magnetic units involved: organic systems, metal-based systems, hetero-bimetallic assemblies, or mixed organic-inorganic systems. The design of molecular magnets has also opened the doors for the unique possibility of designing polyfunctional molecular materials, such as magnets exhibiting second-order optical nonlinearity, liquid crystalline magnets, or chiral magnets. Solubility of molecular magnets, their low density and biocompatibility are attractive features. Being weakly colored, unlike the opaque classic magnets, possibilities of photomagnetic switching can be envisaged. Persistent efforts continue to design the ever-elusive polymer magnets for applications in industry. While providing a brief overview of the field of molecular magnetism, we highlight some recent developments, with emphasis on a few studies from the author's own lab.

Keywords: conjugated polymers; magnetic polymers

1. Introduction

Most molecular materials are organic in nature. Molecular crystals are made up of well-defined molecules, which do not change their geometries appreciably upon entering the crystal lattice. This is because intermolecular interactions are non-covalent in nature, viz. hydrogen bonding, van der Waals interactions, donor-acceptor charge transfer, etc., that are much weaker than the energies of typical chemical bonds, ionic or covalent. This provides an interesting possibility to modify the properties of a molecular solid in a predetermined way by attaching a *function* to the

                    DOI: 10.1002/masy.200450814

molecular building-block (i.e., functionalizing the molecule) to engineer a bulk molecular material with designer characteristics. Synthetic organic chemistry employs ambient temperatures, and is known for imparting immense flexibility in the synthesis of molecule-based compounds. This has given rise to the synthesis of new molecular materials designed to perform several functions originally attributed to the metallic lattices, such as high electric conductivity or photoactivity in polymers, and superconductivity in charge-transfer organic complexes/fullerenes. Quite obviously, there has been a strong urge to develop ferro(ferri)magnetic molecular materials, too, by designing new combinations of interactions between magnetic centres in organic (or polymeric) materials, preferably with p-orbital based spins. Theoretical models did predict the possibility of attaining long-range magnetic order in molecular materials. However, organic compounds are mostly diamagnetic, with closed-shell structures, and even if one (or more) unpaired electrons are maintained stable in an organic molecule, stabilization of a triplet state (parallel alignment of spins) requires that the orthogonality conditions be satisfied (Hund's rules), which is difficult. So, obtaining antiferromagnetic coupling with no spontaneous magnetic moment is preferred. No wonder, the conventional magnetic materials used in present-day technology, such as, Fe, Fe_2O_3, Cr_2O_3, $SmCo_5$, $Nd_2Fe_{14}B$, etc. are all atom-based materials, in which magnetic order arises from co-operative spin-spin interactions between unpaired electrons located in d-orbitals or f-orbitals. Their synthesis typically depends on solid-state chemistry or high-temperature metallurgical routes.

2. Purely Organic Magnets

The first two ferro(ferri)magnetic molecular compounds exhibiting a spontaneous magnetization below a certain temperature, T_c, were reported in 1986 [1,2], and subsequently molecular magnets of many different categories have been synthesized and interest in the field of molecule-based magnets has been growing steadily. The discovery of ferromagnetism involving p-electrons in iron-containing organic-based material [1] was an important step forward because magnetism in metal-free compounds must involve electrons from p-atomic orbitals, which was considered impossible not long ago. Although significant progress has been made in the preparation of π-conjugated oligomers and polymers with large values of spin quantum number S, persistent efforts to design polymer magnets which, when made, would have a huge impact on applications in industry have not borne fruit [3,4]. Most recently however, Rajca et al [5] have reported the observation of magnetic properties comparable to that of insulating spin glasses and blocked superparamagnets (S = 5000, and slow orientation of the magnetization by a small magnetic field of 1 Oe below 10 K) in a highly cross-linked organic π-conjugated polymer, obtained from polyethers. Inspired by recent observations of gate-induced superconductivity in oligomers like anthracene [6] and polymers like polythiophene [7], Arita et al [8] have proposed the possibility of band ferromagnetism in a purely organic polymer structures like PAT [poly(4-amino-1,2,4-triazole], which consists of a chain of five-membered rings. They have proposed that when the flatband of such materials is made half-filled, with appropriate dopings to be realized using a field-effect transistor structure, the ground state is ferromagnetic, as indicated by spin density functional calculations. This is interesting because organics exhibiting (single-)band ferromagnetism have yet to be synthesized, though multiorbital ferromagnets such as TDAE-C_{60} [9] are known, as discussed later.

Typical synthetic approach to design molecule-based magnets consists of choosing molecular precursors, each bearing an unpaired spin (the *function*, as shown in Fig. 1), and assembling them in such a way that there is no compensation of spins at the scale of the crystal lattice [4,10,11].

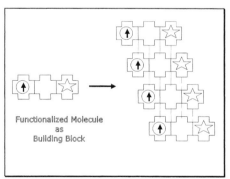

Fig. 1: Generation of a molecular lattice, through self-assembly, employing a functionalized molecule as a building block. The latter could have just one function attached – say, an unpaired electron denoted by an arrow, in which case the three-dimensional lattice is aimed to have a long-range magnetic order. A building block could also have an additional second function to facilitate the design of bifunctional molecular magnets possessing another property, such as chirality, liquid crystallinity, or ferroelectricity, etc. in addition to magnetism.

More recently, the design of molecular magnets has also employed the tools of supramolecular chemistry [12]. The interactions between spin carriers may occur through space, in which case we have a genuine molecular lattice, or, through bond when we are faced with a polymeric or extended structure. In the latter case, the interactions are usually much stronger, particularly so when the bridging ligands are conjugated.

The design of a molecular magnet requires that: (a) all the molecules in the lattice have unpaired electrons, and, (b) the unpaired electrons should have their spins aligned parallel along a given direction. Magnetism being a co-operative effect, the spin-spin interaction must extend to all the three dimensions. Specific occurrence of *spin delocalisation* and *spin polarisation* in molecular lattices, unlike in the case of ionic/metallic compounds, is helpful in bringing about ferromagnetic interaction by facilitating necessary intermolecular exchange interactions.

The delocalization of spin density in certain molecules makes it possible for magnetic interactions to take place across extended bridges between magnetic centres far apart from each other, propagating through conjugated bond linkages, which act as molecular wires. Spin polarization, i.e. the simultaneous existence of positive and negative spin densities at different location within a given radical is crucial for intermolecular exchange interactions to bring about ferromagnetic interaction between organic radicals, as per McConnell's model [13]. Spin density across different regions of the nitronyl nitroxide radical NITR (R=alkyl), a versatile building block with spin $S = 1/2$ ground state (Fig. 2), for instance, shows positive values, equally delocalized between N and O within each N-O group, and a small negative value on the bridging sp^2 carbon, due to spin polarization.

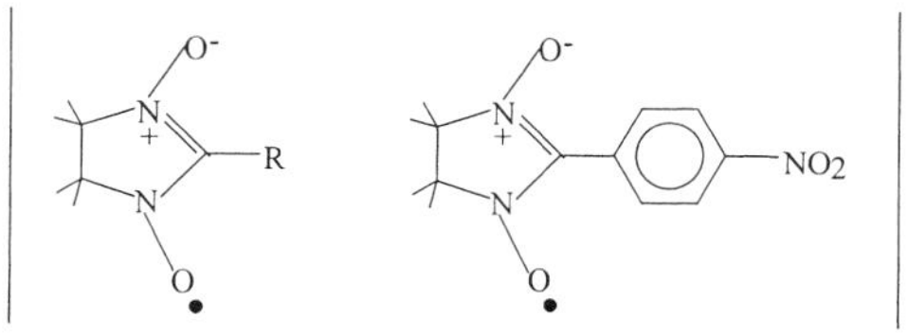

Fig. 2: Chemical structure of NITR; and *p*-NPNN (• indicates an unpaired electron).

By substituting different alkyl groups (like R = benzyl, isopropyl, methyl, ethyl, phenyl, etc.) in NITR, one can tune the single-radical ground state to establish new exchange pathways through varied coordination sites. For instance, ferromagnetism at 0.6 K arises solely from *p* orbital spins in the β-phase of R = phenyl compound, 4-nitrophenylnitronyl nitroxide) (*p*-NPNN, with formula $C_{13}H_{16}N_3O_4$, shown in Fig. 2), a metal-free organic magnet which contains only C, H, N, and O elements [14].

Ferromagnetism has also been obtained for the purely organic fullerene-based charge-transfer material, [tetrakis(dimethylamino)ethene][C_{60}] with T_c of 16.1 K [9]. Fullerene has no intrinsic magnetic moment. For a magnetic moment to exist, an electron must be transferred to C_{60} from a donor molecule. Another example of a fullerene-based ferromagnet was the cobaltocene-doped derivative, which has a T_c of 19 K [15]. The highest ordering temperature reported to date for an organic magnet has been for the β-phase of the (4-cyanotetrafluorophenyl)dithiadiazolyl, a sulfur-based free radical, which was found to be a weak ferromagnet below 35.5 K [16,17]. Under a pressure of 1.6 GPa, this temperature can be raised to 65 K [18]. Most interestingly, but raising some controversy nonetheless, Makarova et al [19] have reported recently strong magnetic signals (including saturation magnetization, hysteresis, and attachment to a magnet at room temperature), in a two-dimensional rhombohedric phase of C_{60}, that resembles graphite. The temperature dependence of saturation magnetization and remanence indicate a T_c of about 500 K, but the magnetic behavior is very sensitive to the preparation conditions. The material is obtained by converting a crystalline state of isolated molecules, held together only by van der Waals forces, into polymeric phases in which the molecules are covalently bonded, using a high-pressure, high-temperature polymerization route.

3. Ferrimagnetic Building Blocks

A powerful strategy to build a molecule-based magnet is based on the use of ferrimagnetic chains containing alternating spins of unequal magnitude $S_A \neq S_B$, and assembling them in such a way that there is a net spin, leading to a long-range magnetic order in the lattice [10,20]. Here, S_A stands for the large spin and S_B for the small spin on two different spin carriers, A and B, such as Mn(II) ions ($S = 5/2$) and Cu(II) ions ($S = 1/2$), respectively, within the same molecular precursor (Fig. 3). A large number of molecular (ferro)ferrimagnets have been assembled using this technique, the spin carriers in them being either two different metal ions [21,22] or a metal ion and an organic radical [23-26], with intervening ligands which serve as effective exchange pathways.

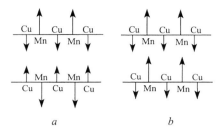

Fig. 3: Assembly of ferrimagnetic chains leading to a net (a) zero or (b) non-zero magnetic moment in bulk.

Heterobimetallic species, in which two different metal ions are bridged by extended bisbidentate ligands such as oxamato [27,28], oxamido [21,29], or oxalato [22], in particular, allow a variety of spin topologies. Using this priniciple, Kahn's group have synthesized several Mn(II)Cu(II) molecular magnets in which the ferrimagnetic interactions is propagated through bisbidentate ligands, viz. MnCu(opba).0.7DMSO which is synthesized by reacting the Cu(II) precursor Cu(opba)$^{2-}$ (Fig. 4), where opba stands for phenylenebis(oxamato), with a divalent ion, Mn(II) in a 1:1 stoichiometry, behaves as an amorphous magnet with a spontaneous magnetization below $T_c = 6.5$ K.

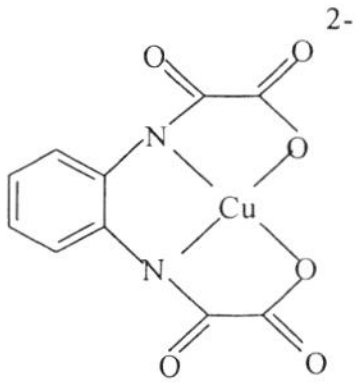

[Cu(opbz)]$^{2-}$ 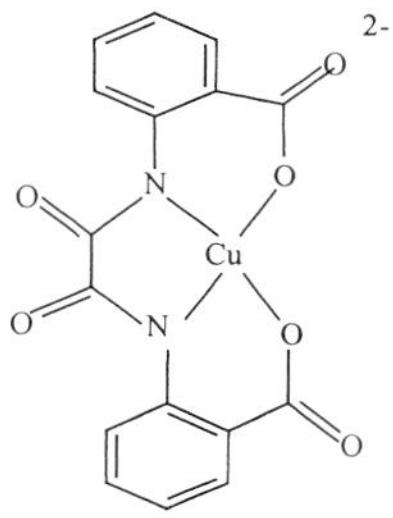   [Cu(obbz)]$^{2-}$

Fig. 4: A schematic of the copper dianion precursors [Cu(opba)]$^{2-}$ and [Cu(obbz)]$^{2-}$, where 'opba' and 'obbz' stands for phenylenebis(oxamato) and oxamidobis(benzoato) ligands, respectively.

Aiming at increasing the Curie temperatures of this class of compounds led to the synthesis of 2:3 Mn(II)Cu(II) compounds A$_2$M$_2$[Cu(opba)]$_3$.$n$solv with two-dimensional character, by employing [Cu(opba)]$^{2-}$ to crosslink the chains in a two-dimensional network, in the presence of two equivalents of non-coordinating cation A^+. Among them is $(NBu_4)_2Mn_2[Cu(opba)]_3.6DMSO.1H_2O$, exhibiting a transition at $T_c = 15$ K towards a ferromagnetically ordered state [27], the T_c value of which rises to 22.5 K when all the solvent molecules are removed [27,30]. Elucidation of structure of 1:2 Mn(II)Cu(II) compound $(NBu_4)_2Mn[Cu(opba)]_2$ by us, recently, revealed a new crystallographic arrangement among the class of 'opba' molecular magnets [31]. It crystallizes in orthorhombic structure (with space group $Pna2_1$) consisting of a zig-zag chain with terminal 'opba' groups. It is a filled structure with most of the space filled by cations; the clefts in the space in-between are filled with two NBu_4 cations (Fig. 5). So far they have been known to exist as one-dimensional chains for the 1:1 proportion, and as planar graphite-like sheet structures for the 2:3 ratio.

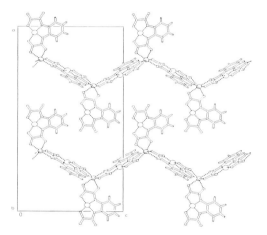

Fig. 5: Structure of 1:2 Mn(II)Cu(II) compound $(NBu_4)_2Mn[Cu(opba)]_2$ consisting of a zig-zag chain with terminal 'opba' groups.

An effective process for assembling spin-bearing precursors is when polymerization is associated with dehydration. When the precursors are linked to each other after polymerization, the interaction involving a spin carrier A from a given unit and a spin carrier B from the adjacent unit becomes large, and the resulting ground state spin of the polymer $[AB]_n$, where n is the number of units, becomes $n(S_A - S_B)$. If n is infinite and the system is three-dimensional, a long-range ferrimagnetic ordering should occur, which can also be considered as a ferromagnetic coupling of the ground state spin $(S_A - S_B)$ of the AB units, which in the case of Mn(II)Cu(II)-

based compounds is $S = 2$. The compound MnCu(obbz).5H$_2$O [21], which was obtained by the polymerization of ferrimagnetic molecular precursor units obtained by reacting the copper dianion [Cu(obbz)]$^{2-}$ (Fig. 4), where obbz is the oxamidobis(benzoato) ligand, with Mn(II) ions, exhibits a minimum at 44 K in its $\chi_M T$ versus T plot, χ_M being the molar magnetic susceptibility and T the temperature, a signature of a one-dimensional ferrimagnet, and a sharp maximum at 2.3 K due to a three-dimensional antiferromagnetic ordering. Upon dehydration it yields the monohydrate, MnCu(obbz).1H$_2$O, which orders ferromagnetically below the critical temperature (T_c) of 14 K, due to the non-compensation of the magnetic moments on Mn(II) and Cu(II) [21,32]. Subsequently, Ni(II)-, Fe(II)- and Co(II)-based bimetallic chain compounds have also been synthesized using the [Cu(obbz)]$^{2-}$ precursor [21,32-35]. Using the mixed metal ion spin-organic radical spin approach, a 46 K magnet was synthesized by the reaction of a trinitroxide radical, with three parallel spins ($S = 3/2$), with bis(hexafluoroacetylacetonato)manganese(II), [Mn(II)(hfac)$_2$] [26].

The Mn(II) ion is not able to prevent the domains from rotating freely under an applied field because it is a magnetically isotropic ion. Most of the Mn(II)Cu(II)-based magnets, though exhibit a high value of T_c (12 to 30 K), are soft ferromagnets, therefore, exhibiting rather narrow magnetic hysteresis loops below T_c with rather weak coercive field values ($H_c < 50$ Oe at 4.2 K) [27,36-38]. The same is generally true for Fe(II)Cu(II)-based magnets (Fig. 6). It is the coercivity of a magnet which confers a memory effect on it. Rather strong coercive fields are expected for Co^{2+}-based molecular magnets where Co^{2+} ion in distorted octahedral environment, being magnetically anisotropic, can assume preferred orientations. Replacing Mn(II), with an orbital singlet state (^{6}A$_1$), by Co(II) with an orbital triplet ground state (^{4}T$_1$) in (cat)$_2$Mn$_2$[Cu(opba)]$_3$ [27,36,37], where 'cat' denotes a cation, results in a dramatic rise in coercivity. For instance, $H_c = 3000$ Oe for (NBu$_4$)$_2$Co$_2$[Cu(opba)]$_3$.3DMSO.3H$_2$O and 3100 Oe for (rad)$_2$Co$_2$[Cu(opba)]$_3$.0.5DMSO.3H$_2$O [38], where "rad" stands for the radical cation 4,4,5,5-tetramethyl-2-(1-methylpyridin-1-ium-4-yl)-4,5-dihydroimidazol-1-oxyl 3-oxide and "opba" denotes phenylenebis(oxamato). The compound [(Etrad)$_2$Co$_2$\{Cu(opba)\}$_3$(DMSO)$_{1.5}$].0.25H$_2$O, having formula C$_{61}$H$_{63.5}$N$_{12}$O$_{23.75}$S$_{1.5}$Co$_2$Cu$_3$, exhibits quite high coercivity, up to 24 kOe at 6 K for a sample of small crystals [39]. Here, Etrad$^+$ deonotes an ethyl radical cation, 2-(1-ethylpyridinium-4-yl)-4,4,5,5-tetramethyl-4,5-dihydroimidazol-1-oxyl 3-oxide. These values are much higher than those for the commercial atom-based materials Fe$_2$O$_3$ or CrO$_2$. The synthetic procedures employed to obtain molecular materials are different from those employed in solid-state chemistry and the molecular crystal lattice is characteristically *soft*, as compared to the ionic or metallic lattices. We have exploited these attributes to demonstrate that for certain molecular magnets, assembled from Co(II)Cu(II)-based ferrimagnetic chains, it is possible to modify the magnetic properties dramatically and reversibly through a mild dehydration-rehydration process, and have named this class of compounds as *molecular magnetic sponges* [33,34,40,41]. This is because they show 'sponge'–like characteristics, viz. a reversible cross-over under dehydration to a polymerized long-range magnetically ordered state with spontaneous magnetization, and transforming back into the isolated units underlying the initial non-magnetic phase by reabsorbing water, i.e. rehydration of both noncoordinated and coordinated water molecules (Fig. 7).

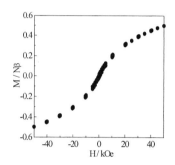

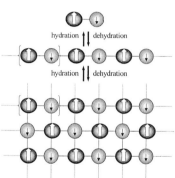

Fig. 6: Magnetization data at 5 K for $Fe^{II}Cu^{II}(obbz).1H_2O$ as a function of applied field.

Fig. 7: The dehydration-polymerization process, typically applicable to the reversible sponge-like behaviour of $CoCu(obbz).nH_2O$.

Coercivity values for these sponges are high and for some of these a colour change, too, occurs reversibly and simultaneously with the change in magnetic properties at the transition temperature corresponding to the dehydration-rehydration process. In the case of $[CoCu(obbz)].nH_2O$, we confirmed that the Co-O bonds could be broken and created without destroying the essence of the molecular architecture. The main features of the four Co(II)Cu(II)-based *molecular magnetic sponges* synthesized by our group, viz. $CoCu(pbaOH)(H_2O)_3.2H_2O$, $CoCu(pba)(H_2O)_3.2H_2O$, $CoCu(obbz)(H_2O)_4.2H_2O$ and $CoCu(obze)(H_2O)_4.2H_2O$ are high values of T_c (38, 33, 25 and 25 K, respectively) and H_c (5.66, 3, 1.3 and 1 kOe, respectively). The symbols pbaOH, pba and obze denote 2-hydroxypropane-1,3-diylbis(oxamato), propane-1,3-diylbis(oxamato) and oxamido-*N*-benzoato-*N*'-ethanoato, respectively.

4. Polycyanometallates

A unique feature of the molecular magnets is that they are usually weakly coloured unlike the opaque classic magnets. Design of molecular ferromagnets of low density that are transparent and have a tunable, high T_c is a cherished goal. Photomagnetic switching has been reported in molecular magnets, especially where hexacyanometallates $[M(CN)_6]^{n-}$ are used as molecular building blocks.

Most metal hexacyanometallates have a cubic structure as shown in Fig. 8. Metal ions are situated at the corners of the cube and they are octahedrally coordinated by the nitrogen or carbon atom of the cyanide group. The cyanide groups are bridging the metal ions along the cube edges. Depending on the number of charges of the N- and C-coodinated metal ions, a certain number of cations can occupy interstitial positions for charge compensation. These interstitial metal ions may be the same as the N- and C-coordinated metal ions, or they may be alkali metal ions. In this context, much interest has been focused recently on the molecule-based cyanides related to the well-known Prussian Blue, $Fe^{III}_4[Fe^{II}(CN)_6]_3.15H_2O$, whose structure was described by Ludi and Gudel [42] as highly disordered cubic cell consisting of alternating ferrocyanide and ferric ions, with linear Fe^{III}-N-C-Fe^{II} bridges.

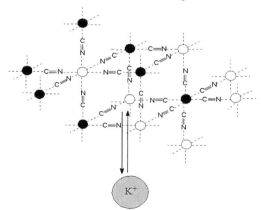

Fig. 8 : Scheme of the structure of metal hexacyanometallates (dark circles denote Fe^{3+}, Co^{3+}, etc. and empty circles denote Cu^{2+}, Ni^{2+}, Fe^{3+}, etc.)

The cyano bridge is known to mediate strong antiferromagnetic or ferromagnetic interactions, as in the case of Prussian Blue analogues, $A_k[B(CN)_6].xH_2O$, where A and B are transition metal ions [43]. T_c values above room temperatures, viz. 376 K [44], have been reported for hexacyanometallates but their high fcc symmetry is often accompanied by inherent disorder among different cationic sites, making it difficult to grow single crystals or to study any magnetic anisotropy. Recently, we have demonstrated the use of a Langmuir monolayer of octadecylamine as a templating agent at the air-water interface for growing oriented crystals of the Prussian-Blue-related metal(II) hexacyanoferrate(III) in a Langmuir-Blodgett trough, where metal ion can be Ni, Co or Cu [45]. X-ray diffraction pattern for one of them is compared with the polycrystalline material in Fig. 9.

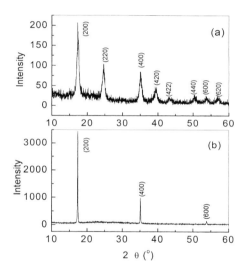

Fig. 9 : CuK_α XRD pattern of (a) bulk randomly oriented $Ni_3[Fe(CN)_6]_2.nH_2O$ powder, and (b) Ni hexacyanoferrate film transferred on glass from the air-water interface. Electron diffraction pattern confirmed that the crystallites have their {100} planes parallel to the monolayer.

An alternative strategy adopted by us to grow single crystals was to employ the heptacyano anion $[Mo^{III}(CN)_7]^{4-}$ as a precursor because its pentagonal bipyramidal coordination sphere is incompatible with a cubic lattice. Moreover, the Mo^{3+} ion in the $[Mo^{III}(CN)_7]^{4-}$ chromophore is low-spin, with a local spin of $S_{Mo} = 1/2$, and the g-tensor associated with the ground Kramers doublet is very anisotropic. Reaction of Mn(II) ions with the heptacyano $[Mo^{III}(CN)_7]^{4-}$ precursor led to lowering of symmetry and high values of Curie temperatures (T_c = 51 K) arising from a ferromagnetic interaction between the low-spin Mo^{3+} and the high-spin Mn^{2+} through the Mo^{III}-C-N-Mn^{II} bridges. The use of a macrocycle reduced the symmetry further by imposing heptacoordination on Mn(II) ion, too. A low T_c (3 K) for the compound $[Mn^{II}L]_6[Mo^{III}(CN)_7][Mo^{IV}(CN)_8]_2.19.5H_2O$, where L is a macrocycle, was attributed to the existence of diamagnetic Mo^{4+} along certain CN bridges, in addition to the paramagnetic Mo^{3+} [46].

More recently, octacyanometallates have also been used as versatile building blocks [47]. In order to examine the role of Mo^{IV} ion exclusively in the propagation of spin-spin interactions in $3d$-based polycyanomolybdates, polycyanoferrates(II) were synthesized. $Fe^{II}_2[Mo^{III}(CN)_7].8H_2O$ orders ferromagnetically below 65 K (Fig. 10), but $Fe^{II}_2[Mo^{IV}(CN)_8].8H_2O$ is paramagnetic down to 2 K.

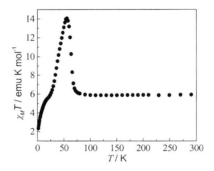

Fig. 10: $\chi_M T$ versus T curve for $Fe_2[Mo(CN)_7].8H_2O$.

5. Role of the Azido Bridge

An end-to-end (e-e) coordination of the azido group (N_3) results in moderate to strong antiferromagnetic coupling whereas an end-on (e-o) coordination gives rise to ferromagnetic coupling between paramagnetic centres (Fig. 11).

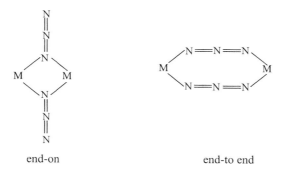

end-on   end-to end

Fig. 11: End-on and end-to-end bridging modes of the azido ligand (N_3^-) connecting two metal ions.

We have recently designed a unique *trans*-coordinated azido-Mn(II) 1-D chain compound of the formula $[Mn(L)(N_3)(PF_6)]_n$ by using a macrocyclic pentaaza ligand, a molecular building block which occupies the equatorial plane of the Mn(II) ion, thus forcing the two incoming N_3^- (azido) ligands to coordinate to the metal in the *trans* position. The crystal structure of $[Mn(L)(N_3)(PF_6)]_n$ is orthorhombic, with space group *Pbca*, and consists of 1-D arrays of Mn(II) units linked by single *trans* e-e azido bridges propagating along the *a* axis (Fig. 12). Magnetic susceptibility measurements indicate antiferromagnetic interactions between the Mn(II) centres (S_{Mn} = 5/2) below 30 K which are mediated through the azide group with a very small value of *J* = -5.0(1) cm^{-1} [48]. Both the EPR and magnetic behaviour confirm that $[Mn(L)(N_3)(PF_6)]_n$ consists of 1-D Mn(II) arrays magnetically well isolated from the neighboring chains.

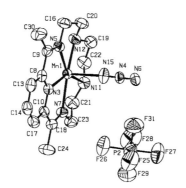

Fig. 12: ORTEP view of $[Mn(L)(N_3)(PF_6)]$ with numbering scheme.

6. The Current Scenario and Future Perspectives

Synthetic activity in molecular magnets has recently led to the development of what are known as *single molecule magnets* (SMMs), such as $Mn_{12}O_{12}(CH_3COO)_{16}(H_2O)_4$, which are magnetic clusters with large spin ground state (e.g. $S = 10$). Crystals of SMMs function as collections of monodisperse nanomagnets, unlike other nanoscale magnetic materials which have size distribution. Having mesoscopic dimensions, these magnetic aggregates are on the border between the classic and quantum regimes, and serve as model systems to study quantum tunneling and quantum coherence etc. [49,50]. Single molecule magnets are exciting new sources of magnetic phenomena relevant to materials such as magnetocaloric effect, QuBits, etc. [51].

The field of molecular magnetism is throwing up other potentially useful materials, such as those exhibiting negative magnetizations [52], spin-crossover materials with large hysteresis effects above room temperature [53], those exhibiting photomagnetic switching [54], and materials whose properties can be modulated electrochemically. Photoinduced magnetization has been demonstrated recently in the organic-based magnet $Mn(TCNE)_x.y(CH_2Cl_2)$ [55]. Solubility of molecular magnets and their low density are attractive features. So is their biocompatibility which makes them useful in drug-targeting or as high-relaxivity magnetic resonance imaging (MRI) contrasting agents. New amphiphilic Gd(III) chelates, capable of self-organizing by forming micelles in aqueous solutions, have been designed with the aim of obtaining a high-relaxivity MRI contrasting agent [56]. Some groups are attempting to utilize the high degree of directionality of hydrogen bonding between open-shell molecules to obtain supramolecular self-organization aimed at achieving new molecular ferromagnets.

The Langmuir-Blodgett technique is a powerful tool to organize molecules in a multilayer architecture. A weak ferromagnetic state has been established in LB films of manganese octadecylphosphonate $[n\text{-}C_{18}H_{37}PO(OH)O]_2Mn$ [57]. New hybrid ferromagnetic LB films based on ferri-ferrocyanide Prussian Blue have been deposited as multilayer lamellar structures in which the magnetic properties of the inorganic sheets are combined with other functions arising from the organic part, such as dimethyldioctadecylammonium bromide (DODA) [58]. LB films of the Mn_{12} acetate cluster (SMM) have also been reported, which exhibit hysteresis [59].

Multilayer LB architectures have been organized from Prussian Blue magnets as well as those of the Mn_{12} acetate SMM.

Magnetic spin ladder systems like bis(piperidinium) tetrabromocuprate(II), $(C_5H_{12}N)_2CuBr_4$ have been reported recently [60]. Apart from the fact that a molecule is the ultimate unit for data storage, the design of molecular magnets has also opened the doors for the unique possibility of designing polyfunctional materials at the molecular level such as those exhibiting ferromagnetic behaviour and second-order optical nonlinearity [61], or liquid crystallinity and magnetism [62]. Recently, we have been involved in the design of a three-dimensional chiral, transparent ferrimagnet $K_{0.4}[Cr(CN)_6][Mn((S)\text{-pn})](S)\text{-pnH}_{0.6}$ with $T_c = 53$ K, where (S)-pn stands for (S)-1,2-diaminopropane [63]. Most recently, Itkis et al [64] have succeeded in obtaining, for the first time, a trifunctional phenalene-based neutral radical, which exhibits magneto-opto-electronic bistability with hysteresis loops centred near room temperature at 335 K, which opens the possibility of a new type of electronic devices, where multiple physical channels can be used for writing, reading, and transferring information. Magnetic bistability, have been reported in the past in molecular systems, such as doped spin-crossover compounds [65] and in the 1,3,5-trithia-2,4.6-triazapentalenyl organic radical [66], for which the magnetic transition is accompanied by a change of color, too.

Acknowledgements.

This article is dedicated to the memory of the late Prof. Olivier Kahn, a pioneer in the area of molecular magnetism, and a great source of inspiration. The author has benefited a great deal from the interaction with Prof. K. Inoue, Dr. J.P. Sutter, Dr. C. Mathoniere, Prof. Helen Stoeckli-Evans, Dr. M.D. Sastry and Prof. M. Tokumoto. The author also places on record his deep appreciation for the contributions of Dr. Sipra Choudhury, and his students S.A. Chavan, Aman K. Sra, Prasanna Ghalsasi and Nitin Bagkar on which this article is largely based.

[1] J.S. Miller, J.C. Calabrese, A.J. Epstein, R.B. Bigelow, J.H. Zang, and W.M. Reiff, J. Chem. Soc., Chem. Commun. (1986) 1026.
[2] Y. Pei, M. Verdaguer, O.Kahn, J. Sletten, and J.-P. Renard, J. Am. Chem. Soc. 108 (1986) 428.
[3] A.A. Ovchinnikov, Theor. Chim. Acta 47 (1978) 297.
[4] K. Itoh and M. Kinoshita, Eds. *Molecular Magnetism*, Gordon and Breach Science, Publishers, Amsterdam (2000).
[5] S. Rajca, J. Wongsriratanakul and S. Rajca, Science 294 (2001) 1503.
[6] J.H. Schoen, S. Berg, C. Kloc and B. Batlogg, Science 287 (2000) 1022.
[7] J.H. Schoen, A. Dodabalapur, Z. Bao, C. Kloc, O. Schenker and B. Batlogg, Nature 410 (2001) 189
[8] R. Arita, Y. Suwa, K. Kuroki and H. Aoki, Phys. Rev. Lett. 88 (2002) 127202.
[9] P. Allemand, K. Khemani, K. Koch, F. Wudl, K. Holczer, S. Donovan, G. Gruner and J.D. Thompson, Science 253 (1991) 301.

[10] O. Kahn, *Molecular Magnetism*, VCH, New York (1993).
[11] J.S. Miller, Inorg. Chem. 39, (2000) 4392.
[12] O. Kahn, Ed. *Magnetism: A Supramolecular Function*; NATO ASI Series C, Vol. 484, Kluwer, Dordrecht. (1996).
[13] H.M. McConnell, J. Chem. Phys. 39 (1963) 1910.
[14] M. Tamura, Y. Nakazawa, D. Shiomi, K. Nozawa, Y. Hosokoshi, M. Ishikawa, M. Takahashi and M. Kinoshita, Chem. Phys. Lett. 186 (1991) 401.
[15] A. Merzel et al. Chem. Phys. Lett. 298 (1998) 329.
[16] A.J. Banister, N. Bricklebank, I. Lavender, J.M. Rawson, C.I. Gregory, B.K. Tanner, W. Clegg, M.R.J. Elsegood and F. Palacio, Angew. Chem. Int. Ed. Engl. 35 (1996) 2533.
[17] F. Palacio et al. Phys. Rev. Lett. 79 (1997) 2336.
[18] M. Mito et al., Polyhedron 20 (2001) 1509
[19] T.L. Makarova, B. Sundqvist, R. Hohne, P. Esquinazi, Y. Kopelevich, P. Scharff, V.A. Davydov, L.S. Kashevarova, and A.V. Rakhmanina, Nature 413 (2001) 716.
[20] O. Kahn, Adv. Inorg. Chem., 43 (1995) 179.
[21] K. Nakatani, J.Y. Carriat, Y. Journaux, O. Kahn, F. Lloret, J.P. Renard, Y. Pei, J. Sletten, and M. Verdaguer, J. Am. Chem. Soc. 111 (1989) 5739.
[22] H. Okawa, M. Mitsumi, M. Ohba, M. Kodera, and N. Matsumoto 1994, Bull. Chem. Soc. Jpn., 67 (1994) 2139.
[23] A. Caneschi, D. Gatteschi, R. Sessoli and P. Rey, Acc. Chem. Res. 22 (1989) 392.
[24] W.E. Broderick, J.A. Thompson, P. Day and B.M. Hoffman, Science 249 (1990) 410.
[25] K. Inoue and H. Iwamura, J. Am. Chem. Soc. 116 (1994) 3173.
[26] K. Inoue, T. Hayamizu, H. Iwamura, D. Hashizume and Y. Ohashi, J. Am. Chem. Soc. 118 (1996) 1803.
[27] H.O. Stumpf, Y. Pei, O. Kahn, J. Sletten, and J.P. Renard, J. Am. Chem. Soc. 115 (1993) 6738.
[28] O. Kahn, Y. Pei, M. Verdaguer, J.P. Renard and J. Sletten, J. Am. Chem. Soc. 111(1988) 7823.
[29] Y. Pei, O. Kahn, K. Nakatani, E. Codjovi, C. Mathoniere and J. Sletten, J. Am.Chem. Soc. 113 (1991) 6558.
[30] S.A. Chavan, R. Ganguly, V.K. Jain and J.V. Yakhmi, J. Appl. Phys. 79 (1996) 5260.
[31] A. Neels, H. Stoeckli-Evans, S.A. Chavan and J.V. Yakhmi, Inorg. Chim. Acta 326 (2001) 106.
[32] S.A. Chavan, J.V. Yakhmi and I.K. Gopalakrishnan, Mol. Cryst. Liq. Cryst. 274 (1995) 11.
[33] J. Larionova, S.A. Chavan, J.V. Yakhmi, A.G. Froystein, J. Sletten, C. Sourisseau, and O. Kahn, Inorg. Chem. 36 (1997) 6374.
[34] O. Kahn, J. Larionova and J.V. Yakhmi, Chem. Eur. J. 5 (1999) 3443.
[35] A.K. Sra, M. Tokumoto and J.V. Yakhmi, Philos. Mag. B (April 2001) to appear
[36] H.O. Stumpf, L. Ouahab, Y. Pei, D. Grandjean, and O. Kahn, Science 261 (1993) 447.
[37] H.O. Stumpf, L. Ouahab, Y. Pei, P. Bergerat, and O. Kahn, J. Am. Chem. Soc. 116 (1994) 3866.
[38] H.O. Stumpf, Y. Pei, C. Michaut, O. Kahn, J.P. Renard, and L. Ouahab, Chem. Mater. 6 (1994) 257.
[39] M.G.F. Vaz et al Chem. Eur. J. 5 (1999) 1486
[40] S. Turner, O. Kahn and L. Rabardel, 1996, J. Am. Chem. Soc. 118 (1996) 6428.
[41] S.A. Chavan, J. Larionova, O. Kahn and J.V. Yakhmi, Philos. Mag. B 6 (1998) 1657.
[42] A. Ludi and H.U. Gudel, Struct. Bonding (Berlin), 14 (1973) 1.
[43] W.R. Entley and G.S. Girolami, Science 268 (1995) 397.
[44] S.M. Holmes and G.S. Girolami, J. Am. Chem. Soc. 121 (1999) 5593.
[45] S. Choudhury, N. Bagkar, G.K. Dey, H. Subramanian and J.V. Yakhmi, Langmuir 18 (2002) 7409.
[46] A.K. Sra, M. Andruh, O. Kahn, S. Golhen, L. Ouahab and J.V. Yakhmi, Angew. Chem. Int. Ed., 38 (1999) 2606.
[47] A.K. Sra, G. Rombaut, F. Lahitete, S. Golhen, L. Ouahab, C. Mathoniere, J.V. Yakhmi and O. Kahn, New J. Chem. 24 (2000) 871.
[48] A.K. Sra, J.-P. Sutter, P. Guionneau, D. Chasseau, J.V. Yakhmi and O.Kahn, Inorg. Chim. Acta 300-302 (2000) 778.
[49] J.R. Friedman, M.P. Sarachik, J. Tejada and R. Ziolo, Phys. Rev. Lett. 76 (1996) 3830.
[50] B. Barbara and L. Gunther, Phys. World (March 1999) 35.
[51] M.N. Leuenberger and D. Loss, Nature 410 (2001) 789.
[52] C. Mathoniere, S.G. Carling, D. Yusheng, and P. Day, J. Chem. Soc., Chem. Commun., (1994) 1551.
[53] J. Krober, E. Codjovi, O. Kahn, F. Groliere, and C. Jay, J. Am. Chem. Soc. 115 (1993) 9810.
[54] K. Hashimoto and S. Ohkoshi, Philos. Trans. R. Soc. Lond. A 357 (1999) 2977.
[55] D.A. Pejakovic, C. Kitamura, J.S. Miller and A.E. Epstein, Phys. Rev. Lett. 88 (2002) 057202.

[56] J.P. Andre, E. Toth, H. Fischer, A. Seelig, H.R. Macke and A.E. Merbach, Chem. Eur. J. 5 (1999) 2977.
[57] C.T. Seip, G.E. Granroth, M.W. Meisel and D.R. Talham, J. Am. Chem. Soc. 119 (1997) 7084.
[58] C. Mingotaud, C. Lafuente, J. Amiell and P. Delhaes, Langmuir 15 (1999) 289.
[59] M.C. Leon, H. Soyer, C. Mingotaud, C.J.G. Garcia, E. Coronado and P. Delhaes, Mol. Cryst. Liq. Cryst. 334 (1999) 669.
[60] B.C. Watson et al., Phys. Rev. Lett. 86 (2001) 5168.
[61] S. Benard, P. Yu, J.P. Audiere, E. Riviere, R. Clement, J. Guilhem. L. Tchertanov and K. Nakatani, J. Am. Chem. Soc. 122 (2000) 9444.
[62] K. Binnemans et al, J. Am. Chem. Soc. 122 (2000) 4335.
[63] K. Inoue, H.Imai, P.S. Ghalsasi, M. Ohba, H. Okawa and J.V. Yakhmi, Angew. Chem. Int. Ed. Eng. 40 (2001) 4242.
[64] M.E. Itkis, X. Chi, A.W. Cordes and R.C. Haddon, Science 296 (2002) 1443.
[65] J. Krober, E. Codjovi, O. Kahn, F. Groliere and C. Jay, J. Am. Chem. Soc. 115 (1993) 9810.
[66] W. Fujita and K. Awaga, Science 286 (1999) 261.

Macromol. Symp. **2004**, *212*, 159-168

Spectroscopic Studies of Charge-Ordering System in Organic Conductors

K. Yakushi,[1] *K. Yamamoto,*[1] *R. Świetlik,*[1] *R. Wojciechowski,*[1] *K. Suzuki,*[1]
T. Kawamoto,[2] *T. Mori,*[2] *Y. Misaki,*[3] *K. Tanaka*[3]

[1] Institute for Molecular Science, Nishigo-naka, Myodaiji, Okazaki, 444-8585, Japan
[2] Department of Organic and Polymeric Materials, Tokyo Institute of Technology, 2-12-1, O-okayama, Meguro-ku, Tokyo 152-8552, Japan
[3] Department of Molecular Engineering, Kyoto University, Yoshida-Honmachi, Sakyo-ku, Kyoto 606-8501, Japan

Summary: Charge localization generates a non-uniform charge distribution in some organic conductors. Phase transitions accompanying such a localization of charge are studied by using infrared and Raman spectroscopy. We first introduce θ-(BEDT-TTF)$_2$MM'(SCN)$_4$ (M=Rb, Cs, Tl; M'=Zn, Co) as typical examples of a charge-ordering system, where BEDT-TTF is bis(ethylenedithio)tetrathiafulvalene. We apply the same spectroscopic technique to α'-(BEDT-TTF)$_2$IBr$_2$, θ-(BDT-TTP)$_2$Cu(NCS)$_2$, and (TTM-TTP)I$_3$, which show the phase transitions from low-resistivity to high-resistivity state, where TTM-TTP is 2,5-bis[4,5-bis(methylsulfonyl)-1,3-dithiol-2-ylidene]-1,3,4,6-tetrathiapentalene.

Keywords: charge transfer; infrared spectroscopy; Raman spectroscopy

Introduction

In molecular conductors, the molecular orbital barely overlaps with those of neighbor molecules. Therefore, the transfer integral (t), which contributes to the delocalization of charge, is usually smaller than the on-site (U) and off-site (V) Coulomb interactions, which contribute to localization of the charges. A strong correlation effect originates from the comparative magnitude of t, U, and V, and thus many molecular conductors are located at the boundary between a metal (delocalized state) and insulator (localized state). A variety of ground states such as spin-density wave (SDW), charge-density wave (CDW), metal, and superconductivity are found in molecular conductors. For example, the various phases of BEDT-TTF salts are theoretically sorted out with the aid of U, V, and t.[1,2] When the charge is localized, it often induces charge disproportionation (CD) and eventually generates an inhomogeneous charge distribution. This localized state is

DOI: 10.1002/masy.200450815

called a charge-ordered (CO) state, since the localized charges often form a new periodic structure. Recently, metal-insulator phase transitions accompanying such CO are found in several organic conductors.[3,4,5,6] It becomes apparent that the charge ordering is widely associated with metal-insulator phase transitions in a variety of organic conductors. The relationship to unconventional superconductivity is suggested from the viewpoint of experiment[7] and theory[8].

It is well known that the frequency of some C=C stretching modes of, for example, BEDT-TTF shows a downshift depending upon the degree of oxidation of the molecule.[9] Therefore the C=C stretching mode can be used as a probe to detect the charge distribution. Using this principle, we have investigated the CO phase transition. In this paper, we will review the previous study and present the application of the infrared and Raman spectroscopy, which have been conducted in our laboratory.

θ-(BEDT-TTF)$_2$MM'(SCN)$_4$ (M=Rb, Co, Tl; M'=Zn, Co)

This system was initiated by Mori et al., and was abbreviated as θ-MM'.[7] θ-RbZn shows a metal-insulator transition at 190 K. Miyagawa et al.[5] and Chiba et al.[10] found the existence of CO below the meta-insulator-transition temperature (T_{MI}) by [13] C NMR measurements. Mori et al.[7] and Watanabe et al.[11] performed the low-temperature x-ray crystal structure analysis, and showed that the unit cell was doubled below T_{MI} to lower the symmetry. On the other hand, θ-CsCo is metal-like down to ~20 K and continuously increases its resistivity below this temperature. In contrast to θ-RbZn, the thermopower of θ-CsCo exhibits no anomaly at around 20 K. The low-temperature x-ray diffraction experiment shows that a three-dimensional short-range ordering occurs at 20 K, which corresponds with a similar doubling of the unit cell to θ-RbZn.[12] The low-temperature phase of θ-CsCo has not been well characterized.

Figure 1 shows the temperature dependence of the 632.8 nm-excited Raman spectra of θ-RbZn and θ-CsCo. θ-RbZn shows a dramatic change in the spectral region of the C=C stretching region. By contrast, θ-CsCo does not show such a drastic change but a continuous change including the growth of the very broad band at around 1250 cm^{-1}. The interpretation of θ-CsCo has not been made. The lowest-temperature spectrum of θ-RbZn was analyzed with the aid of the ^{13}C-substituted compound as shown in Fig. 2.[13] The three C=C stretching modes of BEDT-TTF are classified into IR-active ν_{27} (b$_{1u}$) and Raman-active ν_2 (a$_g$) and ν_3(a$_g$). ν_{27} is an anti-phase

C=C stretching mode in the five-membered rings, while ν_2 and ν_3 are associated with the C=C stretching mode of the central bridge and in-phase mode of the five-membered rings.[14] The bridge and ring C=C stretching modes are almost equally mixed in neutral BEDT-TTF^0. However, they are separated in the (BEDT-TTF)$_2^+$ dimer,[15,16] and ν_2 and ν_3 are respectively assigned to the in-phase ring and bridge C=C stretching modes. Therefore ν_3 is expected to show a large downshift in the ^{13}C-substituted compound.

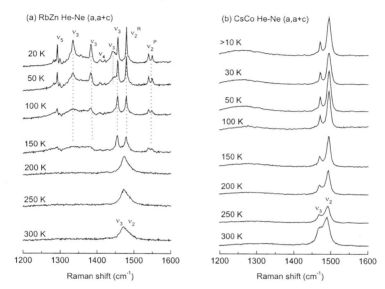

Figure 1. Raman spectra of θ-RbZn and θ-CsCo excited by a 632.8 nm laser.

Using this property, ν_3 is separated from ν_2, ν_4, and ν_5 as shown in Fig. 2. Below T_{MI}, every intra-molecular phonon is four-fold degenerate if the intermolecular interaction is very weak, since the unit cell contains four molecules. However, ν_3 is strongly coupled with the charge-transfer excited state (EMV (electron molecular vibration) coupling).[17] Through this EMV interaction, the degenerate ν_3 is extensively split into four modes. Owing to the EMV interaction, the splitting of ν_3 does not always signify CD. As discussed in ref. 13, CD is proved by the appearance of ν_3^1(A) in the IR E∥a spectrum. Another evidence is the splitting of ν_2 in the same polarization [See (a,a) polarization]. Since ν_2 has a weak EMV interaction, the splitting of ν_2 originates almost purely from the difference in oxidation degree, so that the deviation from +0.5 in (BEDT-

TTF$)^{0.5+\delta}$(BEDT-TTF$)^{0.5-\delta}$ is estimated as $\delta=0.35$ from the separation between ν_2^P (charge-poor site) and ν_2^R (charge-rich site).

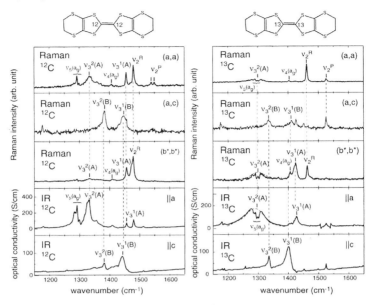

Figure 2. Polarized Raman spectra of natural and ^{13}C-substituted θ-RbZn. For example, (a,c) show that the laser and scattered light are polarized along the a- and c-axes, respectively, in the back scattering geometry.

Another important information from Fig. 2 is the selection rule that the A and B symmetry modes are separately observed in different polarizations. This selection rule proves that the glide-plane symmetry is preserved at this temperature. Therefore, the charge-rich sites connected by the glide-plane symmetry are aligned along the a-axis to form a horizontal stripe. The conclusion is that ν_2 can be used to estimate the CD ratio and ν_3 can be used to examine the symmetry. The interpretation of the spectrum above T_{MI} is more difficult, since the spectrum shows a single broad band with a non-Lorentzian lineshape. ν_2 and ν_3 are well separated and observed, respectively, at 1474 cm^{-1} and 1503 cm^{-1} in a metallic BEDT-TTF compound, κ-(BEDT-TTF)$_2$Cu[N(CN)$_2$]Br.[16] They are not completely separated in the room-temperature spectrum of θ-CsCo, which is more metallic than θ-RbZn. The spectral shape seems to be associated with the fluctuation of charge density. In the case of θ-RbZn, we speculate that the charge is nearly localized above T_{MI} and is extensively fluctuating around $\delta=0$ owing to the incoherent hopping.

We have investigated the θ-TlZn and θ-TlCo using the same method. In this process, we found an orthorhombic phase of θ-TlZn in addition to a monoclinic phase of θ-TlZn. Since the orthorhombic phases of θ-TlZn and θ-TlCo are isostructural to θ-RbZn, the spectral change at the phase transition temperature is almost exactly the same as that of θ-RbZn except for the transition temperature. However, the spectral change of monoclinic θ-TlZn gradually occurs at much higher temperature (~250 K) than T_{MI}~165 K. Probably the charge is more fluctuating than in θ-RbZn above T_{MI}.

α'-(BEDT-TTF)$_2$IBr$_2$

This compound is non-metallic and shows a phase transition at 200 K from low- to high-resistivity state.[18] Although the structural change is investigated by x-ray diffraction, no drastic change has been detected at 200 K.[19,20] Since the band calculation suggests a semimetallic

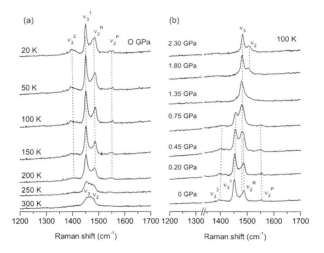

Figure 3. Temperature and pressure dependence of the Raman spectrum of α'-(BEDT-TTF)$_2$IBr$_2$.

band, one of the scenarios for this phase transition is the gap enhancement caused by a small structural change in a narrow-gap semiconductor. However, the spin degrees of freedom are preserved below 200 K. Thus the localization of charge is more plausible for the mechanism of the phase transition at 200 K.

Based on the knowledge of the assignment in θ-RbZn, we applied Raman spectroscopy to this phase transition. Figure 3a shows the temperature dependence of the Raman spectra excited by a 514.5 nm laser. The broad room-temperature spectrum resembles that of θ-RbZn shown in Fig. 1a. At 200 K, the spectrum shows a clear splitting, and the broad band splitting into four or six at 20 K. Based on the assignment shown in Fig. 1a, the disproportionation ratio is estimated as δ~0.4. The molecules are stacked along the a-axis, and the unit cell contains two independent stacks. In each stack, two molecules are connected by inversion symmetry. Therefore the unit cell contains four molecules. If the center of symmetry is preserved, each mode is classified into two in-phase (A_g) and two out-of-phase (A_u) modes, and only the A_g mode is Raman-active. There are two modes v_2 and v_3 in this spectral region, so that four bands should be observed at most if the center of symmetry is preserved at 20 K. Figure 3a suggests the conservation of the center of symmetry below T_{MI}. Figure 3b shows the pressure dependence of the spectrum at 100 K. On increasing pressure, the four bands continuously decrease in intensity, and disappear at 1.35 GPa. Instead, a new band at ~1475 cm^{-1} grows up, and finally the spectrum shows well-resolved v_2 and v_3, which are similar to those of a metallic BEDT-TTF salt. We speculate that the charge distribution is homogeneous at 2.3 GPa. High pressure appears to be more effective in increasing t than U and V.

θ-(BEDT-TTP)$_2$Cu(NCS)$_2$

The CO phase transition was found by Ouyang et al. in this compound, which was synthesized by Misaki et al.[21] The weakly dimerized β-type of BDT-TTP salts such as (BDT-TTP)$_2$SbF$_6$ is metallic down to 2 K and shows a typical Drude-type reflectivity down to 600 cm^{-1}.[22,23] However, the θ-type salt, θ-(BDT-TTP)$_2$Cu(NCS)$_2$, is non-metallic up to 350 K as shown in Fig. 4a. The structure resembles that of monoclinic θ-(BEDT-TTF)$_2$TlZn(SCN)$_4$.[24] This compound shows a sharp increase in resistivity at ~250 K. The optical conductivity obtained from the Kramers-Kronig transformation of the reflectivity spectrum also shows a remarkable change at ~250 K. The spectral weight less than 2000 cm^{-1} shifts to the high-wavenumber region, and thus the optical gap continuously increases near the phase transition temperature. The localization and disproportionation of the charge below the phase transition temperature is proved by the Raman spectrum.

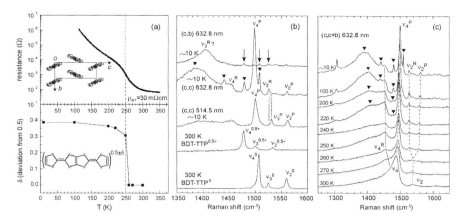

Figure 4. (a) Electric resistance and disproportionation radio δ, (b) comparison of the Raman spectra, and (c) temperature dependence of the Raman spectrum of θ-(BDT-TTP)$_2$Cu(NCS)$_2$.

Before explaining the Raman spectra in Figs. 4b and 4c, let us introduce the normal coordinate analysis reported by Ouyang et al.[25] The five C=C stretching modes are named as Raman-active ν_2,ν_3, and ν_4 and infrared-active ν_{21} and ν_{22}. These modes are rather isolated from other fundamental modes. The five C=C stretching modes are significantly mixed in BDT-TTP^0, whereas they are rather separated in BDT-TTP$^+$. Thus ν_2 and ν_3 in BDT-TTP$^+$ are assigned mainly to the in-phase mode of the C=C stretching of outer rings and bridges, ν_4 is assigned to the C=C stretching of the inner ring, ν_{21} and ν_{22} are assigned to the out-of-phase modes of the C=C stretching of outer rings and bridges, respectively. As shown in Fig. 4b, the three Raman-active modes show a low-frequency shift by 20-25 cm^{-1} from BDT-TTP^0 to BDT-TTP$^{0.5+}$. These modes can become a good probe to detect the oxidation state. As shown in Fig. 4c, the Raman spectrum in the C=C stretching regions dramatically changes below the phase transition temperature. In Fig. 4b, the Raman spectra excited by 632.8 nm are compared with those excited by 514.5 nm. Since 514.5 nm is close to the first excited state of BDT-TTP^0, the Raman modes associated with BDT-TTP^0 are enhanced owing to the resonance effect. If we take the resonance effect into account, the resemblance of (c,c) spectrum excited by 514.5 nm to that of BDT-TTP^0 straightforwardly leads to the view that the charge disproportionation occurs at ~10 K. Based on the temperature dependence shown in Fig. 4c, we assigned the Raman bands at ~10 K as shown in Fig. 4b. Although the ~10 K spectrum is complicated, the spectrum near the phase transition, for example 240 K, is easily interpreted. At this temperature, both v_2 and v_4 are split into the modes corresponding to the charge-poor (v_j^P) and charge-rich (v_j^R) sites. From the separation between v_2^P and v_2^R, the disproportionation ratio δ is estimated as shown in the bottom panel of Fig. 4a. This parameter is regarded as the order parameter of this phase transition.

Additional bands marked by triangles appear below 220 K as shown in Fig. 4c. The appearance of these additional bands signifies the doubling of the unit cell including the short-range order. As indicated by arrows in Fig. 4b, three bands appear in (c,c) as well as in (c,b) polarization. The appearance of these bands indicates that the glide-plane symmetry is broken at ~10 K. We have examined this selection rule at 200 K, and have found the breaking of the selection rule. Therefore, we speculate that the localized charge is aligned along the b-axis.

(TTM-TTP)I$_3$

The high conductivity ($\sigma_{RT}\sim10^3$ S/cm) of (TTM-TTP)I$_3$ attracted attention, because this compound is a half-filled quasi-one-dimensional system. The electric resistivity is almost temperature-independent down to ~160 K and steeply increases below this temperature.[26] Below ~160 K, the lattice is doubled ($2k_F$ modulation) and the electronic system undergoes a spin singlet state.[27,28] Onuki *et al.* proposed through the solid-state NMR experiment that a charge disproportionation such as (TTM-TTP)$^{(1+\delta)+}$ (TTM-TTP)$^{(1-\delta)+}$ accompanied this phase transition.[29] We applied the vibrational spectroscopy to this phase transition if such inhomogeneous charge distribution appeared. Since this molecule has the same π-conjugated skeleton as BDT-TTP, we use the same notation for the C=C stretching modes as BDT-TTP for convenience.

TTM-TTP is stacked along the c-axis. A strong v_{22} mode was observed in the E⊥c reflection spectrum. In the low-temperature optical conductivity, v_{22} showed neither shift nor splitting down to 20 K. Since the frequency of v_{22} is most sensitive to the charge among the five C=C stretching modes, this observation clearly indicates that the charge disproportionation does not occur at the phase transition temperature, ~160 K. At room temperature, the Raman-active modes were observed at v_2=1490 cm^{-1}, v_3=1455 cm^{-1}, and v_4=1433 cm^{-1}, when 514.5 nm laser was used. Figure 5a shows the Raman spectra measured by 785 nm excitation, in which v_4 is missing in

(TTM- TTP)I_3. As shown in this figure, ν_3 shows no change below ~160 K, whereas ν_2 splits into two bands at 1487 cm^{-1} and 1499 cm^{-1} at 155 K. Since the charge disproportionation is not found in ν_{22}, we interpret that this splitting is caused by the asymmetric deformation of the molecule. In other words, the center of symmetry of the molecule is broken below ~160 K due to the structural change ($2k_F$ modulation). When the molecule has a center of symmetry, the C=C stretching modes of the outer five-membered rings are classified into the Raman-active in-phase mode ν_2 and infrared-active out-of-phase mode ν_{21}. The frequencies of ν_2 and ν_{21} are close, because these two C=C bonds are far from each other.[25] If the symmetry is broken and two C=C bonds become non-equivalent, these two modes mix with each other and both modes becomes infrared and Raman active.

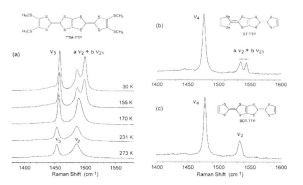

Figure 5. Raman spectra of (TTM-TTP)I_3, (ST-TTP)$_2$AsF$_6$, and (BDT-TTP)$_2$SbF$_6$.

To confirm this speculation, we measured the Raman spectra of the isostructural charge-transfer salts of symmetric BDT-TTP and asymmetric ST-TTP. When we use 785 nm laser, ν_3 is missing in both compounds. As shown in Fig. 5a and 5b, ν_2 shows no splitting in (BDT-TTP)$_2$SbF$_6$, whereas ν_2 shows a clear splitting with the separation 10 cm^{-1} in (ST-TTP)$_2$AsF$_6$. This separation is very close to that (12 cm^{-1}) of (TTM-TTP)I_3 below ~160 K. This observation strongly supports our speculation. Usually the asymmetric environment around the molecule does not have such a big influence on the molecular structure, since the stabilization energy by π-conjugation is much larger than the intermolecular interaction. This long molecule structure may have an inherent instability to the asymmetric distortion.[30-32]

168

[1] H. Kino, H. Fukuyama, *J. Phys. Soc. Jpn.* **1996**, 65, 2158.

[2] H. Seo, *J. Phys. Soc. Jpn.* **2000**, 69, 805.

[3] K. Hiraki and K. Kanoda, *Phys. Rev. Lett.* **1998**, 80, 4737.

[4] D. S. Chow, F. Zamborszky, A. Alavi, D. J. Tantillo, A. Bauer, C. A. Merlic, and S.E. Brown, *Phys. Rev. Lett.* **2000**, 85, 1698.

[5] K. Miyagawa, A. Kawamoto, K. Kanoda, *Phys. Rev. B* **2000**, 62, R7679.

[6] J. Ouyang, K. Yakushi, Y. Misaki, and K. Tanaka, *Phys. Rev. B* **2001**, 63, 054301.

[7] H. Mori, S. Tanaka, T. Mori, *Phys. Rev. B* **1998**, 57, 12023.

[8] J. Merino and R. H. McKenzie, *Phys. Rev. Lett.* **2001**, 87, 237002.

[9] M. E. Kozlov, K. I. Pokhodnia, A. A. Yurchenko, *Spectrochim. Acta* **1989**, 45A, 437.

[10] R. Chiba, H. M. Yamamoto, K. Hiraki, T. Nakamura, T. Takahashi, *Synth. Met.* **2001**, 120, 919.

[11] M. Watanabe, Y. Noda, Y. Nogami, H. Mori, private communication.

[12] M. Watanabe, Y. Nogami, K. Oshima, H. Mori, S. Tanaka, *J. Phys. Soc. Jpn.* **1999**, 68, 2654.

[13] K. Yamamoto, K. Yakushi, K. Miyagawa, and K. Kanoda, and A. Kawamoto, *Phys. Rev. B* **65**, 085110 (2002).

[14] M. E. Kozlov, K. I. Pokhondnia, A. A. Yurchenko, *Spectrochim. Acta* **1987**, 43A, 323.

[15] J. E. Eldridge, Y. Xie, H. H. Wang, J. M. Williams, A. M. Kini, J. A. Schlueter, *Mol. Cryst. Liq. Cryst.* **1996**, 284, 97.

[16] M. Maksimuk, K. Yakushi, H. Taniguchi, K. Kanoda, A. Kawamoto, *J. Phys. Soc. Jpn.* **2001**, 70, 3728.

[17] J. E. Eldridge, K. Kornelsen, H. H. Wang, J. M. Williams, A. V. Strieby Crouch, D. M. Watkins, *Solid State Commun.* **1991**, 79, 583.

[18] M. Tokumoto, H. Anzai, T. Ishiguro, *Synth. Met.* **1987**, 19, 215.

[19] Y. Nogami, S. Kagoshima, T. Sugano, G. Saito, *Synth. Met.* **1986**, 16, 367.

[20] M. Watanabe, M. Nishikawa, Y. Nogami, K. Oshima, G. Saito, *J. Korean Phys. Soc.* **1997**, 31, 95.

[21] J. Ouyang, K. Yakushi, Y. Misaki, and K. Tanaka, *Phys. Rev. B* **2001**, 63, 054301.

[22] T. Nakada, T. Ishiguro, T. Miura, .Y. Misaki, T. Yamabe, T. Mori, *J. Phys. Soc. Jpn.* **1998**, 67, 355.

[23] J. Ouyang, K. Yakushi, Y. Misaki, K. Tanaka, *J. Phys. Soc. Jpn.* **1998**, 67, 3191.

[24] H. Mori, S. Tanaka, T. Mori, A. Kobayashi, and H. Kobayashi, *Bull. Chem. Soc. Jpn.* **1998**, 71, 797.

[25] J. Ouyang, K. Yakushi, T. Kinoshita, N. Nanbu, M. Aoyagi, Y. Misaki, and K. Tanaka, *Spectrochim. Acta Part A* **2002**, 58, 1643.

[26] T. Mori, T. Kawamoto, J. Yamaura, T. Enoki, Y. Misaki, T. Yamabe, H. Mori, and S. Tanaka, *Phys. Rev. Lett.* **1997**, 79, 1702.

[27] M. Maesato, Y. Sasou, S. Kagoshima, T. Mori, T. Kawamoto, Y. Misaki, T. Yamabe, *Synth. Met.* **1999**, 103, 2109.

[28] N. Fujimura, A. Namba, A. Kambe, Y. Nogami, K. Oshima, T. Mori, T. Kawamoto, Y. Misaki, T. Yamabe, *Synth. Met.* **1999**, 103, 2111.

[29] M. Onuki, K. Hiraki, T. Takahashi, D. Jinno, T. Kawamoto, T. Mori, K. Tanaka, and Y. Misaki, *J. Phys. Chem. Solids* **2001**, 62, 405.

[30] S. F. Rak and L. L. Miller, *J. Am. Chem. Soc.* **1992**, 114, 1388.

[31] K. L.ahil, A. Moradpour, C. Bowlas, F. Menou, P. Cassoux, J. Bonvoisin, J.-P. Launay, G. Dive, and D. Dehareng, *J. Am. Chem. Soc.* **1995**, 117, 9995.

[32] K. Pokhodnia, P. Cassoux, J. Bonvoisin, a. Mlayah, L. Brossard, S. Frenzel, oand K. Mullen, *J. Phys. Chem. B* **1997**, 101, 3665.

Macromol. Symp. **2004**, *212,* 169-178

Electrical and Spectral Properties of Organic Salts Formed from BEDT-TTF and Magnetic Anions

Andrzej Graja,[*1] *Andrzej Łapiński,*[1] *Vladimir A. Starodub*[2]

[1] Institute of Molecular Physics, Polish Academy of Sciences, Smoluchowskiego 17, 60-179 Poznań, Poland
E-mail: graja@ifmpan.poznan.pl
[2] Department of Chemistry, Kharkov National Karazin University, 61077 Kharkov, Ukraine
E-mail: vladimir.a.starodub@univer.kharkov.ua

Summary: Spectral and electrical investigations of the semiconducting (BEDT-TTF)$_2$W$_6$O$_{19}$ and the metal-like (BEDT-TTF)$_6$(Mo$_8$O$_{26}$)(DMF)$_3$ salts were performed. The vibrational and electronic spectra of the crystalline samples were analysed and assignment of the vibrational features was proposed. The temperature evolution of the vibrational spectra of the (BEDT-TTF)$_2$W$_6$O$_{19}$ salt was discussed.

Keywords: BEDT-TTF salts; charge transfer; infrared spectroscopy; polyoxometalate anions; UV-vis spectroscopy

Introduction

Recently, much attention was devoted to exploration of the novel lattice architectures and physical properties resulting from the association of organic cation radicals, such as BEDT-TTF (bis(ethylenedithio)tetrathiafulvalene) with bulky anions (e.g. polyoxometallates[1]). One of the reasons for the interest in polyoxometalate-based materials, containing both localized and delocalized electrons is the possibility of formation of compounds where the coexistence of co-operative magnetic and electric properties such as ferromagnetism and superconductivity can be observed.[2] Moreover, Coulomb interactions between organic (cations) and inorganic (anions) subsystems may also induce a partial charge transfer (CT) which can lead to an increasing electrical conductivity through the anionic system.

These hybrid materials are usually built of alternating organic and inorganic layers. The organic layers are created by peculiar arrangement of BEDT-TTF species with short intermolecular contacts. The inorganic ones are formed from polyoxometalate anions and solvent molecules (if they occur). The polyoxometallate structures are made by the association of MO$_6$ octahedra (where M is Mo, W, Nb).[3-9] Both organic and inorganic

 DOI: 10.1002/masy.200450816

building blocks show a structural disorder which is so important for the electron distribution in these blocks.

A detailed spectral investigation of a crystal belonging to this type of compounds, (BEDT-TTF)$_2$Mo$_6$O$_{19}$, has been recently published by Visentini et al.[9] In this salt, the fully oxidised organic species (BEDT-TTF$^+$) overlap in an eclipsed geometry and form quasi-isolated dimers.[4] According to our studies, the crystals show semiconducting properties similar to those of (TTF)$_2$Mo$_6$O$_{19}$.[6]

The crystal structure of the (BEDT-TTF)$_2$W$_6$O$_{19}$ salt, which is isostructural with (BEDT-TTF)$_2$W$_6$O$_{19}$, was described by Triki et al.[5] The inorganic layers of the salt are formed by (W$_6$O$_{19}$)$^{2-}$ polyanions, which are members of a group of discrete isopolymetallates having a Lindquist-type structure.[10] The orthogonalized dimers of (BEDT-TTF)$^+$ cations form the organic layers. The electrical conductivity investigations performed by Kravchenko et al.[11] showed that (BEDT-TTF)$_2$W$_6$O$_{19}$ is a narrow-gap semiconductor (E_a = 0.17 eV). Its conductivity (measured in pellet, at room temperature) is between 0.07 and 0.12 S·cm^{-1}. The preliminary IR spectra of the salt have been also presented.[11]

The electrochemical preparation and characterisation of the (BEDT-TTF)$_6$(Mo$_8$O$_{26}$)(DMF)$_3$ salt have been recently reported by us.[12] Inorganic layers of this salt are formed from (Mo$_8$O$_{26}$)$^{4-}$ anions and dimethylformamide (DMF) molecules. Organic layers are built from planar and non-planar BEDT-TTF units, which are organised in two types of step-chains. The d.c. electrical conductivity of the crystalline sample of (BEDT-TTF)$_6$ (Mo$_8$O$_{26}$)(DMF)$_3$ is about 3 S·cm^{-1} at room temperature, increasing gradually down to 60 K, where it amounts to a value of about 12 S·cm^{-1}. Below this temperature, the conductivity decreases rapidly reaching at 4 K the value five orders lower than at about 300 K.[12] The low-temperature range can be described by a hopping model but the conductivity at $T > 60$ K reveals a metal-like character. IR and Raman spectra of (BEDT-TTF)$_6$(Mo$_8$O$_{26}$)(DMF)$_3$ single crystals were also analysed.[12]

This work reports mainly the extended IR spectral investigations of the semiconducting (BEDT-TTF)$_2$W$_6$O$_{19}$ salt. Its properties are discussed and compared with the data for the hybrid semiconducting (BEDT-TTF)$_2$Mo$_6$O$_{19}$[9] and metal-like (BEDT-TTF)$_6$(Mo$_8$O$_{26}$)(DMF)$_3$[12] salts recently published.

Experimental

Black crystals of (BEDT-TTF)$_2$W$_6$O$_{19}$ salt with submillimetre dimensions were obtained on a smooth platinum wire electrode by anodic oxidation of solution of organic donor BEDT-TTF in the presence of a background salt. More details of preparation of the salt were given by

Kravchenko et al.[11] The procedure of preparation of $(BEDT-TTF)_6(Mo_8O_{26})(DMF)_3$ from DMF solution by anodic oxidation under galvanostatic conditions was described elsewhere.[12]

For d.c. electrical conductivity measurements, bar-shaped crystals of both $(BEDT-TTF)_2W_6O_{19}$ and $(BEDT-TTF)_6(Mo_8O_{26})(DMF)_3$ salts were selected. The conductivity was measured along the long axis of the crystals, using a standard four-probe technique.[12]

Absorption spectra of $(BEDT-TTF)_2W_6O_{19}$ and $(BEDT-TTF)_6(Mo_8O_{26})(DMF)_3$ in KBr pellets were recorded in the range 400 - 45000 cm^{-1} with a Perkin Elmer UV-VIS-NIR Lambda 19 and FT IR Perkin Elmer 1725 X spectrometers, at room temperature. The polarised reflectance spectra for single crystals of $(BEDT-TTF)_2W_6O_{19}$ and $(BEDT-TTF)_6(Mo_8O_{26})(DMF)_3$ were studied in the frequency range 600 - 7000 cm^{-1} with a 1725 X spectrometer equipped with an IR microscope, a suitable polariser and a narrow-band MCT detector. An Oxford Instruments helium cryostat was used for measurements of the reflectivity of the crystals down to 4 K.

Results and discussion

According to Visentini et al.[9], two electronic bands are present at around 4500 and 7000 cm^{-1} in the dimerised salt of $(BEDT-TTF)_2Mo_6O_{19}$. The strong electronic band at 7000 cm^{-1} is associated with an intradimer charge transfer transition, but the medium 4500 cm^{-1} band with a lateral CT transition between dimers. The electronic absorption spectrum of isostructural salt $(BEDT-TTF)_2W_6O_{19}$ is similar to the salt mentioned above. The strong absorption band at ca. 6000 cm^{-1} and the weak band at ca. 4300 cm^{-1} are observed in the polarised transmission spectra recorded for a thin crystalline sample (Fig. 1). It is characteristic that the polarisations of both bands are perpendicular to each other; this corresponds well to electronic bands assignment proposed by Visentini et al.[9] In contrast, the spectrum of $(BEDT-TTF)_6(Mo_8O_{26})(DMF)_3$ is completely different. It consists of a strong band at about 11200 cm^{-1}, which can be attributed to the CT between cations.[12] Assuming that the nearest-neighbour Coulomb repulsion energies are small, the position of this band measures the on-side Coulomb repulsion between charges on the same BEDT-TTF molecule. The relatively weak band at about 7300 cm^{-1} originates from the CT between the neutral and ionic forms of BEDT-TTF. The differences in the electronic absorption spectra of $(BEDT-TTF)_2W_6O_{19}$ and $(BEDT-TTF)_6(Mo_8O_{26})(DMF)_3$ salts reflect the basic differences in their crystal structure and, as a result, their electrical transport properties (semiconductors in contrast to metal-like material).

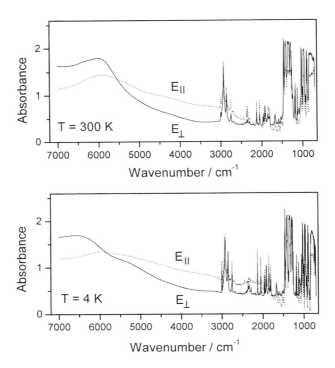

Figure 1. Anisotropy of the electronic and vibrational spectra of a (BEDT-TTF)$_2$W$_6$O$_{19}$ single crystal at 4 and 300 K.

Vibrational spectra of (BEDT-TTF)$_2$W$_6$O$_{19}$ were recorded in three ways: absorption of dispersed powdered sample in KBr matrixes, polarised/unpolarised transmission of thin single crystals and polarised reflection from a well developed face of the crystal. The anisotropy of the transmission spectra (Fig. 1) of a (BEDT-TTF)$_2$W$_6$O$_{19}$ crystal suggests that the electronic transition at about 6000 cm^{-1} occurs mainly along the BEDT-TTF step-chains (for perpendicular polarisation it is much weaker). On the other hand, the much weaker band at about 4300 cm^{-1} is assigned to a lateral CT transition between neighbouring BEDT-TTF species. Besides, the sensitivity of transmission measurements in thin crystalline samples is much higher than that of transmission measurements of absorption in KBr pellet. It is also seen from Figs. 1 and 2 that the region of vibrational modes is extremely rich. Studies of transmission of crystalline samples permit to record numerous very weak bands between 2291 and 1488 cm^{-1}, which are not seen in standard absorption spectra. It seems that they represent higher harmonic and/or combination modes of the BEDT-TTF cation. Vibrational features observed in (BEDT-TTF)$_2$W$_6$O$_{19}$ and (BEDT-TTF)$_6$(Mo$_8$O$_{26}$)(DMF)$_3$ salts are reported in Table 1.

Table 1. Frequencies and assignments of infrared vibrational features of $(BEDT\text{-}TTF)_2W_6O_{19}$ and $(BEDT\text{-}TTF)_6(Mo_8O_{26})(DMF)_3$.

Frequencies[a], cm^{-1}			Assignment[9-17]
$(BEDT\text{-}TTF)_2W_6O_{19}$ from KBr	$(BEDT\text{-}TTF)_2W_6O_{19}$ from single crystal	$(BEDT\text{-}TTF)_6(Mo_8O_{26})(DMF)_3$ from KBr pellet	
2998 m	2995 m		
2972 m	2974 m		
2940 sh	2941 w		
2924 m	2926 m		
2850 vw	2853 w		
2820 vw	2817 w		
2742 vw	2747 m		
2690 vw	2693 vw		
2361 vw	2360 vw		
2337 vw	2333 vw		
	1454 m		
1452 m	1452 w	1451 m	$A_g(\nu_2)$
	1426 w		
1420 sh	1422 vs		
1415 vs	1419 vs	1416 vs	$A_g(\nu_3)$
	1410 sh		
1350 sh	1356 vw		
1343 vs	1344 vs	1341 vs	$A_g(\nu_3)$
1294 vw	1294 w	1293 sh	$B_{3g}(\nu_{57})$ or $B_{1u}(\nu_{29})$
1282 m	1284 m	1282 m	$A_g(\nu_5)$
	1260 vw	1257 vw	$B_{3u}(\nu_{67})$
	1238 vw		
	1185 vw		
1169 w	1174 vw	1169 w	$A_u(\nu_{1u})$
	1127 vw	1127 vw	$B_{1g}(\nu_{21})$
	1059 vw		
1025 w	1025 w	1027 m	$B_{3g}(\nu_{58})$
	1007 vw	1007 vw	$B_{3g}(\nu_{59})$
	993 vw		
974 s	977 w	976 s	W_6O_{19} and $A_g(\nu_6)$
962 s	963 w	963 s	Mo_8O_{26}, W_6O_{19}
	923 m	927 w	$A_g(\nu_7)$
	897 vw	896 w	$B_{2u}(\nu_{49})$
		884 w	$B_{3g}(\nu_{60})$
813 ms		814 sh	W_6O_{19}
803 vs		805 vs	Mo_8O_{26}, W_6O_{19}
		720 w	
		676 w	DMF
		669 w	DMF
		645 w	$A_g(\nu_8)$
585 m		585 m	W_6O_{19}
493 m		493 m	$A_g(\nu_9)$
478 m		478 m	$B_{1u}(\nu_{3u})$
444 s		443 s	W_6O_{19}, $A_g(\nu_{10})$
		406 w	DMF

[a] Intensities: vs – very strong, m – medium, w – weak, vw – very weak, sh – shoulder.

The IR spectra of both investigated salts contain several bands representing normal vibrations of species forming the compound: BEDT-TTF cations, $W_6O_{19}^{2-}$, $Mo_8O_{26}^{4-}$ anions and/or

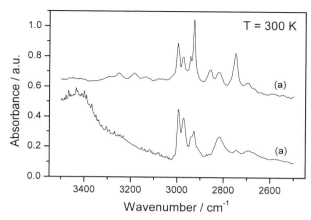

Figure 2. Infrared absorption spectrum of $(BEDT\text{-}TTF)_2W_6O_{19}$ (a) in KBr pellet, (b) transmission spectrum recorded for thin crystalline sample, at room temperature in the range of CH_2 vibrations.

solvent molecules. The strongest and broadest BEDT-TTF bands observed in transmission spectrum of $(BEDT\text{-}TTF)_2W_6O_{19}$ are given by its totally symmetric vibrations activated by the electron-molecular vibration (EMV) coupling: 1452 (v_2), 1419 and 1344 (v_3), 1284 (v_5), and 923 cm^{-1} (v_7). These activated bands are distinctly shifted relative to the corresponding Raman lines of BEDT-TTF molecules. There are also a few weaker bands representing active IR vibrations of BEDT-TTF species. Some bands originate from the vibrations of anionic ($W_6O_{19}^{2-}$ or $Mo_8O_{26}^{4-}$) sublattices. Few weak bands of $(BEDT\text{-}TTF)_6(Mo_8O_{26})(DMF)_3$ can be attributed to solvent vibrations. These bands are not distinctly shifted in comparison with the free solvent vibrations, suggesting weak interaction between DMF and the species forming the salt.

In order to properly analyse the structured bands of the $(BEDT\text{-}TTF)_2W_6O_{19}$, which dominate the IR spectra of the salt, they were decomposed and fitted at each temperature to Gaussian form by the least-square method (Fig. 3). An analysis of T-dependences of the component band frequencies, their bandwidths and integral intensities shows that nearly all bands shift (with no discontinuities) by a few cm^{-1} (0.5 – 0.01 %) towards higher frequencies with decreasing temperature. Usually, the bandwidth decreases distinctly (more than twice) and the integral intensity increases also without discontinuities by tens of % (Fig. 4). This is caused by shortening (strengthening) of molecular bonds in BEDT-TTF cations. Only the band at 1057 cm^{-1} undergoes an extreme intensity increase (more than 14 times). On the other hand, the 1127 cm^{-1} component shows a unique deviation from the above rules. Its frequency slightly decreases (more than 1 cm^{-1}) with temperature but the bandwidth decreases only by about 7 %. This band is assigned to bending vibrations of CCH and SCH bonds of BEDT-TTF cation. A softening of the band is probably caused by stepwise hindering of the CH_2 motions. In contrast to this observation, some electronic parameters evaluated from a Drude-Lorentz analysis of the reflectance spectra of $(BEDT-TTF)_6(Mo_8O_{26})(DMF)_3$ salt show a temperature anomaly at about 180 K.[12] The features in the T-dependences of the plasma frequency and the relaxation rate correlate well with the electric conductivity properties reported elsewhere.[12]

The polarised reflectance spectra of $(BEDT-TTF)_2W_6O_{19}$ single crystals corresponding to two extreme polarisations at two selected temperatures (300 and 4 K), are reported in Fig. 5. The parallel polarization ($E_\parallel$) has been found for the maximum of reflected energy along the long axis of the crystals. For the polarization perpendicular ($E_\perp$) to the above, the reflected energy reaches minimum. The spectra show an anisotropy for both electronic and vibrational excitations. For $E_\parallel$ light polarization, the band at 6000 cm^{-1} as well as its vibrational features

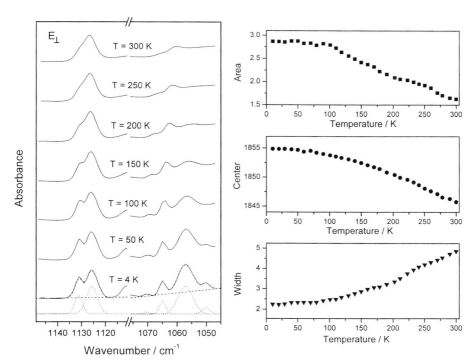

Figure 3. Temperature evolution of the bands in the region of stretching and bending vibrations of CH, SCH and CCH groups of BEDT-TTF. Dotted lines represent the best band decomposition.

Figure 4. Temperature dependences of the frequency, bandwidth and integral intensity of the 1846 cm^{-1} component.

are relatively strong confirming the anisotropy of the crystal mentioned before. The vibronic structure is perfectly visible in both transmission (Fig. 1) and reflection (Fig. 5) spectra of

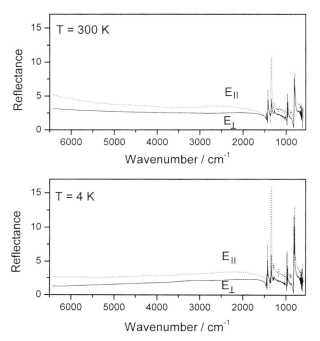

Figure 5. Polarised reflectance spectra of $(BEDT-TTF)_2W_6O_{19}$ single crystals at $T = 300$ K and at $T = 4$ K for parallel and perpendicular polarizations of the IR beam.

$(BEDT-TTF)_2W_6O_{19}$ crystals. The occurrence of developed vibronic features in the IR spectra is typical of organic semiconductors. On the contrary, metallic properties of $(BEDT-TTF)_6(Mo_8O_{26})(DMF)_3$ salt are confirmed by poorly visible structure in the spectra below 1500 cm^{-1}.[12] Its spectral anisotropy is much greater than that reported previously for the semiconducting $(BEDT-TTF)_2W_6O_{19}$ salt. The anisotropy of the salt is adequate to its layered structure and metallic properties well fitted by the Drude-Lorentz dielectric function for $E_\perp$ light polarisation.

The optical conductivity spectra $\sigma(\omega)$ of $(BEDT-TTF)_2W_6O_{19}$ were obtained by the Kramers-Krönig transformation of their reflectivity (Fig. 6). It was performed in the range 600-6500 cm^{-1} by extrapolating the reflectance data to zero frequency with a constant value. For the extrapolation to higher frequencies, we used experimental data for $(BEDT-TTF)_2Mo_6O_{19}$.[9] The conductivity spectra of $(BEDT-TTF)_2W_6O_{19}$ show a distinct vibronic structure below 1500 cm^{-1}. This structure is the consequence of strong coupling of totally symmetric A_g

vibrations of the BEDT-TTF$^+$ cation to suitable electronic excitations. The intensity of vibronic bands increases strongly at low temperature. The $\sigma(\omega)$ spectra of the (BEDT-TTF)$_6$(Mo$_8$O$_{26}$)(DMF)$_3$ salt do not change with temperature and they are qualitatively similar above 2000 cm^{-1}, for a given polarization. For $E_\parallel$ polarization, the intensity of the large vibronic band at about 1300 cm^{-1} increases considerably at 4 K. This band is mainly a consequence of strong coupling of totally symmetric A$_g$ vibrations of C=C bonds with intrastack or interstack charge transfer. For the neutral BEDT-TTF molecule, the two A$_g$ modes observed at A$_g(\nu_3) = 1493$ cm^{-1} and A$_g(\nu_2) = 1551$ cm^{-1} are assigned to C=C stretching.[15] Due to the strong vibrational coupling of these modes, a single, intense and broad vibrational feature is usually observed in salts with partially ionised BEDT-TTF molecules. The situation is completely different for $E_\perp$ polarization, which reveals metal-like properties of the (BEDT-TTF)$_6$(Mo$_8$O$_{26}$)(DMF)$_3$ salt. At room temperature, a strong interband transition is observed at about 2400 cm^{-1}. On the low-frequency wing, a strong and structured band corresponding to vibronic effects is also seen. At about 1500 cm^{-1}, typical antiresonance

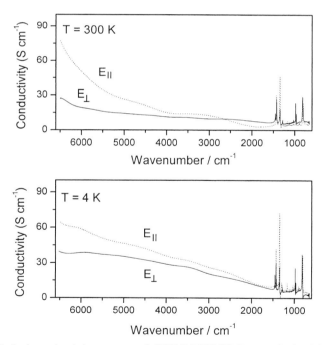

Figure 6. Optical conductivity spectra of (BEDT-TTF)$_2$W$_6$O$_{19}$ as obtained by the Kramers Krönig analysis of the reflectance spectra for parallel and perpendicular polarizations at $T = 4$ K and $T = 300$ K.

is observed, which results from a coupling between the intramolecular vibrations and the interband electronic transition.

Conclusion

The semiconducting $(BEDT-TTF)_2W_6O_{19}$ salt, single crystals and powdered samples, was investigated in a broad spectral range at various temperatures. The salt shows similarities to $(BEDT-TTF)_2Mo_6O_{19}$ but is completely different from the metal-like $(BEDT-TTF)_6(Mo_8O_{26})(DMF)_3$. The electronic absorption spectrum of the salt under study consists of two bands at about 6000 and 4300 cm^{-1} showing anisotropy. The vibrational spectrum of $(BEDT-TTF)_2Mo_6O_{19}$ is dominated by totally symmetric modes of the cation activated by electron-molecular vibration coupling. The spectra recorded for thin single crystals of the salt are very rich and contain several very weak bands, which could be attributed to harmonic and/or combination modes of the $BEDT-TTF^+$ cation.

Acknowledgements

This work was supported by the Polish Committee for Scientific Research (grant 2 PO3B 087 22), NATO (grant PST CLG 972 846) and the Ministry of Education and Sciences of Ukraine (201/2001).

[1] E. Coronado, C.J. Gómez-Garcia, *Chem. Rev.* **1998**, 98, 273.
[2] L. Ouahab, *Chem. Mater.* **1997**, 9, 1909.
[3] S. Triki, L. Ouahab, J. Padiou, D. Grandjean, *J. Chem. Soc., Chem. Commun.* **1989**, 1989, 1068.
[4] S. Triki, L. Ouahab, D. Grandjean, J.-M. Fabre, *Acta Crystallogr., Sect. C* **1991**, 47, 1371.
[5] S. Triki, L. Ouahab, D. Grandjean, J.-M. Fabre, *Acta Crystallogr., Sect. C* **1991**, 47, 645.
[6] D. Attanasio, C. Bellitto, M. Bonamico, V. Fares, S. Patrizio, *Synth. Met.,* **1991**, 41-43, 2289.
[7] L. Ouahab, M. Bencharif, A. Mhanni, D. Pelloquin, J.-F. Halet, O. Peña, C. Garrigou-Lagrange, J. Amiell, P. Delhaes, *Chem. Mater.* **1992**, 4, 666.
[8] C. Bellitto, M. Bonamico, V. Fares, F. Federici, G. Righini, M. Kurmoo, P. Day, *Chem. Mater.* **1995**, 7, 1475.
[9] G. Visentini, M. Masino, C. Bellitto, A. Girlando, *Phys. Rev. B* **1998**, 58, 9460.
[10] I. Lindquist, *Ark. Kemi* **1953**, 5, 247.
[11] A.V. Kravchenko, V.A. Starodub, A.R. Kazachkov, A.V. Khotkevich, *Functional Mater.* **2000**, 7, 693.
[12] A. Łapiński, V. Starodub, M. Golub, A. Kravchenko, V. Baumer, E. Faulques, A. Graja, *Synth. Met.,* submitted.
[13] R. Świetlik, P. Le Maguerès, L. Ouahab, *Adv. Mater. Opt. Electron.* **1997**, 7, 67.
[14] L. Ouahab, S. Golhen, S. Triki, A. Łapiński, M. Golub, R. Świetlik, *J. Cluster Sci,.* accepted.
[15] J.E. Eldridge, C.C. Homes, J.M. Williams, A.M. Kini, H.H. Wang, *Spectrochim Acta A* **1995**, 51, 947.
[16] J.E. Eldridge, C.C. Homes, H.H. Wang, A.M. Kini, J.M. Williams, *Synth. Met.* **1995**, 70, 983.
[17] J.E. Eldridge, Y. Xie, Y. Lin, C.C. Homes, H.H. Wang, J.M. Williams, A.M. Kini, J.A. Schlueter, *Spectrochim. Acta A* **1997**, 53, 565.

Macromol. Symp. **2004**, *212*, 179-190

Novel Tunable Optical Properties of Liquid Crystals, Conjugated Molecules and Polymers in Nanoscale Periodic Structures as Photonic Crystals

Katsumi Yoshino,[1] *Hiroyuki Takeda,*[1] *Masahiro Kasano,*[1] *Shigenori Satoh,*[1] *Tatsunosuke Matsui,*[1] *Ryotaro Ozaki,*[1] *Akihiko Fujii,*[1] *Masanori Ozaki,*[1] *Akira Kose*[2]

[1] Department of Electronic Engineering, Graduate School of Engineering, Osaka University, 2-1 Yamada-Oka, Suita, Osaka 565-0871, Japan
[2] Nihon Koken Kogyo Co. Ltd. Tachikawa, Tokyo 190-0033, Japan

Summary: Tunability of optical properties such as transmission and reflection by temperature and applied voltage has been demonstrated in synthetic opals and inverse opals like three-dimensional photonic crystals infiltrated with liquid crystals, conjugated molecules and polymers, in accordance with theoretical calculation. A new type of tunability based on uncoupled mode in a two-dimensional photonic band gap influenced by the field-dependent anisotropy in liquid crystals has also been demonstrated theoretically. Spectral narrowing and lasing have been observed in these opals infiltrated with conducting polymers and fluorescent dyes like three-dimensional photonic crystal and fluorescent dye-doped cholesteric and ferroelectric liquid crystal as one-dimensional photonic crystal. These lasing wavelength can be controlled by the applied voltage. Laser emission was also realized with conducting polymer on a surface relief grating formed on an azo-polymer film by interference optical beam.

Keywords: conducting polymer; laser, opal; liquid crystal; photonic crystal

Introduction

Recently photonic crystals with three-dimensional periodic structures of the order of optical wavelength have attracted great interest from both fundamental and practical viewpoints because novel concepts, such as photonic band gaps in which some energy range of photons cannot exist, and also various novel applications utilizing unique characteristics of photonic crystals have been proposed.[1,2] A wide variety of applications can be realized utilizing the unique photonic band structure, i.e., utilizing photonic band gap, structure photonic band edge structure and defects in photonic crystals. Even in one-dimensional periodic structure, a photonic band gap corresponding to the stop band appears in one direction.

The photonic band gap is dependent on the refractive index, periodicity, occupation ratio, crystal structure and so on. We have proposed a tunable photonic crystal in which the photonic band structure such as band shape and band gap can be tuned by external applied

© 2004 International Union of Pure and Applied Chemistry

DOI: 10.1002/masy.200450817

field and also by ambient conditions.[3-5] As one of the methods to realize tunable photonic crystals, various materials such as organic molecules, liquid crystals and conducting polymers can be infiltrated in interconnected nanoscale voids in synthetic opals prepared by the sedimentation of silica spheres and also in their inverse opals, which have a three-dimensional periodic structure of the order of optical wavelength and are considered as prototype photonic crystals.

In this paper, we report on the tunability and its dynamic behavior as functions of structures and materials of three-dimensional periodic structures, temperature and applied field. The results are also discussed in terms of theoretical band calculations. A new type of tunability utilizing anisotropy of liquid crystals in photonic crystals but not using photonic band gap is also discussed. Lasing characteristics are examined in opals containing organic molecules and conducting polymers like three-dimensional photonic crystals and also in cholesteric and ferroelectric liquid crystals as one-dimensional photonic crystals; their wavelength tunability is discussed. Lasing in conducting polymer in a surface relief grating formed on an azo-polymer film is also discussed.

Experimental

Synthetic opals were prepared by sedimentation of silica (SiO$_2$) spheres with a diameter of several hundreds nm in water. Inverse opals were prepared by infiltrating various materials such as phenol resin in to the interconnected nanoscale periodic array of voids in the synthetic opal and subsequently removing SiO$_2$ by a chemical treatment with HF solution. As shown in Fig.1, the inverse opals have periodic array of nanoscale voids. Various materials can be infiltrated in to these nanoscale voids of opals and inverse opals.

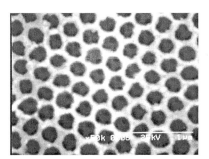

Figure 1. Scanning electron microphotograph of the inverse opal.

Polymer opals containing fluorescent dyes were also prepared by utilizing polymer (latex) nanoscale spheres doped with fluorescent dyes (Uvitex EBF (Chiba)) at 95 °C for two hours.

Upon irradiation of the interference optical beam on the surface of a polymer containing azogroups in the side chain, a surface relief grating was formed. Utilizing this surface grating as a template, i.e., upon casting an other photopolymer on this surface relief of the azopolymer and then removing the surface relief, an inverse surface relief structure was formed, on which a conducting polymer film could be formed by casting chloroform solution of ROPPV (poly[(2.5-dialkoxyphenylene)vinylene]).

Cholesteric and ferroelectric liquid crystals which have helical structure are also studied as one-dimensional photonic crystals. Lasing characteristics have been studied of these liquid crystals, after doping with fluorescent dyes.

For excitation source of lasing experiment, second- or third-harmonic lights of Q-switched Nd:YAG laser (Spectra Physics, Quanta-Ray INDI) whose pulse width and pulse repetition frequency were 8 ns and 10 Hz, respectively, and a second harmonic light of a regenerative amplifier system based on a Ti:sapphire laser (Spectra Physics) whose pulse width, wavelength and pulse repetition frequency were 150 fs, 400 nm and 1 kHz, respectively, were used. The emission spectra of the sample were measured using a CCD multichannel photodetector (Hamamatsu Photonics, PMA-11) having spectral resolution of 3 nm or a spectrograph with CCD (Oriel, 256) with a 0.5 nm resolution.

Results and discussion

Tunability of synthetic opals and inverse opals infiltrated with liquid crystals and conducting polymers

Liquid crystals such as nematic liquid crystals, smectic liquid crystals including ferroelectric liquid crystals and anti-ferroelectric liquid crystals, and cholesteric liquid crystals, were successfully infiltrated in to synthetic opals made of silica spheres and their inverse opals as photonic crystals having a three-dimensional structure with a periodicity of the order of optical wavelength.

The optical stop-band in the transmission spectrum and the peak in the reflection spectrum can be tuned by temperature and also by the applied voltage in the spectral range covering from ultraviolet to infrared. A sharp peak in the reflection spectrum and a dip in the transmission spectrum observed in the synthetic opal shifted drastically upon infiltration of liquid crystals in the nanoscale voids of opals. They also shifted to shorter wavelengths upon increasing temperature, showing a stepwise change at the phase transition point as shown in Fig. 2. These results well coincide with the theoretical analysis.

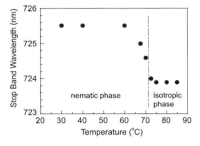

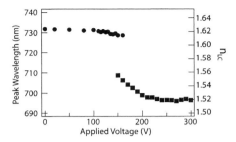

Figure 2. Temperature dependence of the peak wavelength of optical stop-band in a synthetic opal infiltrated with a nematic liquid crystal (ZLI1132, Merck).

Figure 3. Voltage dependence of the peak wavelength of stop-band in the reflection spectrum of the polymer inverse opal infiltrated with 5CB.

The shifts of peaks in reflection spectra and dips in transmission spectra due to infiltration and temperature change were larger in the case of inverse opal, which can be explained by a larger volume fraction of voids in the inverse opal than in the original opal.

Upon voltage application the peak and dip in spectra also shifted to shorter wavelength as shown in Fig. 3. It should be noted that there exists a hysteresis. The tuning of the optical properties by applying a voltage above some threshold voltage has been interpreted in terms of the change of refractive index of infiltrated liquid crystals accompanied by a field-induced reorientation of liquid crystal molecules. The threshold originates from the intense interaction of liquid crystal molecules and the inner surface of the void.

The switching time, in particular the rising time upon field application was much faster than that of a conventional nematic liquid crystal cell and decreased proportionally to the square of the applied field, which suggests that the driving torque for molecular reorientation originates from dielectric force.

In the case of the inverse opal infiltrated with nematic liquid crystals, a similar change of the peak was observed but at a certain voltage the peak disappeared, which can be explained by the coincidence of the refractive index of liquid crystals to that of polymer. At higher fields, the peak again shifts to shorter wavelengths with a higher rate. This suggests that the surface anchoring force also influences dynamic characteristics.

The deformation of sphere voids in the polymer replica also has some effects on the tuning behavior.

Opals and inverse opals infiltrated with smectic liquid crystals and cholesteric liquid crystals also exhibited novel characteristics. Their dynamic behavior is determined by the origin of the driving force of the molecular reorientation of the liquid crystal due to the applied field. In the case of nematic liquid crystals, for example, the response time decreased with increasing the applied voltage proportional by to the reciprocal of the square of the applied voltage. In the case of the photonic crystal infiltrated with ferroelectric liquid crystals with spontaneous polarization P_s, the torque seems to originate in the $P_s E$ torque.

The stop-band and reflection peak in the spectra and the dynamic behavior upon field application at various temperatures were confirmed to depend on the periodicity of photonic crystals and the material forming photonic crystal (opals and replicas).

We have theoretically studied the photonic band structure of photonic crystals infiltrated with liquid crystals and found a new type of tunability in two-dimensional photonic crystals infiltrated with anisotropic liquid crystals. We have calculated the photonic band structure of a two-dimensional photonic crystal made of liquid crystal columns in square lattice, utilizing the wave equation modified by magnetic field.

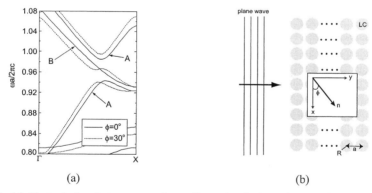

(a)   (b)

Figure 4. (a) Photonic band structures of two-dimensional photonic crystals made of liquid crystal columns as a function of liquid crystal molecular orientation ϕ. A and B represent the symmetric and antisymmetric modes, respectively. (b) Schematic model of plane waves incident on photonic crystals composed of liquid crystals. n and ϕ show the director and molecular orientation of liquid crystals.

By the analysis, it was clarified that the uncoupled modes in two-dimensional photonic crystals, which exist in the case of random orientation of liquid crystals, disappear when anisotropy appears due to the alignment of liquid crystal molecules upon voltage application.

The solid and dotted lines in Fig. 4a indicate the photonic band structures with anisotropic liquid crystals of orientation of $\phi = 0$ ° and $\phi = 30$ °, respectively. Here, ϕ is the angle between the director of liquid crystal molecules and the plane perpendicular to the optical beam as shown in Fig. 4b. For these structures, the optical transmittance when directors of liquid crystals are oriented at $\phi = 0$ ° and $\phi = 30$ ° are shown in Fig. 5. This result indicates that the reorientation of liquid crystal molecules by the applied field results in the transmission change. We are preparing the experiment to support this new idea.

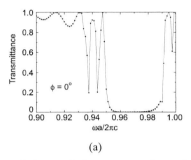
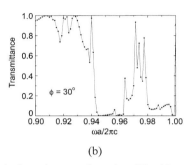

(a)   (b)

Figure 5. Transmission spectra of two-dimensional photonic crystal made of liquid crystal columns as a function of molecular orientation of the liquid crystals, ϕ.

Spectral narrowing and lasing in opals infiltrated with conducting polymers and polymer opals containing dyes

We have already reported observations of lasing in synthetic opals and inverse opals infiltrated with various dyes and conducting polymers. In those cases, lasing was clearly observed when fluorescence wavelengths of the dyes and conducting polymers overlap with the photonic band edge of the opals and inverse opals. For example, blue laser emission was observed in the coumarin-infiltrated opals made of 180 nm silica spheres, on the other hand, green and red laser emissions were observed when RO-PPV and cyanine dye were infiltrated into the green and red opals which were made of 220 nm and 250 nm silica spheres, respectively.

We have also studied lasing with polymer opals. Figure 6 shows the electron microscope images of the surface of plastic opals and surface of broken opals. As evident from this figure, the three-dimensional periodic structure is realized with a regular array of plastic spheres.

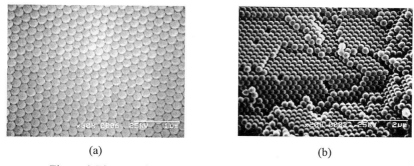

Figure 6. The scanning electron microscope images of plastic opal.

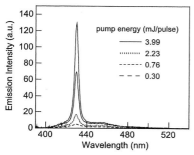

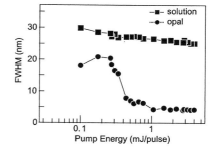

Figure 7. Emission spectra of plastic opal
containing Uvitex EBF as a function of the
pump energy.

Figure 8. Emission peak width (FWHM)
of plastic opal containing Uvitex EBF as a
function of the pump energy.

As shown in Fig. 7, the emission intensity of opals prepared from the spheres containing
Uvitex EBF increases drastically and the spectral width becomes narrower with increasing
excitation intensity of THG (355 nm) of Nd-YAG laser. The spectral narrowing above some
threshold excitation intensity in the dye-doped plastic opal is more clearly shown in Fig. 8.
This can be explained by amplified spontaneous emission and lasing.

Surface relief grating made on azo-polymer film and lasing

Upon irradiating the interferential light beams on the azo-polymer, the trans-cis isomerization
of azobenzene in the side chain induces mass transport and forms the surface relief grating
(SRG) corresponding to the distribution of the intensity of the interfered laser light. The
formed SRG can be transferred onto a photopolymer using the following procedure. A
sandwiched cell was fabricated using an azo-polymer film with the SRG and a glass substrate.
A UV-curable prepolymer was inserted into the sandwiched cell, and cured by the UV

irradiation. After removing the substrates, a flexible polymer film with an SRG was obtained. Figure 9 shows atomic force microscope images of the holographically fabricated SRG on the azo-polymer film and the transferred inverse SRG on the photopolymer.

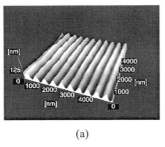

(a)

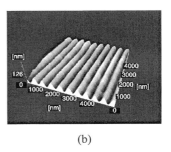

(b)

Figure 9. Scanning electron microscope images of (a) surface relief grating on azo-polymer and (b) the inverse surface relief grating on photopolymer.

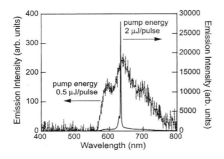

Figure 10. Emission spectra of RO-PPV coated on the inverse surface relief grating as a function of the pump energy.

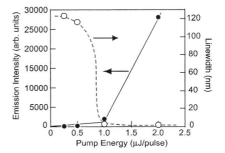

Figure 11. Pump energy dependences of emission intensity and peak width of the emission spectrum of RO-PPV on the inverse surface relief grating.

Figure 10 indicates emission spectra of RO-PPV coated on the photopolymer with the inverse SRG as a function of energy of the excitation light (532 nm). As evident from this figure, upon intense excitation a sharp emission peak was observed compared with the broad emission at low excitation intensity. Figure 11 shows the dependences of the emission intensity and spectral width on the excitation energy. As evident from this figure, above the threshold the emission intensity increases and spectral width decreases drastically. These facts clearly indicate that the mirrorless lasing was realized by the simple method. The periodicity of the inverse surface relief grating can be easily controlled by changing the incident angle of the writing interference light beam and various other light-emitting conducting polymers and also insulating polymers containing light-emitting molecules can be overcoated on the inverse relief. Therefore, the laser device with various emission wavelength can be easily fabricated by this procedure.

Cholesteric liquid crystal and ferroelectric liquid crystal as one-dimensional photonic crystal and lasing

Cholesteric liquid crystals and ferroelectric smectic liquid crystals containing chiral carbon have helical structure with a pitch in the range of several hundreds of nm and several hundreds of μm. Therefore these liquid crystals exhibit properties of one-dimensional photonic crystals. We have incorporated fluorescent dyes in these liquid crystals and observed a laser action upon optical excitation.

Low-molecular-weight cholesteric liquid crystals (CLCs) were sandwiched between glass plates and the lasing can be observed in this sandwiched configuration. However, polymerized cholesteric liquid crystals (PCLC) can be used in a free-standing configuration without glass substrates, i.e. as flexible film.

PCLC film was fabricated utilizing two types of photopolymerizable CLC mixture (Merck KGaA). DCM as fluorescent dye was dissolved in the PCLC.

Figure 12a shows transmission spectrum of the DCM-doped PCLC. The transmittance drop at around 600 nm is due to the selective reflection corresponding to the pitch of the helix. As evident in Fig.12b, the fluorescence spectrum exhibits remarkable narrowing with increasing excitation intensity. Figure 13 shows the dependence of the emission intensity and spectral width on the excitation intensity. These figures clearly indicate that the lasing is performed in the PCLC film. It should be mentioned that the lasing can be observed in the bent film and even focusing of the laser emission was realized utilizing a spherically curved PCLC film upon optical excitation.

Laser action based on a twist defect mode was also demonstrated in a composite film composed of two dye-doped PCLC layers. The twist defect was introduced as a discontinuous jump of the director rotation around the helical axis at an interface of two PCLC layers. At high excitation energy above the threshold, the laser action was observed at the twist defect mode wavelength in the middle of 1-D photonic band gap of the PCLC helical structure.

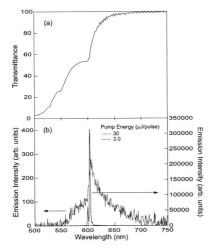

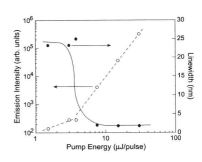

Figure 12. Transmission (a) and fluorescence (b) spectra of the DCM-doped PCLC film.

Figure 13. Pump energy dependences of emission intensity and peak width of the emission spectrum of the DCM-doped PCLC film.

Lasing was also observed in fluorescent-dye-doped ferroelectric liquid crystal (FLC). FLC mixture with a short helical pitch containing a fluorescent dye (Coumarin 500) was sandwiched between two ITO glass plates. We prepared two types of cells with different orientations, homeotropic and homogeneous cells with the helical axis parallel and perpendicular to the glass plate. In both, cases upon optical excitation, laser emission was observed in the directions parallel and perpendicular to the cell.

As evident from Fig. 14, the intensity increases and spectral width decreases with increasing excitation intensity above the threshold intensity. Lasing was observed at the edge of the photonic band gap, i.e., at the stop-band edge.

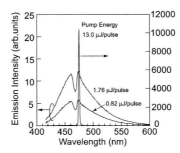

Figure 14. Emission spectra of dye-doped ferroelectric liquid crystal as a function of the pump energy.

It should be stressed that the lasing wavelength shifted drastically with increasing applied voltage as shown in Fig. 15. Therefore, the photonic band gap of the FLC was confirmed to be shifted with the applied voltage as shown in Fig. 16. This clearly supports that the voltage tunability of lasing wavelength originates from the voltage tunability of the photonic band gap.

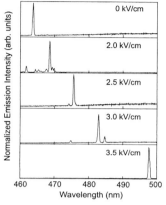

Figure 15. Lasing spectra of a dye-doped FLC as a function of applied electric field.

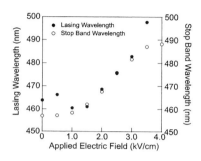

Figure 16. Electric field dependence of helix pitch and lasing wavelength of a dye-doped FLC.

Conclusions

We demonstrated tunable optical properties of liquid crystals, conjugated molecules and polymers with nanoscale periodic structures as photonic crystal. Optical properties such as transmission and reflection were tuned by changing temperature and applying voltage to synthetic opals and inverse opals infiltrated with liquid crystals, conjugated molecules and polymers. Calculations were performed of a novel type of tunability based on uncoupled mode in a two-dimensional photonic band gap influenced by the field-dependent anisotropy in liquid crystals. Spectral narrowing was demonstrated using a plastic opal prepared from the spheres containing a fluorescent dye as three-dimensional photonic crystal. The dye-doped cholesteric and ferroelectric liquid crystals were used as one-dimensional photonic crystal, and the laser actions in these systems were demonstrated. Especially in the latter system, the lasing wavelength could be controlled by the applied voltage. Laser emission was also realized with conducting polymer on a flexible photopolymer with a surface relief grating transferred from the azo-polymer film.

Acknowledgements

The authors would like to acknowledge Prof. F. Kajzar for providing the azo-polymer, Prof. W. Haase for the short pitch ferroelectric liquid crystal and the Merck KGaA for the photopolymerizable cholesteric liquid crystal. This work is in part supported by a grant-in-aid for scientific research from the Japan Ministry of Education, Culture and Sports, Science and Technology.

[1]   E. Yablonovitch, Phys. Rev. Lett. **58**, 2059, 1987.
[2]   S. John, Phys. Rev. Lett. **58**, 2486,1987.
[3]   K. Yoshino, K. Tada, M. Ozaki, A. A. Zakhidov, R. H. Baughman, Jpn. J. Appl. Phys. **36**, L714, 1997.
[4]   K. Yoshino, S.B. Lee, S. Tatsuhara, Y. Kawagishi, M. Ozaki, A.A. Zakhidov, Appl. Phys. Lett. **73**, 3506, 1998.
[5]   K. Yoshino, S. Tatsuhara, Y. Kawagishi, M. Ozaki, A.A. Zakhidov, Z.V. Vardeny, Jpn. J.Appl. Phys. **37**, L1187, 1998.
[6]   K. Yoshino, S. Tatsuhara, Y. Kawagishi, M. Ozaki, A.A. Zakhidov, Z.V. Vardeny, Appl. Phys. Lett. **74**, 2590, 1999.
[7]   S. Sato, H. Kajii, Y. Kawagishi, A. Fujii, M. Ozaki, K. Yoshino, Jpn. J. Appl. Phys. **38**, L1475, 1999.
[8]   K. Yoshino, S. Satoh, Y. Shimoda, Y. Kawagishi, K. Nakayama, M. Ozaki, Jpn. J. Appl. Phys. **38**, L961, 1999.
[9]   K. Yoshino, Y. Shimoda, Y. Kawagishi, K. Nakayama, M. Ozaki, Appl. Phys. Lett. **75**, 932, 1999.
[10]   K. Yoshino, Y. Kawagishi, M. Ozaki, A. Kose, Jpn. J. Appl. Phys. **38**, L784, 1999.
[11]   Y. Shimoda, M. Ozaki, K. Yoshino, Appl. Phys. Lett. **79**, 3627, 2001.
[12]   M. Ozaki, Y. Shimoda, M. Kasano, K. Yoshino, Advanced Materials **14**, 514, 2002.
[13]   T. Matsui, R. Ozaki, K. Funamoto, M. Ozaki, K. Yoshino, Appl. Phys. Lett. **81**, 3741, 2002.
[14]   M. Ozaki, M. Kasano, D. Ganzke, W. Haase, K. Yoshino, Advanced Materials **14**, 306, 2002.
[15]   T. Matsui, M. Ozaki, K. Yoshino, F. Kajzar, Jpn. J. Appl. Phys. **41**, L1386, 2002.

Rough Electrode Surface: Effect on Charge Carrier Injection and Transport in Organic Devices

Sergey V. Novikov

A.N. Frumkin Institute of Electrochemistry, Leninsky prosp. 31, 119071 Moscow, Russia

Summary: The effect of electrode roughness on charge carrier injection and transport is considered. An explicit formula connecting the roughness profile of an electrode with the distribution of the electric field at its surface (and the electrostatic potential in the bulk of transport layer) is derived for the case of smooth roughness, when the typical height of roughness element is small in comparison with its size across the surface (this is a very typical situation). This formula gives us an opportunity to measure the electrode surface profile (e.g., by AFM) and then to calculate various injection properties of this particular electrode for any kind of injection. General properties of the electrode - organic layer interface in the case of significant (not smooth) roughness are considered and a suitable numeric procedure for the calculation of the surface electric field distribution is proposed. It is also shown that rough surface of electrodes generates an additional energy disorder in the bulk of transport layer. This principal result indicates that the electrode roughness affects not only carrier injection but carrier transport as well. Roughness-induced energetic disorder produces a channel-like structure in the vicinity of the electrode, thus providing separation of electrons and holes. Such separation should decrease the charge recombination rate and, hence, the emitting efficiency of light emitting devices. At the same time, the separation is favorable to solar cells.

Keywords: charge injection; charge transport; rough interfaces

Introduction

Morphology of the electrode surface and structure of the interface layer between the electrode and bulk of the organic transport layer significantly affect device performance: its efficiency, stability etc. A major aspect of this influence is the dependence of the carrier injection rate on the structure of electrode surface. Rough surface creates a non-uniform distribution of the electric field and, thus, non-uniform distribution of the injection rate. Usually, organic transport devices have a sandwich geometry (geometry of a flat capacitor): bottom electrode/organic (polymer) layer/top electrode. The bottom electrode is usually deposited on a glass substrate, then a polymer solution is cast on it and dried, and then the top electrode is thermally deposited on the polymer layer. It is

DOI: 10.1002/masy.200450818

well known that these two interfaces have different structures, the bottom one having a rather sharp transition from the metal to organic layer, and the top one having a diffusive structure with the transition layer of thickness of several nanometers [1,2]. This structural asymmetry of electrodes results in different injection properties. In this paper we focus our attention on the interface with the bottom electrode, because it can be approximately described as a rough (random) surface separating a metal (or another conducting material) and organic material. Mathematically, the rough surface may be described by setting its height $h(\rho)$ for a given position $\rho = (x,y)$. In many cases this surface is relatively smooth so that a typical height of the element is small in comparison with its size across the surface [3-7]. In this situation it is possible to calculate distribution of the electric field at the surface knowing the surface profile function $h(\rho)$. In this paper we present the result of such calculation.

Roughness of the electrode surface affects not only injection, but also the charge transport across the organic layer. In the case of ideally flat electrodes, the surface charge accumulated at the surface if a voltage is applied to the device just produces a uniform electric field E_0 in the bulk of the device. In the case of rough surface, electrostatic potential,

$$\varphi(\mathbf{r}) = -E_0 z + \delta\varphi(\mathbf{r}) \tag{1}$$

attains a random component $\delta\varphi(\mathbf{r})$, thus giving an additional contribution $\delta U(\mathbf{r}) = e\,\delta\varphi(\mathbf{r})$ to the carrier random energy. Hence, roughness of the electrode surface induces an additional energetic disorder in the bulk of organic transport layer and certainly affects charge carrier transport. We should expect (and this is indeed the case) that in contrast to the more usual case of intrinsic structural disorder, the magnitude of this additional disorder should be proportional to E_0 decaying when going away from the electrode. Thus this presents a case of spatially inhomogeneous disorder.

Organic layer between rough conducting electrodes: smooth roughness

We model electrode surfaces by two random surfaces $z = h_0(\rho)$ and $z = L + h_L(\rho)$ assuming that $<h(\rho)> = 0$ (here angular brackets denote a statistical average); L is the transport layer thickness. To calculate $\varphi(\mathbf{r})$, we have to solve a Laplace equation $\Delta\varphi(\mathbf{r}) = 0$ inside the layer taking into account the boundary conditions

$$\varphi\big|_{z=h_0(\rho)} = 0 \quad , \quad \varphi\big|_{z=L+h_L(\rho)} = V_0, \tag{2}$$

where V_0 is a voltage applied to the device. It is convenient to make a transformation to new coordinates $X = x$, $Y = y$ and $Z = L(z-h_0)/(L+h_L-h_0)$, so the boundary conditions transform to $\varphi|_{Z=0} = 0$ and $\varphi|_{Z=L} = V_0$. For relatively smooth electrodes, it is sufficient to take into account only terms up to $O(h)$. The actual solution is described elsewhere [8], here we show only the result

$$\varphi / V_0 = \frac{Z}{L} + \frac{1}{4\pi^2 L} \int d\mathbf{k}\, e^{i\mathbf{k}\rho} \left[\left(1 - \frac{Z}{L} - \frac{\sinh k(L-Z)}{\sinh kL} \right) h_0(\mathbf{k}) + \left(\frac{Z}{L} - \frac{\sinh kZ}{\sinh kL} \right) h_L(\mathbf{k}) \right], \tag{3}$$

where $h(\mathbf{k})$ is a Fourier transform of $h(\rho)$. Direct calculation of the correlation function $C(\mathbf{r}, \mathbf{r}') = \langle \delta U(\mathbf{r}) \delta U(\mathbf{r}') \rangle$ (back in the initial coordinates) gives for distances greater than the surface correlation length l

$$C(\mathbf{r},\mathbf{r}') \propto \frac{e^2 E_0^2 l^2 (z+z')}{\left[(z+z')^2 + (\rho-\rho')^2 \right]^{3/2}}, \tag{4}$$

where E_0 is the applied electric field. Thus, the magnitude of the disorder $[C(\mathbf{r}, \mathbf{r})]^{1/2}$ decays as $1/z$ for $z \gg l$. In this universal regime, the distribution of $\delta U(\mathbf{r})$ is approximately Gaussian irrespectively of the distribution of $h(\rho)$, because many surface domains with size l^2 give contributions to $\delta U(\mathbf{r})$. Typical spatial distribution of φ in the vicinity of the rough electrode is shown in Fig. 1.

The rough surface of the electrode creates a channel-like disordered structure in the distribution of electrostatic potential. This structure leads to separation of electrons and holes (electrons tend to move along ridges of the potential, and holes tend to move along valleys). Typical scale of the structure in the direction perpendicular to the electrode plane is surface correlation length l. Such separation should decrease the charge recombination rate and, hence, the emitting efficiency of light emitting devices. At the same time, the separation is favorable to solar cells.

Electric field $E(\rho)$ at the electrode surface at $Z = 0$ is

$$E(\rho) = E_0\left\{1 + \frac{1}{4\pi^2}\int d\mathbf{k}e^{i\mathbf{k}\rho}\frac{k}{\sinh kL}\left[h_0(\mathbf{k})\cosh kL - h_L(\mathbf{k})\right]\right\} \qquad (5)$$

Distribution $h(\mathbf{k})$ decays for $k \gg 1/l$, and behavior of Eq. 5 depends on parameter $\gamma = l/L$. If $\gamma \ll 1$, then

$$E(\rho)/E_0 \approx 1 + \frac{1}{4\pi^2}\int d\mathbf{k}e^{i\mathbf{k}\rho}kh_0(\mathbf{k}) = 1 + O(h/l), \qquad (6)$$

and in the opposite case, $\gamma \gg 1$

$$E(\rho)/E_0 \approx 1 + \frac{h_0(\rho) - h_L(\rho)}{L} = 1 + O(h/L). \qquad (7)$$

Equation 7 corresponds to first terms of the series expansion of equation $E = E_0 L/(L + h_L - h_0)$ which describes simple re-scaling of the electric field in the case of very smooth variation of electrode surfaces. A non-uniform spatial distribution of $E(\rho) = E_0 + \delta E(\rho)$ leads to a significant spatial variation of the injection current J which has a nonlinear dependence on E, be it a Fowler-Nordheim (FN) [9]

$$J_{FN} = J_0(E/E_{FN})^2 \exp(-E_{FN}/E) \qquad (8)$$

or a Richardson-Schottky injection (RS) [10]

$$J_{RS} = J_0 \exp[(E/E_{RS})^{1/2}] \qquad (9)$$

(here E_{FN} and E_{RS} are parameters which depend on temperature and properties of materials and metal/organic interface). In typical cases parameter E_{FN} is large [11], and parameter E_{RS} is small [12] in comparison with E_0; for this reason, even relatively small variation of $h(\rho)$ and $E(\rho)$ leads to a significant spatial variation of injection current (see Fig. 2). Small variations of $E(\rho)$ could even change completely the functional dependence of the total current $<J>$ on E_0. For example, in the case of Gaussian distribution of $h_{0,L}$, the resulting distribution of δE has a Gaussian form with zero mean and variance $<(\delta E)^2> = 2(E_0 h / l)^2$, here $h^2 = <h^2(\rho)>$.

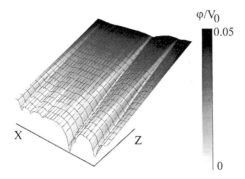

Figure 1. Spatial distribution of the potential in the vicinity of rough electrode (electrode plane is parallel to the XY plane, and the Z axis is directed to the bulk of the transport layer).

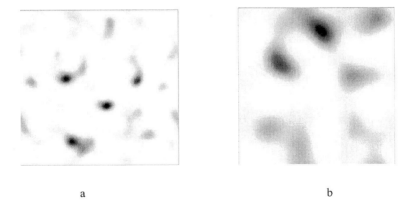

a   b

Figure 2. Spatial distribution of the FN injection rate in the case of Gaussian random surface profiles with the Gaussian surface correlation function $\Omega(\rho) = <h(\rho)h(0)> \propto \exp(-\rho^2/2l^2)$ and $h = 1$ nm: (a) $h/l = 0.125$ and (b) $h/l = 0.062$. We used $E_{FN} = 1\times10^7$ V/cm [11] and $E_0 = 1\times10^6$ V/cm. Darkness is proportional to the injection intensity. Note that in Fig. 2a the maximal intensity is approximately 70 times higher than the intensity for a flat electrode, and in Fig. 2b the maximal intensity is 14 times higher. The size of the area is approximately $0.1\mu \times 0.1\mu$.

A straightforward calculation of $\langle J \rangle$ gives

$$\ln\langle J_{FN} \rangle \propto -\left(\frac{E_{FN}}{E_0}\right)^{2/3} \lambda^{-1/3}, \qquad \lambda \frac{E_{FN}}{E_0} \gg 1, \qquad (10)$$

$$\ln\langle J_{RS} \rangle \propto \left(\frac{E_0}{E_{RS}}\right)^{2/3} \lambda^{1/3}, \qquad \lambda \sqrt{\frac{E_0}{E_{RS}}} \gg 1.$$

Here $\lambda = \langle(\delta E)^2\rangle / E_0^2 = 2(h/l)^2$.

Very rough electrodes

If $h/l \approx 1$, then the perturbation theory solution is not valid. In this case a better approach is numerical solution of the integral equation

$$\int d\rho e^{ik\rho - kh(\rho)} B(\rho) = 4\pi^2 \delta(\mathbf{k}) \qquad \text{or} \qquad \int d\rho' \frac{h(\rho')}{\left[h^2(\rho') + |\rho - \rho'|^2\right]^{3/2}} B(\rho') = 2\pi \qquad (11)$$

which describes the distribution of the charge density $B(\rho) = \sigma(\rho)/\sigma_0$ at the surface of rough electrode (we assume $l \ll L$, so the effect of the second electrode is negligible). Equation 11 was derived by the approach similar to that presented in [13]. Here $\sigma(\rho)$ is the surface charge density and σ_0 is the charge density for a flat electrode. Electric field may be calculated using equation

$$E(\rho) = \frac{E_0 B(\rho)}{\sqrt{1 + (\nabla h)^2}} \qquad (12)$$

A suitable method to solve Eq. (11) is a solution of the corresponding lattice version of the problem, which turns out to be a very large system of linear equations. Resulting spatial field distributions for the Gaussian surface are shown in Fig. 3 and distribution of the magnitude of the electric field $P(E/E_0)$ is shown in Fig. 4. For smooth surface, the electric field magnitude fluctuates around E_0 but with increasing roughness, the electric field is effectively concentrated at the tops of surface peaks while the rest of the surface carries a very weak field. The exponential tail of the distribution for large E/E_0

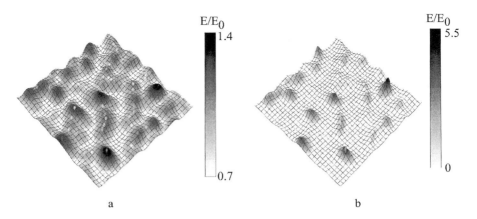

Figure 3. Spatial distribution of the electric field at the surface of rough Gaussian electrodes with different roughness (area of $16l \times 16l$ is shown): (a) $h/l = 0.075$ and (b) $h/l = 0.75$.

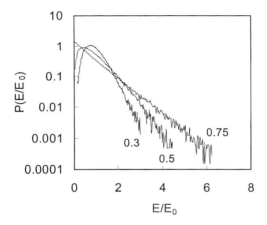

Figure 4. Distribution of the electric field magnitude at the surface of rough Gaussian electrodes with different roughness (values of h/l are indicated with corresponding curves).

$$\ln P = -bE/E_0 \qquad (13)$$

seems to be a fingerprint of the Gaussian distribution of height (preliminary simulation data suggest that the development of the exponential tail does not depend on the functional form of the surface correlation function, but the tail has a different form for a non-Gaussian height distribution). The dependence of parameter b on h/l may be approximated by $b \propto (h/l)^{-\alpha}$ with $\alpha = 1.3 \pm 0.2$ (quality of this approximation can be seen in Fig. 5). The exponential tail of the distribution leads to the dependences: $\ln\langle J_{FN}\rangle \propto -(E_{FN}/E_0)^{1/2}$ for the FN injection and $\ln\langle J_{RS}\rangle \propto E_0/E_{RS}$ for the RS injection.

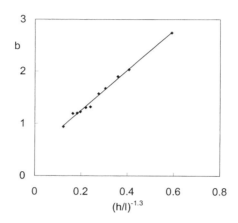

Figure 5. Dependence of parameter b on h/l.

Surface of real ITO samples

Data of AFM scans for different indium tin oxide (ITO) samples indicate that the model of Gaussian random surface may be a good approximation of the ITO electrode surfaces [14-16] (with the surface correlation function having the Gaussian form, too). Statistical properties of ITO electrodes are summarized in Table 1. It is reasonable to expect that the AFM data underestimate the actual roughness of the interface because of the ultrathin water layer effects (such layer frequently covers the sample surface), non-ideal deconvolution of the contribution of cantilever tip shape, non-ideal alignment of the organic material at the electrode surface (possible presence of the empty space), etc. For this reason the calculated values of parameter h/l should be considered as a lower bound on actual values. A comparison of Table 1 and Fig. 4 suggests that we should expect the evolution of the exponential tail of the electric field distribution at least for some of ITO electrodes. Thus, the electric field dependence of the total injection current should significantly differ from the field dependence of the proper microscopic injection rate (see Fig. 6). Quite frequently, such complicated field dependence of the injection current is attributed to the contribution of several injection mechanisms [11]. Our result indicates that it may be attributed to the roughness effect as well.

Table 1. Statistical properties of different ITO samples

ITO sample	h, nm	l, nm	h/l
Cornell group [14]	0.41	38	0.011
Minsk group [15]	1.8	18	0.1
Potsdam group, polished [16]	4.1	24	0.17
Potsdam group, unpolished [16]	4.1	14	0.3

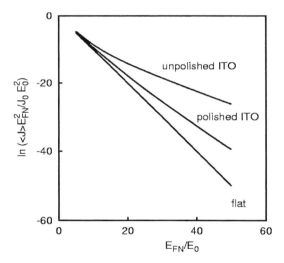

Figure 6. Calculated field dependence of the FN injection rate for the Gaussian electrode having parameters of the Potsdam ITO electrodes. We assume $E_{FN} = 1\times10^7$ V/cm [11].

Conclusion

We presented results of the calculation of the distribution of electrostatic potential in the bulk of organic layer and electric field at the surface of rough electrodes. For very rough electrodes the form of the distribution of E depends on statistical properties of the profile. In the bulk of the layer, the magnitude of the roughness-induced energetic disorder decays as the inverse distance when going away from the electrode, thus providing the case of spatially inhomogeneous disorder. Calculation of the distribution of the electric field offers an opportunity to study various aspects of charge injection.

Acknowledgments

Partial financial support from the Russian Fund for Basic Research (grant 02-03-33052) and the International Science and Technology Center (grant 2207) is acknowledged. This work was partly supported by the University of New Mexico and utilized the UNM-Alliance Roadrunner Supercluster at the Albuquerque High Performance Computing Center. I am greatly indebted to G.G. Malliaras, Y. Shen, O. Stukalov, L. Brehmer, A. Buchsteiner, and J. Stephan for providing ITO samples data and valuable discussions.

[1] M. Abkowitz, J.S. Facci, J. Rehm, *J. Appl. Phys.* **1998**, *83*, 2670.
[2] G.G. Malliaras, J.R. Salem, P.J. Brock, J.C. Scott, *Phys. Rev. B* **1998**, *58*, 3411.
[3] M. Rasigni, G. Rasigni, J.P. Palmari, A. Llebaria, *Phys. Rev. B* **1981**, *23*, 527.
[4] S.J. Fang, S. Haplepete, W. Chen, C.R. Helms, *J. Appl. Phys.* **1997**, *82*, 5891.
[5] S. Lefrant, I. Baltog, M. Lamy dela Chapelle, M. Baibarac, G. Louarn, C. Journet, P. Bernier, *Synth. Met.* **1999**, *100*, 13.
[6] H.C. Lin, J.F. Ying, T. Yamanaka, S.J. Fang, C.R. Helms, *J. Vac. Sci. Technol. A* **1997**, *15*, 790.
[7] Y.S. Kim, M.Y. Sung, Y.H. Lee, B.K. Ju, M.H. Oh, *J. Electrochem. Soc.* **1999**, *146*, 3398.
[8] S.V. Novikov, *Proc. SPIE* **2000**, *4104*, 84.
[9] R.H. Fowler, L.W. Nordheim, *Proc. R. Soc. London, Ser. A* **1928**, *119*, 173.
[10] S.M. Sze, "*Physics of Semiconductor Devices*", Wiley, New York 1981.
[11] P.S. Davis, Sh.M. Kogan, I.D. Parker, D.L. Smith, *Appl. Phys. Lett.* **1996**, *69*, 2270.
[12] A.J. Campbell, D.D.C. Bradley, J. Laubender, M. Sokolowski, *J. Appl. Phys.* **1999**, *86*, 5004.
[13] A. Garcia-Valenzuela, N.C. Bruce, D. Kouznetsov, *J. Phys. D* **1998**, *31*, 240.
[14] Y. Shen, D.E. Jacobs, G.G. Malliaras, G. Koley, M.G. Spencer, A. Ioannidis, *Adv. Mater.* **2001**, *13*, 1234.
[15] O. Stukalov, private communication.
[16] L. Brehmer, A. Buschsteiner, J. Stephan, private communication.

Macromol. Symp. **2004**, *212*, 201-208

Optical Rotatory Power of Biodegradable Polylactic Acid Films

Yoshiro Tajitsu

Department of Polymer Science and Engineering, Faculty of Engineering, Yamagata University, Yonezawa, Yamagata 992-8510, Japan
E-mail: tajitsu@yz.yamagata-u.ac.jp

Summary: Polylactic acid (PLA) is a polymer material on which biodegradability research has been the most advanced. PLA is a chiral polymer in which molecules containing asymmetric carbon atoms have a helical structure. Two optical isomers of PLA exist, PLLA (poly(L-lactic acid)) and PDLA (poly(D-lactic acid)). In this study, using various physical processes, we fabricated various samples such as oriented PLLA film, PLLA fiber, rolled PLLA film and forged PLLA plate. We observed a large optical rotatory power ρ in the cylindrical plate fabricated using a forging process. ρ of forged PLLA plates is 7200°/mm which is approximately 300 times larger than that of α-quartz.

Keywords: biodegradable polymer; chiral polymer; forging process; optical rotatory power; polylactic acid

Introduction

The amount of industrial waste is increasing every year and, as a result, environmental pollution has become more serious than ever. Biodegradable polymers have attracted much attention as they are materials that are expected to alleviate environmental pollution.[1,2] Polylactic acid (PLA) is a polymer material on which biodegradability research has been the most advanced. PLA has been used practically for medical products.[2] For this reason, the research on its mechanical properties and structure, such as its higher-order and crystal structures, has advanced.[3-6] On the other hand, PLA is a chiral polymer in which macromolecules containing asymmetric carbon atoms have a helical structure. Two optical isomers exist in PLA: poly(L-lactic acid) (PLLA) and poly(D-lactic acid) (PDLA). For a long time, chiral polymers have been expected to exhibit optical rotary power, ρ, in their solid state, compared with inorganic low-molecular-weight crystals. A typical substance with large

DOI: 10.1002/masy.200450819

ρ is α-quartz (α-SiO$_2$). The definition of ρ is as follows. The vibration plane of the electric field E of linearly polarized light (LP) revolves clockwise or counterclockwise when LP passes through the substance. This is the mechanism of optical activity. In this case, ρ is defined as the ratio of the rotation angle of the vibration plane of E of LP to unit length. ρ of α-quartz is due to the helical structure formed by molecules. Therefore, it is believed that ρ of a chiral polymer is large. Actually, although Kobayashi et al.[7] revealed huge intrinsic ρ of the PLLA crystal, a PLLA film with large ρ in the fiber axis direction has not been prepared for practical use until now. No PLA film with large ρ has been prepared to date, because it was very difficult to control its higher-order structure.

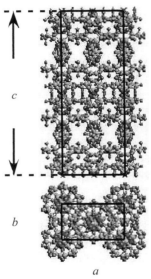

Figure 1. PLLA crystal structure projected onto ($\bar{1}$10) and (001).

Figure 1 shows the PLLA crystal structure [1-7], based on a base-centered orthorhombic unit cell. It contains two 10/3 helical chains arranged along the c-axis; the point group is D$_2$. a, b and c are the lengths of the unit cell. Here we emphasize that PLLA is a chiral polymer and, in the crystal, the polymer molecules form the helical structure. The crystal structure of PLLA is characterized by this helical structure, which gives rise to ρ.

Mathematical representation

In order to analyze the optical activity, we derive the mathematical representation.[8,9] The dielectric displacement D in crystals is induced by E of the LP with wavenumber vector k (k_1, k_2, k_3). It is represented as

$$\left[D \right]_m = \sum_{l=1}^{3} \varepsilon_{ml} E_l + j \left[G \times E \right]_m \quad (m = 1 - 3) \tag{1}$$

with

$$\left[G \right]_m = g_{m1} k_1 + g_{m2} k_2 + g_{m3} k_3 . \tag{2}$$

Here, G is the gyration vector. ε_{ml} is the permittivity tensor and g_{ml} is the gyration tensor. As shown by the first term of Eq. (1), E causes D to occur in one direction. On the contrary, according to the second term, under the existence of G, D with a phase lag of 1/4 period relative to E exists. Then the direction of D becomes perpendicular to the direction of E. This is the reason why the rotation of LP occurs in crystals. In this study, we evaluate g_{ml} of PLLA films.

$$\rho = \frac{\pi G}{\lambda n_0} \tag{3}$$

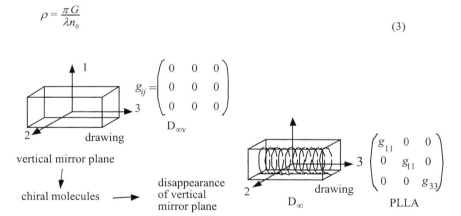

Figure 2. Macrosymmetry change due to physical processing, such as stretching, of polymer samples.

with

$$G = G \cdot k . \tag{4}$$

Equation (3) shows a simple relationship between ρ and G, which was used for evaluation of ρ in this report. This discussion concerns single crystals; next we consider the optical acitivity of a PLLA film.

Optical rotatory power of polymer film

Polymer films of 100% crystallinity cannot be obtained by conventional methods. Amorphous components are always present in complex higher-order structures. Furthermore, no one-to-one correspondence is found between the macrooptical properties and crystal characteristics. For crystalline polymer film, we must consider macroscopic symmetry based on the point group theory, as shown in Fig. 2.[9,10] ρ of an isotropic film

does not exist, even though g_{ml} is present in the crystal state. Also, the point group of a drawn polymer film is $D_{\infty v}$. Although optical activity does not arise, a PLLA molecule has chirality. The mirror plane disappears and the point group becomes D_{∞}. In the case of PLLA, the crystal symmetry differs from the macroscopic symmetry. ρ of the oriented PLLA film used here is based on g_{11} and g_{33}.

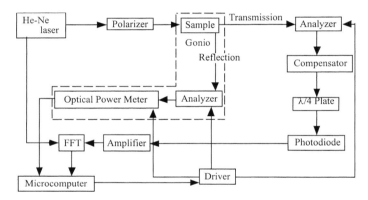

Figure 3. Schematic diagram of the experimental setup for measuring the optical activity.

Measurement System

We developed a new experimental system for measuring the optical activity of polymer film, as shown in Fig. 3.[11] LP is incident on the film and then becomes elliptically polarized due to the dichroism of the polymer film. The elliptic axis rotates around the optical axis (fiber axis) due to ρ. The vector operation of the Stokes parameters with the optical arrangement can be represented as [11-13]

$$S' = P_\phi \cdot R_{\delta,\chi} \cdot T_\theta \cdot P_0 \cdot S \ . \tag{5}$$

Here, S is the Stokes parameter of LP from the He-Ne laser. The terms $P_{\delta,\chi}$ and T_θ are

Mueller matrixes of a linear phase shifter and the azimuth rotator equivalent to that of a polymer film, respectively. The rotation angle θ of the elliptic axis is caused by ρ. The apparent retardation δ of the linear birefringence is due to the dichroism of the polymer film. In practice, we obtain θ and δ using the least-squares method. From these θ and δ, we calculate the gyration tensor g_{ij}.

Sample preparation

We prepared PLLA with a molecular weight of 100,000 ~ 600,000 and D-isomer content of 0.001 ~ 1.000 %. Then, using physical processes, we fabricated various samples such as stretched film, fiber, rolled film and a forged cylindrical plate.[10] To enable measurement, we cut each sample perpendicular to its length using a microtome and polished it. The forging process for metals, in which metal is struck into a die with a hammer, is well known. It is difficult to apply a forging process to polymers, because polymers have poor ductility characteristics. As a result of many considerations of issues such as how to apply pressure, the temperature and the die shape, we were able to fabricate a PLLA forged sample.[14]

Optical rotatory power of PLLA and PDLA samples

Using our measurement system, we obtained ρ of PLLA and PDLA samples. We summarize typical experimental results in Table 1.

Rolled PLLA film (hot rolling) and stretched PLLA film

A heated PLLA sheet was repeatedly passed through a pair of counter-rotating rollers and rolled to a fixed size and shape. The PLLA films obtained here were rolled to thicknesses of 100 μm to 1 mm.

Table 1. Optical rotatory power of various materials (wavenumber: 632.8 nm).

Material	Optical rotatory power (∞/mm)
Stretched PLLA sample	9.5
Rolled PLLA sample	4.5
Fiber PLLA sample	20.2
Forged PLLA sample	7200
Forged PDLA sample	-7110
α-Quartz (α-SiO$_2$)[7]	25
AgGaS$_2$[7]	720
α-HgS[7]	-300

The value of ρ was less than 10°/mm. ρ of the rolled PLLA film increases with increasing temperature of the heating gate and rollers. We performed uni- and biaxial stretching of PLLA films, but the value of ρ was less than 10°/mm even though the conditions during drawing, such as drawing ratio from 4 to 7 and drawing temperature from 70 to 120 °C, were controlled. The value of ρ of less than 10°/mm was almost the same as that of rolled PLLA film.

PLLA fiber

We tried, while spinning PLLA into fibers, to control the conditions of spinning. The value of ρ reached over 20°/mm. These are highly significant results. We also found that ρ increased with increasing drawing ratio. The maximum ρ value of rolled and stretched film was less than 20°/mm, while that of fiber reached 20°/mm. Here, we emphasize that ρ of 20°/mm is almost the same as that for α-quartz which is known to exhibit high ρ. There is a high probability that ρ of a fiber sample fabricated under optimum conditions will increase.

Giant optical rotatory power of forged PLLA and PDLA samples

ρ of the forged PLLA sample was 300 times that of α-quartz. This is an outstanding value.

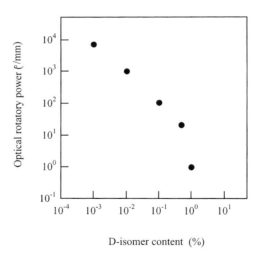

D-isomer content (%)

Figure 4. The D-isomer content dependence of optical rotatory power of forged PLLA sample.

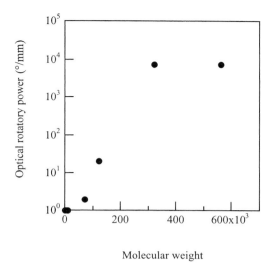

Figure 5. The molecular weight dependence of optical rotatory power of forged PLLA sample.

On the other hand, PDLA films possess dextrorotatory power. This is consistent with the right-handed rotation of helical chain molecules of PDLA. Other important findings are as follows. The factors affecting the value of ρ are the D-isomer content and molecular weight. Figures 4 and 5 show the plots of ρ against D-isomer content and molecular weight. ρ of PLLA increases with increasing molecular weight and decreasing D-isomer content. The higher-order structure of PLLA films is greatly influenced by the D-isomer content and molecular weight. Thus, the unprecedented giant ρ of the PLLA sample is due to its higher-order structure.

Conclusion

We fabricated various PLLA film and fiber samples and developed a new experimental system for measuring the optical activity of polymer films. Using the new system, we measured ρ of PLLA samples. As a result, in the film fabricated using the new physical process, we observed giant ρ. The findings from this study are summarized as follows.

(1) PLLA and PDLA films possess levorotatory power and dextrorotatory power, respectively. ρ of PLLA increases with increasing molecular weight and decreasing D-isomer content. The

208

higher-order structure of PLLA films is greatly influenced by the D-isomer content and molecular weight. Thus, the unprecedented giant ρ of the PLLA sample is due to its higher-order structure.

(2) The maximum ρ value of a rolled and stretched of PLLA film is less than 20°/mm, while that of a PLLA fiber reaches 20°/mm which is almost the same as that for α-quartz. We believe that there is a strong probability that ρ of a PLLA fiber sample fabricated under optimum conditions will increase.

(3) ρ of forged PLLA film is 7200°/mm which is approximately 300 times larger than that of α-quartz. This is an outstanding value.

Acknowledgments

We are grateful to Professor E. Fukada of the Kobayashi Institute of Physical Research and Professor T. Masuko of the Yamagata University for useful comments. We thank Dr. Shikinami of Takiron Co., Ltd., for providing forged PLLA samples. We also thank M. Matsuo, M. Sukegawa and H. Shinoda of Mitsui Chemical, Co., Ltd., for providing pigment powders and PLLA samples, and T. Nakamura and T. Natsui of ThreeBond Co., Ltd., for providing epoxy resins. We thank H. Kowa and K. Muraki of Uniopt Co., Ltd., for technical assistance in the birefringence measurement. This work was supported in part by Grants-in-Aid for Scientific Research (Nos.13650944 and 14022205) from the Ministry of Education, Culture, Sports, Science and Technology, Japan.

[1] G. Scott, in: *Biodegradable Plastics and Polymers*, Elsevier, London 1994, p.79.
[2] C. Gebelein, C. Carraher, in: *Biotechnology and Bioactive Polymers*, Plenum Press, New York 1994, p.103.
[3] E. Fukada, H. Nalwa, in: *Ferroelectric Polymers*, Marcel Dekker, New York 1995, p.393.
[4] P. De Santis, A. J. Kovacs, *Biopolymers* **1968**, 6, 299.
[5] W. Hoogsteen, A. R. Postema, G. Brinke, P. Zugenmaier, *Macromolecules* **1990**, 23, 634.
[6] T. Miyata, T. Masuko, *Polymer* **1997**, 38, 4003.
[7] J. Kobayashi, T. Asahi, M. Ichiki, H. Suzuki, T. Watanabe, E. Fukada, Y. Shikinami, *J. Appl. Phys.* **1995**, 77, 2957.
[8] E. Wahlstrom, in: *Optical Crystallography*, John Wiley & Sons, New York 1969, p.121.
[9] J. F. Nye, in: *Physical Properties of Crystals*, Clarendon Press, Oxford 1985, p.260.
[10] E. Fukada, *IEEE Trans. Ultrasonics, Ferroelectrics and Frequency Control* **2000**, 47, 1277.
[11] Y. Tajitsu, M. Aoki, Y. Kamei, R. Nishina, H. Suzuki, *Ferroelectrics* **1998**, 218, 103.
[12] M. Born, E. Wolf, in: *Principles of Optics*, Pergamon Press, Oxford 1975, p.518.
[13] E. Hecht, in: *Optics*, Addison-Wesley, Reading 1990, p.292.
[14] Y. Shikinami, M. Okuno, *Biomaterials* **1999**, 20, 859.

Macromol. Symp. **2004**, *212*, 209-217

Features of Charge Carrier Concentration and Mobility in π-Conjugated Polymers

Gytis Juška,[1] *Kristijonas Genevičius,*[1] *Kęstutis Arlauskas,**[1] *Ronald Österbacka,*[2] *Henrik Stubb*[2]

[1]Vilnius University, Saulėtekio 9, bd. 3, LT-2040 Vilnius, Lithuania
[2]Åbo Academy University, Porthansgatan 3, FIN-20500 Turku, Finland

Summary: We show the possibilities of experimental investigation of charge carrier mobility and concentration features by extraction methods of equilibrium, photoexcited and injected charge carriers in π-conjugated polymers, where, due to relatively high conductivity, the classic time-of-flight method is inappropriate.

Keywords: charge carrier extraction; conjugated polymers; hole concentration; hole mobility; thin films

Introduction

π-Conjugated polymers are used extensively in light-emitting, field-effect and other molecular electronic devices. The relatively low charge carrier mobility of these polymers causes that the conventional time-of-flight (TOF) method is used for investigation of charge carrier transport features. However, this method can be applied if Eq.1 is fulfilled

$$t_{tr} = \frac{d^2}{\mu U} << \tau_\sigma = \varepsilon\varepsilon_0 / \sigma \qquad (1)$$

where μ is charge carrier mobility, t_{tr} is duration of charge carrier drift through the interelectrode distance d, U is voltage applied on sample electrodes, τ_σ is dielectric relaxation time, ε is dielectric permittivity, and σ is the bulk conductivity of sample. The latter condition means that in the bulk of sample the amount of equilibrium charge ($e p_0 d$) must be less than the amount of charge on sample electrode ($\varepsilon\varepsilon_0 U / d$). In opposite case, after voltage has been applied on sample electrodes, the amount of equilibrium charge is sufficient that the electric field inside of sample, during a time shorter than the charge carrier transit time, will be redistributed. In relatively high-conductivity polymers, the condition that $t_{tr} << \tau_\sigma$ can be achieved by reducing thickness of samples, i.e. by measuring thin layers. However, in such a case, the absorption depth of even UV light is comparable to the

DOI: 10.1002/masy.200450820

thickness of sample. Thus, the bulk charge carrier photogeneration prevails. Moreover, the high capacitance of thin layers causes that only the integral mode of TOF photocurrent transients may be used for investigations. Compared with the current mode in the case of surface charge carrier photogeneration, the shape of photocurrent transient of integral TOF mode is not so expressive, causing very often erroneous interpretation.

In this paper we would like to demonstrate the possibilities of investigation of π-conjugated polymers using charge carrier extraction by the linearly increasing voltage (CELIV) method, for which condition (1) is not necessary, and to discuss errors, which may appear using TOF method for investigation of charge carrier mobility when this condition is ignored.

Polymers and equipment

The π-conjugated polymers used were: poly(*p*-phenylenevinylene) (PPV), regioregular poly(3-hexylthiophene) (RRPHT), regiorandom poly(3-hexylthiophene) (RRaPHT), regioregular poly(3-octylthiophene) (RRPOT; Sigma Aldrich), regioregular poly(3-dodecylthiophene) (RRPDDT), and regioregular poly(3-alkylthiophene) (RRPAT). All layers have been prepared by spin coating or solution casting. The polymer was dissolved in chloroform in a concentration of 10 mg/ml. The solutions were filtered through a 0.20 μm filter before the solution was cast onto pre-patterned ITO-covered glass substrates (Planar International). Finally a 30-nm (semitransparent) aluminium top electrode was evaporated under a pressure of 1 Pa. The films were made in air, but stored and measured in a closed-cycle cryostat (Oxford CCC1104) under vacuum to make temperature measurements possible.

The experimental setup for the CELIV measurement was made up; a variable pulse generator (Stanford DS345) and a memory oscilloscope (Tektronix TDS680B) were used to record the extraction currents. The only restriction imposed by the experimental setup is that at least one contact of the sample should be (partially) blocking. The voltage rise speed, A, used in these experiments, ranged between 10 and 10^6 V/s. For the TOF measurements, a nitrogen laser (Oriel), with a pulse width of 7 ns, energy 3.55 eV and energy per pulse of 500 μJ was used together with the pulse generator and a delay function generator (Stanford DG 535) to ensure a proper delay time between voltage and light pulses. TOF was measured with 50 Ω input impedance on the oscilloscope.

Extraction of equilibrium charge carriers

The linearly increasing voltage pulse applied on sandwich-type sample electrodes, one of which is blocking, raises a current transient, the shape of which is presented in Fig. 1.

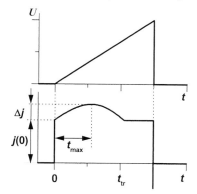

Fig. 1. Schematic view of CELIV method.

From the very initial step of this current $j(0)$, the interelectrode distance d or the dielectric permittivity $\varepsilon\varepsilon_0$ may be estimated:

$$\frac{\varepsilon\varepsilon_0}{d} = \frac{j(0)}{A}.$$ (2)

From t_{max} and Δj, the drift mobility is calculated

$$\mu = \frac{2d^2}{3At_{max}^2}, \quad \text{when } \Delta j \leq j(0), \text{ i.e. } t_{tr} \leq \tau_\sigma,$$ (3)

$$\mu = \frac{\tau_\sigma d^2}{At_{max}^3}, \quad \text{when } \Delta j \gg j(0), \text{ i.e. } t_{tr} \gg \tau_\sigma.$$ (4)

The bulk conductivity can be estimated as

$$\sigma = \frac{3\varepsilon\varepsilon_0}{2t_{max}} \cdot \frac{\Delta j}{j(0)},$$ (5)

and, from calculated μ and σ, the equilibrium charge carrier density p_0 can be found.[1] The shortage of this method is that the above mentioned parameters are measured in electric field varying in time. However, due to the fact that major extraction of charge carriers occurs at t_{max}, the estimated parameters will correspond to the values when electric field is $F = A\, t_{max}/$

d. In Fig. 2, typical μ, σ, p_0 dependences on A show that charge carrier concentration is independent of A and thus on electric field. The p_0 dependences on temperature demonstrate a similar result (Fig. 3). Hence, the mobility dependence on temperature causes activation characteristic of conductivity.

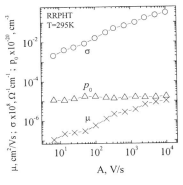 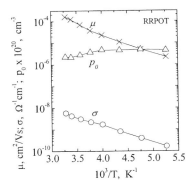

Fig. 2. Dependences of σ, μ, and p_0 on A.   Fig. 3. Dependences of σ, μ, and p_0 on T.

The numerical modelling results in the case of Gaussian distribution of localised states demonstrated regularities of CELIV parameters, which allow, from experimentally measured $\Delta j \sim A^\beta$ and $t_{max} \sim A^\gamma$ power-law dependences, to evaluate the nature of mobility dependence on electric field.[2] If $\mu(F)$ is caused by stochastic transport, then $(\beta - \gamma) = 1$ and $(\beta + \gamma) < 0$ must be valid (see Fig. 4).

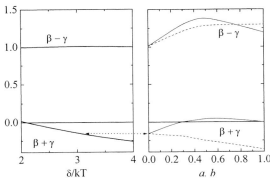

Fig. 4. Numerical modelling of $(\beta + \gamma)$ and $(\beta - \gamma)$ as a function of dispersion parameter δ/kT and of Pool-Frenkel parameters a for mobility (dashed line) and b for retrapping time (full line) using $\delta/kT = 3$.

If micromobility (μ_0) depends on electric field according to the Poole-Frenkel law ($\mu_0 \sim \exp(a\sqrt{F})$) typical of organic polymers then $(\beta - \gamma) > 1$, $(\beta + \gamma) < 1$ and decreases when a increases. Experimentally, a increases when the temperature is decreasing.

When the retrapping probability from localised states depends on electric field in a similar way ($\tau_R^{-1} \sim \exp(b\sqrt{F})$), then $(\beta - \gamma) > 1$ and $(\beta + \gamma)$ increases with b. Experimental investigation results of PPV, RRPHT, RRPOT gave that $(\beta - \gamma) > 1$ and $(\beta + \gamma) > 0$, which demonstrates that the mobility and, herewith, conductivity dependence on electric field is caused by electric field-induced release from the localised states.[2,3]

For low A, i.e. under a weak electric field condition, $\Delta j > j(0)$ has been obtained for a majority of π-conjugated polymers. The latter condition means that $\tau_\sigma \ll t_{tr}$, hence, the main necessary TOF condition is not fulfilled. According to Ref. [4], when the voltage on sample electrodes, $U < U_c = ep_0 d^2/2\varepsilon\varepsilon_0$ (U_c is critical voltage corresponding to the condition $t_{tr} = 2\tau_\sigma$), after equilibrium charge carrier extraction from the depletion depth $l < d$, the drift of charge through the depletion region is obtained and the current is

$$j(t) = Q_0 \sqrt{\frac{2}{t_{tr}\tau_\sigma}} \cdot \exp\left(-\frac{t}{\tau_\sigma}\right). \tag{6}$$

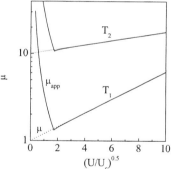

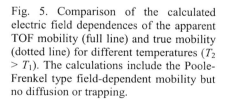

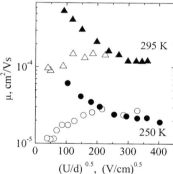

Fig. 5. Comparison of the calculated electric field dependences of the apparent TOF mobility (full line) and true mobility (dotted line) for different temperatures ($T_2 > T_1$). The calculations include the Poole-Frenkel type field-dependent mobility but no diffusion or trapping.

Fig. 6. Electric field dependence of drift mobility in RRPOT measured using TOF (solid dots) and CELIV (open dots) at 295 K and 250 K.

In $\log j \sim \log t$ plot the shape of $j(t)$ transient looks similar to the dispersion-type small charge drift transient. However, the apparent transit time through depletion region is independent of voltage on electrodes, and the calculation using this transit time gives erroneous mobility values, $\mu_{app} = 2U_c/U\ln 2$ (see Fig. 5). In Fig. 6, the mobility dependences on electric field, measured using TOF and CELIV (of the same RRPOT layer), are presented, demonstrating that the apparent mobility (estimated by TOF) decrease with electric field is caused by electric field redistribution due to extraction of equilibrium charge carriers. Since the equilibrium charge carrier concentration is independent of temperature, the value of critical voltage is independent of temperature too, as it is seen in Fig. 6. To evaluate the critical electric field from the conductivity values, it is obvious that not only in RRPOT, but also in PPV, RRPAT [5-8], the measured mobility decrease with electric field is erroneous.

Extraction of photoexcited charge carriers

In CELIV measurement, illumination of sample by continuous light of bulk absorption induces extraction of photoexcited charge carriers (photo-CELIV), similarly to equilibrium charge extraction. More wide-ranging possibilities may be obtained if the pulse of linearly increasing voltage is applied delayed by some time t_d after sample has been illuminated by short pulse of light. From the CELIV and photo-CELIV current transients, the dependences of charge carrier mobility and concentration on delay time can be investigated. A schematic view of the photo-CELIV method is shown in Fig. 7.

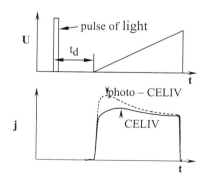

Fig. 7. Schematic view of photo-CELIV.

Experimentally measured hyperbolic mobility (Fig. 8) and photoexcited charge carrier concentration (Fig. 9) dependences on t_d indicate the dispersive character of transport. The drift mobility decrease with increasing t_d may be caused by charge carrier trapping to deeper localised states in the case of dispersive transport. Another reason for such dependence may be the drift mobility dependence on t_d caused by charge carrier density, which decreases when t_d increases. The tentative experimental results point to the latter case. From concentration dependence on t_d (see Fig. 9), the time $\tau_{1/2}$ has been estimated, which corresponds to the time period when the initial amount of photoexcited charge carriers decreases by half. Different $\tau_{1/2}$ values have been obtained for different polymers: 10^{-5} s for RRPOT, and 10^{-2} s for RRPDDT. The increasing of $\tau_{1/2}$ when decreasing temperature indicates the trapping influence on recombination.

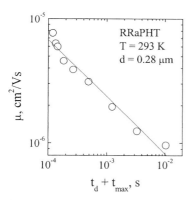

Fig. 8. Dependence of photoexcited charge carrier mobility on delay time.

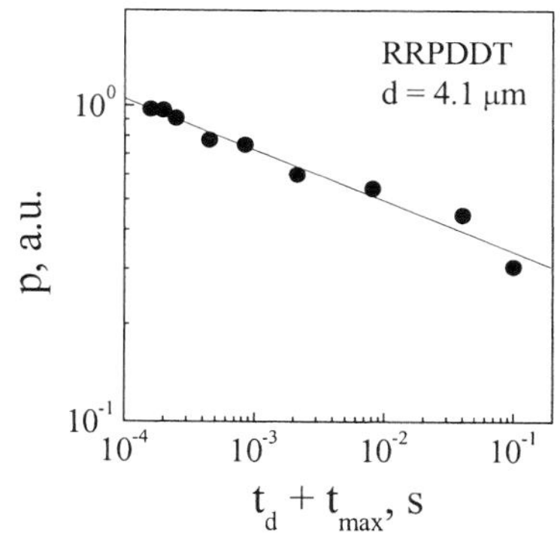

Fig. 9. Dependence of photoexcited charge carrier concentration on delay time.

Extraction of injected charge carriers

In Fig.10, a schematic view of current when the saw-type voltage is applied on the sample electrodes (injection-CELIV) is shown. In the beginning the increasing voltage (region I) in the charge carrier extraction direction is applied, and the capacitance current transient is obtained only. When the decreasing voltage changes its polarity (region III), the charge carrier injection begins. After the voltage changes direction again (region IV), the extraction of injected charge carriers is clearly seen. The dependences of amounts of injected Q_i and extracted Q_e charge carriers on voltage for constant triangle pulse duration are shown in Fig. 11.

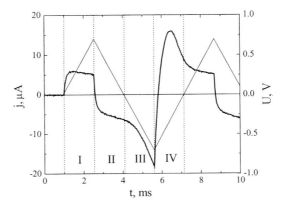

Fig. 10. Schematic view of injection-CELIV. Current transient – bold line.
Voltage – thin line.

These results are compared with charge Q_0 of geometric capacitance of sample. The two experimental results indicate that the space charge limited current condition is fulfilled. First result is that the Q_i depends linearly on voltage, and the second is that in region IV, after the voltage changes direction while polarity remains the same, the current changes its sign (Fig. 10). The equality $Q_i = Q_e$ shows that charge carrier lifetime is longer than duration of extracting pulse t_p (see Fig. 11), and, during injection, charge collects at the blocking electrode. For $t_p \gg t_{tr}$, $Q_e/Q_0 = t_p/3t_{tr}$ is valid, and thus the charge carrier mobility can be evaluated.

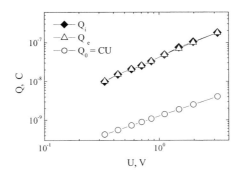

Fig. 11. Dependences of injected (Q_i), extracted (Q_e) charge, and charge of geometric capacitance (Q_0) on triangle-pulse voltage U.

Conclusions

In this paper the possibilities of investigation of features of equilibrium and photoexcited charge carrier transport in π-conjugated polymers of relatively high conductivity, using CELIV methods, have been demonstrated. The following experimental results have been obtained:

- The conductivity dependences on temperature and electric field are caused by drift mobility;
- The dependence of drift mobility on electric field is caused by electric-field release of charge carriers from trapping states;
- In low-electric-field region measured by TOF, a decrease in apparent mobility with increasing electric field is caused by the electric field redistribution;
- A decrease in drift mobility and concentration of photoexcited charge carriers when delay time after photoexcitation increases has been observed, pointing to the influence of trapping in energetically distributed localised states;
- The charge carrier transit time together with the mobility can be evaluated investigating the extraction of injected charge in low-conductivity and not photosensitive materials.

Acknowledgments

We thank Planar International for the ITO substrates. Financial support from the Academy of Finland Grant Nos. 48853 and 5075, Technology Development Centre in Finland (Tekes), and Lithuanian State Science and Studies Foundation Grant No. P-17/01 are acknowledged.

[1] G. Juška, K. Arlauskas, M. Viliūnas, J. Kočka, *Phys. Rev. Lett.* **2000**, *84*, 4946.
[2] G. Juška, K. Arlauskas, M. Viliūnas, K. Genevičius, R. Österbacka, H. Stubb, *Phys, Rev.* B **2000**, *62*, R16235.
[3] K. Genevičius, R. Österbacka, G. Juška, K. Arlauskas, H. Stubb, *Thin Solid Films* **2002**, *403-404*, 414.
[4] G. Juška, K. Genevičius, K. Arlauskas, R. Österbacka, H. Stubb, *Phys. Rev.* B **2002**, *65*, 233208.
[5] S. S. Pandey, W. Takashima, T. Endo, M. Rikukawa, K. Kaneto, *Synth. Met.* **2001**, *121*, 1561.
[6] S. Nagamatsu, S. S. Pandey, W. Takashima, T. Endo, M. Rikukawa, K. Kaneto, *Synth. Met.* **2001**, *121*, 1563.
[7] K. Kaneto, K. Hatae, S. Nagamatsu, W. Takashima, S. S. Pandey, T. Endo, and M. Rikukawa, *Jpn. J. Appl. Phys.* **1999**, *38*, L1188.
[8] S. S. Pandey, W. Takashima, S. Nagamatsu, T. Endo, M. Rikukawa, and K. Kaneto, *Jpn. J. Appl. Phys.* **2000**, *39*, L94.

Macromol. Symp. **2004**, *212*, 219-224

Infrared and Raman Studies of the Charge Ordering in the Organic Semiconductor κ-[Et₄N][Co(CN)₆(ET)₄]·3H₂O

Roman Świetlik,[*1,3] *Lahcène Ouahab,*[2] *Joseph Guillevic,*[2] *Kyuya Yakushi*[3]

[1] Institute of Molecular Physics, Polish Academy of Sciences, ul. M. Smoluchowskiego 17, 60-179 Poznań, Poland
[2] LCSIM, UMR 6511 - CNRS, Université de Rennes I, 35042 Rennes Cedex, France
[3] Institute for Molecular Science, Myodaiji, Okazaki, 444-8585, Japan

Summary: Polarised infrared reflectance and Raman spectra of the charge transfer salt κ-[Et₄N][Co(CN)₆ (ET)₄]·3H₂O were measured as a function of temperature. The salt undergoes a phase transition at 150 K, which is related to a charge ordering inside conducting ET layers. The charge ordering has a considerable influence on vibrational as well as electronic spectra. New vibrational bands related to ET⁺ cations are seen below 150 K. Moreover, formation of a new energy gap (charge gap) in electronic excitation spectrum is observed.

Keywords: charge transfer; infrared spectroscopy; Raman spectroscopy; transitions

Introduction

Three isostructural charge transfer salts yielded by the organic donor bis(ethylenedithio)tetrathia-fulvalene (BEDT-TTF, ET) with diamagnetic (M = Co^{III}) or paramagnetic (M = Fe^{III}, Cr^{III}) hexacyanometalate trianions [M(CN)₆]$^{3-}$ exhibit very similar semiconducting properties and undergo two phase transitions at 240 K and 150 K.[1,2] A specific feature of their crystal structure is that ET molecules in conducting layers are arranged in nearly perpendicular dimers, in a characteristic manner of the so-called κ-phase structure. At room temperature, two kinds of ET dimers can be distinguished (A and B), whereas below 240 K the unit cell is doubled and there exist four kinds of dimers (A, B, C and D). The charge distribution inside ET layers was studied in the Fe^{III} salt by comparison of the bond lengths of ET donors determined from the X-ray structural data. From these data it results that at room temperature all ET molecules possess a charge +0.5 (both in dimers A and B), but below 240 K a charge redistribution takes place leading to the following charge pattern: A(ET^0)₂ B(ET$^{+0.5}$)₂ C(ET$^{+0.5}$)₂ D(ET^{+1})₂. The phase

© 2004 International Union of Pure and Applied Chemistry

DOI: 10.1002/masy.200450821

transition at 150 K has no distinguishable influence on the crystal structure, suggesting a minor role of structural change at this temperature.

Recently, the phase transitions and in particular charge ordering phenomena in Fe^{III} salts were studied by IR and Raman spectroscopy.[3,4] The existence of charge redistribution inside ET layers was unambiguously confirmed by the Raman experiment on powdered crystals dispersed in KBr. The Raman data showed that the charge redistribution is not an abrupt process but develops gradually below 240 K.[3] Subsequently, the IR experiment on Fe^{III} single crystals provided an evidence that significant changes of electronic and vibrational spectral features, related to charge ordering, occur below 150 K.[4] Here we report on IR and Raman investigations of the phase transition at 150 K in single crystals of the Co^{III} salt.

Experimental

Single crystals of the salt κ-[Et$_4$N][Co(CN)$_6$(ET)$_4$]·3H$_2$O were prepared by an electrochemical method, as described elsewhere.[1] The Et$_4$N$^+$ cations incorporated in crystal lattice show a great disorder, which is not determined by X-ray experiment even at low temperatures. Polarised reflectance spectra were recorded from the best-developed crystal face (001) in the frequency range (600 – 10000 cm^{-1}) using FT-IR Nicolet Magna 760 spectrometer equipped with an IR microscope. Single-crystal Raman spectra were measured within the region of C=C stretching vibrations (1200 – 1700 cm^{-1}) with NIR excitation (λ = 785 nm) using Renishaw Ramascope System 1000. The experiments were carried out in vacuum cryostats (T = 40 - 300 K). To avoid problems with sample cracking, due to loss of water molecules from crystals under vacuum, the sample temperature was reduced down to about 270 K before cryostat evacuation.

Results and discussion

The phase transition at 150 K has a drastic influence on Raman spectrum of the κ-[Et$_4$N][Co(CN)$_6$(ET)$_4$]·3H$_2$O salt within the region of C=C stretching bands (Figure 1). At room temperature we observe two bands at about 1495 and 1465 cm^{-1} which can be related to $\nu_2(a_g)$ and $\nu_3(a_g)$ modes of ET, respectively. Positions of these bands in Raman spectrum are strongly dependent on the degree of ionisation of ET and they are often used for determination of charge residing on ET (for neutral ET^0, $\nu_2(a_g)$ = 1552 cm^{-1} and $\nu_3(a_g)$ = 1493 cm^{-1}; for ET$^+$ cation,

$v_2(a_g) = 1455$ cm^{-1} and $v_3(a_g) = 1431$ cm^{-1}).[5] As shown recently, the mode $v_2(a_g)$ is even more valuable for this purpose than $v_3(a_g)$ because of weaker coupling with electrons, which yields an additional shift towards lower frequency.[6] Using this method we estimate that the frequencies 1495 and 1465 cm^{-1} correspond to the molecules with charges +0.58 and +0.45, respectively. The average charge on ET is close to the value +0.5 estimated from the stoichiometry and crystallographic data. Below 150 K, due to charge ordering, additional strong bands appear in the spectrum at frequencies 1452, 1427, 1417 and 1407 cm^{-1}. The bands at 1452 and 1427 cm^{-1} are assigned to $v_2(a_g)$ and $v_3(a_g)$ modes of ET$^+$ cations, respectively. Their positions are very close to the frequencies of Raman bands of ET$^+$ cations reported elsewhere.[5] The bands at 1417 and 1407 cm^{-1} can be also attributed to the $v_3(a_g)$ mode of ET molecules with slightly higher oxidation state than +1 or of the molecules in different environment. An important result is that in the studied Co^{III} salt the charge ordering is not a gradual process as in the Fe^{III} salt but proceeds abruptly at 150 K.

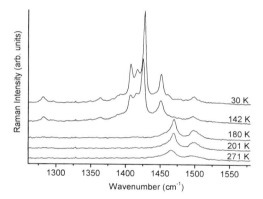

Figure 1. Raman spectra of a κ-[Et$_4$N][Co(CN)$_6$(ET)$_4$]·3H$_2$O single crystal as a function of temperature (NIR excitation: λ = 758 nm).

The IR reflectance spectra were measured for two perpendicular orientations of the electrical vector of polarised light corresponding to the maximum ($E_\parallel$) and minimum ($E_\perp$) of reflected energy. For polarisation $E_\parallel$, the electrical vector of polarised light was parallel to the direction [110], which is the direction of strongest side-by-side S⋯S contacts between neighbouring ET molecules. The reflectance spectra at different temperatures are displayed in Figure 2. Optical conductivity spectra were determined by Kramers-Kronig analysis of the reflectance data (Figure 3). The phase transition at 150 K has a considerable influence on electronic part of the reflectance spectrum. The most striking feature is the appearance of a new electronic band centred at about 7000 cm^{-1} for $E_{\parallel}$ and 7200 cm^{-1} for $E_{\perp}$. This band is assigned to charge transfer mechanism $ET_I^+ + ET_{II}^+ = ET_I^0 + ET_{II}^{2+}$ and provides an evidence of formation of ET^+ cations below 150 K. In a compound containing quasi-isolated dimers $(ET^+)_2$, this charge-transfer transition was observed at 7000 cm^{-1}.[7] Simultaneously, we observe also important modifications of the band related to charge transfer $ET_I^+ + ET_{II}^0 = ET_I^0 + ET_{II}^+$ centred at 2200 cm^{-1} for $E_{\parallel}$ and 3100 cm^{-1} for $E_{\perp}$ (at about 160 K). Below 150 K, the intensity of this band decreases and its maximum shifts towards higher frequencies. This shift is a consequence of opening of a new energy gap in electronic excitation spectrum (charge gap). By fitting to experimental data, the reflectance calculated with a Drude-Lorentz dielectric function (procedure commonly used for semiconductors), the energy gaps below and above 150 K were estimated: 0.27 eV (at 160 K) and 0.30 eV (at 120 K) for polarisation $E_{\parallel}$; 0.40 eV (at 160 K) and 0.48 eV (at 120 K) for polarisation $E_{\perp}$.

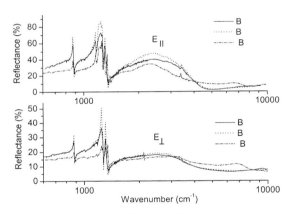

Figure 2. Temperature dependence of the polarised reflectance spectra of the κ-[Et₄N][Co(CN)₆ (ET)₄]·3H₂O crystal for two perpendicular polarisations corresponding to the maximum ($E_{\parallel}$) and minimum ($E_{\perp}$) of reflected energy (note the logarithmic frequency scale).

In vibrational part of the IR spectrum, we find several bands related to the totally symmetric C=C and C-S stretching vibrations of ET molecule activated by strong coupling with the charge transfer. These bands undergo also significant changes due to the phase transition at 150 K. The most important modification is that below 150 K we observe new bands assigned to the C=C modes of ET^+ cations: $v_2(a_g)=1393$ cm^{-1}, $v_3(a_g)=1345$ cm^{-1} and $v_{27}(b_{1u})=1451$ cm^{-1}. These frequencies are in good agreement with analogous bands observed in the salts containing ET radical cations with charge +1.[7]

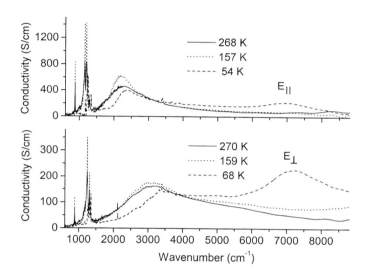

Figure 3. Temperature dependence of the optical conductivity spectra of a κ-[Et$_4$N][Co(CN)$_6$ (ET)$_4$]·3H$_2$O crystal as obtained from the reflectance spectra by Kramers-Kronig transformation.

The modifications of both IR and Raman spectra of Co^{III} crystals caused by the charge ordering are very similar to those recorded previously for Fe^{III} crystals. The most important difference is that the charge redistribution in the Fe^{III} salt proceeds gradually below 240 K, as proved by Raman studies of powders [3] as well as single crystals [8], but in the Co^{III} salt this is an abrupt process, which is directly related to the phase transition at 150 K. Such different behaviour is not clear at the moment. A possible explanation is that disorder in Fe^{III} crystals is larger than in Co^{III} crystals, as suggested by the fact that C=C bands in Fe^{III} Raman spectra are broader than analogous bands in Co^{III} spectra.

Recent investigations demonstrate that the charge-ordering phenomena are frequently observed in nearly uniform one- and two-dimensional conductors. For understanding of the charge-ordered states in quarter-filled quasi-two-dimensional ET salts, it is necessary to consider both on-site and intersite Coulomb interactions, together with the full anisotropy of transfer integrals.[9] In non-dimerised, nearly uniform salts the intersite Coulomb interactions are responsible for the charge ordering giving rise to various stripe-type charge-ordered states. The presence of the relevant values of the intersite Coulomb interactions can be also expected in the dimerised κ-phase salts; nevertheless, the stability of their insulating state is due to the on-site Coulomb repulsion and large dimerisation of conducting ET layers. In κ-phase salts, the charge distribution is uniform among ET molecules and charge ordering is not expected. Therefore, the transition from uniform to non-uniform charge distribution at 150 K in the κ-phase Co^{III} salt is quite surprising. An explanation of this unexpected phenomenon can be the fact that in Co^{III} salt the intra-dimer interaction between ET molecules is weak in comparison with interactions between neighbouring dimers, whereas in typical κ-phase salts, the situation is opposite, i.e. intra-dimer interactions are much stronger than inter-dimer ones.

This work was partially supported by KBN under grant 2 P03B 087 22 and Polonium project No 03273NK.

[1]   P. Le Maguerès, L. Ouahab, N. Connan, C.J. Gómez-García, P. Delhaès, J. Even, M. Bertault, *Solid State Commun.* **1996**, *97*, 27.
[2]   P. Le Maguerès, L. Ouahab, P. Briard, J. Even, M. Bertault, L. Toupet, J. Rámos, C.J. Gómez-García, P. Delhaès, T. Mallah, *Synth. Met.* **1997**, *86*, 1859.
[3]   R. Świetlik, M. Połomska, L. Ouahab and J. Guillevic, *J. Mater. Chem.* **2001**, *11*, 1313.
[4]   R. Świetlik, A. Łapiński, M. Połomska, L. Ouahab and J. Guillevic, *Synth. Met.*, in press.
[5]   J. Moldenhauer, Ch. Horn, K.I. Pokhodnia, D. Schweitzer, I. Heinen, H.J. Keller, *Synth. Met.* **1993**, *60*, 31.
[6]   K. Yamamoto, K. Yakushi, K. Miyagawa, K. Kanoda, A. Kawamoto, *Phys. Rev. B* **2002**, *65*, 085110.
[7]   G. Visentini, M. Masino, C. Bellitto and A. Girlando, *Phys. Rev. B* **1998**, *58*, 9460.
[8]   R. Świetlik, unpublished data.
[9]   H. Seo, *J. Phys. Soc. Jpn.* **2000**, *69*, 805.
[10] K. Kino and H. Fukuyama, *J. Phys. Soc. Jpn.* **1995**, *64*, 2726.

Degradation Effects in Polymer Light Emitting Devices Due to Heat Treatment

Frank Janssen,[1] *Marco Sturm,*[1] *Leo van Ijzendoorn,*[1] *Arnoud Denier van der Gon,*[1] *Herman Schoo,*[2] *Martien de Voigt,*[1] *Hidde Brongersma*[1]

[1] Department of Applied Physics, Eindhoven University of Technology,
  Den Dolech 2, P.O. Box 513, 5600 MB Eindhoven, Netherlands
[2] TNO Institute of Industrial Technology, Eindhoven, Netherlands Dutch
  Polymer Institute, Netherlands

Summary: The characteristics of polymer light emitting diodes (PLEDs) (ITO/PPV/Ca) depend strongly on the conditions during preparation and operation. We studied the effects of heat treatment (during and after preparation) of PLEDs with OC_1C_{10}-PPV as active layer. PLEDs showed a reduction of both the current and the light output to 40 % after annealing for only 30 min at 65 °C. Effects on *I-V* characteristics were studied by measuring single carrier devices (hole- and electron-dominated devices). The current reduction after heat treatment can be ascribed to degradation of the ITO/PPV and the Ca/PPV interfaces.

Keywords: annealing; conjugated polymer; degradation; interfaces; light emitting diodes (LED)

Introduction

Polymer light emitting diodes (PLEDs) are considered promising candidates for full-colour, cheap, flexible displays, which are easy to process [1]. Polymer LEDs consist of an emitting polymer layer (in our case dialkoxy-*p*-phenylenevinylene (OC_1C_{10}-PPV)), which is spin-coated on anode (indium-tin-oxide (ITO) covered glass) and covered with calcium cathode prepared by evaporation.

The temperature history during preparation is considered to be important for the PLED performance. The glass/ITO/PPV layer is usually heated before application of the cathode in order to remove impurities from the PPV. The device will also be heated when a protective Al layer is applied onto Ca, because of the radiation of the evaporator. Furthermore, in commercial applications, some temperature resistance is necessary.

We studied the effects of heat treatment on PLEDs with calcium cathodes by electrical characterisation of PLEDs, hole-only (ITO/PPV/Au and Au/PPV/Au) and electron-only (TiN/PPV/Ca) single-carrier devices.

© 2004 International Union of Pure and Applied Chemistry   DOI: 10.1002/masy.200450822

Preparation of PLEDs

Substrates of glass covered with ITO (100 nm ITO, 30 $\Omega/\square$, Merck) were cleaned in an ultrasonic bath successively with acetone (Uvasol, Merck) and propan-2-ol (Uvasol, Merck). Next, a UV ozone treatment was conducted for 20 min. Subsequently, the samples were transferred without air access to a glove box (O_2 and H_2O < 1 ppm), where an OC_1C_{10}-PPV layer was spin-coated onto the ITO. Then, the samples were transferred to the evaporation chamber which is connected to the glove box. Here a 80-nm thick calcium cathode was evaporated from an effusion cell, at a deposition rate of 0.3 nm/s. The pressure during the evaporation process was ~$1*10^{-5}$ Pa.

Electrical and optical characteristics were measured in the evaporation chamber. Heat treatment was performed in the evaporation chamber with an infrared lamp, temperature was measured with NTC thermistors, which were in contact with the glass of the PLEDs. After the heat treatment, the samples were left to cool down before electrical and optical characteristics were measured.

Hole-dominated single-carrier devices were prepared consisting of glass/ITO/PPV/Au and glass/Au/PPV/Au structures. Electron single-carrier devices were made according to Bozano et al. [2] consisting of glass/TiN/Ca/PPV. For both glass/Au and glass/TiN, the same cleaning procedure as for glass/ITO was performed apart from the UV ozone treatment, which was omitted.

Characteristics of PLEDs

Figure 1 shows the light output against current at 6 V bias for untreated and annealed PLEDs (glass/ITO/PPV/Ca). It is clear from Figure 1 that thermal treatment leads to reduction in current and light output. The power efficiency at 6 V is not influenced much by the heat treatment. It can be seen that the current and light output reduction depends strongly on the treatment temperature.

The reduction of current and light output after evaporation of 50 nm of aluminium onto the top of the calcium is also shown in Figure 1. The temperature of the PLEDs increased during evaporation, because of the radiation of the aluminium evaporator. Temperature measurements on the rear side of the PLED indicated temperatures of 48 °C immediately after evaporation of aluminium.

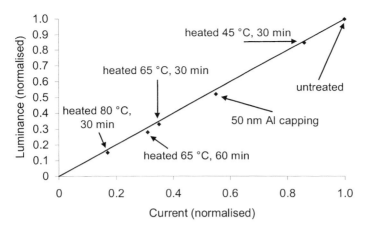

Figure 1. Relative current and brightness at 6 V bias for thermal treatments at different temperatures and time scales. The current and brightness are normalised to the current and brightness measured on the same devices before heat treatment. Experiments were performed with PLEDs with 130 nm OC_1C_{10}-PPV and a calcium cathode. Heat treatment was performed at pressure $\sim 10^{-7}$ Pa. The line indicates points were the relative losses in current and brightness are equal (constant power efficiency). All devices were measured at equal temperatures.

It was verified that the electrical characterisation itself does not cause the degraded PLED performance by comparing I-V and E-V characteristics of annealed PLEDs, which were and were not characterised before the annealing step.

Subsequently, annealing at 65 °C for 30 min was carried out at different stages of the production process to find causes of the current and light output reduction. The results of these experiments are shown in Figure 2. Heat treatment before application of the calcium cathode also led to a reduced current and light output, but the reduction was smaller than for samples treated after application of the cathode. Heat treatment both before and after application of the cathode resulted in almost the same current and light-output reduction as heat treatment after application of the cathode.

Apparently, calcium causes a part of the current reduction and another part is caused by ITO and/or the PPV. In order to study the influence of the calcium and ITO more thoroughly, single-carrier devices were prepared.

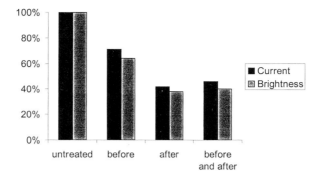

Figure 2. Current and light output at 6 V of ITO/PPV (130 nm)/Ca PLEDs, normalised to the current and light output of untreated devices. Devices were heat-treated for 30 min at 65 °C *before, after* and both *before and after* application of the calcium cathode. If the devices were treated before application of the cathode, they were left to cool down before the calcium was applied.

Hole-only single-carrier devices

In order to find the mechanism that caused the current reduction, hole-only single-carrier devices consisting of ITO/PPV/Au structures were prepared. The high work function of gold blocked the injection of electrons and, therefore, these devices were considered as hole-only single-carrier devices. Heat treatment (30 min at 65 °C) before, after or both before and after application of the Au cathode, all resulted in a reduction of the current to 80 % (at 6 V) compared with untreated devices. In order to separate the effects of the degradation of the ITO/PPV interface and the PPV itself, hole-only devices with a gold anode and a gold cathode were prepared (Au/PPV/Au). These devices did not show any reduction of the currents after annealing at 65 °C for 30 min. It can be concluded that the ITO caused the current reduction in ITO/PPV/Au devices and that the heat treatment of PPV itself did not cause the decreased PLED performance.

Electron-only single-carrier devices

Next, measurements were performed on electron-only single-carrier devices. In these devices the ITO anode was replaced with TiN in order to block the holes [2]. The TiN/PPV/Ca did not show light output in the used voltage range (0-6 V) and impedance spectroscopy confirmed that these samples can be described as single-carrier electron-only devices.

Heat treatment before application of the calcium cathode did not result in reduction of the current; moreover, the current-voltage characteristics of treated and untreated devices were equal. If the devices were heated after (or before and after) application of the calcium cathode, the current at 6 V decreased to 40 % and the shape of the current-voltage characteristics changed. We concluded that heat treatment influences the electron currents; this occurred only when the calcium was present. The PPV did not cause the decreased electron current, because heat treatment before application of the cathode did not have any effect on the performance.

Modification of the Ca/PPV interface

In the previous paragraph, we concluded that calcium caused the decreased electron current. In order to get a better understanding of the effects of electron current reduction on PLED performance, we tried to modify the Ca/PPV cathode without performing heat treatment.

It can be expected that, by heating the device, calcium atoms diffuse into the PPV and modify the Ca/PPV interface. An alternative approach, to modify the Ca/PPV interface, changes the evaporation rate of the calcium. If the evaporation rate is very low, the probability of clustering of calcium particles on the PPV surface is very small, as shown in ref. [3] for other metals on polymers. Free calcium atoms are therefore considered to diffuse easier into the PPV.

Devices with calcium cathodes deposited at two different evaporation rates (0.3 and 0.003 nm/s) were prepared by evaporating calcium at two different temperatures (510 °C to 380 °C) and characterized (Figure 3). It can be seen that devices with cathodes deposited at 0.003 nm/s have a worse characteristics; the current was reduced by 45 % and the brightness by 50 % at 6 V.

The time needed to evaporate the cathode is obviously different for the two evaporation rates, which could lead to different temperatures during evaporation of the devices due to the radiation of the evaporator. However, the temperatures measured on the glass of the PLED immediately after evaporation are equal (38 °C) for both deposition rates.

In the case of slow evaporation, the calcium at the PPV interface is exposed longer to impurities, (e.g. oxygen), which might be present in vacuum. Oxidation of the cathode leads to decreased performance [4]. Our mass spectrometer did not detect impurities in vacuum, which means that the partial pressures of oxygen and water vapor were below 10^{-7} Pa. Furthermore, in ref. [4], it is shown that oxidation of calcium during evaporation leads to a decreased power efficiency. The power efficiency, however, was (nearly) unchanged in our experiments, so it can be concluded that impurities did not influence our experiments.

Therefore, following [3], we expect that a low evaporation rate induces calcium diffusion into the PPV, which apparently reduces electron injection as indicated by a lower current and light output.

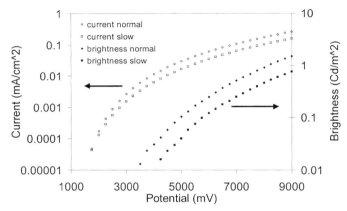

Figure 3. Current and brightness characteristics of ITO/PPV (250 nm)/Ca PLEDS. The first 20 nm of calcium is deposited onto the PPV at different rates. For the 'normal' devices, the evaporation rates were 0.3 nm/s and for the 'slow' devices it was 0.003 nm/s

Conclusions

Annealing of PLEDs (ITO/PPV/Ca) at various stages of preparation leads to reduction of current and the light output. This reduction is caused by the ITO/PPV and the Ca/PPV interfaces. Changes in the PPV itself (if present) do not influence the performance. It was shown that modification of the Ca/PPV interface by lowering the evaporation rate of the calcium resulted in worse device characteristics. Probably more calcium diffuses into the PLED, which apparently leads to worse PLED performances due to, example, the formation of (electron) traps. Heating PLEDs after application of the calcium cathode can also be assumed to induce calcium to diffuse into the PPV and cause a part of the observed current reduction.

[1] J. H. Burroughes, D.D.C. Bradley, A.R. Brown, R.N.Marks, K. Mackey, R.H. Friend, P.L. Burn, A.B. Holmes, *Nature* **1990,** 347, 539.
[2] L. Bozano, S.A. Carter, J.C. Scott, G.G.Malliaras, P.J. Brock, *Appl. Phys. Lett.* **1999** 74, 1132.
[3] F. Faupel, V. Zaporojtchenko, T. Strunskus, J. Erichsen, K. Dolgner, A. Thran and M. Kiene, "Metallization of polymers 2", *ACS Symp. Ser.,* **2001.**
[4] G.G. Andersson, M.P. de Jong, F.J.J. Janssen, J.M. Sturm, L.J. van IJzendoorn, A.W. Denier van der Gon, M.J.A. de Voigt, H.H. Brongersma, *J. Appl. Phys.* **2001,** 90, 1376.

Macromol. Symp. **2004**, *212*, 231-237

Space Charge Distribution in Luminescent Conjugated Polymers

Dieter Geschke,*[1] Frank Feller,[1] Andy P. Monkman[2]

[1] University of Leipzig, Faculty of Physics and Earth Sciences, Institute of Experimental Physics I, Polymer Research Group, Linnéstraße 5, 04103 Leipzig, Germany
[2] University of Durham, Department of Physics, Organic Electroactive Materials Research Group, South Road, Durham DH1 3LE, UK

Summary: We have used for the first time the laser intensity modulation method (LIMM) to resolve the depth profile of space charges in films of poly[(2-(2-ethylhexyl)-5-methoxy-1,4-phenylene)vinylene] (MEH-PPV), poly(pyridine-2,5-diyl) (PPY) and poly(fluorene) (PFO). The results demonstrate that in conjugated polymers space charges can not only be created but also stored permanently.

Keywords: conjugated polymers; light-emitting diodes

Introduction

Impressive scientific and technological progress has recently been achieved in the area of organic light-emitting diodes. An equal fundamental research motivation has been the desire to better understand and control charge injection into, charge migration through, and radiative recombination in macromolecular solids. Charge injection is influenced by space charge distribution[1] and therefore it is of fundamental interest to know whether the current in the device is injection-limited or transport-limited[2]. If the barrier height ϕ between the electrode work function and the HOMO or LUMO levels of the polymer is lower than about 0.3 eV in most conjugated polymers, the injection rate becomes higher than the transport rate in the bulk of the polymer, thus charge is injected faster than it can be transported in low-mobility materials. The result is an accumulation of space charge near the injecting contact, which shields the external electric field and hinders further charge injection. On the other hand, if the injection barrier is high (about 1 eV), the injection rate becomes so low that every injected charge can be led away fast enough. Thus a very small and homogeneously distributed space charge is expected in such systems.

We used for the first time a well established thermal wave method (laser intensity modulation method, LIMM)[3,4] to resolve the depth profile of space charge and the distribution of the internal electric field in 3-4 μm films of poly[(2-(2-ethylhexyl)-5-

 DOI: 10.1002/masy.200450823

methoxy-1,4-phenylene)vinylene] (MEH-PPV), poly(pyridine-2,5-diyl) (PPY) and poly(fluorene) (PFO).

Experimental

MEH-PPV, PPY, and PFO were synthesised as described previously[5]. 4-5 μm films were prepared by spin coating from a polymer solution onto glass substrates that carry a 100-nm gold electrode (rear electrode). A second 100-nm gold electrode (front electrode) was evaporated on top of the film and covered with a 20-nm bismuth layer to enhance the heat absorption of the laser beam.

The LIMM experiment is explained in more detail elsewhere[6] and only briefly described here. The set-up (Fig. 1) consists of a HeNe laser, its beam being intensity-modulated by an acoustic-optical modulator with a frequency f supplied by a frequency generator. The laser beam hits the front electrode. Heat is absorbed and transformed into a heat wave that penetrates into the sample. The pyroelectric response of the sample is detected as a small AC current between the gold electrodes, which is amplified by a current-to-voltage converter (CVC) using a gain of 10^5 V/A and measured with a lock-in amplifier.

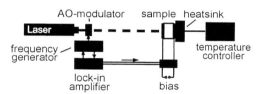

Fig. 1: Experimental set-up for the laser intensity modulation method (LIMM).

In addition, a galvanic decoupler has been used to allow the application of a DC bias to the very sensitive AC current measurements across the sample electrodes.

Depending on the modulation frequency f of the incident laser beam and on certain thermal parameters of the material, the heat wave establishes a dynamic temperature profile $T(z)$ with respect to the coordinate z along the film normal. The resulting inhomogeneous thermal expansion modifies the space charge distribution $r(z)$ and gives rise to a small pyroelectric current (with A irradiated area, L film thickness)

$$I_p(f) = 2\pi i \cdot f \cdot \frac{A}{L} \int_0^L r(z)\ T(f,z)\ dz. \qquad (1)$$

After measuring $I_p(z)$, the distribution function $r(z)$ has to be inferred via inversion of the integral Equation 1 using the temperature profile $T(f,z)$, that can be calculated for the specific thermal parameters. This is, however, known to be an ill-posed inverse problem. In order to solve such ill-posed inverse problems, special mathematical methods, i.e. regularisation methods, are necessary.

Results

Figure 2 shows a typical LIMM spectrum of a 4-µm thick PPY sample between Au electrodes. The sample was poled using a DC electric field of 25 MV/m for several minutes and measured under electric field.

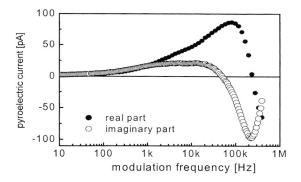

Fig. 2: LIMM spectrum of a Au/PPY/Au sample with laser incident on the front electrode. Measurements were performed at room temperature using an applied electric field of 25 MV/m.

The lower panel of Fig. 3 shows the space charge profile in the PPY film. Directly behind the front electrode, a pronounced space charge layer is detected, which has the same polarity as the nearby electrode and is therefore denoted as a homo-charge layer. Deeper in the bulk of the film, at about 500 nm, a second weaker space charge layer is found that is of opposite polarity (denoted as hetero-charge layer). The upper panel shows the electric field profile in the film, demonstrating how the accumulated space charge distorts the internal field. We assign the positive layer to be due to the injection of positive charge carriers from the anode into near-surface traps and the second hetero-charge layer to intrinsic charge carriers that have been separated within the external electric field and accumulated in the trapping zone.

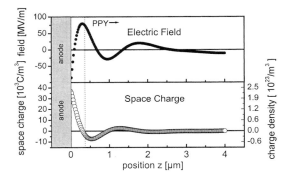

Fig. 3: Distribution of electric field (upper panel) and space charge (lower panel) obtained from the LIMM spectrum in Fig. 2 via regularisation.

The stability of the accumulated space charge was investigated by measuring the pyroelectric signal for different values of the applied electric field. Figure 4 shows such pyroelectric current - voltage curves.

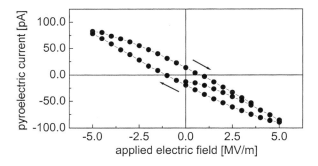

Fig. 4: Pyroelectric current-voltage characteristics of a 4-μm thick Au/PPY/Au sample. To account for the slow response of the pyroelectric signal to changes of the external bias, the sample was allowed to settle for 10 min before each bias reading. Measurements were performed at room temperature using a modulation frequency of 33 kHz.

Obviously the pyroelectric current shows clear hysteresis behaviour. Thus, the polymer can be charged by application of an electric field. A significant amount of the accumulated charge remains stored in the material and can only be removed by an electric field. An assessment of the magnitude of the charge density profile allows to estimate the density of permanently stored charge to be as high as 2×10^{21} m^{-3}. In PPY space charges can not only be created but also stored.

We have compared the pyroelectric response of PPY with that of other conjugated polymers, i.e. with MEH-PPV and PFO. The LIMM spectra for the three polymers using equal external electric field and electrode configuration are depicted in Fig. 5.

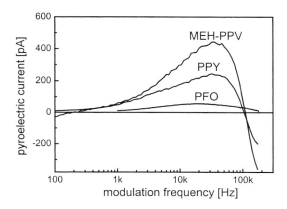

Fig. 5: LIMM spectra of the three conjugated polymers under investigation represented by their real part. All three polymers where subjected to an electric field of 25 kV/m and measured at room temperature.

As expected, enhancement of the space charge accumulation is found in MEH-PPV corresponding to the lower injection barriers for holes ($\phi \approx 0.3$ eV). On the other hand, it is noticeable that the pyroelectric current in PFO is much weaker than in PPY, although hole injection should also be improved due to a lower barrier height (ϕ(Au-PFO) ≈ 0.6 eV and ϕ(Au-PPY) ≈ 0.9 eV). Furthermore, neither in MEH-PPV nor in PFO, a considerable poling effect could be measured. After removing the external bias, the pyrocurrent returned to zero or to a constant, very small value.

The three polymers were subjected to an instantaneous switching of the external bias and the temporal pyroelectric response was observed. This reveals another difference in the charge accumulation and transport behaviour between the polymers. While, in PPY accumulation and annihilation of space charge occurs in the time scale of several minutes, PFO and MEH-PPV films seem to instantaneously react to an external bias. A more detailed analysis of the temporal pyroelectric current may provide information about the nature and depth of the relevant traps and will be subject of a future work.

Discussion

The results show significant accumulation of space charge in the three conjugated polymers under study with no definite correlation to the injection barrier heights. Especially in PPY, where the difference between the Au electrode work function and the valence/conduction band (HOMO/LUMO) of the polymer is in the range of 1 eV, only small space charge should be expected in the material because of the very low injection rate. We therefore suggest the following model that is illustrated in Fig. 6 to explain the emergence of space charges in these devices.

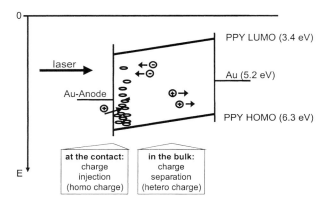

Fig. 6: Proposed model to explain the pyroelectric results in PPY. Charge carriers are injected directly into deep traps at the surface of the polymer layer. Simultaneously, mobile charged states in the bulk are separated by the influence of the external electric field.

We presume an increased concentration of deep traps near the front surface of the sample. These traps can have various origins (intrinsic lattice defects introduced during the casting of the film, extrinsic impurities introduced by the ambient atmosphere before the evaporation of the top electrode, etc.). An injection of carriers is then conceivable by direct transfer from the electrode into the deep traps at the surface rather than over the barrier into the transport level. As a result, deeply trapped charge is accumulated as space charge near the front electrode. Any spatial redistribution of the space charges requires detrapping of the charge carriers and, depending on the depth of the traps, the response to an external bias can be instant or very slow. Detrapping of charge from deep traps in the time scale of minutes has also been reported by Campbell and Bradley[7]. A more detailed analysis of the pyroelectric current transients will be covered in a later publication.

More difficult to understand is the smaller hetero-charge layer deeper in the bulk of the sample. A possible explanation could be that intrinsic charge arising from charged defects form mobile states (polarons) which subsequently separate in the external electric field. There are, however, also some questions left by this explanation.

Conclusions

In PPY and MEH-PPV, charge storage has a significant influence on the internal electric field in the sample and films can be charged permanently depending on their bias history. The space charge in the polymer cannot follow changes of the external bias instantaneously - the samples are still redistributing internal charges after minutes. This behaviour should be kept in mind when making measurements of electronic properties of these polymers. PFO differs more significantly from the other polymers, showing a very low concentration of deep traps. In addition, the charges in PFO can be redistributed in less than one second since only shallow traps are addressed in this material.

[1]   A. J. Campbell, M. S. Weaves, D. G. Lidzey, D. D. C. Bradley, *J. Appl. Phys.* **84**, 6737 (1998).
[2]   P. W. M Blom, M.C. J. M. Vissenberg, *Mater. Sci. Eng.* **27**, 53 (2000).
[3]   F. Feller, D. Geschke, A. P. Monkman, *Appl. Phys. Lett.* **79**, 779 (2001).
[4]   S. B. Lang, D. K. Das-Gupta, *J. Appl. Phys.* **59**, 2151 (1986).
[5]   S. Dailey, M. Halim, E. Retourt, L. E. Horsburgh, I., D. W. Samuel, A. P. Monkman, *J. Phys.-Condensed Matter* **10**, 5171 (1998).
[6]   N. Leister, D. Geschke, *Liq. Cryst.* **24**, 441 (1998).
[7]   A. J. Campbell, D. D. C. Bradley, *Proc. SPIE* **3797**, 326 (1999).

Macromol. Symp. **2004**, *212*, 239-244

Time-Resolved Luminescence of Terbium Complexes with Poly(2- and 4-vinylpyridine *N*-oxide)s in Aqueous Solution

Drahomír Výprachtický,[1] *Věra Cimrová,*[1] *Yoshi Okamoto,*[2] *Rudolf Kotva*[1]

[1] Institute of Macromolecular Chemistry, Academy of Sciences of the Czech Republic, Heyrovský Sq. 2, 162 06 Prague 6, Czech Republic
E-mail: vyprach@imc.cas.cz,
E-mail: cimrova@imc.cas.cz
[2] Polymer Research Institute, Polytechnic University, Six MetroTech Center, Brooklyn, New York 11201, USA
E-mail: yokamoto@duke.poly.edu

Summary: Terbium complexes with polymer ligands of poly(2- and 4-vinylpyridine *N*-oxide)s (P2VPNO, P4VPNO) in aqueous solution were prepared and characterized. Multi-exponential decays of the $^5D_4 \rightarrow ^7F_5$ terbium transition at 545 nm of [P2VPNO-Tb^{3+}] and [P4VPNO-Tb^{3+}] complexes were measured. The non-linearity of semi-logarithmic plots of time-resolved luminescence was more pronounced in [P4VPNO-Tb^{3+}] than in [P2VPNO-Tb^{3+}], being reduced by addition of salts such as sodium formate or acetate. We assume that multi-exponential decays of Tb^{3+} in the complexes are caused by a back metal-to-ligand energy transfer via triplet state of *N*-oxide polymer ligand. By carrying out separate experiments in water and deuterium oxide, the number of coordinated water molecules in the [P4VPNO-Tb^{3+}] complex was estimated as 4–5, assuming that the Tb^{3+} aqua complex contains nine water molecules.

Keywords: lanthanide complexes; poly(vinylpyridine N-oxide); terbium metal; time-resolved luminescence

Introduction

Lanthanide metal ions such as Tb^{3+} and Eu^{3+} exist as stable trivalent ions exhibiting characteristic luminescence in aqueous solution, which is known to be strongly influenced by their immediate coordinated environment [1]. The luminescence intensity of the ions is normally quite weak in aqueous solution since the coordinated water molecules serve as efficient quenchers of the emission [2]. Recently we have observed that luminescence intensities of the Tb^{3+} ion are greatly enhanced upon binding to polycarboxylates,[3] poly(vinylpyridine)s,[4] or poly(vinylpyridine *N*-oxide)s.[5] The results indicated that Tb^{3+} ions were strongly bound to the polymer ligands and

DOI: 10.1002/masy.200450824

that some of their inner-coordinated solvent molecules (water, methanol) were displaced upon binding. The luminescence decays were measured and the number of coordinated water (polycarboxylates) or methanol (polyvinylpyridines) molecules was determined using the deuterium isotope effect.[6] Here we investigated the Tb^{3+} complexes with poly(2-vinylpyridine N-oxide) (P2VPNO) and poly(4-vinylpyridine N-oxide) (P4VPNO) in aqueous or deuterium oxide solutions by time-resolved fluorescence technique.

Experimental

Materials and Synthesis. Terbium chloride hexahydrate was purchased from Rhone-Poulenc Basic Chemicals Co. and anhydrous terbium chloride from Aldrich; both were used as received. Syntheses and characterization of poly(2- and 4-vinylpyridine N-oxide)s were reported.[5]

Ultraviolet spectra. A Carry 2300 UV-vis-near IR spectrometer was used to determine the absorption spectra in water. The spectra of both polymers have a maximum at 255 nm; the molar absorption coefficients of P2VPNO and P4VPNO, ε_{255}, are 6 550 and 10 200 L mol^{-1}cm^{-1}, respectively.

Steady-state fluorescence spectra. A Perkin Elmer LS50B fluorescence spectrometer was used for steady-state fluorescence measurements. The slit width was 5 nm for both excitation and emission monochromators and the fluorescence intensity was reported in arbitrary units. The samples in a quartz cuvette (1 x 1 x 4 cm) were measured in the L-format arrangement.

Time-resolved luminescence. The time-resolved luminescence decays in the millisecond time scale were measured with a Perkin Elmer LS50B instrument in the phosphorescence mode. The Tb^{3+} luminescence intensity was measured at twenty different delay points following the excitation (xenon lamp equipped with a chopper). The luminescence decay, $F(t)$, was analyzed by triple-exponential fit ($i = 3$) given by

$$F(t) = C + \sum_i B_i \exp(-t/\tau_i),$$ (1)

where B_i is a preexponential factor representing the fractional contribution to the time-resolved decay of the component with a lifetime τ_i [rel $B_i = (B_i \ \tau_i / \Sigma \ B_i \ \tau_i) \times 100 \ \%$], C is background and t is time. Least-square analysis of luminescence decay curves was used for evaluation of τ_{1-3}.

Results and Discussion

For [P4VPNO-Tb^{3+}] complexes, the multi-exponential decays (Eq. 1) in water or deuterium oxide were measured (Figure 1a).

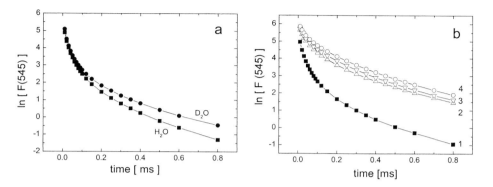

Figure 1: Time-resolved luminescence of the [P4VPNO-Tb^{3+}] complex at 545 nm (λ_{ex} = 320 nm) in (a) water (■) or deuterium oxide (●); (b) in water (■) and after addition of 0.0244 M (△) or 0.134 M (▽) of sodium acetate or 0.153 M (○) of sodium formate. [P4VPNO] = 1.16 × 10^{-2} M, [TbCl$_3$] = 1.0 × 10^{-3} M.

When solvents containing O-H groups are coordinated to lanthanide ions, efficient non-radiative deactivations take place via vibronic coupling with the vibrational states of the O-H oscillators (3500 cm^{-1}). If the O-H oscillators are replaced by the low-frequency O-D oscillators (2800 cm^{-1}), the vibronic deactivation pathway becomes much less efficient. By carrying out separate experiments in H$_2$O and D$_2$O solutions, the number of coordinated water molecules (n) in lanthanide – ligand complexes can be calculated [6] from the experimental excited-state lifetimes measured (in ms) in water (τ_{OH}) or deuterated water (τ_{OD}) using the equation: $n = q\,(1/\tau_{OH} - 1/\tau_{OD})$, where q = 4.2 was determined for Tb^{3+}. For single-exponential decays, the n could be evaluated; we determined the values 3.6, 3.4, and 2.4 for poly(acrylic acid), syndiotactic poly(methacrylic acid), and isotactic poly(methacrylic acid) ligands, respectively.[3,7] This approach to calculation of n was not directly applicable in the case of multi-exponential decays. Nevertheless, skipping the coinciding strongly nonlinear points for $t < 0.1$ ms, the linear regression of the points with $t > 0.1$ ms gave τ_{OH} = 0.18 ms and τ_{OD} = 0.22 ms (regression coefficient ~ 0.98). From these values we can roughly estimate $n = 4\text{-}5$. If it is assumed that the Tb^{3+} aqua complex contains nine water molecules[8], then about 4-5 water molecules were displaced upon the [P4VPNO-Tb^{3+}] complex formation and the coordination increased the luminescence intensity of Tb^{3+} (Figure 2).

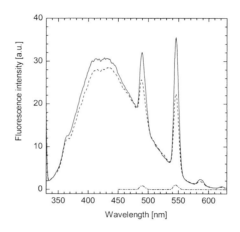

Figure 2: Emission luminescence spectra of $TbCl_3$ ($\lambda_{ex} = 353$ nm, dash-and-dott line) in water and of the [P4VPNO-Tb^{3+}] complex ($\lambda_{ex} = 320$ nm) in water in the presence (solid line) or absence (dashed line) of sodium formate; [P4VPNO] = 1.16×10^{-2} M, [$TbCl_3$] = 1.0×10^{-3} M, [HCOONa] = 0.153 M.

The terbium luminescence increase may be accounted for by two effects: the replacement of some of the water molecules inner-coordinated to Tb^{3+} ion upon complex formation and a possible energy transfer from the P4VPNO ligand to the Tb^{3+} ion. The ligand-to-metal energy transfer was proved by experiments in the presence and absence of terbium metal.[5] We assume that the multi-exponential decays at 545 nm are caused by back metal-to-ligand non-radiative energy transfer (quenching). The non-radiative deactivation of Tb^{3+} via the triplet state of N-oxide-containing branched macrocyclic ligands was described.[6] The N-oxide polymers have no ionizable groups, but charge separation in the N^+–O^- bond results in a large dipole moment. The quenching is likely associated with a too compact arrangement of N^+–O^- groups in the coordination sphere of the metal ion. We have found out that the curvature of multi-exponential decay of the [P4VPNO-Tb^{3+}] luminescence can be reduced by addition of salts such as sodium acetate or sodium formate

(Figure 1b). Addition of the salt increases the luminescence lifetime (Table 1) and intensities of terbium $^5D_4 \rightarrow {}^7F_6$ and $^5D_4 \rightarrow {}^7F_5$ transitions at 490 and 545 nm, respectively (Figure 2). In a competitive process with the polymer ligand, formate or acetate anions probably optimize the ligand-metal distances in the energy transfer processes. Such explanation was further supported by time-resolved luminescence decays of the [P2VPNO-Tb^{3+}] and [P4VPNO-Tb^{3+}] complexes at 545 nm (Figure 3a).

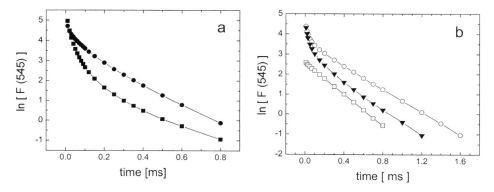

Figure 3: Time-resolved luminescence in aqueous solutions at 545 nm (λ_{ex} = 320 nm) of (a) [P2VPNO-Tb^{3+}] (●) and [P4VPNO-Tb^{3+}] (■), [ligand] = 1.16×10^{-2} M, [TbCl$_3$] = 1.0×10^{-3} M; (b) 1.0×10^{-2} M TbCl$_3$ (□), 1.0×10^{-2} M TbCl$_3$ + 1.16×10^{-3} M P4VPNO (○), 1.0×10^{-3} M TbCl$_3$ + 1.16×10^{-3} M P4VPNO (▼).

The complex with the P2VPNO ligand was quenched less, possessing a higher content of the longer-lifetime component (Table 1) than that with the P4VPNO ligand. Steric reasons seem to play an important role. In P2VPNO, the N$^+$–O$^-$ group is more shielded by its polymer backbone and cannot occupy such tight arrangement in the metal sphere as that of P4VPNO. Finally, direct excitation of terbium leads to a single-exponential decay of its $^5D_4 \rightarrow {}^7F_5$ luminescence at 545 nm. In the [P4VPNO-Tb^{3+}] complex, because of absorption coefficients of the ligand and metal at the excitation wavelength, the overall terbium luminescence was acquired via ligand-to-metal energy transfer. With increasing ligand-to-metal ratio, the number of ligands in coordination sphere of the metal increases and a non-linearity (quenching) of time-resolved decays increases (Figure 3b). It was verified that the quenching was not caused by the chloride counterion after

TbCl$_3$ dissociation. Experiments of time-resolved luminescence of [P4VPNO-Tb^{3+}] and [P2VPNO-Tb^{3+}] performed with terbium triflate or nitrate provided similar results to those with terbium chloride.

In conclusion, with increasing ligand concentration, the ligand-to-metal energy transfer increases and one observes a higher emission intensity of Tb^{3+} ion at 490 and 545 nm; simultaneously, back metal-to-ligand quenching increases, which leads to multi-exponential decays of the $^5D_4 \rightarrow ^7F_6$ and $^5D_4 \rightarrow ^7F_5$ emissions of terbium.

Table 1: Luminescence lifetimes (τ_{1-3} in triple-exponential fit; Eq. 1) calculated by least-square analysis from decay curves 1-5 shown in Figures 1b and 3a.

Luminescence decay	Luminescence lifetime τ_i (μs) and rel B_i (%)		
	τ_1 (rel B_1)	τ_2 (rel B_2)	τ_3 (rel B_3)
1	13 (18 %)	43 (41 %)	167 (41 %)
2	19 (8 %)	70 (33 %)	258 (59 %)
3	20 (7 %)	75 (28 %)	233 (65 %)
4	20 (5 %)	72 (25 %)	235 (70 %)
5	12 (3 %)	69 (27 %)	200 (70 %)

Acknowledgment

We thank the Grant Agency of the Czech Republic (grant No. 203/01/0512) and the Grant Agency of the Academy of Sciences of the Czech Republic (grant No. AVOZ 4050913) for support.

[1] R. K. Gallagher, *J. Phys. Chem.* **1964**, *41*, 3036.
[2] J. L. Kropp, M. W. Windsor, *J. Phys. Chem.* **1967**, *71*, 477.
[3] S. Okamoto, D. Výprachtický, H. Furuya, A. Abe, Y. Okamoto, *Macromolecules* **1996**, *29*, 3511.
[4] D. Výprachtický, K. W. Sung, Y. Okamoto, *J. Polym. Sci., A: Polym. Chem.* **1999**, *37*, 1341.
[5] Y. Okamoto, T. K. Kwei, D. Výprachtický, *Macromolecules* **1998**, *31*, 9201.
[6] N. Sabbatini, M. Guardigli, J.-M. Lehn, *Coord. Chem. Rev.* **1993**, *123*, 201.
[7] H. Luján-Upton, Y. Okamoto, A. D. Walser, *J. Polym. Sci., A: Polym. Chem.* **1997**, *35*, 393.
[8] W. D. Horrocks, D. R. Sudnick, *J. Am. Chem. Soc.* **1979**, *101*, 334.

Macromol. Symp. **2004**, *212*, 245-250

Physical Properties and Structure of Thin Conducting Ion-Beam Modified Polymer Films

Margarita Guenther,[1] *Gerald Gerlach,*[1] *Gunnar Suchaneck,*[1] *Dieter Schneider,*[2] *Bodo Wolf,*[3] *Alexander Deineka,*[4] *Lubomir Jastrabik*[4]

[1] Dresden University of Technology, Institute for Solid State Electronics, Mommsenstr. 13, 01062 Dresden, Germany
[2] Fraunhofer-Institute for Material and Beam Technology, Winterbergstrasse 28, 01277 Dresden, Germany
[3] Dresden University of Technology, Institute for Crystallography and Solid State Physics, Zellescher Weg 16, 01062 Dresden, Germany
[4] Institute of Physics, Academy of Sciences of the Czech Republic, Na Slovance 2, 182 21 Prague 8, Czech Republic

Summary: In this work, a complex investigation of the film surface composition, chemical bonding, conductivity, optical properties, density, hardness and Young's modulus of ion-beam-modified polyimide films was carried out. It was shown that the partial destruction of chemical bonding under ion bombardment leads to the formation of graphite-like, amorphous carbon islands, which increase the surface film conductivity by several orders of magnitude, from an insulating to a semiconducting region. Strong enhancements of both the conductivity and the optical absorption coefficient occur when the fraction of amorphous carbon clusters dispersed in a polyimide matrix reaches 40 %. The values of hardness, Young's modulus and density at high irradiation doses reach the values typical of a hydrogenated amorphous carbon.

Keywords: conductivity; density; ion implantation; polyimide; XPS

Introduction

Thin films of aromatic polymers such as polyimides (PI) find extensive use in electronics. For particular sensor applications, physical properties of these films must be tailored. In this case, PI films are used as water vapor-uptaking layers for bimorphic humidity sensors. It was shown that ion implantation improves sensoric properties of polyimides.[1-3] In this work, the changes in structure and physical properties of polymer films are discussed in dependence on the dose and energy of implanted ions and irradiation conditions for PI films, which lead to optimal sensor properties.

© 2004 International Union of Pure and Applied Chemistry

DOI: 10.1002/masy.200450825

Experimental

The polyimide considered in this study is DuPont's polyimide PI2566 obtained from 4,4'-(1,1,1,3,3,3-hexafluoropropan-2-yl)bis(phthalic anhydride) and 4,4'-oxydianiline, Figure 1a). The samples were obtained by spin coating of Si substrates with PI precursor solutions (polyamic acids / N-methyl-2-pyrrolidone). The PI films were soft-baked at 90 °C for 20 min to remove the solvent and then they were cured in nitrogen at 400 °C for 1 h. The final film thickness was typically between 500 and 600 nm. The PI films were irradiated with a B^+ ion beam (energy 180 keV, dose 10^{13} - 10^{16} B^+/cm^2).

a)

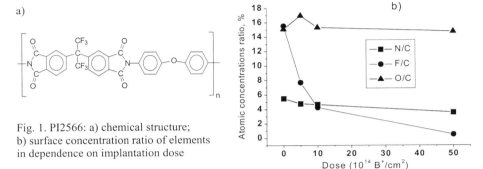

Fig. 1. PI2566: a) chemical structure;
b) surface concentration ratio of elements
in dependence on implantation dose

The chemical structure of the modified polyimide layers was investigated by X-ray photoelectron spectroscopy (XPS) using a Physical Electronics PHI 5702 electron spectrometer. The sampling depth of this technique amounts to 2 - 5 nm at a take-off angle of 45°. The atomic ratios of the elements were estimated using tabulated sensitivity factors. The spectra were referenced to the C 1s line at 284.8 eV arising from C-C or C-H bonds. The core level peaks have been analyzed by means of a Gauss-Lorentz best-fit computer program with background subtraction. The curve-fitting quality was evaluated by chi-convergence. Volume and surface conductivity measurements were performed using electrode configurations consisting of an evaporated NiCr bottom electrode, pressure-measuring and guard electrodes as described in IEC 93.[4] Refractive index and extinction coefficient spectra were calculated from the main ellipsometric angles Δ and Ψ measured in the spectral range 400-900 nm with a Woollam spectral ellipsometer working in the rotating analyzer mode. Hardness and Young's modulus E of modified PI films were investigated by the Hysitron depth-sensing nanoindentation technique with a Berkovich indenter. The nanoindentation analysis was carried out by the Oliver and Pharr method.[5] Density and E modulus were determined by a surface-acoustic-waves (SAW) technique as described in [6].

Results and discussion

The changes in the polymer surface composition after ion implantation (determined by XPS) are shown in Fig. 1b. The N/C, F/C and O/C atomic composition ratios were estimated from the ratios of integrated net intensities of the N 1s (at binding energy 400.15 eV), F 1s (688.26 eV) and O 1s (532 eV), respectively, to C 1s (285 eV) spectra with a correction for the atomic sensitivity factor. The chemical modifications observed at ion doses higher than $5 \cdot 10^{14}$ B$^+$/cm^2 are caused essentially by the drastic depletion of F and N, the modification of the residual O and the enrichment of C. The virgin C 1s core level spectrum deconvoluted into Gaussian components consists of five distinct peaks. The lowest binding energy peak at about 284.9 eV, which is the most intense, is attributed to emission from aromatic C atoms not linked to O or N.[7,8] The peak at 286.8 eV is due to C atoms linked to nitrogen (C-N) and oxygen (C-O-C), with a possible extra contribution of C atoms of the 6FDA ring.[7-9] The peak at 288.3 eV is due to carbonyl bonds (C=O) of the imide groups[7-9] and CF$_3$-C-CF$_3$ bonds[10]. The peaks at 291 eV and 292.8 eV are respectively assigned to an energy loss effect (shake-up satellite due to the $\pi \rightarrow \pi^*$ transitions of the benzene rings), very common in aromatic structures[7,11], and to CF$_3$ bonds[10,11]. The O 1s band is resolved into two components at 532 eV and 533.1 eV. They are assigned to imide carbonyls C=O and C-O-C linkages[7,12,13]. The relative quantity of each bonding component can be estimated from the area it covers divided by the total area of the C 1s and O 1s spectrum. Changes of the C 1s and O 1s peak areas after ion bombardment are depicted in Fig. 2. Figure 2a shows the decrease in the aromatic C component at 284.9 eV, the C-N bonding component at 286.8 eV and the CF$_3$ bonding component at 292.8 eV indicating that the aromatic, imide and C(CF$_3$)$_2$ groups are partly destroyed. The broken C-N and C-C bonds between benzene and imide rings result in the formation of polar amide groups (-CO-N-) at low doses up to 10^{15} ions/cm^2. The further fragmentation process leads to the formation of nonpolar phenyl groups starting at doses higher than 10^{15} B$^+$/cm^2.[7] Simultaneously, the low binding energy peak appears at 284.2 eV, in a position which can be assigned to graphite-like, amorphous carbon (a-C) rings.[7,9,14] The residual oxygen is present as ether (C-O-C), carboxylate (O-C=O at 534.17 eV [11]),

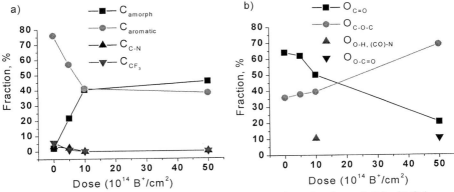

Fig. 2. Changes of XPS peak fraction in dependence on implantation dose: a) C 1s; b) O 1s

hydroxy or amide group (O-H and -CO-N- at 531.4 eV[10,15]), while the imide carbonyl (C=O) component becomes negligible (Fig. 2b). With increasing irradiation dose, the formation of amorphous and graphite-like structures is promoted which increases the surface conductivity σ_s by several orders of magnitude from an as-prepared value of $10^{-15}\Omega^{-1}$ to a value of $10^{-7}\Omega^{-1}$ at a dose of 10^{16} B[+]/cm^2 (Fig. 3). The ion-beam irradiation destroys the anisotropy of the refractive index of polyimide layers.[16-18] This isotropization of the surface-near part of the polymer layers and the increasing values of their optical constants also indicate the formation of a-C islands due to the ion bombardment. In Fig. 3, the absorption coefficient k at a wavelength of 670 nm is presented in dependence on the a-C fraction. A strong enhancement of σ_s and k occurs when the fraction of a-C clusters dispersed in the PI matrix reaches 40 %. This value corresponds to the percolation threshold for the appearance of three-dimensional hopping conductivity in the thin surface-near part of the polymer layer.[8,19] In addition, the carbonization and partial graphitization and the radiation-induced crosslinking play an important role in the enhancement of electric conductivity.[19] The crosslinking is responsible for the formation of three-dimensionally-connected, rigid networks, and improves the carrier mobility, which is one of the most important parameters that govern the conductivity in implanted polymers. The degree of crosslinking of the polymer matrix can be estimated by the values of hardness and E-modulus, because the E-modulus of polymer is directly proportional to the crosslink density (or inversely proportional to the average molecular weight beetween crosslinks).[20] The radiation-induced crosslinking results in an enhancement of hardness and Young's modulus of the polymer layers up to 10 and 6 times, respectively (Fig. 4a). Here, it can be concluded that the strong enhancement of hardness and E modulus occurs at ion doses higher than 10^{15} B$^+$/cm^2 in consequence of increasing crosslinking. The modification of the polymer microstructure during irradiation with energetic ions involves the emission of volatile molecules.[2,19]

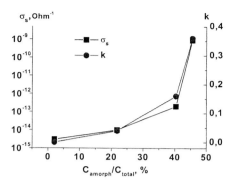

Fig. 3. Conductivity and optical extinction coefficient in dependence on the amorphous carbon fraction

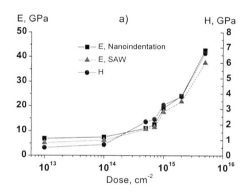 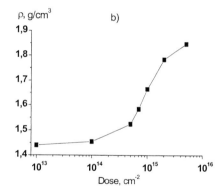

Fig. 4. Impact of ion implantation: a) on hardness and elastic modulus; b) on density

The thickness of the polymer films decreases due to the loss of material as volatile decomposition products and due to polymer shrinkage at ion doses higher than $5 \cdot 10^{14}$ B$^+$/cm^2.[18] The outgassing increases the density of the modified layer in comparison with the virgin PI from a value of 1.44 g/cm^3 (as-prepared) up to 1.85 g/cm^3 (Fig. 4b). The values of hardness, E modulus and density at high irradiation doses reach those typical of a hydrogenated amorphous carbon.[21,22]

Conclusions

All surface changes of ion-beam-modified PI films strongly depend on the implantation parameters. Electric conductivity, values of optical constants, density, hardness and E modulus of PI layers increase with increasing boron ion energy and dose. The PI2566 layers have the maximal moisture uptake after ion implantation with a dose of 10^{15} B^+/cm^2 at ion energy 180 keV, as it was shown in [3]. The observed increase in moisture uptake up to a dose of 10^{15} B^+/cm^2 is related to the formation of polar hydroxy and amide groups at the surface and correlated with the destruction of the hydrophobic $C(CF_3)_2$ groups. The small wettability of a-C islands and the increasing crosslinking result in a decrease in the moisture uptake of PI layers at ion doses higher than 10^{15} B^+/cm^2. Thus, irradiation at medium doses between 10^{14} and 10^{15} B^+/cm^2 results in optimum sensoric properties of ion-beam-modified polyimide films in humidity sensors.

Acknowledgments

The authors gratefully acknowledge support of this work by the Deutsche Forschungsgemeinschaft (grant Ge 779/6-2).

[1] G. Gerlach, K. Baumann, R. Buchhold, and A. Nakladal, German Patent DE 198 53 732, **1998**.
[2] R. Buchhold, *"Bimorphe Gassensoren"*, Dresden Univ. Press, Dresden, and Munich 1999, p.111.
[3] M. Guenther, K. Sahre, G. Suchaneck, G. Gerlach, K.-J. Eichhorn, *Surf. Coat. Technol.* **2001**, 142-144, 482.
[4] Standard IEC 93: Methods of test for insulating materials for electrical purposes; Volume resistivity and surface resistivity of solid electrical insulating materials, 1980.
[5] W. C. Oliver, G. M. Pharr, *J. Mater. Res.* **1992**, 7, 1564.
[6] D. Schneider, Th. Witke, Th. Schwarz, B. Schöneich, B. Schultrich, *Surf. Coat. Technol.* **2000**, 126, 136.
[7] G. Marletta, C. Oliveri, G. Ferla and S. Pignataro, *Surf. Interface Anal.* **1988**, 12, 447-454.
[8] J. Davenas, G. Boiteux, *Adv. Mater.* **1990**, 2, 521-527.
[9] D. Xu, X. Xu, S. Zou, *Appl. Phys. Lett.* **1991**, 59, 3110-3112.
[10] G. Suchaneck, M. Guenther, B. Adolphi, G. Gerlach, K. Sahre, K.-J. Eichhorn, A. Deineka, L. Jastrabik, Proc. of ISPC 15, Orleans, July 9-13, **2001**, 2485-2490.
[11] K.-W. Lee, A. Viehbeck, *IBM J. Res. Develop.* **1994**, 38, 457-474.
[12] K. C. Yung, D. W. Zeng, T. M. Yue, *Appl. Surf. Sci.* 2001, 173, 193-2102.
[13] Z. Qin, J. Zhang, H. Zhou, Y. Song, T. He, *Nucl. Instrum. Methods Phys. Res., Sect. B* **2000**, 170, 406-412.
[14] H. Song, O. J. Ilegbusi, *Thin Solid Films* **2001**, 388, 114-119.
[15] K.C.Yung, D.W. Zeng, *Surf. Coat. Technol.* **2001**, 145, 186-193.
[16] K. Sahre, K.-J. Eichhorn, F. Simon, D. Pleul, A. Janke, G. Gerlach, *Surf. Coat. Technol.* **2001**, 139, 257.
[17] G. Gerlach, M. Guenther, G. Suchaneck, K. Sahre, K.-J. Eichhorn, A. Deineka, L. Jastrabik, MRS Symp. Proc. 2001, 661, KK2.8.
[18] M. Guenther, G. Gerlach, G. Suchaneck, K. Sahre, K.-J. Eichhorn, B. Wolf , A. Deineka, L. Jastrabik: Ion-beam induced chemical and structural modification in polymers, *Surf. Coat. Technol.* **2002**, in press.
[19] E. H. Lee, in *"Polyimides"*, M. Ghosh, and K. Mittal, Eds., Marcel Dekker, New York 1996, p.471.
[20] A. Charlesby, *Radiat. Phys. Chem.* **1992**, 40, 117-120.
[21] Y.Q. Wang, *Nucl. Instrum. Methods Phys. Res., Sect. B* **2000**, 161-163, 1027-1032.
[22] B.Schultrich, H.-J. Schiebe, G. Grandremy, D. Schneider, P. Siemroth, *Thin Solid Films* **1994**, 253, 125.

Macromol. Symp. **2004**, *212,*251-256

Ultraweak Spectrally Resolved Thermoluminescence in Polymers

Ewa Mandowska, * *Arkadiusz Mandowski, Józef Świątek*

Institute of Physics, Pedagogical University, ul. Armii Krajowej 13/15, 42-200 Częstochowa, Poland
E-mail: e.mandowska@wsp.czest.pl

Summary: Spectrally resolved thermoluminescence (TL) measurements provide a lot of information concerning trapping and recombination states in various dielectric materials. As long-lived luminescence is typically weak, this technique requires a very sensitive apparatus. This paper describes the measurement system that is used for studying spectrally resolved TL and phosphorescence decay in polymers. Exemplary investigations are presented for poly(N-vinylcarbazole) (PVK) thin films. The obtained spectra clearly indicate the influence of solvents (dioxane, chlorobenzene) and excitation wavelengths on TL in PVK samples.

Keywords: luminescence; phosphorescence; spectrally resolved thermoluminescence (TL); poly(N-vinylcarbazole) (PVK); thin films

Introduction

The thermoluminescence (TL) method was proposed by Urbach to study trapping processes in crystalline phosphors.[1] First theoretical explanation and a method of analysis of TL curves was proposed by Randall and Wilkins.[2] During TL measurement a sample at an appropriately low temperature is excited by a high-energy (e.g. UV or X-ray) radiation. When the sample is heated, the energy is released in the form of light. A series of peaks appearing on TL glow curve is usually attributed to trap levels characterised by different activation energies. Recently, many authors have pointed out the virtues of making spectral measurements during TL. The possibility to construct a three-dimensional representation of the intensity of emission as a function of wavelength and temperature adds materially to the ability to interpret the behaviour of the studied materials. The temperature peaks correspond to traps from which charge carriers are released whereas the emission bands carry information about the charge carrier recombination levels.

Much research has been done with poly(N-vinylcarbazole) (PVK, Figure 1).[3-5] It contains charge-transporting carbazole pendant groups. Though the polymer is not electrically

 DOI: 10.1002/masy.200450826

conductive it has good photoconductivity properties with the emission peak located in the violet-blue region. In this paper we present some results of spectrally resolved TL measurements performed in PVK thin films. Particularly, the influence of excitation wavelength and the influence of solvents used for the preparation of PVK samples on TL features are presented.

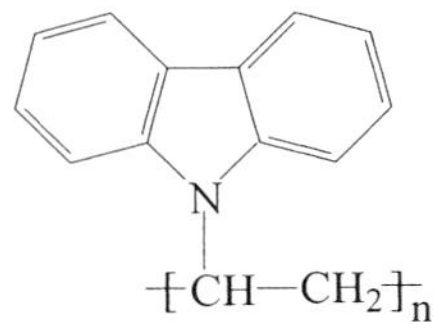

Figure 1. Poly(*N*-vinylcarbazole) (PVK).

Experiment

PVK (trade name Luvican, provided by BASF) was purified by dissolving in chloroform and reprecipitating into methanol six times. The powder was then redissolved in a solvent. Thus two kinds of samples were produced - PVKc (dissolved in chlorobenzene) and PVKd (dissolved in dioxane). Layers 50-100 μm thick were cast onto thick (0.3 mm) copper disk with diameter 20 mm. The samples were dried at room temperature for 24 h surrounded by vapour of the solvent.

The TL measurements were carried out in a cryostat that enabled sample temperature to be controlled between 78 K and 700 K. The sample was cooled with liquid nitrogen. A linear temperature rise, with respect to time, was provided by an autotuning temperature controller. The samples were excited with a 75 W xenon arc lamp installed in a high-intensity illumination system with a monochromator. The luminescence intensity was monitored using the spectrograph connected to a LN/CCD-1024E camera (chip format 1024×256, spectral range 190 - 1080 nm). The CCD camera was cooled with liquid nitrogen. It was connected to the temperature controller, which is necessary to stabilise working temperature of the camera (from 143 K to 203 K). A schematic diagram of this arrangement is shown in Figure 2. The sample was cooled down to 82 K and then excited with UV light at 323 nm or 363 nm. TL spectrum was recorded while heating the sample from 82 K to 300 K at a constant heating rate of 0.8 K/s. Temperature- and wavelength-resolved TL signals were obtained using custom-made software. The resolution of the system was varied from 10 to 20 nm depending on the of spectrograph grating, width of the spectrometer slit and the binding used for CCD camera chosen according to the intensities of the measured samples. The temperature resolution was about 3 K. The spectra were numerically calibrated with respect to wavelength and intensity.

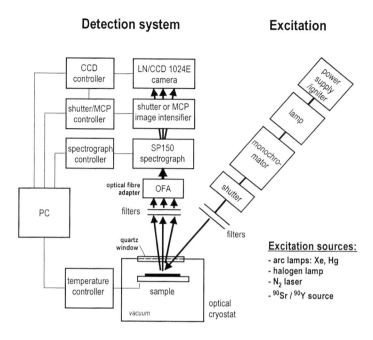

Figure 2. Block diagram of the measurement system.

Results

TL in PVK appears in the wavelength range 300-850 nm. Figures 3 and 4 demonstrate spectrally resolved TL spectra of PVKc excited at 323 nm (PVKc323) and 363 nm (PVKc363), respectively. Emission spectra of PVKc323 and PVKc363 are different. Figure 5 clearly shows the difference at the maximum intensity of TL in PVKc323 and PVKc363. When PVK was dissolved in dioxane, the TL spectra were the same for both excitations: 323 nm (PVKd323) and 363 nm (PVKd363) (Figure 6).

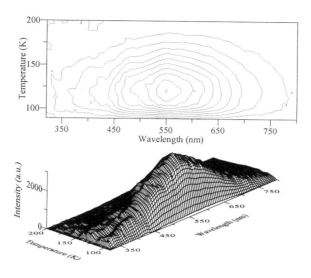

Figure 3. Isometric and contour plots of TL emission from PVKc323 sample. The spectra were recorded at 82 K - 300 K, after excitation at 323 nm.

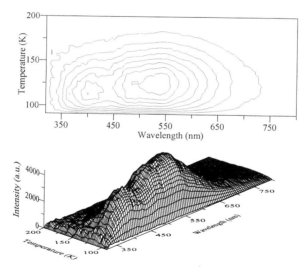

Figure 4. Isometric and contour plots of TL emission from PVKc363 sample. The spectra were recorded at 82 K – 300 K, after excitation at 363 nm.

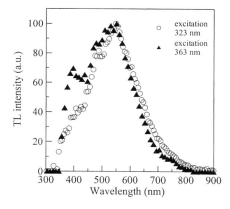

Figure 5. TL spectra of a PVKc thin film, at the maximum intensity of TL peak. Temperature range was 110 K-120 K (for PVKc323) and 123 K-133 K (for PVKc363).

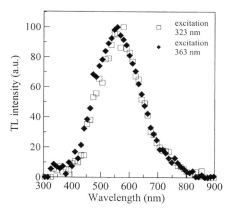

Figure 6. TL spectra of a PVKd thin film at the maximum intensity of TL peak. Temperature range was 110 K-120 K (for PVKd323) and 104 K-114 K (for PVKd363).

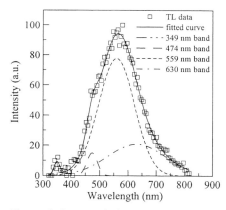

Figure 7. Deconvolution of TL emission spectrum in the temperature range 113-123 K (at the peak maximum) in PVKc363 to four bands.

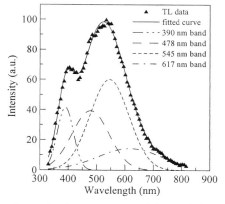

Figure 8. Deconvolution of TL emission spectrum in the temperature range 110-120 K (at the peak maximum) in PVKd323 to four bands.

As one can see, the TL spectra of PVKc323 and PVKd323 are not identical. The kind of the solvent used for preparation of the sample is also important. To characterise the emission spectra quantitatively, PVKc and PVKd were deconvoluted to several Gaussian bands. Figures 7 and 8 show fitting for PVKc363 and PVKd323, respectively. Table 1 presents detailed results of deconvolution for all samples.

Table 1. Emission spectra deconvoluted with four Gaussian bands in PVK at the maximum intensity (λ - wavelength, I - intensity of the emission spectra).

PVKc323		PVKc363		PVKd323		PVKd363	
λ (nm)	I (a.u.)	λ (nm)	I (a.u.)	λ (nm)	I (a.u.)	λ (nm)	I (a.u.)
382	13	390	47	352	9	371	6
490	49	462	35	477	19	473	17
570	41	538	72	558	80	552	88
622	33	619	18	636	25	648	16

Conclusions

Spectrally resolved thermoluminescence is a new powerful technique providing a lot of information relating to trap and recombination centres in dielectric solids. The measurements were made for PVK thin films to study the influence of a solvent on luminescence properties of this material. Emission spectra of PVK prepared in chlorobenzene and dioxane were deconvoluted for several nearly identical bands with significantly different intensities. This may suggest different structural arrangement of PVKc and PVKd. Due to various topological configuration some of the radiative transitions are more preferred but the other less. The same applies to luminescence quenching. Further studies are still required to explain physical basis of the reported phenomena.

Acknowledgements

We are grateful to Dr S. Tkaczyk for his help in preparation of the samples. This work was partly supported by grant No. 4T11F00922 from the Polish State Committee for Scientific Research.

[1] F. Urbach, *Wien. Ber. II A* **1930**, *139*, 363.
[2] J. T. Randall, M. H. F. Wilkins, *Proc. R. Soc. A* **1945**, *184*, 366.
[3] J. Sanetra, D. Bogdał, S. Nizioł, P. Aramatys, J. Pielichowski, *Synth. Met.* **2001**, *121*, 1731.
[4] G. Z. Li, N. Minami, *Chem. Phys. Lett.* **2000**, *331*, 26.
[5] J. Sanetra, F. Bai, M. Zheng, G. Yu, D. Zhu, *Thin Solid Films* **2000**, *363*, 118.
[6] D. B. Romero, M. Schaer, M. Leclerc, D. Ades, A. Sioce, L. Zuppiroli, *Synth. Met.* **1996**, *80*, 271.

Recombination in Complex Systems – An Analytical Approach

Arkadiusz Mandowski

Institute of Physics, Pedagogical University, ul. Armii Krajowej 13/15, 42-200 Częstochowa, Poland
E-mail: arkad@wsp.czest.pl

Summary: A new analytical model is proposed to describe the kinetics of trapping and recombination of charge carriers in complex systems with an arbitrary spatial distribution of traps and recombination centres. The structural properties of a material are described by two functions Γ_m and Γ_n irrespective of the thermal history of the sample. A simple method is proposed to determine the function Γ_m from simultaneous thermoluminescence (TL) and thermally stimulated conductivity (TSC) measurements.

Keywords: Monte Carlo simulation; recombination; thermoluminescence; thermally stimulated conductivity; traps

Introduction

Classic theories of trapping and recombination of charge carriers in dielectrics relate to the two analytically described cases. The first relates to uniform distribution of traps.[1] The second relates to pairs of traps and recombination centres placed close to each other.[2] None of the cases is likely to prevail in complex organic solids. Under such circumstances it is usually not possible to formulate a set of analytical equations describing charge carriers kinetics in such processes as phosphorescence, thermoluminescence (TL), thermally stimulated conductivity (TSC) and others. So far, no theory has been constructed that takes into account arbitrary spatial distribution of traps and recombination centres. Some peculiarities of the kinetics that may take place in non-homogeneous systems (e.g. 1-D structures or clusters) were shown in a number of papers in last few years.[3-5] Examples include apparently composite structures of monoenergetic peaks[4] and additional 'displacement' peaks.[6-8] The results were obtained numerically using Monte Carlo algorithms.[3] The numerical results were hardly compared with the experiment due to the lack of an appropriate analytical theory. In this paper new equations are presented for charge carrier trapping in a system with an arbitrary spatial distribution of traps and recombination centres. It is shown that the distribution could be characterised by two functions Γ_n and Γ_m.

DOI: 10.1002/masy.200450827

These functions depend only on structural properties (e.g. spatial distribution, energy barriers, etc.) of metastable states in a solid. The Γ function has to be determined separately for traps and recombination centres (RC). However, in most cases this process could be significantly improved by utilizing special symmetry properties. Examples of Γ functions and a method of determining Γ's from experimental data are presented.

Theory - two standard models

Commonly accepted explanation of long-lasting phosphorescence and thermoluminescence phenomena is based on the assumption of metastable levels (traps and recombination centres) situated within the energy gap. Although direct transition from trap to a recombination centre is possible, most of the transitions take place through excited states. This is shown schematically in the case of 'active' electron traps in Figure 1.

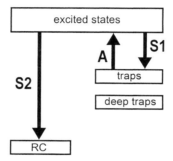

Figure 1. Energy diagram for a system consisting of trap levels, one kind of recombination centres (RC) and a number of deeper traps. **A** denotes thermally activated transitions (detrapping). **S1** and **S2** denote trapping and recombination, respectively. These are structural transitions depending on spatial configuration of traps and RC.

The diagram comprises several cases, including the model where electrons are excited to the conduction band, and the model of localized transitions. Only for these two cases it is possible to formulate a set of kinetic equations describing charge carriers kinetics. To simplify the following equations, we will assume that, within a given temperature range, only one trap level and one type of recombination centres are 'active'. All transitions to a set of deeper (thermally disconnected) traps are neglected. Thus, in the framework of the conduction model, the charge carrier kinetics obeys the following set of equations:

$$-\dot{n} = n\nu\exp\left(\frac{-E}{kT}\right) - n_c A(N - n), \tag{1a}$$

$$- \dot{m} = Bmn_{\mathrm{c}},\tag{1b}$$

$$m = n + n_{\mathrm{c}} + M,\tag{1c}$$

here E stands for the activation energy, N, n, and m denote the concentrations of trap states, electrons trapped in 'active' traps and holes trapped in recombination centres, respectively. n_{c} denotes the concentration of electrons in the conduction band. M stands for the number of electrons in the thermally disconnected traps (deep traps), i.e., traps that are not emptied during the experiment. A and B stand for the trapping and recombination probabilities, respectively, and ν is the frequency factor.

Another situation was considered by Halperin and Braner[2]. They assumed that traps and recombination centres are closely correlated in space, forming pairs that can be considered as independent units - i.e., the whole charge transfer takes place within groups of one kind, each having one trapping state, one excited state and one recombination centre. In this case the kinetic equations are the following (in a slightly modified form with respect to[2]):

$$- \dot{n} = n\nu\exp\left(\frac{-E}{kT}\right) - \overline{A}n_{\mathrm{e}},\tag{2a}$$

$$- \dot{m} = \overline{B}n_{\mathrm{e}},\tag{2b}$$

$$m = n + n_{\mathrm{e}},\tag{2c}$$

where n_{e} denotes the concentration of electrons in the excited state. Because the transport of charge carriers does not take place through the conduction band, the TL peak should not be accompanied by TSC. It is assumed that $\overline{A}$ and $\overline{B}$ are constants.

Theory - generalized equations

The simple energy configuration shown in Figure 1 comprises many more systems than those conceived by Eqs (1 and 2). Typical examples of this kind are presented in Figs 2 and 3. Figure 2 shows a set of clusters of traps and RC's. The clusters are separated by a distance and/or energy barriers. Each cluster has the energy configuration shown in Fig. 1. Figure 3 illustrates two one-dimensional systems of traps. It is assumed that charge carriers thermally released from traps are able to move along the chain. These exemplary arrangements are much closer to what one can find in such complex structures as molecular crystals or polymers. Properties of thermally stimulated relaxation processes in such systems were studied numerically using the Monte Carlo technique. Many new interesting features of this 'unusual type of kinetics' were discovered. Unfortunately, the results were hardly compared with experiment due to the lack of an appropriate analytical model of the phenomena. It will be shown that such a theory is possible to construct.

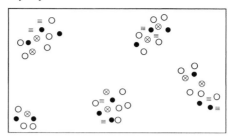

Figure 2. Groups of spatially correlated traps and recombination centres. ● - filled traps, ⊗ - empty traps, ≡ - thermally disconnected (deep) traps, O - recombination centres.

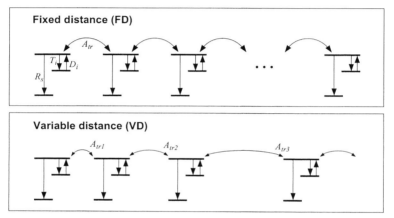

Figure 3. Exemplary one-dimensional models of a system of traps and RC. A carrier released from a trap can move along the chain. A_{tr}, A_{tr1}, ... A_{tr3} denote transitions between excited states. T_i, D_i and R_s denote trapping, detrapping and recombination, respectively.

Let us define the variable

$$\theta(t) = \int_0^t n_e(t')dt' \tag{3}$$

From simple probabilistic arguments[3,5] it is obvious that the total concentration of free holes should depend especially on $\theta(t)$ that is proportional to the total time spent by electrons in the excited state, and also on some other parameters W_k - dependent, e.g., on the initial occupation of traps. Therefore:

$$m(t) = F_m[\theta(t), W_k] \tag{4}$$

where F_m is an unknown function. As W_k does not depend on time it is obvious that

$$\frac{dm}{dt} = \frac{\partial F_m}{\partial \theta} n_e \tag{5}$$

Using similar arguments, in a special case when transitions S2 and A (Fig. 1) do not occur,

$$\frac{dn_c}{dt} = -\frac{dn}{dt} = \frac{\partial F_n}{\partial \theta} n_e \tag{6}$$

where F_n is another function. The equations allow to characterise transitions S1 and S2. The above partial derivatives are not constant, but depend on the actual concentrations of carriers in a destination trap level. Therefore we denote:

$$\Gamma_m(m) = -\frac{\partial F_m}{\partial \theta} \tag{7}$$

$$\Gamma_n(n) = -\frac{\partial F_n}{\partial \theta} \tag{8}$$

A comparison with previous sets of the kinetic equations allows to write finally:

$$-\dot{n} = n\nu \exp\left(\frac{-E}{kT(t)}\right) - \Gamma_n(n)n_e, \tag{9a}$$

$$-\dot{m} = \Gamma_m(m)n_e, \tag{9b}$$

$$m = n + n_e + M, \tag{9c}$$

Therefore we claim that the set of Eqs (9) describes the charge carrier kinetics in an arbitrary system consisting of traps and RC's, irrespective of its thermal history. As the shape of the Γ's depends solely on spatial distribution of traps and RC's, we conclude that the functions Γ_m and Γ_n describe structural properties of the system.

Properties of Γ functions

To show some basic properties of Γ_m and Γ_n, we calculate the functions using the previously described Monte Carlo method. Γ_m in the case of $M=N$ is shown in Fig. 4. The cases $n_0=1$ and $n_0=10^6$ correspond to the localised transitions and the conduction band model, respectively. In these cases Γ_m is represented by a straight line. The same result was obtained for Γ_n. Both Γ_m and Γ_n for the same case $M=N$ are presented in Fig. 5. Only for two standard cases (localised transitions and STM) the Γ's are linear. In all other cases the Γ's are nonlinear monotonic functions. The nonlinearity results in distorted (apparently composite[4,5]) TL peaks. However, the striking feature is the symmetry between Γ_m and Γ_n. This property may facilitate experimental determination of Γ functions.

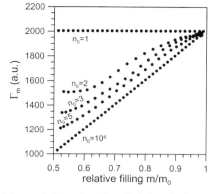

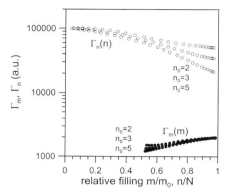

Figure 4. The function $\Gamma_m(m)$ calculated for $M=N$ and various occupancy of a single cluster n_0.

Figure 5. The functions $\Gamma_m(m)$ and $\Gamma_n(n)$ calculated for $M=N$ and $r=A/B=100$. n_0 and m_0 denote initial concentrations of trapped electrons and holes, respectively.

Conclusions

A new analytical model is proposed to describe TL kinetic properties for an arbitrary spatial distribution of traps and recombination centres. The structural properties of a material are described by two functions $\Gamma_m(m)$ and $\Gamma_n(n)$ irrespective of the thermal history of the sample measured. Any modification of the spatial arrangement or traps does not influence the structural function for recombination centres and *vice versa*. Thus the basic question concerns the possibility of determining Γ's from experimental data. Surprisingly, the answer is very simple, at least in the straightforward case, when TL and TSC intensities are characterized by common definitions: $J_{TL} \propto -\dot{m}$ and $J_{TSC} \propto n_e$. Applying the relations to Eq. (9b), one gets

$$\Gamma_m\left(m_0 - \alpha \int_0^t J_{TL}(t')dt' \right) \propto J_{TL}/J_{TSC} \tag{10}$$

Therefore, the information on Γ_m could be achieved from simultaneous TL/TSC measurements. Any assumptions with respect to Γ_n can be made using symmetry arguments.

[1] R. Chen, S. W. S. McKeever, *"Theory of Thermoluminescence and Related Phenomena"*, World Scientific, Singapore 1997.
[2] A. Halperin, A. A. Braner, *Phys. Rev.* **1960,** *117*, 408.
[3] A. Mandowski, J. Świątek, *Philos. Mag. B,* **1992,** *65*, 729.
[4] A. Mandowski, J. Świątek, *Radiat. Prot. Dosim.* **1996,** *65*, 25.
[5] A. Mandowski, J. Świątek, *J. Phys. III (France)* **1997,** *7*, 2275.
[6] A. Mandowski, J. Świątek, *Synth. Met.* **2000,** *108*, 203.
[7] A. Mandowski, *Radiat. Meas.* **2001,** *33*, 747.
[8] A. Mandowski, *J. Electrostatics* **2001,** *51-52*, 585.

Macromol. Symp. **2004**, *212*, 263-268

Effect of Spatial Irregularities on the Temperature and Field Dependence of the Mobility in Liquid-Crystalline Conjugated Polymer Films

Simon J. Martin, Agapi Kambili,* Alison B. Walker

Department of Physics, University of Bath, Claverton Down, Bath BA2 7AY, UK

Summary: We have performed simulations of time-of-flight measurements via the Monte Carlo approach for films made of conjugated polymers in the liquid-crystalline phase. In spatially regular films with a distribution of on-site energies the mobility is affected by the interplay between electrostatic and thermal energy. When the width of the on-site energy distribution is comparable to the thermal energy, the mobility increases with temperature at low fields, but shows the opposite behaviour at larger fields. However, when spatial irregularities in the arrangement of the film are introduced, the mobility is enhanced at all temperatures, as the electrostatic energy plays less of a role in charge transport than thermal activation over energy barriers.

Keywords: charge transport; conjugated polymers; light-emitting diodes; Monte-Carlo simulation; spatial arrangement

Introduction

Conjugated polymers[1] have promising applications to solar cells and photodetectors[2-4]. In polymer light-emitting diodes[5] detailed investigation of the materials has resulted in rapid progress[6]. To commercialise polymer devices, high efficiencies, brightness, and carrier lifetimes are required[7]. It is, therefore, essential to fully understand the fundamental physics of electrical transport through conjugated polymers; for a recent review, see[8].

Previous theoretical studies have dealt with strongly disordered organic materials, with charge transport mainly attributed to hopping[9-11], and low carrier mobilities that depend strongly on temperature and electric field. However, large carrier mobilities are generally required. Recent measurements on aligned polymer films[12-14] have demonstrated enhanced carrier mobilities, varying only weakly with the electric field. To our knowledge, this weak field dependence has not yet been explained.

    DOI: 10.1002/masy.200450828

This realization has motivated us to investigate the transport properties of liquid-crystalline conjugated polymer films with the chains nematically aligned perpendicular to the direction of transport. In a previous work[15] we looked into the character of transport through such polymer films, and we showed that it is possible to obtain non-dispersive transport, in agreement with time-of-flight experiments conducted on liquid-crystalline polyfluorene (PFO) films[12, 13]. In the present work the interplay between electric field and temperature on the transport characteristics is examined in combination with the competing effect of spatial irregularities in the arrangement of the polymer chains.

The Model

In order to investigate the transport properties of liquid-crystalline polymer films, we have performed numerical simulations of the time-of-flight technique. The film is composed of conjugated polymer chains of length $L = 100$ nm which are nematically aligned perpendicular to the direction of the electric field, chosen as the x direction, with periodic boundary conditions applied along the other two directions. Based on the extended backbone conjugation of liquid-crystalline polymers, such as PFO, that makes them stiff, and on bond vibrations being of very high frequency and low amplitude, we have assumed that the polymer chains can be described as rigid rods. The thickness of the film in the direction of transport has been taken equal to $d = 1$ μm.

Hopping motion of the charge carriers under the influence of the electric field is assumed, which in general can be either intra- or inter-chain, and is described by a simple Monte Carlo model. The charges are taken to be negative here, but positive charge can be equally treated. At the start of the simulation, the charges are placed on the chains adjacent to the injecting electrode. The unnormalized probability of hopping between two sites i and j is equal to

$$p_{ij} = \gamma \exp(-\frac{\varepsilon_j - \varepsilon_i - e\vec{E}\vec{r}_{ij}}{k_B T}) \tag{1}$$

$\vec{E}$ is the electric field, $\vec{r}_{ij}$ is the relative position vector, k_B is the Boltzmann factor, and T is the temperature. γ denotes the electronic wavefunction overlap, and in the following we have considered only nearest-neighbouring hopping within a cut-off distance equal to 10 Å. The on-site energies ε_i are taken from a Gaussian distribution whose width σ determines the degree of energetic disorder present in the film. Since liquid-crystalline conjugated polymers are generally characterized by a high degree of chemical regularity, a small value of σ is adequate. Here, we have taken $\sigma / \kappa_B T = \alpha$, with α being of the order of 1, and $T = 300$ K.

Results and Discussion

First, we examine the case of equidistant arrangement of the chains, with all inter-chain distances equal to 10 Å (spatially regular case). In Figure 1 we present a current transient from our numerical simulations for electric field $E = 3 \times 10^5$ V/cm at room temperature, and for disorder $\alpha = 1$. The calculated current exhibits the typical behaviour of a time-of-flight signal, with a plateau followed by a decaying tail, indicative of non-dispersive transport[16]. From the current transient we extract the transit time t_T by using the current integration mode, which determines the transit time as the point at which the current has fallen to half its value in the plateau region.

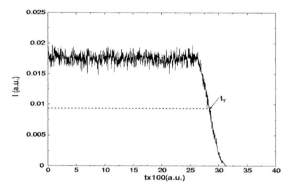

Figure 1. Current transient for $\alpha = 1$, $E = 3 \times 10^5$ V/cm, and $T = 300$ K, for regularly spaced chains. The transit time t_T is indicated by the arrow.

By identifying the transit times from the current transients for various values of the external electric field and for different temperatures, we have calculated the mobility via the relation $\mu = d/(Et_T)$. Figure 2 shows $\ln\mu$ as a function of $E^{1/2}$ for temperatures in the range from 100 K to 350 K. The left panel corresponds to $\alpha = 0.5$, and the right panel to $\alpha = 1$. For $\alpha = 0.5$, as the field increases for a given temperature, $\ln \mu$ decreases, since at large fields the electrostatic energy forces the charge carriers to follow shorter paths, causing the transit time to saturate. $\ln \mu$ decreases with temperature for all values of E, as seen in crystalline semiconductors. On the other hand, when $\alpha = 1$, at a given temperature, $\ln \mu$ always increases with the electric field. The effect is more pronounced at small T, as the thermal energy is not sufficient to enable the carriers to surmount the energy barriers. At a given field, there is a crossover between the behaviour at low fields, where $\ln \mu$ increases with temperature, and the behaviour at high fields, where the mobility decreases with T. A similar crossover has also been observed in other types of polymers within a Master equation approach[17]. However, in that case the crossover appears to occur at the same field for all temperatures considered. At room temperature, our mobility increases by a factor of 1.2, in approximate agreement with experimental data[12].

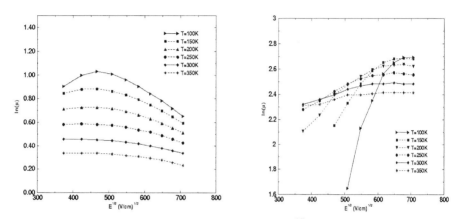

Figure 2. The mobility μ (arbitrary units) as a function of $E^{1/2}$ and for various temperatures, for regularly spaced chains. The left graph is for energetic disorder $\alpha = 0.5$, and the right graph for $\alpha = 1$.

The observed crossover with temperature does not appear in strongly disordered polymeric materials, in which the mobility always increases with T [6]. In Figure 3 we show $\ln \mu$ versus $E^{1/2}$ for various temperatures, but for energetic disorder $\alpha = 1.5$. The crossover that was previously seen now appears only in a small range of the field values, in particular at large E, and around room temperature. Thus, as the energetic disorder increases, the thermal energy aids the motion of the charge carriers towards the collecting electrode.

Figure 4 shows $\ln \mu$ as a function of $E^{1/2}$ and T, for $\alpha = 0.5$ (left) and $\alpha = 1$ (right), respectively, for a film in which the inter-chain distances vary randomly between 7 Å and 13 Å (irregularly spaced chains). Hopping is only allowed to nearest neighbours within the same cut-off distance as before. For small α, ln μ behaves in the same way as for the case of regular chain spacing in that at constant temperature it decreases with the field and for a fixed field it is lowered with temperature. For $\alpha = 1$, the mobility is enhanced with temperature for all values of the electric field, as was shown in recent measurements of PFO films[18]. This behaviour resembles that of strongly disordered polymer films[6]. In our case the increase in ln μ with temperature arises from the spatial irregularities in the arrangement of the chains within the film, since the distribution in the relative position vector leads to variations in the electrostatic energy, and thermal activation dominates. In all cases, however, μ varies only weakly with the electric field.

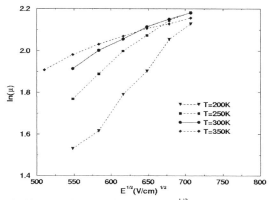

Figure 3. The mobility μ (arbitrary units) as a function of $E^{1/2}$ and for various temperatures, for regularly spaced chains and energetic disorder $\alpha = 1.5$

Summary

In this paper, we have presented calculations of the mobility of charge carriers through liquid-crystalline polymer films. By employing the Monte Carlo technique, we have investigated the electric field and temperature dependence of the mobility for films with energetic disorder comparable to the thermal energy. At room temperature the mobility varies only weakly with the electric field, and for energetic disorder parameter $\alpha = 1$, our predictions are in qualitative agreement with experimental data on PFO films[12]. For regularly spaced chains and for $\alpha = 1$, the mobility decreases with temperature for small fields and increases with T for larger fields. However, for random inter-chain distances the mobility increases with temperature, in agreement with recent experimental findings[18].

268

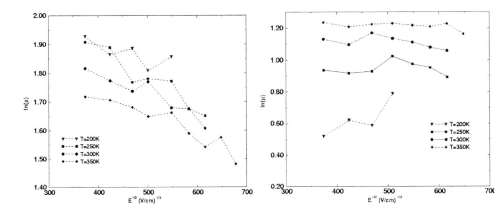

Figure 4. The mobility μ (arbitrary units) as a function of $E^{1/2}$ and for various temperatures, for irregularly spaced chains. The left graph is for energetic disorder $\alpha = 0.5$, and the right graph for $\alpha = 1$.

[1]   M. Pope and C. E. Swenberg, *"Electronic Processes in Organic Crystals and Polymers"* Oxford University Press **1999**
[2]   P. Gattinger, H. Rengel, D. Neher, *J. Appl. Phys.* **1998**, *84*, 3731
[3]   P. Peumans, V. Bulović, S. R. Forrest, *Appl. Phys. Lett.* **2000**, *76*, 2650
[4]   F. Zhang, M. Svensson, M. R. Andersson, M. Maggini, S. Bucella, E. Menna, O. Inganäs, *Adv. Mater.* **2001**, *13*, 1871
[5]   J. H. Burroughes, D. D. C. Bradley, A. R. Brown, R. N. Marks, K. Mackay, R. H. Friend, P. L. Burn, A. B. Holmes, *Nature* **1990**, *347*, 359
[6]   G. Hadziioannou and P. F. van Hutten, *"Semiconducting Polymers, Chemistry, Physics and Engineering"*, Wiley-VCH, Weinheim, 2000
[7]   R. H. Friend, R. W. Gymer, A. B. Holmes, J. H. Burroughes, R. N. Marks, C. Taliani, D. D. C. Bradley, D. A. Dos Santos, J. L. Brédas, M. Lögdlund, W. R. Salaneck, *Nature* **1999**, *397*, 121
[8]   A. B. Walker, A. Kambili, S. J. Martin, to be published in *J. Phys.: Condens. Matter.* **2002**
[9]   H. Bässler, *Phys. Status Solidi* **1993**, *175*, 15
[10] S. V. Novikov, D. H. Dunlap, V. M. Kenkre, P. E. Parris, A. V. Vannikov, *Phys. Rev. Lett.* **1998**, *81*, 4472
[11] S. V. Rakhmanova, E. M. Conwell, *Synth. Met.* **2001**, *116*, 389
[12] M. Redecker, D. D. C. Bradley, M. Inbasekaran, E. P. Woo, *Appl. Phys. Lett.* **1998**, *73*, 1565
[13] M. Redecker, D. D. C. Bradley, M. Inbasekaran, E. P. Woo, *Appl. Phys. Lett.* **1999**, *74*, 1400
[14] H. C. F. Martens, O. Hilt, H. B. Brom, P. W. M. Blom, J. N. Huiberts, *Phys. Rev. Lett.* **2001**, *87*, 086601
[15] A. Kambili, A. B. Walker, *Phys. Rev. B* **2001**, *63*, 012201
[16] J. C. Scott, B. A. Jones, L. T. Pautmeier, *Mol. Cryst.* **1994**, *253*, 183
[17] Z. G. Yu, D. L. Smith, A. Saxena, R. L. Martin, A. R. Bishop, *Phys. Rev. B* **2001**, *63*, 085202
[18] D. Poplavskyy, T. Kreouzis, A. J. Campbell, J. Nelson, D. D. C. Bradley, Paper P1.4, *Materials Research Society, 2002 Spring Meeting, MRS Proceedings* (Eds: G. E. Jabbour, N. S. Sariciftci, S. T. Lee, S. Carter, J. Kido) **2002**, 725

Luminescence Properties of Epoxy Resins Modified with a Carbazole Derivative

Ewa Mandowska,[1] *Wojciech Mazela,*[2] *Piotr Czub,*[2] *Arkadiusz Mandowski,*[1] *Jan Pielichowski,*[2] *Józef Świątek*[1]

[1] Institute of Physics, Pedagogical University, ul. Armii Krajowej 13/15, 42-200 Częstochowa, Poland
E-mail: e.mandowska@wsp.czest.pl
[2] Department of Chemistry and Technology of Polymers, Cracow University of Technology, ul. Warszawska 24, Kraków, Poland

Summary: Luminescence properties of a new material - epoxy resin with added 9-(2,3-epoxypropyl)carbazole (REPK) were studied. Absorption and photoluminescence (PL) spectra of REPK are compared with those of poly(*N*-vinylcarbazole) (PVK). PL in REPK is shifted to shorter wavelengths. Its intensity is higher than in PVK. REPK emits light in the range from 330 nm to 470 nm. PL spectrum of REPK could be well deconvoluted for four emission bands.

Keywords: carbazole; epoxy resins; light emitting devices (LED); luminescence; polymers

Introduction

Much research has been done with organic light emitting devices (LED), especially with polymers containing carbazole groups, which have high application potential in flat panel displays.[1-3] Unfortunately, these materials suffer from inappropriate mechanical properties and their usually complex synthesis. In our study epoxy resin was used as a base material. The epoxy resin with carbazolyl groups was prepared by adding glycidyl derivative of carbazole - 9-(2,3-epoxypropyl)carbazole (EPK) (Figure 1) - to epoxy resin (R). This way we obtained a homogenous composition, which was crosslinked in the next step. In curing process we obtained a homogenous polymer with the carbazolyl group chemically bonded to the resin (REPK). Such polymer has excellent mechanical properties, high thermal stability and good chemical resistance.[4] In this paper we present some results of the luminescence measurements (absorption and photoluminescence (PL)) performed on thin layers of REPK.

DOI: 10.1002/masy.200450829

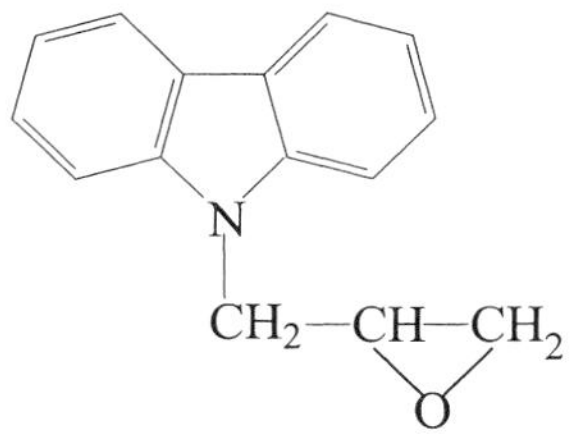

Figure 1. 9-(2,3-epoxypropyl)carbazole (EPK).

Experimental

The sample was prepared as a mixed composition, which contains 10 wt. % of EPK (R10EPK), using low-molecular-weight commercial epoxy resin – Ruetapox 0162 (Bakelite AG). Glycidyl derivative of carbazole was obtained by the reaction of carbazole with epichlorohydrin.[5] The modifier was dried to constant weight and powdered in a ceramic mortar. In the next step, components were mixed mechanically and then kept at 333 K for 24 h. Isophoronediamine (Euredur 46-Ciba) was used as a curing agent. Thin films (1.300 μm thick layers) were deposited from acetone solution on a glass substrate for absorption spectra and on a copper substrate for luminescence spectra. The obtained layers were then cured at room temperature and post-cured at 363 K for 24 hours. The finished material is homogeneous, amorphous, with 2-3% content of crystalline phase.

The PL and absorption measurements were carried out in an optical cryostat. The samples were excited using a 75 W xenon arc lamp installed in a high-intensity illumination system. Luminescence intensity was monitored using an SP150 spectrograph connected to the LN/CCD-1024E camera (chip format 1024x256, spectral range 190-1080 nm). The CCD camera is cooled with liquid nitrogen to reduce the dark current. Temperature of CCD chip is stabilized with a controller between 143 K to 203 K. The samples during PL measurements were excited by UV light (at 282 nm and 286 nm) in air at room temperature. The wavelength resolution of the measurements was about 0.4 nm. The obtained spectra were numerically calibrated with respect to wavelength and intensity.

Results

The absorption spectrum of R10EPK is presented in Figure 2. Its shape is similar to the absorption spectrum of poly(N-vinylcarbazole) (PVK).[6] The absorption edge for this material is ca. 360 nm. PL in R10EPK appears in the wavelength range from 330 nm to 470 nm. PL and absorption spectra are symmetrical, crossing at about 348 nm (Figure 3). Therefore the distance between singlet states S_{00} and S_{10} is about 3.4 eV. The intensity of PL for R10EPK is dependent on the thickness of the sample. It reaches a maximum for about 5 μm of the sample thickness, becoming constant for the sample thickness larger than 16 μm. Analogous measurements were performed for PVK for the same size and the same thickness of samples. The PL intensity of R10EPK is higher than intensity of PVK (Figure 4).

PL emission spectrum of R10EPK has a complex structure. To describe its shape more quantitatively, it was deconvoluted numerically giving four emission bands: 352 nm (maximal intensity), 366 nm, 377 nm and 395 nm. It was assumed that the bands have Gaussian shape. The deconvolution is shown in Figure 5.

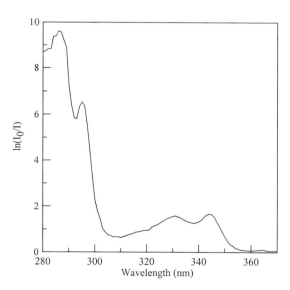

Figure 2. Absorption spectrum of an R10EPK thin film at room temperature.

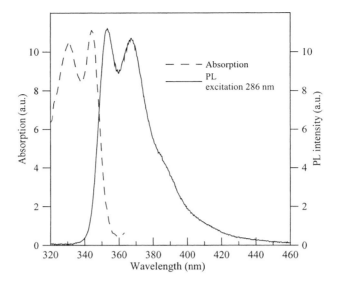

Figure 3. Absorption and photoluminescence spectra of R10EPK thin film at room temperature. PL spectrum measured after excitation at 286 nm.

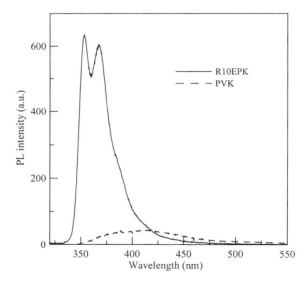

Figure 4. Photoluminescence spectra of R10EPK and PVK thin films at room temperature. Excitation at 282 nm.

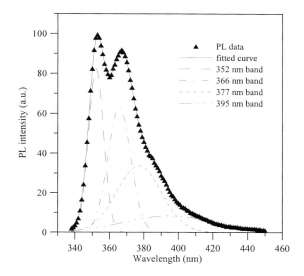

Figure 5. PL spectrum of an R10EPK thin film at room temperature excited at 286 nm. The spectrum is fitted with four emission bands.

Conclusions

Luminescence properties of a new material – R10EPK – were studied and compared with the known PVK polymer. Although absorption spectra of these two materials are similar, PL in R10EPK has greater intensity than in PVK. It is also shifted to shorter wavelengths with an effective range from 330 nm to 470 nm. The distance between singlet states S_{00} and S_{10} was found to be about 3.4 eV. PL emission spectrum of R10EPK has a complex structure. It can be represented as a sum of four Gaussians having maxima at 352 nm, 366 nm, 377 nm and 395 nm. Presumably these emission bands are related to vibrational levels of the singlet state S_0 of EPK. A very high PL efficiency of this material together with its excellent mechanical properties makes it a promising candidate for LED applications. This will be the aim of our further studies.

[1] J. Sanetra, D. Bogdał, S. Nizioł, P. Aramatys, J. Pilichowski, Synth. Met. **2001**, 121, 1731-1732.
[2] G.Z. Li, N. Minami, Chem. Phys. Lett. **2000**, 331, 26-30.
[3] J. Sanetra, F. Bai, M Zheng, G. Yu, D. Zhu, Thin Solid Films **2000**, 363, 118.
[4] Bryan Ellis, Chemistry and Technology of Epoxy Resins, Chapman & Hall, London 1994, p. 146.
[5] W. Mazela, J. Pielichowski, P.Czub, J. Sanetra, Prace Naukowe Instytutu Technologii Organicznej i Tworzyw Sztucznych Politechniki Wrocławskiej **2001**, 50.
[6] Y Qiu, L . Duan, X. Hu, D. Zhang, M. Zheng, F. Bai, Synth. Met. **2001**, 123, 39-42.

Macromol. Symp. **2004**, *212*, 275-280

Carbon Papers with Conductive Polymer Coatings for Use in Solid-State Supercapacitors

Andrew White, * *Robert Slade*

Chemistry, University of Surrey, Guildford, Surrey GU2 7XH, United Kingdom

Summary: Conductive polymer-coated carbon papers have been fabricated through polymerisation of pyrrole-based monomers oxidised with various heteropolyacids. Smooth surfaces are obtained when multiple coatings are applied to the carbon surface and give good contact with the Nafion® electrolyte. Cyclic voltammetry was used to study the electrodes and a.c. impedance and charge / discharge cycling was used to study membrane electrode assemblies (MEA). MEAs were fabricated using a hot-press technique.

Keywords: conducting polymers; electrochemical capacitors; heteropolymetallates; polypyrroles; supercapacitors

Introduction

There is a requirement for more efficient and cleaner ways of producing energy and delivering power by replacing old technologies such as the Carnot-limited internal combustion engine. Viable alternatives include battery and fuel cells, which readily provide large amounts of energy for proposed automotive, portable and power utility applications. [1,2] The major disadvantage of these systems is low power capability caused by thermodynamic or electrode polarisation problems.[3] Supercapacitors (sometimes referred to as ultracapacitors or electrochemical capacitors) solve the power issue when placed in parallel. A hybrid system is high both in energy (a battery or fuel cell component) and in power (a supercapacitor component).[4] There are batteries, such as insertion lithium ion, which have improved power capabilities, but at the time of writing, these are expensive and suffer from temperature problems.

The supercapacitors under consideration in this study incorporate conductive polymer electrodes which store electrical charge faradaically in a three-dimensional system in a similar way to RuO_2-based systems.[5] Polymer electrodes previously studied for use within supercapacitors include polyaniline, polypyrrole and polythiophenes with various structures.[6,7]

DOI: 10.1002/masy.200450830

In this work polymer-coated carbon papers were chemically fabricated and tested as electrodes within a supercapacitor test cell. Polymer coatings were polypyrroles or poly(N-methylpyrrole)s doped with heteropolymetallates. The electrolyte is a Nafion® membrane, which was used to form a solid-state device with the coated electrodes. Nafion® could be considered as a safe alternative to corrosive aqueous acids or flammable and toxic organic-based electrolytes.

Single electrode fabrication and results

Carbon paper was pre-treated by washing in acetone and then hydrogen peroxide (30%) to remove grease and any other impurity. Multiple coatings of the polymer were grown chemically using aqueous heteropolyacids as oxidant solutions until full coverage of the carbon paper was obtained. The monomers, polypyrrole and N-methylpyrrole, were reacted separately with the heteropolyacids. (dodecamolybdophosphoric acid, dodecamolybdosilicic acid and dodecamolybdo-divanadophosphoric acid).

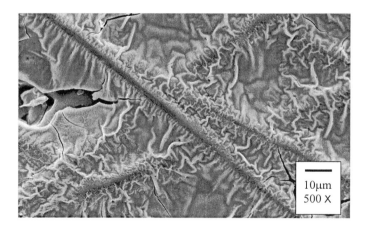

Figure 1. Scanning electron micrograph (secondary electron image) of carbon paper multiply coated with polypyrrole doped with dodecamolybdophosphoric acid.

Black conductive polymer coatings were obtained in both cases, which, when viewed using SEM (Figure 1), were all found to be smooth with a skin-like (pellicular) morphology. The wrinkled surface observed arises from the drying process in which the polymer coating contracts. The smooth surface is necessary for contact with Nafion® electrolyte, reducing any contact resistances that may occur. The polymer coatings adhered well to the surface of the carbon paper, having typical coating thickness 5 - 15 μm.

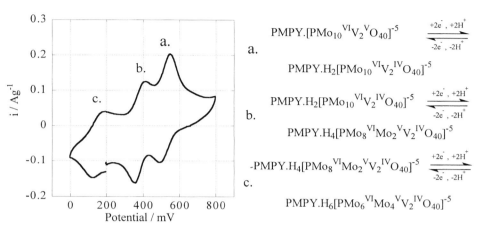

Figure 2. Carbon paper with a poly(*N*-methylpyrrole) coating doped with dodecamolybdo-divanadophosphoric acid (0.1 M H_2SO_4 against Ag / AgCl), with the accompanying mechanism for the electron transfer reactions.

Cyclic voltammetry of the electrodes revealed peaks associated with the three heteropolyacid electron transfers involving $2e^-$ exchanges which completely dominated the polypyrrole and poly(*N*-methylpyrrole) peaks (Figure 2). Lira-Cantú and co-workers have previously observed these peaks associated with encapsulated heteropolyacid.[8,9] The overall charge of the heteropolyacid is maintained as the polymer stabilises *via* proton transfer from the backbone.[10]

Lowering of the scan rate resulted in an increase of the reversibility of the system; for example, the polypyrrole electrode doped with the dodecamolybdophosphoric acid had peak separations, ΔE_p, of ≈ 200 mV at a scan rate of 100 mV s^{-1} which reduced to ≈ 40 mV at 1 mV s^{-1}. Capacitances for the electrodes were calculated from the division of the mean of the anodic and cathodic current peaks (I in A) by the scan rate (s, in mV s^{-1}) (Equation 1).

$$C = \frac{i}{s} \tag{1}$$

The highest capacitance obtained was given by the carbon paper coated with the 10-molybdo-2-vanadophosphoric acid-doped polypyrrole at 1 mV s^{-1}. The values were 776 F g^{-1} (per mass of active polymer) and 0.79 F cm^{-2} (of electrode area).

Fabrication and results from membrane electrode assemblies (MEA)

Disks were cut (1.3 cm in diameter) from the polymer-coated carbon papers and then pre-treated by boiling in deionised water for 30 minutes. Nafion$^®$ 115 was cut into squares (3 cm x 3 cm) boiled successively in hydrogen peroxide (30%), nitric acid (4 mol dm^{-3}) and then repeatedly refluxed in deionised water. The treated Nafion$^®$ was then placed between two electrode disks and pressed at 75 °C for 15 minutes under 5 tonnes using a KBr disk press and thermocouple-controlled drum heaters. The MEA was cut from excess Nafion$^®$ and then re-hydrated to a required water content using either 100% relative humidity (placed in a sealed container over a dish of water) or wet (re-hydrated in a dish of deionised water).

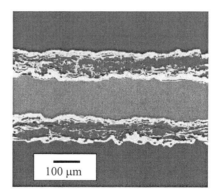

Figure 3. Scanning electron microscopy (secondary electron image) of the cross-section of a typical membrane electrode assembly. With carbon paper electrodes coated with polypyrrole doped with dodecamolybdophosphoric acid.

Cross-sectional analysis (Figure 3) revealed good contact between the electrodes and the Nafion$^®$ with an MEA thickness of 300 μm. The polymer coat ($\approx$15 μm thick) did not appear to penetrate the internal surfaces of the carbon paper. The hot pressing technique reduced the thickness of the carbon weave by 30%, from ≈ 90 to 60 μm. Nafion was reduced to 115 μm, which is lower than the thickness obtained from uncompressed Nafion at 0% relative humidity.

MEAs were placed into an in-house built test cell, placed under a torque of 5 N m and tested using a.c. impedance analysis and charge / discharge cycling techniques. Charge / discharging cycling was observed over 200 cycles at ±1 mA. The a.c. impedance of the MEA was taken *in situ* at the beginning and end of the 200 galvanic cycles.

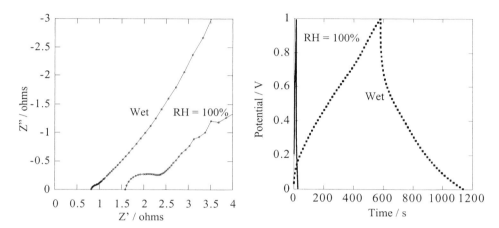

Figure 4. A.c. impedance plot and charge / discharge curves of MEA. Electrode coating: polypyrrole doped with dodecamolybdo-divanadophosphoric acid. Run under wet and RH = 100% conditions.

The resistance relating to the Nafion® electrolyte, the interfacial contact between the membrane and the electrode resistance all increased on decreasing relative humidity or after galvanic cycling (Figure 4). The MEAs which were run under wet conditions were found to give higher values for specific capacitance, specific power and specific energy when compared with the MEA run at RH = 100% (Figure 4). Improved discharge characteristics were observed from the MEA, which incorporated polypyrrole electrodes.

Conclusion

Electrodes were made which showed a suitable morphology for good contact with Nafion®
electrolyte. Cyclic voltammetry showed the strong influence of the heteropolyacid encapsulated
in the electrodes, responses which dominated the CVs. The heteropolyacid dodecamolybdo-
divanadophosphoric acid combined with polypyrrole led to high values of capacitance per unit
mass of active polymer. Charge / discharge and a.c. impedance tests on MEA devices showed
that the most favoured systems were run under wet conditions and employed polypyrrole doped
with heteropolyacids.

Acknowledgements

We like to thank Mr A. Hook for the test cell construction and Regenesys Technologies Limited
for providing part funding and materials for this work. The work was also funded by the
Engineering and Physical Science Research Council (grant GR/R22483/01).

[1] A. Burke, *J. Power Sources*, **2000**, 91, 37.
[2] G. Ariyoshi, K. Murata, K. Harada and K. Yamasaki, *IEICE Trans. Fundam.*, **2000**, E38-A, 1014.
[3] L. P. Jarvis, T. B. Atwater and P. J. Cygan, *J. Power Sources*, **1999**, 79, 60.
[4] D. J. T. Tarnowski, H. Lei, C. Peiter and M. Wixom, 200th Meeting of the Electrochemical Society, Inc. and the 52nd Meeting of the International Society of Electrochemistry.
[5] R. A. Huggins, *Philos. Trans. R. Soc. London, Ser. A*, **1996**, 354, 1555.
[6] F. Fusalba, P. Gouérec, D. Villers and D. Bélanger, *J. Electrochem. Soc.*, **2001**, 148, A1.
[7] J. P. Ferraris, M. M. Eissa, I. D. Brotherston and D. C. Loveday, *Chem. Mater.*, **1998**, 10, 3528.
[8] P. Gómez-Romero and M. Lira-Cantú, *Adv. Mater.*, **1997**, 9, 44.
[9] P. Gómez-Romero, M. Lira-Cantú and N. Casañ, *Solid State Ionics*, **1997**, 101-103, 875.
[10] M. Barth, M. Lapkowski and S. Lefrant, *Electrochim. Acta*, **1999**, 44, 2117.

Electroluminescence and Charge Photogeneration in Poly(9,9-dihexadecylfluorene-2,7-diyl) and Its Blends

Věra Cimrová,* Drahomír Výprachtický

Institute of Macromolecular Chemistry, Academy of Sciences of the Czech Republic, Heyrovský Sq. 2, 162 06 Prague 6, Czech Republic
E-mail: cimrova@imc.cas.cz, vyprach@imc.cas.cz

Summary: Photoluminescence, electroluminescence (EL), charge photogeneration and transport were studied in poly(9,9-dihexadecylfluorene-2,7-diyl) (PFC16), poly[(2,5-dihexadecyl-1,4-phenylene)(1,4-phenylene)] (PPPC16) and their blends. Blending of PFC16 with PPPC16 led to a significant improvement of the EL efficiency and stability compared with the devices fabricated from the neat polymers. Efficient blue and white light-emitting devices (LEDs) were fabricated using the blends. The increase in the EL efficiency was attributed to modification of the charge injection, transport and recombination properties in the blend.

Keywords: electroluminescence; LEDs; poly(1,4-phenylene)s; polyfluorenes; polymer blends

Introduction

Recently, an important issue under investigation has been the development of efficient and stable blue light-emitting materials. In this area, conjugated polymers with large bandgap such as poly(1,4-phenylene)s (PPP) and polyfluorenes (PF) are of particular interest due to their high photoluminescence (PL) efficiency in the blue spectral region.[1-7] The performance of polymer LEDs can be improved by optimizing the parameters of the active medium through polymer blending. We have shown that the aggregation of poly[2,5-bis(isopentyloxy)-1,4-phenylene] (SPPP) molecules could be avoided in blends composed of SPPP and poly[methyl(phenyl)silanediyl] (PMPSi).[6] Using polymers blends, the EL efficiency and time stability can be improved.[6,7]

In this paper we have shown that an increase in the EL efficiency and stability of blue emission in PF-based LEDs can be achieved using polymer blends composed of two electroluminescent polymers - poly(9,9-dihexadecylfluorene-2,7-diyl) (PFC16) and poly[(2,5-dihexadecyl-1,4-phenylene)(1,4-phenylene)] (PPPC16). We have shown that such blending can significantly improve the LED performance.

DOI: 10.1002/masy.200450831

Experimental

Chemical structures of the polymers under study are shown in Figure 1. PFC16 was synthesized by the Yamamoto coupling reaction [8] from 2,7-dibromo-9,9-dihexadecylfluorene using zinc, the nickel(II)/triphenylphosphine complex as a catalyst, and 2,2′-bipyridine as a coligand. SPPPC16 were prepared by the Suzuki coupling of bis(boronate) with the corresponding dibromo derivative in the presence of a palladium catalyst. [8]

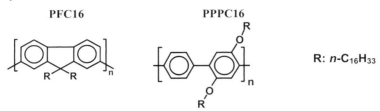

Figure 1. Chemical structures of the polymers under study.

Thin polymer films were prepared by spin-coating from toluene solutions. All films exhibited a good optical quality. Polymer LEDs with a hole-injecting indium-tin oxide (ITO) electrode and an electron-injecting aluminium (Al) electrode were fabricated. The ITO electrode was further optimized by coating with a thin layer of a conductive polymer, poly[3,4-(ethylenedioxy)thiophene]/poly(styrenesulfonate) (PEDT-PSS), to improve its hole-injecting properties. The 50-nm thick PEDT-PSS layers were also prepared by spin-coating. Two-layer ITO/PEDT-PSS/polymer/Al devices were studied. 60-80 nm thick Al electrodes were vacuum-evaporated on the top of polymer films. Typical active areas of the LEDs were 4 mm^2. Samples for charge transport and photogeneration measurements (film thicknesses from 700 to 1300 nm) were prepared by spin coating of the polymer solution in toluene onto stainless steel substrates. All polymer films were dried in vacuum (10^{-3} Pa) at 323 K for 6 h. Cyclic voltammetry was measured using a PA4 polarographic analyzer (Laboratory Instruments, CZ) and ionization potential was determined as described in Ref.[9], where more details are also given concerning the experimental setup. UV-vis spectra were measured on a Perkin-Elmer Lambda 20 spectrometer and EL and PL spectra using a home–made spectrofluorometer with single photon-counting detection (SPEX, RCA C31034 photomultiplier). LEDs were from a Keithley 237 source measuring unit, which served for simultaneous recording of the current passing through the sample. Current-voltage and EL intensity-voltage characteristics were recorded simultaneously using a Keithley 237 source measuring unit and a silicon photodiode with integrated amplifier (EG&G HUV-4000B) for detection of total light output. The measurements of the photogeneration efficiencies were carried out using the method of photoinduced surface potential discharge decay (xerographic).[10] Discharge experiments were performed with a home-built setup as described in Refs[9,11]. A polymer sample was charged to the initial surface potential with a corona and then discharged upon irradiation. The measurements were performed with irradiation incident on the positively charged surface. To ensure emission-limited conditions, the discharge was measured with photon fluxes as low as possible (10^{17}-10^{18} photon $m^{-2} s^{-1}$). Transport properties were studied by measurements of the dark surface potential decay.[12]

Results and Discussion

Typical absorption and PL of thin spin-coated films of PFC16 and PPPC16 on a fused silica substrate are shown in Figs 2a and 2b. Thin films of both PFC16 and PPPC16 exhibited efficient blue PL with the maxima at 425 nm and 415 nm, respectively. Also shown is the electroluminescence (EL) spectrum. In polyfluorene chains with long alkyls, the aggregate formation is not so pronounced as in polyfluorenes with shorter alkyls (octyl), where a red-shifted emission is observed already on freshly prepared films. But the EL efficiency and stability of the single-component PFC16 or PPPC16 devices is not high enough; a higher EL efficiency was detected in PPPC16 devices.

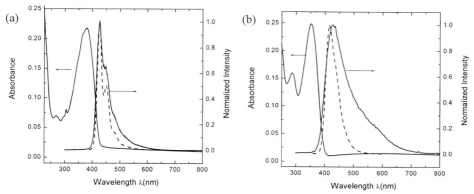

Figure 2. Absorption and photoluminescence (dashed line) spectra of thin polymer films and an electroluminescence (solid line) spectrum of ITO/PEDT:PSS/polymer/Al devices for polymers: (a) PFC16 and (b) PPPC16.

Using blends made of PFC16 and PPPC16, more efficient blue LEDs with the EL maximum at about 425 nm have been fabricated. An example of the PL and EL spectra for a blend LED is shown in Fig. 3a. The PL emission spectra of neat polymers and their blends indicated that PL emission from PFC16 dominates in blends. The EL onset and driving voltages of neat PFC16 and blend devices were nearly the same, even lower values were observed in the blend devices. Performance of the devices is shown Fig. 3b. Blending led to a significant improvement of the EL efficiency and stability compared with the devices made of neat EL polymers. In the blend LEDs, the EL efficiency and EL spectra depended on the contents of the components (see Fig. 3). The EL efficiency in the blends was up to one order of magnitude higher than that in PFC16 LEDs. The EL results can be mainly explained by modification of charge injection, transport and recombination in the polymer blend layer.

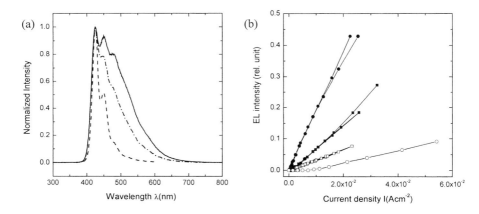

Figure 3. (a) PL spectrum (dashed line) of a thin film of a polymer blend and EL spectrum (solid line) of the ITO/PEDT:PSS/polymer blend/Al device for polymer blends composed of PFC16 and PPPC16 (7:3 - solid line, 1:1 – dash-and-dot line). (b) Spectrally integrated EL intensity as a function of the current density of the LEDs: ITO/PEDT:PSS/PFC16/Al (open circles), ITO/PEDT:PSS/PPPC16/Al (open squares), ITO/PEDT:PSS/PFC16+PPPC16 (1:1)/Al (solid squares) and ITO/PEDT:PSS/PFC16+PPPC16(7:3)/Al (solid circles).

From cyclovoltammetric measurements the value of ionization potential (I_P) of PFC16 and PPPC16 close to 5.8 eV and 5.5 eV were determined, respectively. The I_P value determined for PPPC16 is lower than that for PFC16, so the PPPC16 molecules can facilitate hole injection into PFC16. Transport properties of thin films were studied from the dark decays of the surface potential. Typical dark decay curves are shown in Fig. 4a. The decay curves were analyzed by a model of trap-controlled hopping.[12] The effective hole mobility and other model parameters such as the capture and release rates of charge carriers in a deep trap were determined by curve fitting. The fastest dark decay (curve 1 in Fig. 4a) was observed in PFC16, which corresponded to the highest effective mobility, i.e. the mobility of charge carriers controlled by energetically shallow traps in thermal equilibrium with the transport states. It is significantly higher than that in PPPC16 (curve 2). Depending on the contents of the components in the blend, the effective mobility could be modified (see curve 3 and 4). Due to the lower ionization potential, the PPPC16 molecules act as traps for holes in blend layers, which results in the charge transport modification, i.e. in a lower effective mobility.

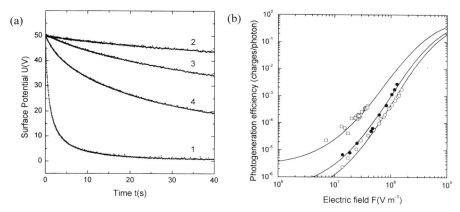

Figure 4. (a) Dark decay curves measured on thin films of (1) PFC16, (2) PPPC16, (3) PFC16+PPPC16(1:1) and (4) PFC16+PPPC16(7:3). (b) The electric field dependences of the photogeneration efficiency in PFC16 (open squares), PPPC16 (open circles), and PFC16+PPPC16(7:3) (solid circles). Solid lines were calculated using the Onsager theory with the Gaussian distribution of electron-hole pair radii with the parameters given in Table 1.

Charge photogeneration was studied to get information about charge dissociation and/or recombination in the electric field. The charge photogeneration efficiency, η_{ph}, was measured as a function of the electric field. η_{ph} in PFC16 was more than one order of magnitude higher than that in PPPC16 and the blends. The results fitted by the Onsager theory of geminate recombination [13] with the Gaussian distribution function of electron-hole (e-h) pair radii $g(r) = (\pi^{-3/2}\alpha^{-3}) \exp(-r^2/\alpha^2)$,[14,11] which gave the best description of the experimental data, are shown in Fig. 4b. The full lines represent the best theoretical fits to our experimental data with parameters given in Table 1. From the distribution parameters, the most effective separation distances of e-h pairs r_{le} and r_{he} at low and high electric fields, respectively, were determined. Compared with PFC16 in the blend, shortening of the separation distances without a change of the primary quantum yield $\eta_{0\alpha}$ was observed. This indicates a decrease in the geminate recombination quenching at the electric field, which is favorable for EL. On the other hand, compared with PPPC16, the separation distances have not changed or only slightly increased in the blend, which indicates a similar charge dissociation or recombination in PPPC16 and the blends.

In conclusion, the EL efficiency and time stability of polymer LEDs were improved using polymer blends made of PFC16 and PPPC16 compared with the single-component LEDs made of the EL polymers. The EL results can be explained mainly by modification of charge injection, transport and recombination, which was demonstrated by the charge transport and photogeneration measurements.

Table 1. The parameters α and $\eta_{0\alpha}$ obtained from fitting of the experimental dependences η_{ph} vs. F (see Fig. 4) using the Onsager model with the Gaussian distribution of e-h pair radii. The most effective separation distances of e-h pairs, r_{le} and r_{he}, at low and high electric fields, respectively.

Polymer	α (nm)	$\eta_{0\alpha}$	r_{le}	r_{he}
PPPC16	0.60	1.0	1.20	0.60
PFC16	0.80	1.0	1.60	0.80
PFC16+ PPPC16(7:3)	0.65	1.0	1.30	0.65
PFC16+PPPC16(1:1) (3:7)	0.60	1.0	1.20	0.60

We would like to acknowledge support of the Grant Agency of the Czech Republic (grants No. 102/98/0696 and No. 203/01/0512) and of the Grant Agency of Academy of Sciences of the Czech Republic (grant No. AVOZ 4050913).

[1] V. N. Bliznyuk, S. A. Carter, J. C. Scott, G. Klärner, R. D. Miller, and D. C. Miller, Macromolecules **32**, 361 (1999).
[2] K.-H. Weinfurtner, H. Fujikawa, S. Tokito, Y. Taga, Appl. Phys. Lett. **76**, 2502 (2000).
[3] M. Leclerc, J. Polym. Sci., Part A: Polym. Chem. **39**, 2867 (2001).
[4] I. Prieto, J. Teetsov, M.A. Fox, D. A. van den Bout, A. J. Bard, J. Phys. Chem. A **105**, 520 (2001).
[5] D. Sainova, T. Miteva, H.G. Nothofer, U. Scherf, I. Glowacki, J. Ulanski, H. Fujikawa, D. Neher, Appl. Phys. Lett. **76**, 1810 (2000).
[6] V. Cimrová, D. Výprachtický, J. Pecka, R. Kotva, Proc. SPIE **3939**, 164 (2000).
[7] V. Cimrová, D. Neher, M. Remmers, I. Kmínek, Adv. Mater. **10**, 676 (1998).
[8] D. Výprachtický, V. Cimrová, L. Machová, V. Pokorná, Collect. Czech. Chem. Commun. **66**, 1473 (2001).
[9] D. Výprachtický and V. Cimrová, Macromolecules **35**, 3463 (2002).
[10] I. Chen, J. Mort, J.H. Tabak, IEEE Trans. Electron Devices **19**, 413 (1972).
[11] V. Cimrová, I. Kmínek, S. Nešpůrek, W. Schnabel, Synth. Met. **64**, 271 (1994).
[12] V. Cimrová, S. Nešpůrek, H. von Berlepsch, J. Phys. D: Appl. Phys. **24**, 1404 (1991).
[13] L. Onsager, Phys. Rev. **54**, 554 (1938).
[14] A. Mozumder, J. Chem. Phys. **60**, 4300 (1974).

Macromol. Symp. **2004**, *212*, 287-292

Polypyrrole Electrodeposition on Inorganic Semiconductors CuInSe₂ and CuInS₂ for Photovoltaic Applications

Sergei Bereznev,[1] *Igor Konovalov,*[2] *Julia Kois,*[3] *Enn Mellikov,*[3] *Andres Öpik*[1]

[1] Department of Basic and Applied Chemistry, Tallinn Technical University, Ehitajate tee 5, Tallinn 19086, Estonia
E-mail: bereznev@edu.ttu.ee
[2] Institut für Solartechnologien, Im Technologiepark 7, 15236 Frankfurt (Oder), Germany
E-mail: ikono@ist-ffo.de
[3] Department of Materials Technology, Tallinn Technical University, Ehitajate tee 5, Tallinn 19086, Estonia
E-mail: enn@edu.ttu.ee

Summary: Thin polypyrrole (PPy) layers with an average thickness of about 0.5 µm were deposited, using potentiostatic and galvanostatic techniques, on CuInSe₂ (CISe) structures prepared electrochemically on glass/ITO substrates and on CuInS₂ (CIS) structures fabricated on Cu tape substrates. The polymer layer of p-type is considered as an alternative to the traditional buffer layer and window layer in the conventional cell structure. The deposition proceeded from an aqueous solution containing sodium naphthalene-2-sulfonate as a dopant. In order to prepare stable PPy films of high quality with a good adherence to the surface of inorganic semiconductors CIS and CISe, the optimal concentrations of reagents, current densities and electrodepositing potentials were selected experimentally. Electrochemical polymerization of pyrrole to PPy on CIS surfaces is faster under white light irradiation and the polymerisation starts at lower potential than in the dark. Significant photovoltage and photocurrent of the fabricated CISe/PPy and CIS/PPy structures have been observed under standard white light illumination.

Keywords: CIS; conducting polymers; electrodeposition; photovoltaic structure; polypyrroles

Introduction

At present, the potentials of photoenergy conversion using electrically conducting polymers (ECP) are under serious investigation due to producing low-cost, large-area and flexible photodiodes and solar cells.[1,2] Although a number of ECP were synthesized, the polypyrrole (PPy) was one of the most intensively studied polymers during the last decade.[3] PPy shows several application advantages like easy synthesis in aqueous media with a wide range of possible dopants, relatively high stability of electric conductivity and good mechanical properties. As noted, the combination of ECP with inorganic semiconductors, e.g. CuInSe₂

DOI: 10.1002/masy.200450832

(CISe) and CuInS$_2$ (CIS), is attractive for use in thin-film photovoltaic cell structures.[4] CIS and CISe are polycrystalline semiconductors with very high optical absorption coefficients, presently intensively studied for photovoltaic applications.[5,6]

Among various techniques of preparation of these structures with PPy, electrodeposition of the polymer on the surface of CIS and CISe deserves special attention because it can be easily controlled by regulating the polymerization current or potential and time for producing high-quality PPy films with predictable properties.[7] Moreover, electrodeposition is an inexpensive, low-temperature and relatively non-polluting method. However, the process of PPy electrosynthesis on the surface of the semiconductors with high electrical resistance is far from being easily controllable due to non-linear electrical processes in the semiconductors. PPy films with good quality are most often polymerized using certain sulfonic acids and their salts (naphthalenesulfonates, toluenesulfonates, etc.) as dopants with planar aromatic groups. These dopants induce structural regularity in the polymer film. Also, they can absorb more easily on the electrode surface and thus facilitate the first steps of the polymerization, leading to improved adhesion of PPy with high conductivity of the film.[3,7]

PPy always has p-type conductivity when doped with sulfonates, but the type of conductivity in CIS or CISe strongly depends on the composition and defects and may be p- or n-type.[6] The conductivity of n-type in the semiconductors is necessary for the formation of a rectifying heterojunction between PPy and CISe layers. The purpose of our work is to determine the optimal conditions of electrochemical preparation of high-quality PPy films with a good adherence to CIS and CISe and to investigate the obtained structures. In order to prepare these photovoltaic structures, the optimal concentrations of reagents, current densities and electrodepositing potentials were selected experimentally.

Experimental

Electrodeposition of n-CISe thin films with the thickness of about 1 μm onto ITO/glass substrates in potentiostatic mode at −900 mV using a Wenking LT 87 potentiostat have been described in a previous paper.[8] Aqueous solutions containing CuSO$_4$/In$_2$(SO$_4$)$_3$/SeO$_2$ in concentrations of 2/8/11 and 3/7/11 mM, respectively, were used for CISe(2-8-11) and CISe(3-7-11) films deposition. Thermal annealing of as-deposited CISe films was performed at 400 °C for 20 min in vacuum. Subsequently, the CISe films were etched in 10% KCN aqueous solution for 5 min to remove possible secondary phases.

CIS films have been prepared on copper tape substrate using the CISCuT method.[9] The idea of the method is a fast roll-to-roll sulfurization in S_x+N_2 gas atmosphere at 500-600 °C of 0.8 μm indium precursor on the copper tape substrate. Fast chemical reaction of indium and copper from the substrate with sulfur from the gas phase results in formation of a stack of several layers at the surface. The top layer of the stack has been shown to be a low-doped $CuInS_2$ phase. Prior to PPy deposition, the structure was etched in aqueous KCN solution and annealed in vacuum at 280 °C for 10 min.

Thin doped PPy films were synthesized on the surface of glass/ITO/CISe and Cu/CIS substrates galvanostatically or potentiostatically in the usual three-electrode electrochemical cell configuration using a VoltaLab™ 32 potentiostat/galvanostat in the dark and under white light illumination of 100 mW/cm^2. Pyrrole (Py) monomer (Aldrich) was distilled in vacuum prior to use. Polymerization was performed in the presence of sodium naphthalene-2-sulfonate (NSA) (Aldrich) as a dopant in de-aerated aqueous solutions of Py. The film thickness was controlled by the charge transferred during electrodeposition using the Faraday law.[10] Experimentally determined optimal values for the current density, applied potential and concentration of reagents are presented in Table 1.

Table 1. Optimal parameters for PPy electrodeposition.

Substrate	PPy deposition	Py/NSA mol/l	Current density mA/cm^2	Potential vs SCE mV	PPy film thickness μm	Deposi-tion time s
ITO/CISe	galvanostatical[a]	0.3/0.1	0.25-0.5	550-700	0.5-2.0	160-1280
Cu/CIS	galvanostatical[a]	0.025/0.009, 0.05/0.017	0.125-0.25	450-900	0.5-1.0	320-640
Cu/CIS	galvanostatical[b]	0.3/0.1	1.0-2.0	60-190	0.5-2.0	40-320
Cu/CIS	potentiostatical[b]	0.3/0.1	0.05-0.3	-200-0	0.25-0.5	400-1600

[a] In the dark.
[b] Under white illumination of 100 mW/cm^2.

Thermal annealing of PPy films at 100 °C for 6 h in air has markedly improved its adherence to CIS and CISe. The metal contact (Ag) was evaporated onto the polymer layer for the structures based on CISe. All investigated photovoltaic structures were fabricated in the sandwich configurations glass/ITO/CISe/PPy/Ag and Cu/CIS/PPy/i-ZnO/n-ZnO respectively. The ZnO:Al window layers were DC-sputtered, the i-ZnO layer was deposited in the presence of oxygen. The thickness of the i-ZnO layer was about 100 nm, the overall thickness of the ZnO window layers was about 1 μm.

Current-voltage characteristics were measured using a VoltaLab™ 32 potentiostat/galvanostat and Keithley 2400 SourceMeter. White light with an intensity of 100 mW/cm² from a tungsten-halogen lamp or xenon lamp was used for irradiation. The glass/ITO/CISe/PPy samples were illuminated from the glass substrate side. The area of the CISe/PPy/Ag and CIS/Ppy/ZnO junctions was about 1 and 0.1 cm², respectively. In the case of Cu/CIS/PPy, the polymer layer side was illuminated since CIS was deposited on the Cu substrate.

Results and Discussion

It is well known that electrodeposition of PPy is remarkably sensitive to the conditions at the surface of the substrate, e.g., to the adsorption of the monomer or oligomers at the electrode surface, the distribution of nucleation centers and the charge and potential distribution at the electrode/electrolyte interface. Frequently, the nucleation sites are sparsely occupied on the surface, resulting in a semicircular and insular morphology.[7] It has been found that NSA is highly suitable as a dopant for the deposition of smooth and adhesive PPy films onto the CIS and CISe layers. Figure 1 shows the electrode potential vs. time during galvanostatical deposition of PPy onto glass/ITO/CISe and Cu/CIS substrates in the presence of NSA as a dopant in the dark and under white light illumination.

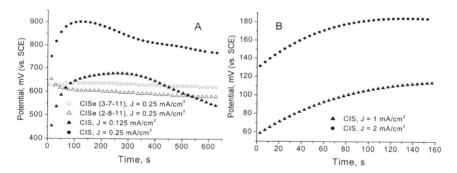

Figure 1. Potential vs SCE during galvanostatic polymerization of Py galvanostatically at glass/ITO/CISe and Cu/CIS substrates: A - in 0.3M Py/0.1M NSA and 0.025M Py/0.009M NSA solutions for the CISe and CIS, respectively, in the dark; B – in 0.3M Py/0.1M NSA solution for the CIS under illumination (100 mW/cm²).

The influence of illumination on PPy electrodeposition on CIS was investigated by cyclic voltammetry (Figure 2A). The cyclic voltammograms and Figure 1B show the significant role of light in polymerisation of Py on the CIS surface. At lower potentials, the anodic current corresponding to Py oxidative polymerization is substantially larger under illumination than in the dark. The synthesis of PPy film begins at –200 mV vs. SCE (Figure 2A) and the galvanostatic curves start at low potential when the current density is relatively high: 1 and 2 mA/cm^2 (Figure 1B). The illumination has been found to play two roles during polymerization: (i) primarily it affects the early stages of polymerization by creating high oxidative potential on the CIS surface relative to the substrate due to the so-called "as-grown cell" formed by the CISCuT process and (ii) it seems to enhance the nucleating process of PPy on CIS. These two effects give the possibility to produce high-quality PPy films on the CIS layers at low potentials and to avoid the danger of CIS degradation due to oxidation. Also, relatively high current densities could be achieved. It should be noted that for deposition of PPy films with the thickness less than 0.5 µm the potentiostatic method results in a better lateral homogeneity of the film, in a good agreement with the literary data.[7]

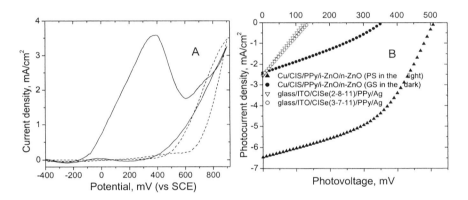

Figure 2. A – Cyclic voltammograms with scan rate of 20 mV/s of PPy doped with NSA deposited on CIS in 0.3M Py/0.1M NSA aqueous solution in the dark and under white light illumination (100 mW/cm^2). B – *I-V* characteristics of the ITO/CISe/PPy/ and Cu/CIS/PPy/i-ZnO/n-ZnO structures under irradiation of 100 mW/cm^2 (GS – PPy layer with the thickness of about 1 µm was deposited galvanostatically at 0.125 mA/cm^2 from 0.025M Py/0.009M NSA aqueous solution in the dark. PS – PPy layer with the thickness of about 0.25 µm was deposited potentiostatically at -200 mV vs. SCE under white light illumination of 100 mW/cm^2). The PPy layers on glass/ITO/CISe substrates were prepared galvanostatically at 0.25 mA/cm^2 in 0.3M Py/0.1M NSA aqueous solution in the dark.

Figure 2B shows photovoltaic properties of the obtained structures under 100 mW/cm^2 white light illumination. The incident light produced a short-circuit photocurrent density I_{sc} = 6.5 mA/cm^2 and an open-circuit voltage V_{oc} = 510 mV for the structure based on the CIS and PPy. The thickness of PPy film has noticeably affected the I_{sc} and V_{oc} because of the light absorption in PPy. In further studies, thinner CISe and PPy films could be used to improve the parameters of both types of the structures. The linearity of I-V characteristics of glass/ITO/CISe/PPy/Ag structures suggests the possibility that the photovoltaic behaviour could be fully controlled by the shunt resistance in the structure. Nevertheless, I_{sc} and V_{oc} are remarkably high and prove the value of the chosen approach.

Conclusions

The optimal conditions for electrodeposition of high-quality PPy films with a good adherence to CIS and CISe were determined in the dark and under illumination. The studies showed that the potentiostatic deposition of thin PPy film onto the CIS under white light illumination is preferable for producing the photovoltaic structures. The best structure so far showed an open circuit voltage of V_{oc} = 510 mV and a short circuit current of I_{sc} = 6.5 mA/cm^2 under xenon lamp white light illumination of 100 mW/cm^2 intensity.

Acknowledgements

The financial support from the Estonian Science Foundation (Grant 4857) is acknowledged.

[1] C. J. Brabec, N. S. Sariciftci, J. C. Hummelen, *Adv. Funct. Mater.* **2001**, *11*, No. 1, 15.
[2] K. Gurunathan, A. V. Murugan, R. Marimuthu, U. P. Mulik, D. P. Amalnerkar, *Mater. Chem. Phys.* **1999**, *61*, 173.
[3] A. G. MacDiarmid, *Synth. Met.* **1997**, *84*, 27.
[4] P. J. Sebastian, S. A. Gamboa, M. E. Calixto, H. Nguyen-Cong, P. Chartier, R. Perez, *Semicond. Sci. Technol.* **1998**, *13*, 1459.
[5] A. Catalano, *Proc. 1st IEEE World Conf. on Photovoltaic Energy Conversion (Hawaii)* **1994**, 52.
[6] H. W. Schock, A. Shah, *Proc. 14th European Photovoltaic Solar Energy Conf. (Barcelona)* **1997**, 2000.
[7] E. Kupila, J. Kankare, *Synth. Met.* **1995**, *74*, 241.
[8] S. Bereznev, J. Kois, E. Mellikov, A. Öpik, D. Meissner, *Proc. 17th European Photovoltaic Solar Energy Conf. (Munich)* **2002**, *1*, 160.
[9] J. Penndorf, M. Winkler, O. Tober, D. Röser, K. Jacobs, *Sol. Energy Mater. Sol. Cells* **1998**, 53, 285.
[10] M. Schirmeisen, F. Beck, *J. Appl. Electrochem.* **1989**, *19*, 401.

Effects of Mechanical and Thermal Stresses on Electric Degradation of Polyolefins and Related Materials

Shaval Mamedov,*[1,2] Vilayet Alekperov,[1,2] Nursel Can,[1] Faruk Aras,[3] Gunes Yilmaz[4]

[1] Yildiz Technical University, Department of Physics, 34010, Istanbul, Turkey
[2] Institute of Physics of the Azerbaijan Academy of Sciences, Baku, H. Javid St. 33, Azerbaijan
[3] Kocaeli University, Technical Education Faculty, Electrical Education Department, 41100, Kocaeli, Turkey
[4] Ar-Ge Turk Pirelli, Mudanya, Bursa, Turkey

Summary: Degradation under the simultaneous effects of mechanical stress and temperature in polyolefins (PE, PP), composites on their basis (PE+PP fibre, PP+PP fibre, PP+glass fibre) and radiation low-density polyethylene (X-LDPE) used in high-voltage cables obeys the thermofluctuation theory of Zhourkov (in certain σ and τ_0 intervals) based on the theory of Arrhenius is presented in the following form:

$$\tau_\sigma = \tau_0 \exp\left[(U_0 - \gamma\sigma)/RT\right] \quad (1)$$

where τ is durability. τ_0 is a constant (10^{-12}-10^{-13}s) equal to period of vibrations of atoms around equilibrium position, U_0 is the activation energy of the mechanical destruction process (at $\sigma = 0$), γ is a structure-sensitive parameter, T is absolute temperature and R is universal gas constant. Electric degradation under the effects of electric field and temperature in the materials mentioned above obeys the equation:

$$\tau_E = \tau_0 \exp\left[(W_0 - \chi E)/RT\right] \quad (2)$$

Here, τ_E, W_0 and χ are analogous to τ_σ, U_0 and γ, respectively.
It is assumed that the following equation is valid under the simultaneous effects of E, σ and T:

$$\tau_{\sigma,E} = \tau_0 \exp\left[(U_0 - (\gamma\sigma + \chi E))/RT\right]. \quad (3)$$

Keywords: electric degradation; mechanical stress; polyolefins; termal stress; thermofluctuation theory

Introduction

Polymers, polymer composites and crosslinked polymers are used extensively in all types of electrical power networks, devices and equipment, presenting attractive features such as excellent dielectric properties and good thermomechanical behaviour. However, being exposed to the external factors, such as electrical strength (E), mechanical stress (σ), and temperature (T), all the materials degrade and become unusable. There are some studies about the individual effects of these factors in the literature but the simultaneous effects are complex and their mechanisms are still being discussed. [1-5] Bagirov et al. have given an invalid equation in terms of units for polyethylene (PE) and polycaprolactam. Also the mechanism of the degradation is unclear. [6] In addition, although the experimental results may be correct, the mathematical expressions of the parameters in the equation presented are not examined.[7] Electric degradation of polymer materials, in particular those used in electric cables, are still subject to discussion. As to the mechanisms, exponential equations are given for electric degradation:

$$\tau_E = \tau_0 \left(\frac{E_0}{E} \right)^m, \tag{4}$$

and for temperature dependence

$$\tau_E = \tau_0\, exp\left(\frac{\Delta W}{kT} \right). \tag{5}$$

Arrhenius equation is given for either the sum of the two or for individual mechanisms.

Experimental

In experiments, PP+PP fiber (0, 10, 20, 30, 40 and 50 %), LDPE (low-density polyethylene)+PP fiber (0, 10, 20, 30, 40 and 50 %), PP+glass fiber (0.1, 0.3, 0.5, 0.7, 1 and 5 %) and unaged XLPE transmission power cables having radiation-crosslinked low-density polyethylene (X-LDPE) insulation of 22 mm thickness are used. Composites samples of 50-100 μm thicknesses are prepared by hot pressing at 453 K and 15 MPa for 10 min. The dynamic mechanical characteristics (σ) are measured by a lever mechanism that keeps σ constant as the cross-section of the sample changes[1]. In order to measure the simultaneous effect of σ and E, and electric degradation of XLPE cable samples, an electric cell is added to the system. One of the electrodes (ϕ 30 mm) is grounded, while the other (ϕ 10 mm) is connected to the voltage source of 0 - 15 kV. The sample is sandwiched between two electrodes shaped in such a way that no electric discharge is likely to arise through the edges of the electrodes. X-LDPE samples of thicknesses 120 - 200 µm are cut with a microtome from the cables. Since the sample thicknesses and strengths are high, the measurements have been made after placing the samples between the electrodes immersed in transformer oil. The structure variations have been observed using a Mattson 1000 Fourier transform spectrometer and polarization microscope.

Experimental results and discussion

Typical graphs for PP+PP fiber, LDPE+PP fiber, PP+glass fiber and X-LDPE cable samples are given (Figures 1-5). Most of them contain PP+20 % PP, because the best results for all properties of the composites are obtained at this composition.

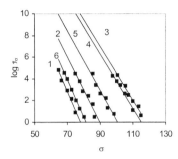
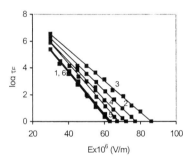

Figure 1. The dependence of log τ_σ on σ (at 183 K) for PP + PP fiber composites (1 - neat PP; 2 – PP + 10 % PP fiber; 3 – PP + 20 % PP fiber; 4 – PP + 30 % PP fiber; 5 – PP + 40 % PP fiber; 6 – PP + 50 % PP fiber).

Figure 2. The dependence of log τ_E on E (at 293 K) for PP + PP fiber composites (1 - neat PP; 2 – PP + 10 % PP fiber; 3 – PP + 20 % PP fiber; 4 – PP + 30 % PP fiber; 5 – PP + 40 % PP fiber; 6 – PP + 50 % PP fiber).

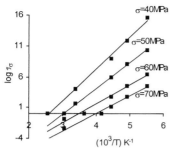

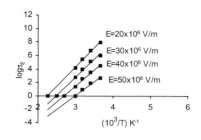

Figure 3. The dependence of $\log \tau_\sigma$ on $1/T$ for neat PP sample.

Figure 4. The dependence of $\log \tau_E$ on $1/T$ for neat PP sample.

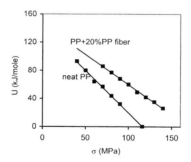

Figure 5. The dependence of activation energy of degradation on σ.

As can be seen in the figures, it is possible to write the following equations for the electric degradations similar to the equation for mechanical degradations:

$$\tau_\sigma = \tau_0 \exp[(U_0 - \gamma\sigma)/RT] \text{ an } \tau_E = \tau_0 \exp[(W_0 - \chi E)/RT].$$

The equations are supported by the following results:

(1) The graphs $\log \tau_\sigma - f(\sigma)$, $\log \tau_E - f(E)$, $\log \tau_{\sigma,E} - f(1/T)$ are linear in certain σ and E intervals; (2) τ_0 and U_0 values are constants for all cases, i.e., the thermofluctuation theory is valid; (3) while the mechanical stress (σ) and electric strength (E) are increasing, τ decreases and also structure-

sensitive parameters (γ and χ) increase. The equation for degradation of the polymer materials mentioned above, occurring while two or more factors change simultaneously is

$$\tau_{\sigma, E} = \tau_0 \exp[(U_0 - (\gamma\sigma + \chi E))/RT]$$

The equation may be expressed as the products of two exponents

$$(\tau_{\sigma, E}) = \tau_0 \exp\left(\frac{W_0 - \chi E}{RT}\right)\exp\left(\frac{U_0 - \gamma\sigma}{RT}\right);$$ probably $W_0 \cong U_0$ if assumed that the mechanisms of

electric and mechanical degradations are similar, one of them belongs to degradation under a long-term electric field and the other belongs to the degradation caused by mechanical stress. In this case, the temperature dependence may be written as the Arrhenius equation. In long-term degradation, variations in the composite structure can be observed. This result is observed both in microphotographs and IR spectra (Figures 6a, b and 7), in which the intensities of the peaks at 1680 cm^{-1} (C = C vibrations) and at 1715-1780 cm^{-1} (C – O and C = C vibrations) belong to amorphous regions. As can be seen from Figure 7, burning of the material and then deterioration of the fibrous structure arise in the broken area. The mechanism of breaking in an electric field following a long-term degradation can be based on the above results. In this process, the role of electric degradation in the instantaneous breaking is a decrease in the breaking potential energy (ΔU).

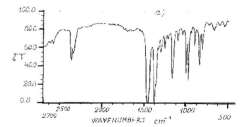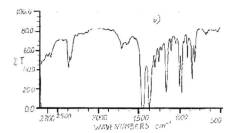

Figure 6. The changes in the IR spectrum of PP+20 % PP composite during long-term degradation in electric field. (a) $t = 0$ and $U = 0$, (b) $t = 5$ h and $U = 9$ kV.

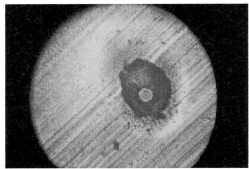

Figure 7. a) Carbonization region; b) complete disorder of fiber structure; pure fibrous structure in the ordered region.

[1] V.R. Regel, A.L. Slutsker, E.Y. Tomashevskii. *The Kinetic Nature Solidity of Solid State* (in Russian), Nauka, Moscow, **1974**, p. 560.
[2] A.A. Tager. *Physical Chemistry of Polymers*, Mir Publishers, **1978**, 653.
[3] A.Ya. Goldman. Translated and Edited by M.Shelef and R.A.Dickie, Washington, DC: *American Chemical Society*, **1994**, 349.
[4] H.H. Kausch. *Polymer Fracture*, Springer, Berlin, **1978**, 456.
[5] A.J. Kinloch, R.J. Young. *Fracture Behaviour of Polymers*, Elsevier, **1988**, 496.
[6] M.A. Bagirov, S.A. Abasov, Ya.H. Ragimov, T.F. Abbasov, E.S. Safiyev, I.M. Ismailov. *Izv. Akad. Nauk Azerb. SSR*, **1982** (1), 74.
[7] Ya.H. Ragimov, Sh.V. Mamedov, S.A. Abasov, V.A. Alekperov. *Dokl. Akad. Nauk Azerb. SSR*, **1988**, 44 (5), 40.

Growth of Organic n-Conductors on Thin Polymer Films for Use in Organic Field Effect Transistors

W. Michaelis,[*1] *C. Kelting,*[2] *A. Hirth,*[2] *D. Wöhrle,*[2] *D. Schlettwein*[1]

[1] Physical Chemistry 1, University of Oldenburg, P.O.Box 2503, 26111 Oldenburg, Germany

[2] Institute of Organic and Macromolecular Chemistry, University of Bremen, P.O.Box 330440, 28334 Bremen, Germany

Summary: Thin films of different polymers - poly(styrene) (PS), poly(methylmethacrylate) (PMMA), poly(vinylcarbazole) (PVCz), poly(vinylchloride) (PVC) and poly(vinylidene fluoride) (PVDF) - were deposited by spin-coating or by vapor deposition. On these polymers, thin films of (hexadecafluorophthalocyaninato)-oxovanadium ($F_{16}PcVO$) were prepared by physical vapor deposition. The growth of these films was monitored in situ by optical spectroscopy. The optical absorbance spectra were analyzed based on the coupling of transition dipoles to obtain information on the intermolecular arrangement of chromophores in the films. In all of these samples, the molecules are oriented with their molecular plane preferentially perpendicular to the substrate surface. This gives the desired overlap of the π-systems for electric conductance parallel to the substrate. Differences in the interactions were detected when deposition temperatures below or above the glass transition temperature of a given polymer were compared. The morphology of the polymer films and the deposited semiconductors were investigated by atomic force microscopy and scanning electron microscopy. The influence of the chosen substrate on the film structure is determined. The optical and electric properties of the films could thereby be influenced and the applicability of such films as active layers in organic thin film transistors is discussed.

Keywords: molecular semiconductor; n-conduction; phthalocyanines; polymer thin films

Introduction

The growth of molecular semiconductor films on polymer substrates is investigated because the polymer can also be used as gate insulators in devices following an all-organic approach. Such an approach allows to avoid high temperatures and thereby ensures a high flexibility with regard to the chosen substrates. Also, it opens the way towards low production costs of envisaged electronic devices.

DOI: 10.1002/masy.200450834

The intermolecular arrangement of organic semiconductors in thin films is decisive for their use in electronic devices, such as organic field-effect transistors or diode structures, since it should determine both the charge carrier mobility and optical properties of such films.[1-3] Recently, molecular n-conducting materials like $F_{16}PcVO$ have attracted special interest.[4-9]

Optical absorbance spectroscopy is used as a method to probe the chromophore interactions of the semiconductor molecules.[10] In the present experimental setup these spectra were measured in situ during the growth of the semiconductor films. Aside from the intermolecular arrangement, the morphology of films determines the effective mobility of charge carriers and is probed by atomic force microscopy (AFM) and scanning electron microscopy (SEM).

Experimental

The films of PS, PMMA, PVC and PVCz were prepared by spin-coating from a methylcyclohexanone solution on an ITO substrate. The film thickness was determined as 270-300 nm (PMMA, PVCz), 600 nm (PVC) and 700 nm (PS) by using a DEKTAK profilometer. The films of PVDF were prepared by thermal decomposition of PVDF in a crucible, evaporation of the fragments and re-polymerization on the substrate, a process performed in analogy to PVD at a pressure of 10^{-3} Pa and an evaporation rate of 0,5 nm/min on a glass support. The thickness of these films was determined as 110 nm.

$F_{16}PcVO$ was synthesized as described earlier.[11] This compound was purified by temperature gradient sublimation in a three zone-oven (Lindberg).

Thin films of this material were then prepared by PVD under high vacuum conditions (10^{-3} Pa). The evaporation rate was 0.5 - 1 nm/min. The films were grown at two temperatures (at about 323 K or 393 K). Optical absorbance spectroscopy was performed with a light beam perpendicular to the substrate surface using a diode array spectrometer, Ocean Optics PC 2000 with a tungsten lamp HL 2000 LL, which was coupled to the deposition vacuum chamber by optical fibers. To heat the substrates, they were mounted on ITO coated glass, which was used as a resistive heater. The spectra were measured in situ during the film growth.

After deposition and optical analysis, these films were characterized ex situ by AFM (MT-MDT Smena A) in resonant mode and by SEM (Zeiss DSM 940).

Results and Discussion

AFM-measurements of the polymer films made by spin coating showed very smooth surfaces with an average roughness in the range of only a few nanometers (Fig. 1). The PVDF films prepared by PVD had a higher roughness of about 30 nm.

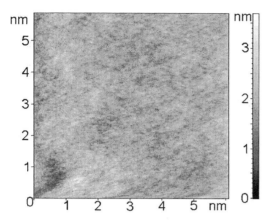

Figure 1: AFM-measurement of a spin-coated PVCz-film on ITO

When $F_{16}PcVO$ was deposited on these polymer films, two different characteristics could be observed depending on the temperature of the substrate during the evaporation. On cold substrates (~ 323 K), the smooth surface of the polymer substrate could be widely preserved during deposition of $F_{16}PcVO$, with a roughness in the range of about 20nm. Generally, these films are too smooth to be measured in SEM in contrast to the films that were prepared at elevated temperature (~ 393 K; Fig.2).

For these films of $F_{16}PcVO$ film morphology could be detected by SEM. In the case of PVCz, small islands have been seen. The same behavior has been found for films prepared on glass at a comparable temperature. On PMMA and PVC, needle-like structures were found that are similar to those reported earlier for films on SiO_2.[12] The film on PS showed a texture that, aside from crystals as on PVC, suggests a growth of $F_{16}PcVO$ that often started from new marked crystallization seeds. A widely homogeneous surface of PS with, however, a few marked defects was thereby indicated when compared with the other polymer surfaces. The formed islands are not so well defined as on PMMA or PVC, but they showed a similar crystalline shape.

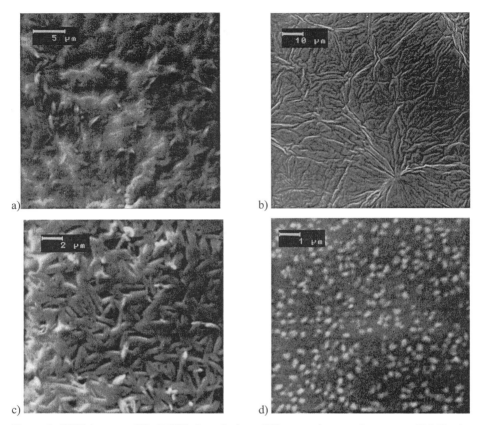

Figure 2: SEM images of $F_{16}PCVO$ deposited on different polymer substrates at 393 K, a) on PMMA, b) on PS, c) on PVC, d) on PVCz

Optical absorption spectra are used to study relative molecular interactions in the films. Spectra of 40 nm $F_{16}PcVO$ on glass, PMMA, PVC, PS, PVCz, and PVDF are depicted in Figure 3. These films were prepared at 323 K. The development of all bands in the spectra was continuous. All spectra were dominated by an absorption band shifted to the blue ($\sim$ 670 nm) relative to that of the individual uncoupled molecules in solution (707 nm). A slightly red-shifted band at about 745 nm accompanied this. On glass at elevated temperature (393 K) an additional band appeared at about 830 nm (Fig. 4). This band did not appear on PS and was generally weaker on the other polymer substrates when compared with the film on glass. A transition of the polymers to their visco-elastic state seemed to suppress this band of F_{16}PcVO since it was not detected on the polymer of the lowest glass transition temperature (PS) but it was most clearly detected on those polymers, which were at 393 K still well below their glass transition temperature.

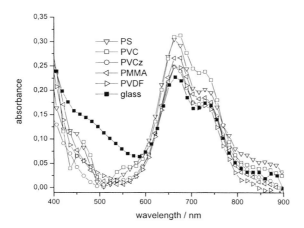

Figure 3: Absorbance spectra of 40 nm F_{16}PcVO films prepared at 323 K

Although small spectral shifts can occur due to different polarizable environments in solids when compared with solvents, the present shifting and splitting of the absorption bands relative to the position in solution could be interpreted within a model based upon the dipole-dipole interaction of the transition dipoles.[10,13] In this model the red shift means a head-to-tail arrangement of the chromophores (J-aggregate), the blue shift a cofacial arrangement (H-aggregate). So it is clear that in all studied films a cofacial orientation of adjacent molecules was dominating. This was accompanied by smaller contributions from a head-to-tail interaction. Since we have observed the individual bands separately in earlier experiments,[14] and since their relative ratio changes even in the present study, we conclude that a mixture of different crystalline phases of F_{16}PcVO was present in the films. The shoulder above 800 nm is caused by a band at 830 nm that has been detected earlier for unsubstituted phthalocyanines and stands for a strongly coupled crystalline phase in which a head-to-tail-arrangement clearly dominates [13].

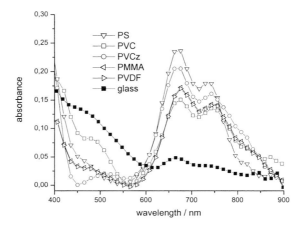

Figure 4: Absorbance spectra of 40 nm $F_{16}PcVO$ films prepared at 393K (on glass only 10 nm $F_{16}PcVO$ on glass)

All films showed a low integral absorbance when normalized to the deposited amount of $F_{16}PcVO$ and when compared with earlier results on alkali halide (100) surfaces [14]. The observed small absorbance for the electric field vector of the incoming light (parallel to the surface) leads to the conclusion that the molecules were essentially standing upright on the surface, because the main transition dipole lies in the molecular plane, so that there was only a weak interaction between the transition dipole and the electric field vector of the light.

Conclusions

Smooth polymer films were obtained by spin coating as well as by vapor deposition that were vacuum-compatible and could serve as a suitable substrate for the growth of thin films of organic semiconductors. The polymers provided a uniform surface for the deposition of phthalocyanines, which then led to the formation of smooth uniform layers of the latter. This was shown in detail for the organic n-conductor $F_{16}PcVO$. By variation of the deposition temperature relative to the glass transition temperature of the polymer, details of the intermolecular coupling and film morphology could be adjusted.

The molecular orientation in the semiconductor films was found in an "edge-on" or "standing" orientation with the molecular plane perpendicular to the substrate surface. This is a highly desired orientation since it offers the possibility of good intermolecular overlap of the aromatic π-systems and thereby provides a fundamental prerequisite for an attractive mobility within the semiconductor film. Regarding possible gate insulators, such a "standing" orientation of $F_{16}Pc$ has been found before only for depositions on amorphous SiO_2. In the present work it is shown that suitable polymer films also provide an interaction that promotes this orientation. They represent promising insulator layers because they can be easily processed, causing minimum thermal stress on device substrates, and thereby allow an all-organic approach to thin film transistors.

Acknowledgement

The authors are grateful to the DFG (Schl 340/ 4-1 and Ja 346/27) for financial support and to S. Martyna (University of Oldenburg) for technical assistance in the SEM analysis.

[1] Law, K.-Y. *Chem.Rev.* **1993**, *93*, 449.
[2] Schmidt, A.; Chau, L.K.; Back, A.; Armstrong, N.R.; in Leznoff, C.C.; Lever, A.B.P. (Ed.): "Phthalocyanines IV", VCH Weinheim **1996**, 307-341.
[3] Forrest, S.R. *Chem. Rev.* **1997**, 97, 1793-1896.
[4] Isoda, S.; Hashimoto, S.; Ogawa, T.; Kurata, H.; Moriguchi, S.; Kobayashi, T. *Mol.Cryst.Liq.Cryst.* **1994**, *247*, 191-201.
[5] Bao, Z.; Lovinger, A.J.; Brown, J. *J.Am.Chem.Soc.* **1998**, *120*, 207-208.
[6] Schlettwein, D.; Graaf, H.; Meyer, J.-P.; Oekermann, T.; Jaeger, N.I. *J.Phys.Chem B* **1999**, *103*, 3078-3086.
[7] Meyer, J.-P.; Schlettwein, D.; Wöhrle, D.; Jaeger, N. I. *ThinSolid Films* **1995**, *258*, 317.
[8] Meyer, J.-P.; Schlettwein, D. *Adv. Mater. Opt. Electron.* **1996**, *6*, 239.
[9] Hashimoto, S.; Isoda, S.; Kurata, H.; Lieser, G.; Kobayashi, T. *J.Porphyrins Phthalocyanines* **1999**, *3*, 585-591.
[10] Kasha, M.; Rawls, H.R.; El-Bayoumi, M.A. *Pure Appl. Chem.* **1965**, *11*, 371.
[11] Hiller, S.; Schlettwein, D.; Armstrong, N.R.; Wöhrle, D. *J.Mater.Chem* **1998**, *8*, 945.
[12] Hesse, K.; Dissertation, University of Bremen 2002.
[13] Chau, L.-K.; England, C.D.; Chen, S.; Armstrong, N.R. *J.Phys.Chem.* **1993**, *97*, 2699-2706.
[14] Schlettwein, D.; Tada, H.; Mashiko, S. *Langmuir* **2000**, *16*, 2872.
[15] Schlettwein, D; Hesse, K.; Tada, H.; Mashiko, S.; Storm, U.; Binder, J. *Chem. Mater.* **2000**, *12*, 989-995.

Macromol. Symp. **2004**, *212*, 307-314

Oxidation State and Proton Doping Level in Copolymers of 2-Aminobenzoic Acid and 2-Methoxyaniline

Ida Mav,[1] *Majda Žigon,*[1] *Jiří Vohlídal*[*2]

[1]Laboratory for Polymer Chemistry and Technology, National Institute of Chemistry, Hajdrihova 19, POB 660, SI-1001 Ljubljana, Slovenia
E-mail: ida.mav@ki.si
[2]Department of Physical and Macromolecular Chemistry, Laboratory of Specialty Polymers, Charles University, Albertov 2030, CZ-128 40 Prague 2, Czech Republic

Summary: UV-vis spectra of homopolymers and copolymers of 2-aminobenzoic acid (OAB) and 2-methoxyaniline (OMA) were analyzed in order to obtain information about the oxidation state and proton doping level of these polymers. Dimethyl sulfoxide (DMSO) was used as a solvent in which protonated forms of polyanilines are preserved and a mixture of *N*-methyl-2-pyrrolidone and triethylamine (0.5 %) as a solvent (NMP/TEA) in which polyanilines are assumed to be non-protonated. Polymers were prepared in the emeraldine salt form, externally doped with HCl. It was found that only external doping is eliminated in NMP/TEA while internal doping by carboxylate groups bound in OAB units remains operative. Since doped quinoid units do not contribute to the quinoid band (Q-band at 630 nm), the intensity ratio of the Q-band and benzenoid band (B-band at 320 nm) cannot be simply correlated with the oxidation state of poly(OMA-*co*-OAB) copolymers in contrast to poly(OMA) and polyaniline. Spectra of copolymers with less than 60 % of OMA units as well as those of poly(OAB) in DMSO and NMP/TEA are almost identical due to internal doping, which is proposed to lead to structures in which main-chain protons are coulombically bound with immobile carboxylate anions. In the spectra of copolymers with less than 60 % of OMA units, a well-resolved band occurs at 500 nm, which can be ascribed to alternating or close-to-alternating sequences of OMA and OAB units.

Keywords: conducting polymers; UV-vis spectroscopy

Introduction

Functional properties of synthetic metals, including polyanilines (PANI), are closely related to their doped states, in which the number of electrons associated with the macromolecule backbone is either decreased (p-doping) or increased (n-doping) as compared with their number associated with the backbone of the parent nonmetallic polymer molecule [1]. This doping is achieved by chemical or electrochemical redox processes involving parent

DOI: 10.1002/masy.200450835

polymers. Unlike other conducting polymers, PANIs also exhibit proton doping consisting in proton exchanges between PANI main chains and proton acid molecules (or acid groups), HA, in which the number of π-electrons associated with a PANI molecule is conserved. As a result, functional properties of PANIs depend on the extent of both oxidation and protonation of their macromolecules.

Oxidation states of PANI are described [2] by the general formula shown in Scheme 1a, where y is the mole fraction of benzenoid units in PANI chains. If $y = 1$, PANI is fully reduced (leucoemeraldine form, here leucoemeraldine base, LB); if $y = 0.5$, it is half-oxidized (emeraldine form, here emeraldine base, EB), and if $y = 0$, PANI is fully oxidized (pernigraniline form, here pernigraniline base, PB). In other words, the quantity $(1 - y)$ is the degree of oxidation of a PANI chain.

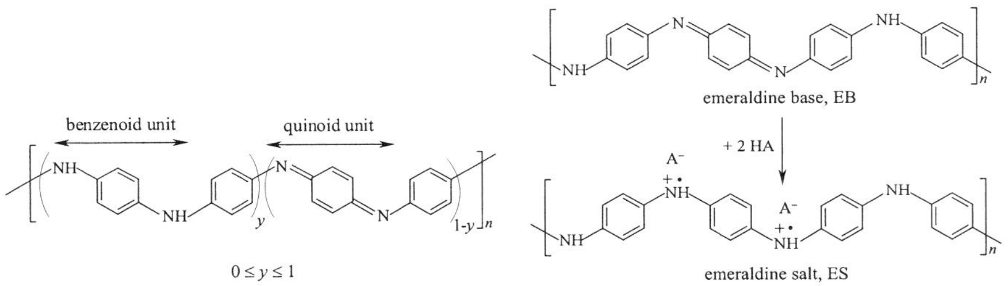

benzenoid unit   quinoid unit

$0 \leq y \leq 1$

emeraldine base, EB

+ 2 HA

emeraldine salt, ES

(a) General formula of PANI base forms   (b) Proton doping of PANI emeraldine base

Scheme 1.

The oxidized PANI base forms easily undergo proton-exchange reactions with HA (Scheme 1b), in which they are transformed to corresponding emeraldine (ES) and pernigraniline salts (PS). These salts are synthetic metals because their main chains comprise mobile electrons, the formation of which is closely related to the presence of quinoid units in PANI chains (Scheme 1b). The ES form is the most important form of PANIs since it shows not only high electrical conductivity but also high stability in air as well as under working conditions [2], which is not the case of PS.

For obtaining high-quality PANI samples, it is essential to achieve a precise control of the polymer oxidation state and level of proton doping. PANIs are generally prepared by diverse synthetic procedures using various reaction conditions and modes of crude PANI post-treatment, which makes the prediction of y value on the basis of synthetic path practically impossible. Typically, the oxidation state of as-prepared PANI ranges from emeraldine to pernigraniline form [3] and the true value of y must be determined by an independent method. The UV-vis spectroscopy of PANI solutions in N-methyl-2-pyrrolidone (NMP) was shown [4] to be a suitable and sufficiently accurate method for the determination of y values of PANI samples.

Since PANI is only sparingly soluble, more soluble polymers of substituted anilines have become a subject of scientific interest. Among substituted PANIs, those with acid pendent groups such as SO_3H [5-10] and COOH [11] are of special interest, because these groups also act as internal proton dopants. Although the conductivity of ES forms of these self-doped PANIs is lower than that of unsubstituted PANI, it is pH-independent at pH $\leq$ 4, which is advantageous from the application point of view. However, the substitution of PANI brings about some drawbacks. First, attached acid groups affect not only the proton-doping level but also the degree of oxidation of PANI chains as it was shown, e.g., for copolymers of aniline-3-sulfonic acid with aniline [9] or 2-methoxyaniline (OMA) [10]. Second, preparation of self-doped PANIs by homopolymerization of corresponding aniline acids (ANIA) meets with difficulties since this reaction usually does not take place under mild conditions. Therefore, self-doped PANIs are typically prepared by copolymerization of ANIA with aniline (ANI) or a reactive substituted aniline bearing alkyl or alkoxy groups [5-10]. Apart from chemical, also electrochemical copolymerization was utilized in preparation of self-doped PANIs, e. g., electrolytic copolymerization of N-(3-sulfopropyl)aniline with N-methylaniline [12] and aniline-2-sulfonic acid with 3-methylaniline and 3-ethylaniline [13].

We have recently reported [14-16] on a chemical copolymerization of OMA with anilinesulfonic acids and aminobenzoic acids. We determined the monomer reactivity ratios by '*in situ*' ^{1}H NMR measurements and characterized the composition of copolymers and oxidation level of emeraldine and leucoemeraldine forms of poly(OMA-*co*-aniline-3-sulfonic acid) also by NMR spectroscopy [10]. In the present paper, we report on the oxidation state and proton doping of as-prepared poly(OMA-*co*-OAB) (OAB stands for 2-aminobenzoic acid) as a function of the copolymer composition. OMA is used as the main comonomer because methoxy groups on benzene rings considerably increase copolymer solubility compared to that of analogous copolymers with unsubstituted aniline. OAB is advantageous because it undergoes homopolymerization under mild conditions unlike other aniline acids.

Experimental

Chemicals. 2-Methoxyaniline (OMA, Aldrich), 2-aminobenzoic acid (OAB, Fluka), ammonium peroxodisulfate ($(NH_4)_2S_2O_8$, Fluka), hydrochloric acid (HCl, Merck), *N*-methyl-2-pyrrolidone (NMP, Aldrich), triethylamine (TEA, Aldrich), methanol (MeOH, Aldrich), dimethyl sulfoxide (DMSO, Aldrich) and DMSO-d_6 (Aldrich) all of AR grade were used as supplied.

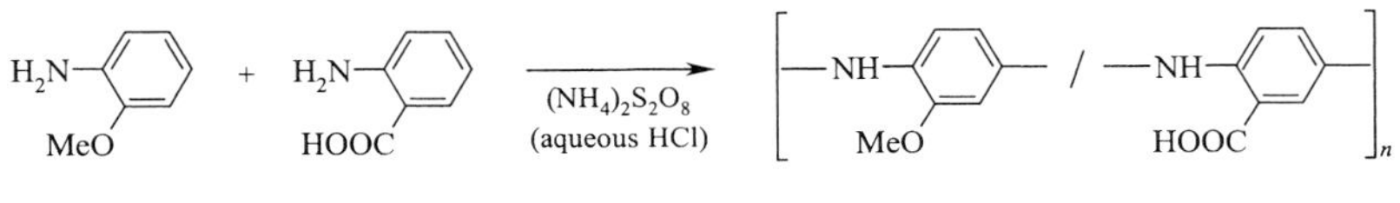

Scheme 2.

Synthesis of copolymers (Scheme 2*).* A solution of $(NH_4)_2S_2O_8$ (0.1 mol) in aqueous HCl (1M, 100 mL) was added to a constantly stirred aqueous HCl solution (1M, 700 ml) of a mixture of OMA and OAB (sum of both monomers 0.1 mol) and the mixture was allowed to react for a given time up to 20 h. After ca 10 – 15 min, the color of the reaction mixture changed from light yellow-red to dark purple and, finally, a continuous precipitation of a dark green polymer occurred. After a chosen reaction time, the polymer precipitate was filtered off, extensively washed with aqueous HCl (1M, 1 L) until a colorless filtrate was obtained, in order to remove soluble oligomers and residual monomers, then with methanol (250 mL) and, finally, the purified polymer was dried in vacuum at 40 °C for 48 h. The feed mole fraction of OMA in the monomer mixture, f_1, varied from 0 to 1 and other reaction conditions were set as follows: oxidant-to-monomer mole ratio Ox/mon = 1, $[HCl]_0$ = 1M, T = 20 °C and t = 20 h.

Analyses, measurements. Nitrogen content was determined on a CE440 LeemanLabs (CHN) analyzer. Mole fraction of OMA units in a copolymer, F_1, was determined from integral intensities of ^{1}H NMR signals of aromatic (around 7 ppm) and methoxy protons (around 3.7 ppm), respectively. NMR spectra were recorded on a Varian VXR 300 MHz spectrometer using a 4.5 degree pulse, a relaxation delay of 3 s, an acquisition time of 4 s and 500 repetitions (1 h). For NMR measurements, copolymers were dissolved in a mixture of NMP, TEA and DMSO-d_6 (70:10:20); solutions of concentration 10 % w/v were used. TEA was added to increase the solubility of copolymers by transforming them from the ES into the EB form. The ^{1}H NMR spectra were referenced to tetramethylsilane internal standard and the ^{13}C spectra to the DMSO-d_6 signal at 39.5 ppm. UV-vis absorption spectra of PANI solutions

(0.02 %) were recorded on a Hewlett Packard 8452 diode-array instrument using quartz cuvettes of optical path of 0.2 cm. A mixture of NMP and TEA (0.5 %) was used as the solvent in which PANIs are in the unprotonated state, whereas DMSO was used as a solvent in which protonated forms of PANIs are preserved [17]. The spectra were measured using as-prepared samples, which were not subjected to any modification so that their oxidation state corresponds to that one resulting from synthetic procedure.

Results and Discussion

Optical spectra of dissolved PANI EB forms consist of two bands of different origin [4,18, 19]. The first band with maximum at about 320 nm, which is the only band of LB form, is due to intrachain $\pi \rightarrow \pi^*$ transition within benzenoid units and, therefore, it is called B-band. The second band with maximum at about 630 nm is due to electron transitions from the benzenoid HOMO to quinoid LUMO states (Q-band of exciton origin). It is contributed to by intrachain as well as interchain processes and its absorption coefficient is in general concentration-dependent, but concentration-independent in dilute solutions ($c < 0.1$ %). The Q/B band intensity ratio is closely correlated with the degree of oxidation of PANI chains as it was shown for unsubstituted as well as for partly and fully ring-substituted PANIs [4,9,19,20]. If EB is oxidized to the PB form, the Q band is replaced by another band with maximum at about 550 nm, which is attributed to a Peierl's gap typical of fully conjugated systems [4]. The Q band is not present in UV-vis spectra of PANI ES forms, in which new bands occur in the regions from 400 to 450 nm and from ca 700 nm to the NIR region. These bands are ascribed to the polaron transitions associated with charged conjugated chains [5,21].

UV-vis spectra of as-prepared (i.e., protonated by HCl) samples of poly(OMA-co-OAB) obtained using various feed ratios f_1, poly(OMA), and poly(OAB) dissolved in NMP/TEA mixture and in DMSO are shown in Fig. 1. The spectrum of poly(OMA) in NMP/TEA (Fig. 1a, curve a) is practically identical with that of the poly(OMA) EB form dissolved in neat NMP [4,18], which proves that the as-prepared poly(OMA) undergoes effective deprotonation in NMP/TEA. The B/Q band intensity ratio of 2 indicates that the poly(OMA) sample is slightly under-oxidized ($1 - y = 0.47$) as compared to the ideal EB state (see, e.g., Fig. 5 in Ref. [4]), which corresponds to a slightly lower than stoichiometric amount of oxidant used in the sample preparation (we used Ox/mon = 1 and reached an OMA conversion of 84 % while the ideal Ox/mon ratio is 1.25 for total monomer conversion [3]). The spectrum of protonated poly(OMA) (Fig. 1b, curve a) contains two polaron transition bands at 450 nm and 820 nm, but an unresolved Q band, as expected.

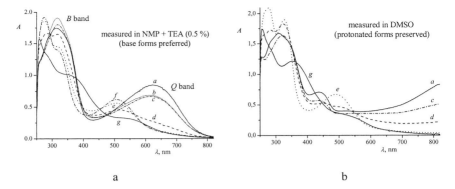

Figure 1. UV/vis spectra of poly(OMA-*co*-OAB) and corresponding homopolymers dissolved in NMP/TEA, in which base forms are preferred, and in DMSO, in which protonated forms are preserved, as a function of the mole fraction of OMA, f_1, in the starting mixture of monomers as well as mole fraction of OMA units, F_1, in the formed copolymer. Polymerization conditions: $[OMA]_0 + [OAB]_0 = 0.1$ M, $T = 20$ °C, reaction time $t = 20$ h, $[HCl]_0 = 1M$, Ox/mon = 1, f_1 (F_1): *a* 1 (1), *b* 0.80 (0.86), *c* 0.66 (0.76), *d* 0.5 (0.65), *e* 0.33 (0.54), *f* 0.20 (0.45), *g* 0 (0).

The second homopolymer, poly(OAB), exhibits substantially different spectroscopic behavior. First, a distinct Q band is not observed in the spectrum of as-prepared (externally doped with HCl) poly(OAB) in basic NMP/TEA solution (Fig. 1a, curve *g*). The spectrum can be described as a broad band decreasing in intensity continuously from the region of $\pi - \pi^*$ transitions below 300 nm to ca 800 nm, on which two shoulders corresponding to maxima at ca 400 and 550 nm occur. Second, the spectrum of DMSO solution of poly(OAB) is very similar to that obtained with NMP/TEA, showing only slightly stronger absorption at 400 nm. This indicates that (i) the band at 400 nm is partly associated with external doping, which disappears in NMP/TEA, and (ii) the internal doping with COOH groups is reduced almost negligibly. Accordingly, bands occurring at 400 and 550 nm should be ascribed to transitions in the doped poly(OAB). In general [22-24], these transitions are associated with polarons and/or bipolarons (PANI chain distortions associated with a quinoid unit in which one (in polaron) or both (in bipolaron) nitrogens are protonated, see Scheme 1b). Theoretical calculations for externally doped PANIs suggest the co-existence of polarons and bipolarons in short chains and predominance of polarons in long chains [24]. However, in a self-doped polymer such as poly(OAB), a high population of bipolarons can be expected due to attractive Coulombic interactions between main-chain cations and immobile COO⁻ counter-anions linked directly to main-chain benzene rings, which makes the dissociation of a bipolaron to two independent polarons difficult. Nevertheless, the question whether the bands observed in poly(OAB) spectra are associated with polarons or bipolarons or both cannot be solved on the basis of the results available and it needs further experimental and theoretical investigation.

Spectra of OMA-rich ($F_1 \geq 0.75$) poly(OMA-co-OAB) copolymers (Fig. 1a, curves b and c, and Fig. 1b, curve c) resemble those of poly(OMA), from which they differ mainly by lowered intensity of both Q (in NMP/TEA) and polaron (in DMSO) bands. The B/Q band intensity ratio evaluated according to Fig. 6 in Ref. [4] provides a value of $(1 - y) \cong 0.4$ for both above-mentioned OMA-rich copolymers containing about 80 % of OMA units, which indicates a close correlation between $(1 - y)$ and F_1 values in OMA-rich copolymers. This is in agreement with block arrangement of OMA units in these copolymers, which is supported by a large difference in reactivity ratios found for the OMA/OAB copolymerization system (21.8 for OMA and 0.064 for OAB [14]). Thus, it can be concluded that an experimental value of the B/Q band intensity ratio provides information on the oxidation state of OMA unit blocks in a given OMA-rich poly(OMA-co-OAB) copolymer only, but not information concerning the entire copolymer molecules. The Q band of OAB units is not observed in UV spectra of NMP/TEA solutions of these copolymers because, as it was shown above, the internal doping does not vanish in this solvent and protonated segments do not exhibit the Q band absorption.

A dramatic change occurs in UV-vis spectra when the OMA unit fraction F_1 decreases to 0.65 or to a lower value (Figs 1a and 1b, curves d, e and f). The Q band (Fig. 1a) as well as polaron bands (Fig. 1b), typical of poly(OMA), are strongly reduced ($F_1 = 0.65$, curve d) or practically disappear ($F_1 < 0.55$, curves e, f) and new distinct bands with maxima at 500 nm and 275 nm appear in the spectra. It is notable that UV spectra measured in both the solvents used become almost identical, similarly to the case of poly(OAB). On the other hand, distinct bands at 500 nm and 275 nm are not present in the poly(OAB) spectra (Fig 1, curve g), which clearly indicates that these bands are not associated with long sequences of OAB units.

We also considered the possibility of the EB form oxidation to the PB form. However, UV band of PB occurs at 530 nm for PANI [25] and at 560 nm for poly(OMA) [4] and the oxidation of EB to PB is accompanied by only a small decrease in the Q band intensity, whereas we observe band at 500 nm and disappearance of the Q band. Neither NMR nor IR spectra of copolymers provide an evidence for the copolymer oxidation to the pernigraniline state. The observed differences can be explained by changes in the proton doping extent of copolymers of diverse composition. These observations suggest that bands at 500 nm and 275

314

nm are closely associated with self-doped copolymers, in which alternating or close-to-alternating hetero-unit sequences of OMA and OAB units are present in high concentrations.

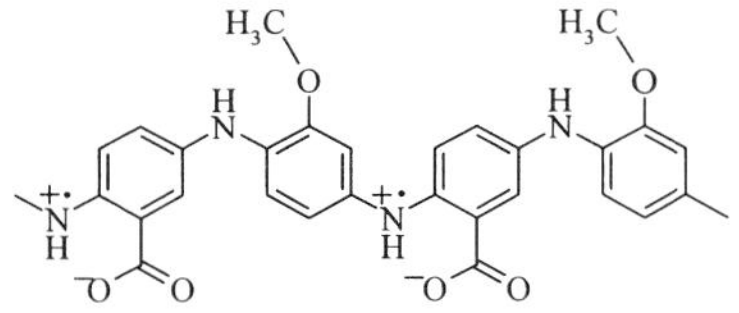

Scheme 3. Self-doped alternating sequences.

The decreased mobility of cations due to Coulomb interactions with immobile COO^- anions is expected to take place in these self-doped sequences (Scheme 3), which provides an explanation for a rather low intensity of the corresponding absorption bands.

Acknowledgements

Financial support by the Czech Republic – Slovenia program KONTAKT (project No. 2001/012), the Grant Agency of the Czech Republic (project No. 202/00/1152), the Ministry of Education, Science and Sport of the Republic of Slovenia (project No. Z2-3334-104), and Ministry of Education, Youth and Sports of the Czech Republic (project MSMT 113100001) is gratefully acknowledged.

[1] A.G. McDiarmid, *Angew. Chem., Int. Ed.* **2001**, *40*, 2581.
[2] S. Kaplan, E.M. Conwell, A.F. Richter, A.G. MacDiarmid, *J. Am. Chem. Soc.* **1988**, *110*, 7647.
[3] a) J. Stejskal, P. Kratochvíl, A.D. Jenkins, *Collect. Czech. Chem. Commun.* **1995**, *60*, 1747; b) J. Stejskal, P. Kratochvíl, A.D. Jenkins, *Polymer* **1996**, *37*, 367.
[4] J.E. Albuquerque, L.H.C. Mattoso, D.T. Balogh, R.M. Faria, J.G. Masters, A.G. MacDiarmid, *Synth. Met.* **2000**, *113*, 19.
[5] J. Yue, Z.H. Wang, K.R. Cromack, A.J. Epstein, A.G. MacDiarmid, *J. Am. Chem. Soc.* **1991**, *113*, 2665.
[6] K.G. Neoh, E.T. Kang, K.L. Tan, *Synth. Met.* **1993**, *60*, 13.
[7] J.Y. Lee, C.Q. Cui, *J. Electroanal. Chem.* **1996**, *403*, 109.
[8] I. Mav, M. Žigon, A. Šebenik, *Synth. Met.* **1999**, *101*, 717.
[9] I. Mav, M. Žigon, A. Šebenik, J. Vohlídal, *J. Polym. Sci. Polym. Chem.* **2000**, *38*, 3390.
[10] I. Mav, M. Žigon, *Polym. Bull.* **2000**, *45*, 61.
[11] M.T. Nguyen, A.F. Diaz, *Macromolecules* **1995**, *28*, 3411.
[12] A. Malinauskas, R. Holze, *Electrochim. Acta* **1998**, *43*, 521.
[13] P.A. Kilmartin, G.A. Wright, *Synth. Met.* **1997**, *88*, 163.
[14] I. Mav, M. Žigon, *Polym. Int.* **2002**, *51*, 1072.
[15] I. Mav, M. Žigon, *J. Polym. Sci. Polym. Chem.* **2001**, *39*, 2482.
[16] I. Mav, M. Žigon, *Synth. Met.* **2001**, *119*, 145.
[17] E. Erdem, M. Sacak, M. Karakisla, *Polym. Int.* **1996**, *39*, 153.
[18] W.S. Huang, A.G. MacDiarmid, *Polymer* **1993**, *34*, 1833.
[19] D. Yang, B.R. Mattes, *Synth. Met.* **2002**, *129*, 249.
[20] M. Angelopoulos, R. Dipietro, W.G. Zheng, A.G. MacDiarmid, A.J. Epstein, *Synth. Met.* **1997**, *84*, 35.
[21] B.C. Roy, M.D. Gupta, L. Bhoumik, J.K. Ray, *Synth. Met.* **2002**, *130*, 27.
[22] M. Kuwabara, Y. Shimoi, S. Abe, *J. Phys. Soc. Jpn.* **1998**, *67*, 1521.
[23] A.J. Heeger, S. Kivelson, J.R. Schrieffer, W.P. Su, *Rev. Mod. Phys.* **1988**, *60*, 781.
[24] Z.T. de Oliviera Jr., M.C. dos Santos, *Solid State Commun.* **2000**, *114*, 49.
[25] J.M. Leng, R.P. McCall, K.R. Cromack, Y. Sun, S.K. Manohar, A.G. MacDiarmid, A.J. Epstein, *Phys. Rev. B* **1993**, *48*, 15 719.

Macromol. Symp. **2004**, *212*, 315-320

Poly[(3,3'-dimethyl-1,1'-biphenyl-4,4'-diyl)ethynediyl(2,5-dioctyl-1,4-phenylene)ethynediyl] and a Related Copolymer – New Blue EL Polymers

Sung-Hoon Joo, Jung-Il Jin*

Division of Chemistry and Molecular Engineering and Center for Electro- and Photo-Responsive Molecules, Korea University, Seoul 136-701, Korea
E-mail: jijin@korea.ac.kr

Summary: Electroluminescence (EL) properties of a new poly(aryleneethynylene) and a related copolymer were studied. They are poly[(3,3'-dimethyl-1,1'-biphenyl-4,4'-diyl)ethynediyl(2,5-dioctyl-1,4-phenylene)ethynediyl] (PPEBE) and related copolymer (PPEBE-*co*-mP) containing 20 mole % of 1,3-phenylene units. Both polymers are blue-light emitters; the former was found to perform better than the latter when the light-emitting diode (LED) device had the configuration of ITO/PEDOT/polymer/Li:Al. The device constructed with the former polymer exhibited the external quantum efficiency of 0.05 and the maximum brightness higher than 400 cd/m^2 with its EL spectrum showing maxima at $\lambda = 445$ and 472 nm. The performance of the device constructed with the copolymer was about one fifth of the device fabricated with the homopolymer.

Keywords: blue emitter; EL; PLED; poly(phenyleneethynylene)

Introduction

Since the Cambridge report[1] on the electroluminescence (EL) properties of poly(*p*-phenylenevinylene) (PPV), interest in the EL phenomena in a wide variety of polyconjugated polymers[2] has been intensified due to the possible development of new display devices based on those polymers. PPV is a green-light-emitting material and polyfluorenes[3,4] are blue-light-emitting polymers. The performance of polymer light-emitting diodes (PLED) devices not only depends on the chemical structure of the polymers but also on many factors such as the nature of electrodes, utilization of additional carrier transporting layers, and so on.

We recently reported highly efficient green-light-emitting PPV derivatives bearing carbazole[5], phenyloxadiazole[6], and fluorene pendants[7] directly bonded to the PPV backbone. They, in general, revealed[8] a better balance in the mobility of the carriers. Moreover, some of the polymers were found to form new intragap states upon contact with the calcium electrode, which appears to be the reason for a lowered threshold electric field for the polymers[9].

Among the polyconjugated polymers, various poly(*p*-phenyleneethynylene) (PPE) derivatives

                     DOI: 10.1002/masy.200450836

also have been studied in PLED applications[10]. We find the possibility that one may be able to reduce the wavelength of the emitted light by replacing the double bonds in, for example, PPV with triple bonds, which increases the bandgap energy. In fact, there are several reports[11,12] claiming that a proper design of a PPE derivative exhibits blue emission.

In this investigation we prepared the following new PPE derivatives and their EL properties were studied by examining the device performance of ITO/PEDOT/polymer/Li:Al.

PPEBE   PPEBE-*co*-mP

Experimental

Polymer Synthesis

PPEBE: 1,4-diethynyl-2,5-dioctylbenzenene (428 mg, 1.22 mmol), 4,4'-diiodo-3,3'-dimethyl-1,1'-biphenyl (580 mg, 1.34 mmol), [Pd(PPh$_3$)$_4$] (84 mg, 0.07 mmol), and CuI (29 mg, 0.15 mmol) were combined in toluene (30 mL) and diisopropylamine (8 mL). The reaction mixture was then stirred at 70 °C. Ammonium iodide salts were formed immediately after starting the reaction and the mixture became highly fluorescent. After a total reaction time of 28 h, the reaction mixture was cooled to room temperature and added dropwise to rapidly stirred acetone (500 mL). After stirring for 2 h, the precipitate was collected and washed with acetone, hot ethanol, and then dissolved in chloroform and filtered through Florisil®(60-100 mesh) to remove residual palladium. The filtrate was dropwise added to cold acetone. The fluorescent precipitate was filtered off and dried overnight in vacuo. PPEBE was obtained as a greenish yellow solid (400 mg, 96 %), ^{1}H NMR (CDCl$_3$, δ ppm) 0.88-2.86 (m, 40H, Ar-CH_2-(CH_2)-CH_3, Ar-CH_3), 7.41-7.60 (m, 8H, Ar-H), Anal. Calcd for C$_{26}$H$_{38}$: C 90.85, H 9.15 %. Found : C 91.60, H 8.40 %

PPEBE-*co*-mP: This polymer was prepared by the same procedure. A mixture of 1,4-diethynyl-2,5-dioctylbenzene (470 mg, 1.34 mmol), 1,3-diethynylbenzene[13] (43 mg, 0.34 mmol), 4,4'-diiodo-3,3'-dimethyl-1,1'-biphenyl (803 mg, 1.85 mmol), [Pd(PPh$_3$)$_4$] (115 mg, 0.10 mmol), and CuI (38 mg, 0.20 mmol) were combined in toluene (35 mL) and diisopropylamine (8.5 mL). PPEBE-*co*-mP was obtained as light yellow solid (410 mg, 97 %), ^{1}H NMR (CDCl$_3$, δ ppm) 0.87-

2.86 (m, 40H, Ar-CH$_2$-(CH$_2$)-CH$_3$, Ar-CH$_3$), 7.41-7.60 (m, 8H, Ar-H), 7.76 (s, 0.01H, Ar-H), Anal. Calcd for C$_{26}$H$_{38}$: C 92.26, H 7.74 %. Found : C 92.95, H 7.05 %

Measurements

GPC analysis was conducted with a Wyatt Dawn EOS system equipped with Ultra-I-stragel columns using THF at a flow rate of 0.5 mL/min at 40 °C with polystyrene as the calibration standard. The luminescence spectra for the polymers were recorded on an AMINCO-Bowman Series 2 luminescence spectrometer at room temperature. The current and luminescence intensity as a function of applied field were measured using an assembly consisting of PC-based dc power supply (HP 6623A) and a digital multimeter (HP 34401). Also a light power meter (Newport Instruments, model 818-UV) was used to measure the device light output in microwatts. Luminance was measured with a Minolta LS-100 luminance meter.

Results and Discussion

Synthesis of polymers: PPEBE and PPEBE-*co*-mP were prepared as described in Experimental (Scheme 1) Polymerization proceded homogeneously and insoluble particles, diisopropylammonium iodide, were formed as the reaction progressed. Both polymers are readily soluble at room temperature in chloroform and tetrahydrofuran. Their number-average molecular weights ($\overline{M}_n$) determined by GPC were found to be 15500 and 10200, respectively. Polydispersity indices were 1.5 and 1.6.

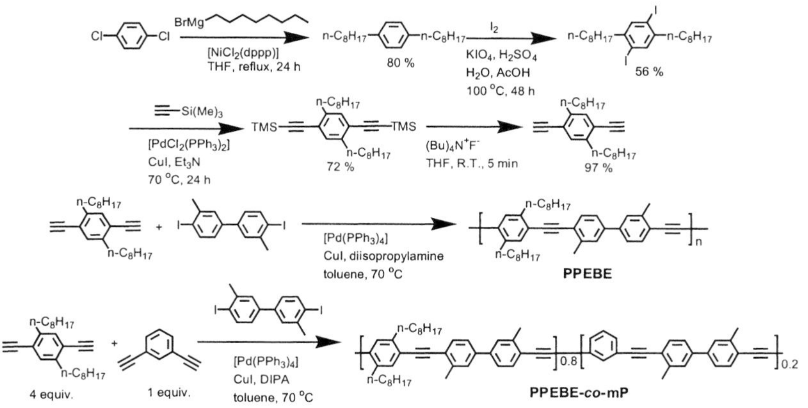

Scheme 1. Synthesis of PPEBE and PPEBE-*co*-mP.

UV-vis absorption and photoluminescence (PL) spectra: Figure 1(a) compares the UV-vis absorption and PL spectra of the two polymers in thin films. We note that as we include the meta-phenylene unit in the copolymer, UV-vis absorption as well as PL spectra in the longer wavelength region is shifted toward blue side. The absorption in 375-450 nm corresponds to $\pi \rightarrow \pi^*$ transitions of the polymers. The PL spectrum of the PPEBE polymer exhibits finer vibronic details with two distinct maxima located at 451 and 485 nm with a weaker shoulder at about 525 nm. In contrast, the PL spectrum of the polymer, PPEBE-*co*-mP, shows a dominant peak maximum at 465 nm with two weaker shoulders in the longer wavelength region. The optical bandgap (E_g) estimated from the absorption edge (450 nm) for both the polymers is 2.7 eV, which is higher than that of PPV (2.4 eV).

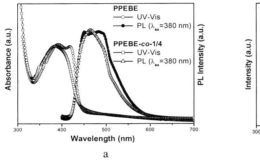 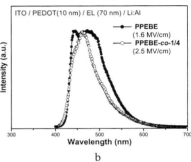

a   b

Figure 1. (a) Absorption and emission studies of PPEBE and PPEBE-*co*-mP, (b) their EL spectra (film thickness 70 nm).

EL device performance

We constructed PLED devices having the configuration of ITO/PEDOT (10 nm)/polymer (70 nm)/Li:Al. Here, PEDOT stands for poly[3,4-(ethylenedioxy)thiopene-2,5-diyl] doped with sulfonated polystyrene obtained from Bayer. The room temperature conductivity of the polymer was 10 S/cm. Figure 1b shows the EL spectra of the two polymers obtained at the operating voltages 1.6 and 2.5 MV/cm, respectively. The overall feature of the spectra is very similar to their corresponding PL spectra given in Figure 1a, although there are some differences in details.

The copolymer emits a better blue light of the desired wavelength. Figure 2 compares the characteristics of the devices fabricated with the polymers. We note several important differences: (1) The threshold electric field of PPEBE is significantly lower (1.4 MV/cm) than that (2.5 MV/cm) of the copolymer. (2) Maximum light output of the PPEBE device is much higher than that of the PPEBE-*co*-mP device. (3) HOMO-LUMO energy levels, measured by cyclovoltammetry and optical band gap of the homopolymer (5.9-3.1 eV) are higher than those of the copolymer (6.1-3.3 eV). (4) Device stability is much greater for the PPEBE device than for the other. Needless to say, the phenomena described in (1) – (4) are correlated.

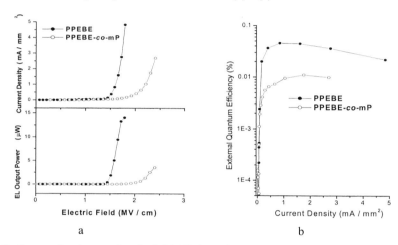

Figure 2. Current density vs. electric field, EL intensity vs. electric field, and external quantum efficiency vs current density curves for EL devices (ITO/PEDOT (10 nm)/polymer (70 nm)/Li:Al).

The maximum brightness observed for the PPEBE device was 410 cd/m^2, which is much higher than the values reported for other poly(*p*-phenyleneethynylene) derivatives[10] and comparable with those reported for polyfluorenes[3].

The external quantum efficiencies of the two devices are shown in Figure 2b. The maximum quantum efficiency observed for the PPEBE device was 0.05 %, decreasing steadily as the current density or the applied electric field increased. The situation for PPEBE-*co*-mP was similar. The maximum efficiency attainable was 0.01 %. Therefore, it is concluded that the inclusion of the *m*-phenylene unit lowers the device performance, although the wavelength of the emitted light is close to the ideal blue light. In the light of the very low current flow in the copolymer device, it is conjectured that inclusion of the *m*-phenylene unit reduces the carrier mobility, which has to be confirmed by future work.

Nevertheless, we believe that a proper modification of the structure of PPEBE and better construction of the PLED device may lead to a successful development of blue-light emitting displays based on poly(phenyleneethynylene)s.

Acknowledgement

This work was supported by the Korea Science and Engineering Foundation through the Center for Electro- and Photo-Responsive Molecules, Korea University. S.-H. Joo is a recipient of the Brain Korea 21 Fellowship supported by the Ministry of Education and Human Resources, Korea.

[1] J. H. Burroughes, D. D. C. Bradley, A. R. Brown, R. N. Marks, K. MacKay, R. H. Friend, P. L. Burn, A. B. Holmes, *Nature* **1990**, 347, 539.

[2] H.-K. Shim, J.-I. Jin, *Adv. Polym. Sci.* **2002**, 158, 193.

[3] A. W. Grice, D. D. C. Bradley, M. Bernius, M. Inbasekaran, W.W. Wu, E. P. Woo, *Appl. Phys. Lett.* **1998**, 73, 629.

[4] M. Bernius, M. Inbasekaran, J. O'Brien, W.Wu, *Adv. Mater.* **2000**, 12, 1737.

[5] K. Kim, Y.-R. Hong, S.-W. Lee, J.-I.Jin, Y.Park, B.-H.Sohn, W.-H. Kim, J.-K. Park, *J. Mater. Chem.* **2001**, 11, 3023.

[6] D. W. Lee, K. Y. Kwon, J.-I. Jin, Y. Park, Y. R. Kim, I. W. Hwang, *Chem. Mater.* **2001**, 13, 565.

[7] B.-H. Sohn, K. Kim, D.-Soo. Choi, Y.-K. Kim, S.-C. Jeoung, J.-I. Jin, *Macromolecules* **2002**, 35, 2876.

[8] J.-I. Jin, C.-E. Lee, J.-S. Joo, Y. Park, Ed. Sasabe, in *Nanotechnology toward the Organic Photonics*, GooTech Press, Chitose, **2002**, p. 253.

[9] V. E. Choong, Y. Park, R. Hsieh, Y. Gao, in *Photonic Polymer Systems - Fundamentals, Methods, and Application*, Marcel Dekker, New York, **1998**, Chap. 5.

[10] A. Montali, P. Smith, C. Weder, *Synth. Met.* **1998**, 97, 123.

[11] G. Brizius, N. G. Pschirer, W. Steffen, K. Stitzer, H.-C. Loye, U. H. F. Bunz, *J. Am. Chem. Soc.* **2000**, 122, 12435

[12] W. Y. Huang, W. Gao, T. K. Kwei, Y. Okamoto, *Macromolecules* **2001**, 34, 1570.

[13] M. Itoh, M. Mitsuzuka, K. Iwata, K. Inoue, *Macromolecules* **2001**, 27, 7917.

[14] A. W. Grice, D. D. C. Bradley, M. T. Bernius, M. Inbasekaran, W. W. Wu, E. P. Woo, *Appl. Phys. Lett.* **1998**, 73, 629.

Macromol. Symp. **2004**, *212*, 321-326

Kinetics of Refractive Index Changes in Polymeric Photochromic Films

Ewelina Ortyl, Stanisław Kucharski*

Institute of Organic and Polymer Technology, Technical University of Wrocław,
50-370 Wrocław, Poland
E-mail: kucharski@itots.ch.pwr.wroc.pl

Summary: Methacrylate copolymers containing, in side chain azobenzene groupings with heterocyclic sulfonamide substituents: *N*-(2,6-dimethylpyrimidin-4-yl)sulfamoyl (sulfisomidine) and *N*-(5-methylisoxazol-3-yl) sulfamoyl (sulfamethoxazole) were investigated. The materials undergo reversible *trans-cis* isomerisation during illumination with light. This results in changes of dipole moment, polarizability and refractive index. Ellipsometric measurements showed a distinct decreasing refractive index during illumination with light corresponding to absorption band (*ca.* 450 nm). Depending on the polymer, the change of real part of refractive index in spin-coated films was between 0.016 and 0.031. The dynamics of growth and decay of refractive index changes, was described by biexponential function approach.

Keywords: azobenzene polymers; ellipsometry; photochromic polymers; refractive index modulation; *trans-cis* isomerisation

Introduction

Polymers containing the azobenzene moiety in side chain or built in the main chain have been attracting much attention as active materials in optical data storage and holographic applications [1]. The ability of the azobenzene grouping to reversibly *trans-cis* (or E-Z) izomerize has been known for a long time and described in numerous papers [2-4]. Under illumination with visible or UV light, corresponding to the maximum of absorption band, the more stable *trans* form undergoes transformation to the less stable *cis* form. The reverse *cis-trans* reaction takes place as a thermal relaxation and is relatively slow. The photoisomerisation of the azobenzene grouping is particularly interesting because its consequence is a change of refractive index.

DOI: 10.1002/masy.200450837

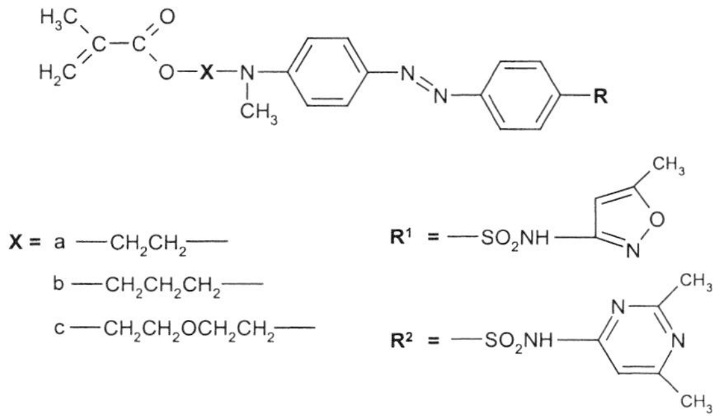

CODE	X	R
MET-1	a	R^1
MET-2	b	R^1
MET-3	c	R^1
IZO-1	a	R^2
IZO-2	b	R^2
IZO-3	c	R^2

Fig. 1. Chromophoric monomers.

Results and discussion

We calculated changes of some molecular properties which take place under light irradiation of chromophoric monomers (Fig. 1). Syntheses of monomers were described previously [5]. We started with optimization of geometrical structures of the isolated molecules of the compounds with GAUSSIAN 98 [6] program using *ab initio* RHF (restricted Hartree-Fock) options with a split-valence 3-21G basis set. The optimized geometry was used to calculate dipole moment (μ), static polarizability (α^0) and first hyperpolarizability (β^0), for both *trans* and *cis* form of the monomers, and potential energy difference, ΔE, between them (Table 1).

For both sulfisomidine and sulfamethoxazole azobenzene derivatives the values of μ, α^0 and β^0 were higher for *trans* form. The differences between dipole moments of *cis* and *trans* isomers increased with increasing length of the aliphatic spacer joining nitrogen atom with methacryloyl group. For monomer IZO-3, which has the longest spacer, the dipole moment difference was the highest, 6.41 D. This difference was considerably lower for monomers with short spacers, IZO-1 and MET-1, 1.88 D and 4.69 D, respectively. In both homologous series of monomers, the decrease in static polarizability and first hyperpolarizability was more pronounced with increasing length of spacer. The differences in potential energy, ΔE, between *trans* and *cis* form of the compounds also increase with increasing length of aliphatic spacer. For IZO-3 monomer, ΔE has the highest value equal to 104.6 kJ/mol.

Table 1. Dipole moments, polarizability, first hyperpolarizability and potential energy difference (ΔE) of the *trans* and *cis* structures calculated by GAUSSIAN.

Compound	μ (D)		$\alpha^0 \times 10^{24}$ (cm^3)		$\beta^0 \times 10^{30}$ (esu)		ΔE (kJ/mol)
	cis	trans	cis	trans	cis	trans	
IZO-1	8.34	10.22	43.12	47.02	10.17	40.80	64.27
IZO-2	5.25	10.58	43.85	49.65	3.88	49.27	92.18
IZO-3	3.98	10.39	45.40	51.65	3.39	52.65	104.61
MET-1	5.52	10.21	40.40	43.92	13.94	45.29	81.23
MET-2	4.24	8.79	41.39	46.42	10.46	53.49	81.58
MET-3	3.73	9.22	41.92	48.21	2.42	50.83	92.45

One of the consequences of the *trans-cis* isomerisation of the azobenzene grouping is the refractive index change. But investigation of these photoinduced phenomena is not easy as there is a problem with isolation of the pure form of *cis* isomer, which is unstable at room temperature, where the measurements are usually carried out. For these reasons we calculated refractive indices for the isolated molecules of their pure *trans* and *cis* forms. We utilized two models, which relate the index of refraction to the polarizability of the molecule [7-9]. In the first model we used the transformated Lorentz-Lorenz equation (Eq. 1):

$$n(\omega) = \sqrt{\frac{3 + 8\pi N \alpha(\omega)}{3 - 4\pi N \alpha(\omega)}} \tag{1}$$

Equation (1) may be approximated by the Lorenz form. In the second model we used the Lorenz expression (2):

$$n(\omega) = \sqrt{1 + 4\pi N \alpha(\omega)} \tag{2}$$

In both equations (1) and (2), N is average number of molecules per unit volume and $\alpha(\omega)$ is the isotropic average polarizability of the isolated molecule, is equal to $\frac{1}{3}(\alpha_{xx} + \alpha_{yy} + \alpha_{zz})$. The volume occupied by an isolated molecule was calculated using Cerius2 program. The values of refractive indices calculated according to Eq. (1) and (2) and differences between both isomers (Δn) are given in Table 2. The most interesting for us were the values of Δn. The highest value of Δn was observed for IZO-3 monomer which was 0.0392.

Table 2. Refractive indices and refractive index changes calculated utilizing two models: Lorenz and Lorentz-Lorenz.

Comp.	Molecule volume (Å^3)		n_{Lorenz}		Δn_{Lorenz}	$n_{\text{Lorentz-Lorenz}}$		$\Delta n_{\text{Lorentz-Lorenz}}$
	cis	trans	cis	trans		cis	trans	
IZO-1	627.53	639.37	1.3649	1.3870	0.0221	1.4871	1.5280	0.0409
IZO-2	652.42	668.88	1.3580	1.3901	0.0321	1.4748	1.5338	0.059
IZO-3	686.3	691.3	1.3531	1.3923	0.0392	1.4659	1.5381	0.0722
MET-1	551.01	558.48	1.3860	1.4099	0.0239	1.5261	1.5724	0.0463
MET-2	581.53	582.12	1.3762	1.4148	0.0386	1.5078	1.5823	0.0745
MET-3	602.35	616.98	1.3690	1.4076	0.0386	1.4945	1.5680	0.0735

Our aim was to compare the calculated refractive index change for monomers with the measured ones for polymers containing the same chromophoric groups. We used ellipsometry for the determination of film thicknesses and refractive index changes during illumination. Ellipsometry measurement yields two quantities Δ and Ψ, which are sensitive to optical parameters of the sample [10]. From the measured values, it is possible to calculate the complex dielectric function expressed by complex refractive index (3):

$$n = n_{\text{r}} + i\,k \tag{3}$$

The refractive index change, Δn_{r}, we calculated according to equation (4):

$$\Delta n_{\text{r}} = n_{\text{r}}^0 - n_{\text{r}}^{\text{photostat.}} \tag{4}$$

where n_{r}^0 is the real part of complex refractive index (3) measured before illumination with light and $n_{\text{r}}^{\text{photostat.}}$ is the real part of the refractive index measured in the photostationary state. Natansohn et al. have observed a biexponential kinetic process in their azobenzene systems [11]. We have found that biexponential functions also well describe the dynamics of growth and decay of refractive index changes in our polymers.

The biexponential functions are

$$\Delta n_{\text{r}} = A\{1 - \exp(-k_{\text{a}}t)\} + B\{1 - \exp(-k_{\text{b}}t)\} \tag{5}$$

for photoinducing, and

$$\Delta n_{\text{r}} = C\exp(-k_{\text{c}}t) + D\exp(-k_{\text{d}}t) + E \tag{6}$$

for the relaxation process in the absence of illumination. In Eqs. (5) and (6) Δn_r is the change of real part of refractive index observed at time t; k_a, k_b, k_c and k_d represent the rate constants with the amplitudes of A, B, C and D respectively. E is the fraction of refractive index modulation conserved for a long time. All values of the constants are given in Tables 3a and 3b. The rate constants of the fast-growth process and stability of the photoinduced refractive index changes (E) depend on the kind of chromophore side group as well as on the kind of the comonomer used: butyl 2-methacrylate (MB) and 2-ethylhexyl acrylate (AI).

But for IZO-3/20MB polymer the refractive index change, Δn_r, was the highest (0.031).

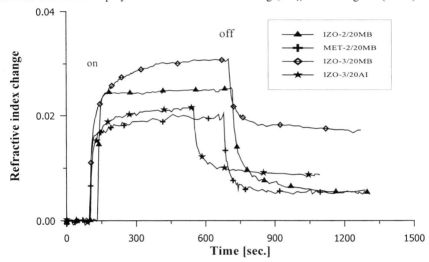

Fig. 2. Typical growth and decay processes of photoinduced refractive index modulation. The light source used for sample illumination was white light of ca. 2 mW cm^{-2} power (from 410 nm on).

Table 3a. Kinetic data for the photoinduced refractive index modulation – growth process.

Polymer	A	k_a	B	k_b	Δn_r
IZO-3/20MB	0.0180	0.0875	0.0127	0.0103	0.031
IZO-3/20AI	0.0176	0.085	0.0036	0.007	0.021
IZO-2/20MB	0.0225	0.085	0.0026	0.007	0.025
MET-2/20MB	0.0168	0.085	0.003	0.006	0.020
MET-3/20AI	0.0144	0.1308	0.0015	0.001	0.016

Table 3b. Kinetic data for the photoinduced refractive index modulation – decay process.

Polymer	C	k_c	D	k_d	E
IZO-3/20MB	0.00842	0.1476	0.00492	0.01436	0.01751
IZO-3/20AI	0.0085	0.1180	0.0041	0.0093	0.0087
IZO-2/20MB	0.01125	0.09444	0.00773	0.00924	0.00541
MET-2/20MB	0.00746	0.2305	0.00748	0.02800	0.00544
MET-3/20AI	0.01417	0.05467	0.00083	0.01347	0.00178

Conclusions

The methacrylate polymers containing azobenzene grouping with sulfonamide substituents in side chains undergo a reversible *trans-cis* isomerisation during illumination with light corresponding to their maximum absorption band. This photoinduced transformation results in change of physicochemical properties of materials, such as refractive index.

The observed decrease in refractive indices of polymers during illumination was in fair agreement with the values calculated for monomers. The refractive index modulation, Δn_r, measured by ellipsometry was in the range of ca. $0.016 - 0.031$. Biexponential functions well described the kinetics of refractive index change in these polymers: they give good fits to the growth curves as well as to the relaxation ones.

[1] S. A. Jenekhe and K. J. Wynne, Eds., *Photonic and Optoelectronic Polymers*; *ACS Symp. Ser.* **1995**, *672*, 236.
[2] Z. Sekkat, W. Knoll, *Photochem.* **1997**, *22*, 117.
[3] S. Kucharski, R. Janik, H. Motschmann and Ch. Radüge, *New J. Chem.*, **1999**, *23*, 765.
[4] R. Janik, S. Kucharski, A. Kubaińska and B. Łyko, *Pol. J. Chem.*, **2001**, *75*, 241.
[5] E. Ortyl, R. Janik and S. Kucharski, *Eur. Polym. J.*, **2002**, *38*, 1871.
[6] M. J. Frisch, W. Trucks, H. B. Schlegel, P. M. W. Gill, B. G. Johnson, M. A. Robb, J. R. Cheeseman, T. Keith, G. A. Petersson, J. A. Montgomery, K. Raghavachari, M. A. Al.-Laham, V. G. Zakrzewski, J. V. Oritz, J. B. Foresman, J. Cioslowski, B. B. Stefanov, A. Nanayakkara, M. Challacombe, C. Y. Peng, P. Y. Ayala, W. Chen, M. W. Wong, J. L. Andres, E. S. Replogle, R. Gomperts, R. L. Martin, D. J. Fox, J. S. Binkley, D. J. Defrees, J. Baker, J. P. Stewart, M. Head-Gordon, C. Gonzalez and J. A. Pople, *Gaussian 98, Rev. A.8*, Gaussian, Inc., Pittsburgh PA, **1998**.
[7] C. J. F. Böttcher, *Theory of Electric Polarization*, 2nd ed., Elsevier, Amsterdam, **1973**, Vol. 1.
[8] K. O. Sylvester-Hvid, P.-O. Åstrand, M. A. Ratner and K. V. Mikkelsen, *J. Phys. Chem. A*, **1999**, *103*, 1818.
[9] K.O. Sylvester-Hvid, K.V. Mikkelsen and M.A. Ratner, *J. Phys. Chem.* A, **1999**, 103, 8447.
[10] R. M. A. Azzam and N. M. Bashara, *Ellipsometry and Polarized Light*, North-Holland, Amsterdam 1979.
[11] A. Natansohn, P. Rochon, J. Gosselin and S. Xie, *Macromolecules*, **1992**, 25, 2268.

Macromol. Symp. **2004**, *212*, 327-334

Quantum Chemical Study of Oxidation Processes in Cu–Phthalocyanine

Petr Toman,[1] *Stanislav Nešpůrek,*[1] *Kyuya Yakushi*[2]

[1] Institute of Macromolecular Chemistry, Academy of Sciences of the Czech Republic, Heyrovský Sq. 2, 162 06 Prague 6, Czech Republic
[2] Institute for Molecular Science, Myodaiji-cho, Okazaki, Aichi 444-8585, Japan

Summary: Quantum chemical calculations reproduced quite well the experimental infrared spectra of CuPc and $CuNO_3Pc \cdot HNO_3$. The agreement in the changes of line intensities during the oxidation supports the idea of ligand oxidation. This result is in agreement with the Mulliken population analysis.

Keywords: electron density; infrared spectroscopy; oxidation; phthalocyanine; quantum chemistry

Introduction

Phthalocyanine (Pc) molecules are stacked one-dimensionally in a metal-over-metal stacking mode in the charge-transfer salts. Owing to this characteristic structure, these compounds have two channels - the metal and ligand ones - in the same molecular column, in which charge carriers are doped by introducing counter anions. The iodine doping makes NiPc and CuPc metallic at room temperature [1,2]. Generally, there are two oxidation possibilities: during the doping a ligand-oxidized or metal-oxidized channel can be formed. Martinsen et al. ascribed the distinct properties of CoPcI to the difference in the doped channel: CoPcI has a metal-oxidized channel, whereas the NiPcI has a ligand-oxidized channel [3]. The differences can be found not only for different metals in phthalocyanine skeleton, but also for different acceptors. Yakushi et al. [4] found on the basis of IR measurements that the ligand was oxidized in $CoPc(AsF_6)_{0.5}$ in contrast to the unoxidized ligand of CoPcI. Their conclusion was drawn based on the several infrared-active vibrational bands whose intensities are sensitive to the oxidation state of Pc [5,6]. Meanwhile, Hiejima and Yakushi conducted the high-pressure experiment of infrared spectrum and found that the high pressure induced the electron transfer from the 3d band to the partially filled π-band in $NiPc(AsF_6)_{0.5}$ and $CoPc(AsF_6)_{0.5}$ [7]. This pressure-induced d-π charge transfer

DOI: 10.1002/masy.200450838

was concluded from the intensity change of the charge-sensitive vibrational bands under high pressure. The intensity changes were related to the differences in electron distributions in Pc^{unox} and Pc^{ox} skeletons, which was supported for NiPc and CoPc by molecular orbital calculations [8]. Because the situation has not been fully clear for CuPc we have aimed our studies to the oxidization process in CuPc molecule and formation of monocation $CuPc^+$.

Experimental and calculation details

Copper-phthalocyanine (CuPc) from Tokyo-Kasei was purified by sublimation into temperature gradient three times. The IR spectrum was measured by Perkin-Elmer FTIR Paragon 1000 PC spectrometer using a KBr pellet of the powdered sample.

Quantum chemical calculations were performed using ab initio Becke's three-parameter hybrid method with the correlation functional of Lee, Yang, and Paar (B3LYP). Computer program Gaussian 98 [9] was used. The B3LYP method is known to give usually very accurate conformational geometries and IR spectra, better than Hartree-Fock one. All results were calculated using the basis set of atomic orbitals containing the 6-31G* basis functions for hydrogen, carbon, and nitrogen and the 6-311+G* basis functions for the central metal (copper). The inclusion of polarization and diffuse functions is important for obtaining correct IR spectra. CuPc molecules were assumed to be isolated. Molecular conformations were determined by means of minimization of the total energy, no symmetry was taken into account by definition, however the resulting molecular conformations were found to retain $\mathcal{D}_{4h}$ symmetry. According to X-ray data taken from literature [10, 11] the CuPc molecule is not perfectly $\mathcal{D}_{4h}$ symmetrical but the distortion is minor. The B3LYP calculated Cu–N bond lengths are equal to 1.967 Å. This value is close to the value 1.94 Å reported for the $\mathcal{D}_{4h}$ averaged geometry by Schaffer et al. [12]. The unoxidized CuPc molecule was treated as neutral, the oxidized molecule was assumed to have the total charge +1e.

The open shell systems can be in principle studied using the restricted or unrestricted wavefunctions. The latter approach allows wider class of the wavefunctions and gives slightly lower total energy. A usual disadvantage of the unrestricted wavefunction is that it is not an eigenfunction of the spin operator S^2. When the deviation of the mean value of the spin operator $<S^2>$ from $S(S+1)$ is substantial, the calculated molecular conformation is suspicious (see e.g.

[13]). In the presented open shell calculations, the values of the square of the total spin $<S^2>$ are as follows (the S(S+1) values are in parenthesis): $CuPc^0$ – 0.753 (0.750), singlet $CuPc^+$ – 1.031 (0.000), triplet $CuPc^+$ – 2.021 (2.000). Thus, the only molecule with rather highly spin-contaminated wavefunction is singlet $CuPc^+$. After the annihilation of the first spin contaminant, the value of $<S^2>$ lowers from 1.031 to 0.251. We do not expect any important influence of this spin contamination on the presented results. For these reasons we used UB3LYP method for the calculation of all presented results.

Results and discussion

Regarding the multiplicity and spin pairing of electrons, the unoxidized CuPc contain one unpaired electron; CuPc molecule is in the doublet state. The oxidized $CuPc^+$ molecule possesses even number of electrons, so one can expect a closed-shell system. However, we found in the previous work [8] that the closed-shell B3LYP wavefunction of $CoPc^+$ is unstable. The stable electronic configuration possesses two unpaired electrons occupying different orbitals with the symmetry a_{1u} and a_{1g}; both the singlet and triplet configurations are possible. The Hartree-Fock (HF) and B3LYP calculations of $CuPc^+$ show similar results: the closed shell wavefunction is also unstable with respect to the becoming unrestricted one; it means the system is going to be open shell. Furthermore, the closed shell B3LYP wavefunction does not fulfill the Aufbau principle. Thus, stable electronic configuration of $CuPc^+$ is an open shell and possesses two unpaired electrons occupying different orbitals with the symmetry a_{1u} and b_{1g} (see Fig. 1); both the singlet and triplet configurations are possible. The B3LYP total energies of these open shell configurations are about 0.05 a.u. lower than the total energy of the unstable closed shell configuration. The difference in the total energy between the triplet and singlet open shell configuration is very low (about 5×10^{-4} a.u.). These facts allow us to believe that the open shell configuration (singlet or triplet) is very probable. Therefore, the further discussion will be concentrated to the open shell configuration but, for the comparison, the closed shell case (in which a_{1u} is a fully occupied orbital and b_{1g} orbital is empty) will be discussed. There are two another possible open shell configurations [10]. We have found that these configurations have theoretical infrared spectra completely different from the experimental ones. Let us try to solve the problem of the oxidation of the metal or Pc ligand. Homborg found diagnostic bands in the fingerprint region of the experimental IR spectrum to distinguish the oxidized ligand, Pc^{ox}, from the unoxidized one, Pc^{unox} [5,6]. The absorption bands at about 1520 and 1290 cm^{-1} appear strongly in non-oxidized Pc ligand (MgPc and CoPc) and disappear in oxidized ligands (MgClPc and CoClPc) whereas 1460 and 1365 cm^{-1} bands appear in the latter and disappear in the former case. This complementary relationship of the intensities was found in the theoretical spectra of CuPc and $CuPc^{+}$ (open shell) but not CuPc and $CuPc^{+}$ (closed shell) - see Fig 2.

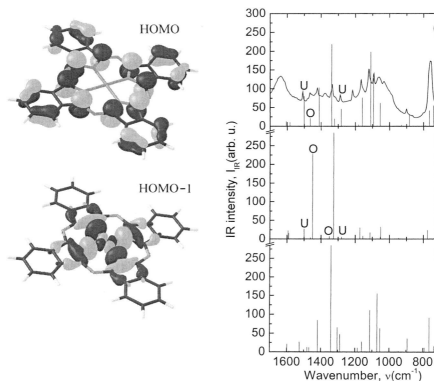

Figure 1. Singly occupied molecular orbitals of $CuPc^{+}$: HOMO (a_{1u} symmetry) and HOMO-1 (b_{1g} symmetry).

Figure 2. (a) Experimental IR pattern of CuPc (line) together with the calculated spectrum (columns), (b) calculated spectrum of $CuPc^{+}$ –open shell, triplet, and (c) calculated spectrum of $CuPc^{+}$ – closed shell. "U" and "O" denote the characteristic diagnostic bands of unoxidized and oxidized Pc ligand, respectively.

Table 1. Experimental and theoretical IR wavenumbers ($\nu_{exp.}$ and $\nu_{theor.}$) and IR intensities ($I_{exp.}$ and $I_{theor.}$) of neutral and oxidized CuPc. Wavenumbers and theoretical intensities are in cm^{-1} and km/mol, respectively. Experimental intensities are labeled by generally used symbols (from weakest to strongest line: vw, w, wm, m, ms, s, vs). The transitions No. 2 and 4 are out-of-plane vibrations with the symmetry A_{2u}. All other transitions are in plane vibrations with the E_u symmetry. The experimental data of the oxidized CuPc$^+$ (measured on CuNO$_3$Pc $\bullet$ HNO$_3$) are taken from the Homborg paper [5].

| No. | neutral CuPc^0 | | | | oxidized CuPc$^+$ | | | | | |
	$\nu_{exp.}$	ν_{theor}	$I_{exp.}$	I_{theor}	$\nu_{exp.}$	ν_{theor} open shell	ν_{theor} closed shell	$I_{exp.}$	I_{theor} open shell	I_{theor} closed shell
1	669	629.6	wm	5	–	627.0	629.3	–	0.04	0.3
2	722	723.6	ms	180	735	729.5	697.2	s	225	156
3	761	741.1	vs	55	–	738.4	736.6	–	1	15
4	shoulder	766.5	shoulder	41	788	776.3	764.9	wm	24	91
5	900	883.4	wm	32	–	882.3	893.1	–	1	35
6	1034	1002.3	w	7	–	1002.8	1001.2	–	1	0.5
7	1061	1055.0	m	62	1065	1051.7	1054.1	wm	32	63
8	1092	1093.6	m	128	–	–	1071.0	–	–	155
9	1121	1110.6	ms	198	1125	1114.4	1113.2	wm	17	111
10	1167	1158.8	ms	76	–	1155.4	1160.2	–	8	27
11	–	1181.7	–	4	1180	1173.4	1185.0	w	30	11
12 "U"	1288	1281.8	wm	45	–	1278.9	1287.9	–	2	47
13	shoulder [a]	1321.1	shoulder [a]	20	1310	–	1303.3	vw	–	66
14	1335	1337.8	m	218	1330	1325.9	1340.7	s	660	360
15 "O"	–	–	–	–	1360	1351.3	–	wm	7	–
16	1422	1412.1	wm	103	–	1404.9	1419.9	–	2	84
17 "O"	1466	1464.6	vw	20	1450	1450.8	1469.3	m	228	12
18 "U"	1508	1500.4	wm	80	1510	1500.5	1526.2	w	27	26

[a] This transition is a shoulder of the peak at 1335 cm^{-1} from the side of higher wavenumbers.

The experimental and theoretical values of these peaks are summarized in Table 1. The calculated CuPc$^+$ spectra are compared with the experimental spectrum of CuNO$_3$Pc $\bullet$ HNO$_3$ taken from the Homborg paper [5]. The IR spectrum of the CuPc$^+$ (open shell, singlet) is not presented in Fig. 2 and Table 1 because it is very similar to the CuPc$^+$ (open shell, triplet) one. The theoretical spectrum of the neutral CuPc reproduces well the experimental results. Only the experimental medium intensity peak at 1216 cm^{-1} (see Fig. 1a) is missing in the theoretical spectrum. Similarly to NiPc and CoPc, there are important differences in the line intensities of the experimental IR spectra between the unoxidized and oxidized CuPc and only minor differences in the peak positions. The experimental intensity changes of peaks No. 14, 15, 17 and 18 agree well with the theoretical changes for open shell case of $CuPc^+$. "U" and "O" diagnostics suggests that very probably in the $CuPc^+$ the ligand is oxidized and the molecule is in open shell configuration. Open shell configuration is supported also by the changes of other transitions given in Table 1. The closed shell configuration is less probable, mainly because of changes in the intensities of peaks 7, 8, and 16.

Figure 3 gives the calculated normal modes of $CuPc^+$ for the diagnostic bands denoted by "U" and "O" in Fig. 2. There are no important differences in the character of the vibrational modes between the neutral and cationic states of CuPc. Thus, only the case of $CuPc^+$ is presented in Fig. 3. The normal modes at 1279 ("U") and 1451 cm^{-1} ("O") are localized at the outer benzene rings, whereas those at 1351 ("O") and 1500 cm^{-1} ("U") are rather localized at the inner pyrrole rings. The changes in the peak intensities of the IR spectra reflect the change of the charge distribution upon the oxidation which takes place in a whole macrocycle as it follows from Table 2.

Table 2. Calculated Mulliken atomic charges and spin densities. The numbering of atoms is given in the figure below the table.

Atom	Mulliken charges			Spin densities			
	$CuPc^0$	$CuPc^+$ open shell	$CuPc^+$ closed shell	$CuPc^0$	$CuPc^+$ open shell triplet	$CuPc^+$ open shell singlet	$CuPc^+$ closed shell
Cu	0.65	0.70	0.74	0.578	0.586	0.581	0.000
N1	-0.58	-0.59	-0.60	0.104	0.054	0.151	0.000
C1	0.38	0.42	0.41	-0.005	0.145	-0.154	0.000
N2	-0.43	-0.42	-0.39	0.002	-0.063	0.067	0.000
C2	0.07	0.08	0.08	0.005	-0.018	0.028	0.000
C3	-0.19	-0.17	-0.17	0.000	0.044	-0.045	0.000
H3	0.15	0.18	0.17	0.000	-0.002	0.002	0.000
C4	-0.13	-0.12	-0.13	-0.001	0.012	-0.014	0.000
H4	0.13	0.17	0.17	0.000	-0.001	0.001	0.000
Total	0.00	1.00	1.00	1.000	2.000	0.000	0.000

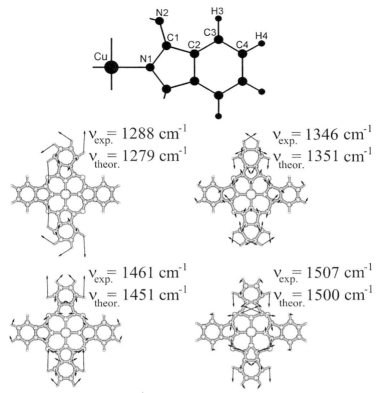

Figure 3. IR vibrational modes of CuPc⁺ (open shell, triplet) denoted as "U" and "O" in Fig. 2.

Table 2 shows the Mulliken charge and spin density distributions in CuPc neutral molecule and its three cations (open shell – triplet, open shell – singlet, and closed shell). The atomic charges of the open shell – singlet and open shell – triplet are the same, so only one common column is shown in the table. Atomic spin densities of the closed shell CuPc⁺ are zero on all atoms by definition. The increase of the positive charge of the all cations in comparison with the neutral molecule is predominantly distributed in the ligands. The charge on the Cu atom in the oxidized state is only slightly more positive (about 0.05 e resp. 0.09 e) than that in the neutral molecule. However, the change of the spin density on Cu atom depends on the type of the cation mentioned in Table 2. Both open shell cations have almost the same spin density on the central Cu atom as the neutral CuPc, it supports the idea that the oxidation occurs mainly in the ligand channel. On the other hand, the difference of spin densities on the central metal in neutral CuPc and $CuPc^+$ (closed shell) is more than 0.5. It means the unpaired electron (in the b_{1g} orbital), which is taken away during the oxidation, is located on both metal and ligand. However, after the ionization the electron density is redistributed in such a way that the atomic charge on Cu atom in $CuPc^+$ (closed shell) in comparison with the neutral molecule raises only about 0.09 e after reaching the equilibrium state – also in this case the ligand is oxidized. These results strongly support the predominant ligand oxidation in CuPc; regardless of which of these cations is actually realized. It seems from our theoretical calculations that Homborg diagnostic peaks indicate only the oxidation from the a_{1u} orbital.

Acknowledgements

This work was supported by the Projects No. KJB1050301 and AV0Z4050913 of the Grant Agency of the Academy of Sciences of the Czech Republic, and the Project No. ME 558 of the Ministry of Education, Youth and Sports of the Czech Republic. The computer time in the academic supercomputing centers Metacenter (Prague, Brno) and Joint Supercomputing Center at the Czech Technical University (Prague) is gratefully appreciated.

[1] T. E. Phillips, R. P. Scaringe, B. M. Hoffman, J. A. Ibers, *J. Am. Chem. Soc.* **1980**, *102*, 3435.
[2] J. Martinsen, S. M. Palmer, J. Tanaka, R. C. Greene, B. M. Hoffman, *Phys. Rev. B* **1984**, *30*, 6269.
[3] J. Martinsen, J. L. Stanton, R. L. Greene, J. Tanaka, B. M. Hoffman, J. A. Ibers, *J. Am. Chem. Soc.* **1985**, *107*, 6915.
[4] K. Yakushi, H. Yamakado, T. Ida, A. Ugawa, *Solid State Commun.* **1991**, *78*, 919.
[5] H. Homborg, *Z. anorg. allg. Chem.* **1983**, *507*, 35.
[6] H. Homborg, W. Kalz, *Z. Naturforsch. B* **1984**, *39*, 1490.
[7] T. Hiejima, K. Yakushi, *J. Chem. Phys.* **1995**, *103*, 3950.
[8] P. Toman, S. Nešpůrek, K. Yakushi, *J. Porphyrins Phthalocyanines* **2002**, *6*, 556.
[9] M. J. Frisch et al.: Computer program Gaussian 98, Revision A.7, Gaussian, Inc., Pittsburgh PA, 1998.
[10] S. Carniato, G. Dufour, F. Rochet, H. Roulet, P. Chaquin, C. Giessner–Prettre, *J. Electron Spectrosc. Relat. Phenom.* **1994**, *67*, 189.
[11] C. J. Brown, *J. Chem. Soc. A* **1968**, *2*, 2488.
[12] A. M. Schaffer, M. Gouterman, E. R. Davidson, *Theor. Chim. Acta* **1973**, *30*, 9.
[13] I. N. Levine, *"Quantum chemistry"*, Prentice–Hall, Englewood Cliffs 1991, p.461.

Atmospheric Aging of Poly[methyl(phenyl)silanediyl] Monitored by FTIR Spectroscopy

Oto Meszáros, * *Pavel Schmidt, Jan Pospíšil, Stanislav Nešpůrek, Ivan Kelnar*

Institute of Macromolecular Chemistry, Academy of Sciences of the Czech Republic, 162 06 Prague, Czech Republic
E-mail: meszaros@imc.cas.cz

Summary: Poly[methyl(phenyl)silanediyl] films were irradiated in ATLAS Weather-O-Meter Ci 3000+ for ca 3 h. Changes in the region of siloxane, carbonyl, hydroxyl, and C-H (aromatic) bands were monitored by FTIR. Main attention was paid to changes in siloxane formation, which could be reduced in the presence of phenolic UV absorber **1** and oxalanilide **3**.

Keywords: formation of siloxane groups; photooxidation; poly[methyl(phenyl)silanediyl]

Introduction

Poly[methyl(phenyl)silanediyl] (PMPSi) is a typical representative of polysilanes (PSi), a group of substituted σ-conjugated polymers with homoatomic backbone made exclusively of Si atoms. Cumulation of the Si atoms allows an extensive electron delocalisation, resulting in unusual electronic and photochemical properties[1-4]. The PSi chain arrangement may be linear, cyclic, branched or crosslinked. PSi with various substituents such as alkyls, cykloalkyls, aralkyls, aryls, functionalised aryls and heterocycles were prepared. PSi consists of Si chains with interacting sp^3 orbitals. Rezonance integral between two sp^3 orbitals localized on the adjoining Si atoms, β_{vic}, is responsible for the formation of Si-Si σ bond. Electron delocalization along the Si chain is a result of an interaction of sp^3 orbitals on the neighbouring Si atoms. The degree of electron delocalization in the chain is a function of the β_{vic}/β_{gem} ratio, where β_{gem} is the resonance integral between two sp^3 orbitals localized on one Si atom. The electron delocalization is optimal when the ratio is unity. The ionization energy of the Si-Si bond is lower in comparison with π-electrons in olefins. Hence the character of PSi is similar to unsaturated carbon polymers[2,5]. Various applications of PSi were published. Among others, they are β-SiC precursors, impregnating agents for strengthening ceramics, photoresists, waveguides and heat sensors. In the last couple of years, a remarkable attention was paid to the use of Psi-based systems as photoconductors, charge transport media materials, charge-dissipating coatings components, non-linear optical materials and photoinitiators of radical and cationic reactions[1-3].

Phototriggered degradation (photolysis and photooxidation) of the PSi backbone during aging, accounting for breaking the σ-conjugation and formation of secondary reaction products[6,7] is considered as a negative effect on electric properties of PSi. Irreversible formation of oxygenated moieties, such as siloxane (Si-O-Si), hydroxy (OH) and carbonyl (>CO) functions in aged PSi was monitored in this study.

Materials and Methods

Polymer: Poly[methyl(phenyl)silanediyl] (PMPSi) was synthesized by Wurtz coupling of purified dichloro(methyl)phenylsilane in a mixture with sodium dried (reflux, 8 h) toluene and 15 % of n-heptane[8]. A sodium metal dispersion was prepared by vigorous stirring and heating of sodium at 110-115 °C (mild reflux) in the solvent mixture.

The reaction was carried out in an inert atmosphere (dry nitrogen) in a three-necked 1000 ml flask equipped with an electromagnetic stirrer and reflux condenser. At the reflux, the monomer was added dropwise to the sodium dispersion and the temperature was kept at 110-117 °C for 3 h. At the end of the reaction the color of the reaction mixture was dark violet and its viscosity increased. After cooling to room temperature, excess of sodium metal was decomposed by successive addition of isopropyl alcohol, methanol and water. The organic layer containing PMPSi was separated, washed with distilled water and dried with anhydrous magnesium dichloride for 24 h. Purification: Insoluble part (crosslinked PMPSi) was separated by centrifugation (10000 rpm, 2 h), PMPSi was precipitated with excess of methanol, filtered off and dried in vacuum (room temperature, 24 h, and 40 °C, 2 h). Low-molecular-weight portion (mostly cyclic oligomers) was removed by boiling PMPSi in diethyl ether (2 h), PMPSi was filtered off and dried in vacuum (room temperature, 24 h).

Characterization of PMPSi: (a) Molecular weight was determined by GPC [modular arrangement: Constametric 3500 pump (LDC Analytical), RIDK 101 and UV-VIS LCD 2563 detectors (Laboratory Instruments Praha), 8x600 mm, SDV 10^4 Å column, (PSS Mainz), PS standards (Merck), THF Chromasolv as the mobile phase, DATAAPEX hardware and software for the data calculations]: $M_w = 23240$, $M_n = 4557$. (b) UV spectral characterization with UV/VIS spectrophotometer Perkin Elmer LAMBDA 20 (1 cm quartz cell), the data were calculated using UV-WINLAB software. (c) For the measurement of IR spectra of neat and aged PMPSi, an FTIR spectrometer Bruker IFS 55 was used and the results were calculated using Opus 3.0 software. The characteristic absorption bands selected according to[9] are listed in Table 1.

Table 1. Characteristics IR absorption bands.

Wavenumber [cm^{-1}]	1122	1111	1098	1800-1650	3725-3100	3100-3000, 800-600	3000-2700
Assignment	Si-O-Si	Si-O-Si	Original skeleton	C=O	OH	C-H aromatic	C-H aliphatic

Additives: UV absorbers were used as delivered. 2-(4,6-Diphenyl-1,3,5-triazin-2-yl)-5-(hexyloxy)phenol (**1**, Tinuvin 1577, Ciba Specialty Chemicals, CAS 147315-50-2), dimethyl [(4-methoxyphenyl)methylidene]propanedioate (**2**, Hostavin PR 25, Clariant, CAS 7443-25-6), N-(5-*tert*-butyl-2-ethoxyphenyl)-N'-(2-ethylphenyl)oxalamide (**3**, Sanduvor EPU, Clariant, CAS 35001-52-6). Characteristic UV absorption spectra of these absorbers were measured in tetrahydrofuran (THF) and are given in Figure 1.

1   **2**   **3**

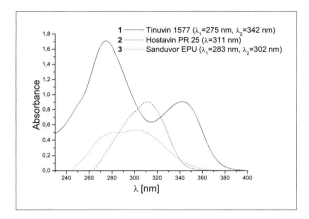

Figure 1. Absorption spectra of the used UV absorbers measured in THF at concentration of 2.10^{-4} mol.l^{-1}

Preparation of PMPSi samples: a 10 wt.% solution of PMPSi was prepared by dissolving the PMPSi powder in toluene under intensive shaking for 60 min and subsequent centrifugation (8000 rpm, 45 min). Thin films of PMPSi on dry crystalline potassium bromide backings were prepared by spin-coating (400 rpm, 5 s and 800 rpm, 60 s) and dried in air (room temperature, 24 h) and in vacuum (45 °C, 4 h). The samples were protected from direct light during all the manipulations. Film thickness varied between 20 and 60 μm.

Preparation of PMPSi samples with UV absorbers: all doped samples were prepared by the method described above, adding stabilizers **1-3**, dissolved in toluene to the polymer solution before spin-coating. For stabilizer **1**, two concentrations of 1.5 wt.% a 5 wt.% were used. After interpretation of the effect of UV absorber **1**, only the samples with 5 wt.% of stabilizers **2** and **3** were prepared. This concentration was decided to be the upper limit for all further experiments.

Irradiation of PMPSi films: the Weather-O-Meter ATLAS Ci 3000+ device equipped with xenon lamp and external and internal borosilicate filters was used. The irradiation was carried out under dry conditions at 60 °C of the black panel temperature (bpt) in air atmosphere. The samples were irradiated in intervals increasing by 2 min from 1 min to 23-27 min. The total irradiation time was longer than 3 h in most cases. The IR spectrum was measured immediately after each irradiation and was compared with those of unstabilized PMPSi. The manipulation time between two irradiations was as short as possible and the samples were protected during this period from direct light.

Results and Discussion

Films of neat and UV absorbers containing PMPSi were irradiated in ATLAS Weather-O-Meter. Spectral changes in the photooxidized material were monitored by FTIR spectroscopy. Changes in characteristic absorption bands listed in Table 1 were monitored.

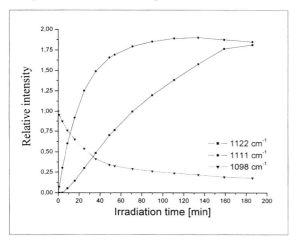

Figure 2. PMPSi without additives exposed in WOM. Formation and separation of three characteristic bands in the 1200-1000 cm^{-1} region.

Figure 2 shows an increase in intensity of the two absorption bands characteristic of Si-O-Si groups (1122cm^{-1}, 1111cm^{-1}) formed in the PMPSi backbone when irradiated in the absence of photostabilizers, and a decrease in intensity of the 1098 cm^{-1} band characteristic of the original skeleton destructed during irradiation. The time evolution of the overall intensity integral in the 1200-955 cm^{-1} region for the stabilized and unstabilized samples is shown in Figure 3.

Photodegradation of pure PMPSi and conversion to siloxane moieties was very rapid (Figure 2). PMPSi films containing 5 wt% of stabilizer **2** was quickly degraded as well (Figure 3). The rate of Si-O-Si groups formation was similar to that of additive-free PMPSi. Stabilizer 2 having absorption maximum at 312 nm was virtually ineffective as the degradation retarder (Figure 3). PMPSi film with 1.5 wt.% of stabilizer **1** degraded in the monitored region 1200-

955 cm^{-1} noticeably slower (Figure 3). The used concentration of **1** was assumed to be insufficient in PMPSi protection. The degradation was markedly retarded in the presence of 5 wt.% of stabilizer **1** in comparison with neat PMPSi and PMPSi with 1.5 wt.% of stabilizer **1** (Figure 3). This indicates an effective retardation of photodegradation by phenolic UV absorber **1** showing an absorption maximum in the UV-A region at 342 nm, i.e. close to that of PMPSi (332 nm). Degradation of the PMPSi film containing 5 wt.% of the non-phenolic stabilizer **3** showing λ_{max} at 302 nm (i.e., in the UV-B region) was characterized by increased intensity in the siloxane region. The increase in the oxygenated functions content was, however, retarded after prolonged irradiation (Figure 3). On the basis of this result, we assume that the photostabilization effect of UV absorbers in PMPSi is less dependent on the characteristic UV absorption maximum of the absorber than on its inherent durability in the substrate and, in particular, on the mechanism of activity. Involvement of the excited state intramolecular proton transfer mechanism[10] in **1** and probably in **3** as well plays certainly an important role. The elucidation of the stabilization mechanism is under study.

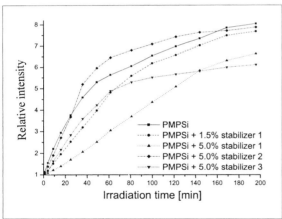

Figure 3. Comparison of the overall integrals of the bands in the 1200-955 cm^{-1} region.

In all samples a noticeable increase in the carbonyl intensity (Figure 4) and that of hydroxyl (Figure 5) group absorption was found. It is assumed that terminal silanol (Si-OH) groups and oxygenated transformations products formed on methyl groups accout for this absorption. Stabilized non-irradiated PMPSi films show non-zero values of the intensity of C=O and OH absorption bands. This may result either from the presence of these functional groups in the added UV absorbers, or from adventitious trace impurities or from occluded humidity.

Interesting changes were found in the 800-650 cm^{-1} region (aromatic C-H out-of-plain deformations). These changes are caused by structural changes in PMPSi due to the generation of Si-O-Si species in the Si backbone.

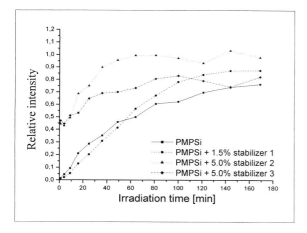

Figure 4. Increase in the intensity of characteristic carbonyl group band in the 1795-1640 cm^{-1} region.

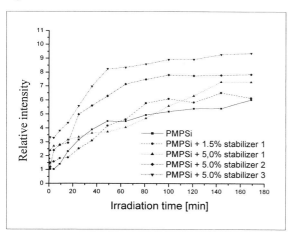

Figure 5. Increase in the intensity of characteristic hydroxy group band in the 3720-3100 cm^{-1} region.

Conclusions

Irradiation of poly[methyl(phenyl)silanediyl] films in ATLAS Weather-O-Meter Ci 3000+ changes the IR spectra in siloxane, carbonyl, hydroxyl and aromatic regions. IR spectral analysis indicates that the phototriggered irreversible degradation of PMPSi consists of several parallel chemical reactions, including dual photoprocesses - photolysis and photooxidation - followed by the conversion of the primary species. Elucidation of effects of environmental conditions and additives on PMPSi degradation is in progress. Main attention was paid to the formation of siloxane groups breaking the Si-Si conjugation. Their formation was retarded by phenolic UV absorber **1** and oxalanilide **3**.

Acknowledgement

Financial support by grant A1050901 of the Grant Agency of the Academy of Sciences of the Czech Republic is gratefully appreciated. The authors thank Ciba Specialty Chemicals, Basle and Clariant, Hunengue, for samples of UV absorbers, Ms. M. Brunclíková for technical cooperation and M. Pekárek, MSc for measurement of UV spectra.

[1] R. D. Miller, J. Michl, Chem. Rev. **1989**, *89*, 1359.

[2] R. West, in: *"The Chemistry of Organic Silicon Compounds"*, S. Patai, Z. Rappoport, Eds., J. Wiley & Sons, New York 1989, p. 1207ff.

[3] S. Hayase, in: *"The Polymeric Materials Encyclopedia"*, J. C. Salamone, Ed., CRC Press, Boca Raton, 1996, p. 6734ff.

[4] S. P. Sawan, S.A Ekhorutomven, in: *"The Polymeric Materials Encyclopedia"*, J. C. Salomone, Ed., CRC Press, Boca Raton, 1996, 6722ff.

[5] S. Nešpůrek, V. Herden, W. Schnabel, A. Eckhards, Czech. J. Phys. **1998**, *48*, 477.

[6] J. Pospíšil, S. Nešpůrek, S. - i. Kuroda, 1st International Conference MoDeSt, Palermo, September 2000.

[7] P. P. C. Satoratto, C. U. Davanzo, I. V. P. Yoshida, Eur. Polym. J. **1997**, 33, *81.*

[8] R. D. Miller, D. Thompson, R. Sooriyakumaran, G. N. Fickes, J. Polym. Sci, Part A: Polym. Chem. **1991**, *29*, 813.

[9] G. Socrates, *"Infrared and Raman Characteristic Group Frequencies. Tables and Charts"*, John Wiley & Sons, Chichester, 2001, p.241.

[10] J. Pospíšil, S. Nešpůrek, Prog. Polym. Sci. **2000** 25, 1261.

Hybrid Organic-Inorganic Coatings and Films Containing Conducting Polyaniline Nanoparticles

Milena Špírková,[1] *Jaroslav Stejskal,*[1] *Jan Prokeš*[2]

[1] Institute of Macromolecular Chemistry, Academy of Sciences of the Czech Republic, 162 06 Prague 6, Czech Republic
[2] Faculty of Mathematics and Physics, Charles University Prague, 121 16 Prague 2, Czech Republic

Summary: Hybrid organic–inorganic coatings and free-standing films made from [3-(glycidyloxy)propyl]trimethoxysilane, amino-terminated poly(oxypropylene) (Jeffamine D-230), colloidal polyaniline nanoparticles and, in some cases, colloidal nanosilica were prepared and characterized. HCl or 4-methylbenzene-1-sulfonic acid were used as catalysts for the sol–gel process and pH tuning, water–propan-2-ol mixture as a solvent. Electrical and mechanical properties and surface morphology of films were studied. The coatings were blue and non-conducting, or green and conducting, depending on preparation conditions. They have a smoother surface than *in-situ* polymerized polyaniline films.

Keywords: atomic force microscopy; films; hybrid networks; mechanical properties; polyaniline dispersion particles

Introduction

Organic–inorganic (O–I) hybrid materials with an *in-situ* formed inorganic phase rank among nanocomposites with interesting properties. They can be prepared by the sol–gel process consisting in the hydrolysis and subsequent polycondensation of alkoxysilanes. The polycondensation reaction results in a multiplicity of structures ranging from monodisperse silica particles to polymer networks depending on the reaction conditions used.[1–3] Another way of their preparation consists in the use of alkoxysilanes bearing polymerizable functional groups[4,5] (glycidyl, amino, isocyanate, methacryloyl, etc.). Then the final product is obtained as a result of two independent polymerization reactions: inorganic structures are formed by sol–gel process of alkoxysilane groups and the organic polymer network by classic polymerization reactions (polyaddition, radical polymerization, etc.).

© 2004 International Union of Pure and Applied Chemistry   DOI: 10.1002/masy.200450840

We have prepared and characterized a series of hybrid O–I coatings and free-standing films made from [3-(glycidyloxy)propyl]trimethoxysilane (GTMS), amino-terminated poly(oxypropylene) (Jeffamine D-230) and submicrometer colloidal polyaniline (PANI) particles. In some cases, colloidal silica (SiO_2) was added as well. The O–I networks were formed by two independent reaction mechanisms: by the sol–gel process and epoxy–amine addition reaction. The influence of the PANI concentration and reaction conditions on properties of hybrid O–I polymeric products is reported in the present study.

Experimental

Materials. GTMS (Aldrich, 98 %) and Jeffamine D-230 (D230, Huntsman Corp; M~230, total amine content 8.45 mequiv g^{-1}) were used as received. PANI particles (37 wt.% PANI, 63 wt.% silica, average particle diameter 430 nm) were prepared by dispersion polymerization of aniline hydrochloride[6] in the presence of nanocolloidal silica[7] (Ludox AS-40, Aldrich; average particle size 35 nm). Propan-2-ol (Lachema, Czech Republic), hydrochloric acid (Lachema, Czech Republic) and 4-methylbenzene-1-sulfonic acid (TsOH, Fluka, Switzerland) were used as received.

Preparation of free-standing films and coatings. GTMS was mixed with water, propan-2-ol, a dispersion of PANI–SiO_2 particles and, in some cases, with colloidal silica (SiO_2) particles at laboratory temperature. Acid conditions (pH 4) for the hydrolysis of alkoxy groups were adjusted with dilute hydrochloric acid or TsOH; D230 was then added. Due to the alkaline character of Jeffamine, pH of the resulting reaction mixture increased up to 8–9. This means that the polycondensation and thermal curing took place under alkaline conditions. Several films were also prepared under acid conditions (after addition of D230, pH was adjusted to 3–4 by addition of HCl or TsOH). The reaction mixture was spread on a glass or polypropylene sheet with a ruler of adjustable 50–500 μm thickness and heated for 2 h at 80 °C and for 1 h at 105 °C.

Methods of characterization. *Static mechanical properties* were measured with an Instron model 6025 (Instron). Specimens of the size 25×6×0.1 mm were tested at laboratory temperature at a test speed of 3.33×10^{-2} mm s^{-1}. All reported values are averages from five independent measurements. *Dynamic mechanical properties* of the films were studied using an ARES apparatus (Rheometric Scientific). A specimen of the size 17×7.5×0.1 mm was measured by oscillatory shear deformation at a constant frequency of 1 Hz and the rate of heating 3 °C min^{-1} to obtain temperature dependences of storage and loss shear moduli, G' and G'', from -100 °C to $+100$ °C. *Conductivity* was measured by the two-probe method with a Keithley 6517 electrometer after deposition of gold electrodes on both sides of films. *Atomic force microscopy* under ambient conditions used a commercial atomic-force microscope (MultiMode Digital Instruments NanoScope Dimension III) in contact mode with Olympus oxide-sharpened silicon nitride probe (OMCL TR-400). The normal force of the tip on the sample was reduced not exceeding 10 nN.

Results and discussion

Tensile properties. The ultimate goal was the preparation of mechanically resistant durable coatings, either electrically conducting (acid polycondensation) or non-conducting (alkaline polycondensation). PANI-free hybrid polymers were tested as well[8] (Table 1, samples 1 and 2). Static mechanical characteristics, i.e., the strain at break ϵ_b, stress at break σ_b, Young modulus E and toughness w of free-standing films were determined from stress–strain dependences.

It was found that the introduction of PANI–SiO$_2$ and SiO$_2$ particles influence tensile properties after alkaline polycondensation. If we choose, e.g., toughness as the criterion of tensile properties, the order of decreasing toughness was: silica-containing films > PANI–SiO$_2$ composite > silica-free products (Table 1, samples 1–3). When the polycondensation is carried out in acid medium, the strain at break is the only parameter comparable with alkaline polycondensation. Other tensile characteristics were reduced (Table 1).

Dynamic Mechanical Analysis. Properties of networks depend strongly on temperature and change mainly in the region close to the glass transition temperature, T_g. The dynamic mechanical properties were measured for sample 7 in Table 1 in the temperature range from -100 to $+100$ °C. Figure 1 shows temperature dependences of the storage and loss shear moduli, G' and G'', and the loss factor, tan δ. The sample is characterized by two glass-transition temperatures T_g; the first at ca -75 °C corresponds to T_g of poly(oxypropylene), while a more distinct transition at *ca* $+10$ °C reflects the glass transition of the organic phase immobilized by inorganic structures.

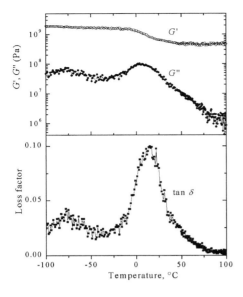

Figure 1. The storage and loss shear moduli, G' and G'', and the loss factor, tan δ, of a typical O–I composite comprising PANI–SiO$_2$ and SiO$_2$ nanoparticles (sample 7 in Table 1).

Table 1. Composition of reaction mixtures, electrical conductivity σ, the strain at break ε_b, the stress at break σ_b, Young modulus E and toughness w (the energy per unit cross-section necessary to break the sample).

No.	PANI–SiO$_2$/GTMS/ SiO$_2$/D230[a], wt.%	Catalyst/ medium[b]	σ S cm^{-1}	ε_b %	σ_b MPa	E MPa	w kJ m^{-2}
1	0/80.4/0/19.6	HCl/B	2.7×10^{-12}	2	8.4	892	1.3
2	0/68.4/15.0/16.6	HCl/B	2.3×10^{-12}	2.1	15.3	1060	3.2
3	7.4/74.5/0/18.1	HCl/B	2.5×10^{-12}	3	7.2	570	2.6
4	7.4/74.5/0/18.1	HCl/A	5.8×10^{-4}	1.3	1.1	75	0.7
5	6.5/64.4/15.0/14.1	HCl/A	3.8×10^{-4}	2.6	2.5	92	1.3
6	7.6/75.9/0/16.5	TsOH/A	2.4×10^{-6}	2.2	0.7	56	0.4
7	6.5/64.4/15.0/14.1	TsOH/A	6.4×10^{-4}	3.5	3.7	156	1.5

[a] The molar ratio of reactive NH end-groups /epoxy, $r = 0.9$–1.
[b] Polycondensation and thermal curing at A: pH 3–4; B: pH 8–9.

Atomic-Force Microscopy. The surface morphology was investigated by atomic-force microscopy using a contact mode. The products based on organic–inorganic hybrid matrix included (a) colloidal PANI–SiO$_2$ particles and (b) PANI–SiO$_2$ and SiO$_2$ particles. For comparison, a PANI film prepared in situ on glass by dispersion polymerization of aniline in the presence of nanocolloidal silica[9] was also characterized. The surfaces of individual products differ substantially (Fig. 2): while the coating containing only PANI–SiO$_2$ particles embedded in the matrix has a very flat and regular surface (not shown here), the coatings of type (b) containing both PANI–SiO$_2$ and small SiO$_2$ nanoparticles can be clearly observed on the surface (Fig. 2a). Profile height differences are units of nm in both cases. The films grown during the dispersion polymerization of aniline have rough surface (Fig. 2b) due to adhering individual PANI–SiO$_2$ particles on the substrate.

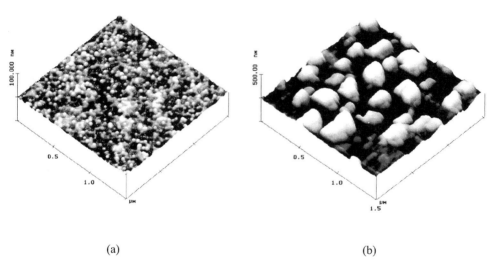

(a)   (b)

Figure 2.   3D images of sample surfaces (contact mode): (a) hybrid O–I coating containing PANI–SiO$_2$ and SiO$_2$ nanoparticles (sample 7 in Table 1), (b) coating prepared by direct polymerization of aniline in the presence of SiO$_2$ nanoparticles. z-Scales are (a) 100 nm and (b) 500 nm.

Electrical properties. The preparation under alkaline conditions (Table 1, samples 1–3) yields blue and non-conducting products as expected from the presence of PANI base. Green and conducting coatings comprising protonated PANI are obtained under acid reaction conditions (samples 4–7). The conductivity of the order of 10^{-4} S cm^{-1} can be regarded as reasonably high as the film contains only ca 3 wt.% of PANI and the system is well below a typical percolation limit of 16 vol.%.

Conclusions

A series of electrically conducting and non-conducting hybrid O–I coatings containing colloidal polyaniline particles was prepared and characterized. Conducting films are produced when the polycondensation and thermal curing proceed under acid conditions. Their mechanical properties can be tuned by conditions of preparation.

Acknowledgements

The authors wish to thank the Grant Agency of the Czech Republic (203/01/0735) and the Grant Agency of Academy of Sciences of the Czech Republic (K4050111) for financial support. Dynamic mechanical measurements were kindly performed by Dr. A. Strachota from the Institute.

[1] C.J. Brinker, C.W.Scherer, "*Sol-Gel Science*", Academic Press, San Diego 1990.
[2] L. Matějka, K.Dušek, J. Pleštil, J. Kříž, F. Lednický, *Polymer* 1999, *40*, 171.
[3] D.A. Loy, B.M. Baugher, C.R. Baugher, D.A. Schneider, K. Rahimian, *Chem. Mater.* 2000, *12*, 3624.
[4] L. Matějka, O.Dukh, J. Brus, W.J. Simonsick, B. Meissner, *J. Non-Cryst. Solids* 2000, *270*, 34.
[5] M.W. Daniels, L.F. Francis, *J. Colloid Interface Sci.* 1998, *205*, 191.
[6] J. Stejskal, *J. Polym. Mater.* 2001, *18*, 225.
[7] J. Stejskal, P. Kratochvíl, S.P. Armes, S.F. Lascelles, A. Riede, M. Helmstedt, J. Prokeš, I. Křivka, *Macromolecules* 1996, *29*, 6814.
[8] M. Špírková, J. Brus, D. Hlavatá, H. Kamišová, L. Matějka, A. Strachota, *Surf. Coat. Int.*, in press.
[9] A. Riede, M. Helmstedt, V. Riede, J. Zemek, J. Stejskal, *Langmuir* 2000, *16*, 6240.

Macromol. Symp. **2004**, *212*, 349-356

Electrical Conductivity of Polycrystalline Films of *p*-Sexiphenyl

Stanislaw Tkaczyk,* Jozef Świątek

Institute of Physics, Pedagogical University, Al. Armii Krajowej 13/15, 42-200 Czestochowa, Poland

Summary: In this work the results of DC conductivity measurements of polycrystalline *p*-sexiphenyl thin films are presented. The investigations concerned the effect of temperature, film thickness and electric field on the DC conductivity mechanism. The thickness of the investigated material varied from 0.2 to 2.5 µm. The measurements were carried out for different electrode polarities of the 0 –100 V voltage and at temperatures ranging from 15 to 325 K. Thin films of *p*-sexiphenyl were obtained by controlled vacuum sublimation on BK-7 glass substrate with gold and aluminium electrodes. Analyzing the obtained results we conclude that injection of the charge carrier from electrodes into the investigated material proceeds by thermionic emission and field emission and it is dependent on temperature and external electric field. The charge carrier transport is controlled by localized states (traps) in the forbidden energy gap. The activation energy calculated from formula $\ln I = f(1/kT)$ varied from kT for low temperature up to 1.0 eV.

Keywords: DC conductivity; field emission; *p*-sexiphenyl; thin films; tunneling

Introduction

Polycrystalline thin films contain many structural defects, which perturb the internal potential of crystalline area (grains) so much, that this decides on electric properties of layer. By changing the defects concentration it is possible to obtain a material with hopping or tunneling between neighboring localized states as the dominant mechanism of carrier transport. In disordered materials there is a big number of local energy levels in the forbidden energy gap. In this connection, energetic diagrams of polycrystalline and amorphous structures are very similar. The polycrystalline materials are a convenient object for hopping conductivity investigations in a wide range of temperatures and electric fields. In many cases one can used a variable-range hopping mode [1,2]. There are same difficulties to define the mechanism of the charge carrier phenomena on the basis of $\ln \sigma = f(T^{-1/4})$ or $\ln I = f(T^{-1/4})$ characteristics only. The following properties of the material determine charge carrier transport and decide whether two- or three-dimensional hopping is observed:

(1) structural non-homogeneity of the investigated material,

(2) the sample thickness,

(3) energetical distribution of localized states,

(4) the existence of multiphonon processes,

(5) the existence of Coulomb interaction during charge carrier transfer.

At high electric fields (10^7 V/m) or in the case when the electrode is not an ideal injecting contact, the current flow through the sample is controlled by contact as well as by bulk phenomena appearing in the material.

The interesting object of the present investigation was p-sexiphenyl – a representative of the para-oligophenyl homologous series.

Experimental

The material under study was p-sexiphenyl produced by Aldrich, with structure formula $C_{36}H_{26}$ and molecular mass 458.6194. The p-sexiphenyl molecule is constructed from six flat benzene rings spread into two dimensions (Fig. 1). The C-C distance between two benzene rings is 1.48×10^{-10} m and the C-C distance inside benzene rings is 1.42×10^{-10} m.

Figure 1. Formula of p-sexiphenyl

The p-sexiphenyl thin films were obtained by vacuum sublimation on BK-7 glass using thin gold and aluminium electrodes. This process was controlled with respect to:

(1) film sublimation rate,

(2) temperature condition of sublimation process ($T_3 = 402$ K),

(3) vacuum condition ($p = 6.7 \times 10^{-5}$ Pa).

The size of crystalline grains in the obtained films depended on the sublimation rate, temperature and final thickness of the film.

The DC conductivity measurement was performed by measuring the current flow through the film (material bulk) between gold and aluminium electrodes. The measurements were carried out in the temperature range from 15 K up to 325 K. Such extensive range of temperature for p-sexiphenyl was used for the first time. The DC current conductivity measurement equipment is described in Ref. [3]

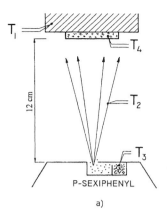
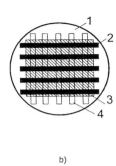

a)   b)

Figure 2. Schematic arrangement of the production of p-sexiphenyl thin films
a) T_1, T_2, T_3, T_4 – copper-constantan thermocouples measuring temperatures of substrate, sublimation chamber, evaporator and sample, respectively
b) 1 – substrate (glass – BK-7)
2 – upper electrode (aluminium)
3 – p-sexiphenyl film
4 – bottom electrode (gold)

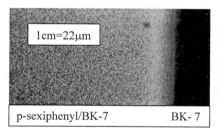

Figure 3. Polarizing microscope pictures taken for transmitting light with crossed polarizer and analyzer. p-Sexiphenyl sample thickness of – d = 1.376 μm, magnification 440, T_1 = 301.6 K, T_2 = 305 K, T_3 = 402 K, T_4 = 302 K

Results and discussion

Analysis of the obtained results for the DC conductivity of p-sexiphenyl polycrystalline thin films suggests that the charge carrier transfer through the material bulk may be described by many different conductivity mechanisms. The shape of the current-voltage characteristics $I=f(U)$ presented in Fig. 4 confirms that non-Ohmic conductivity exists. Analyzing these characteristics we can see that for different p-sexiphenyl thicknesses 0.73 μm film thickness in Fig. 4a and 2 μm in Fig. 4b), the shape of curves is the same. The shape of current-voltage characteristics for *p*-sexiphenyl is similar for semiconductor diodes. The charge carrier transport in the conductance direction exists for (+) polarity of Au electrode and the charge carrier transfer in reverse direction exists for (-) polarity of this electrode. This phenomenon is observed for the temperature range from 15 K up to 325 K.

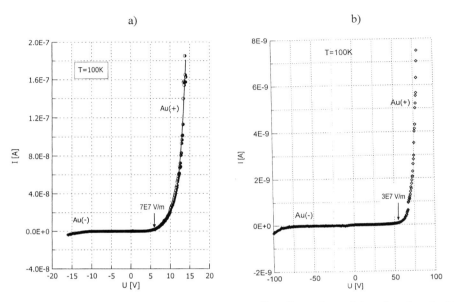

Figure 4. Current-voltage dependence for polycrystalline films of *p*-sexiphenyl equipped with Au-Al electrodes with various electrode polarities. *p*-Sexiphenyl film thickness and sample temperature: a) d=0.73 μm, T=100 K; b) d=2.033 μm, T=100 K

By analysis of the ln $I = f(1/kT)$ relation, where k is the Boltzman constant, T temperature, I current intensity, the activation energy of DC conductivity was determined for various polarizing voltages. Plotting the dependences ln $I = f(1/kT)$, one can identify some rectilinear regions with different slopes. The E_3 region (Fig. 5) relates to non-activated hopping conductivity. The E_2 region gives energy about kT and higher and relates to thermal-activated hopping conductivity with participation of the Poole-Frenkel effect. The E_1 region relates to spontaneous conductivity of *p*-sexiphenyl with activation energy from 0.25 eV up to 0.98 eV (Table 1).

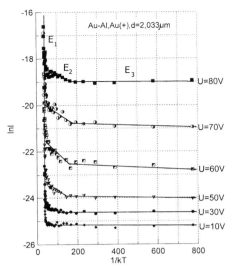

Figure 5. The dependence ln $I=f(1/kT)$ for p-sexiphenyl films - sample thickness 2.033 μm, Au-Al, Au (+)

Table 1. The influence of the voltage on activation energy

U[V]	20	30	40	50	60	70	80
E_1[eV]	0.415	0.563	0.375	0.250	0.420	0.620	0.800
E_2[eV]	0.042	0.043	0.016	0.050	0.117	0.097	0.053

The charge carrier transport at low temperatures takes place through hopping between localized states over the potential barriers that are lowered due to the electric field according to the Poole-Frenkel effect. The hopping is confirmed by straight-line characteristics of $\ln I = f(T^{-1/3})$ and $\ln I = f(T^{-1/4})$. As it is seen from the characteristics (Fig. 6), non-active three-dimensional hopping exists for film thicknesses $d > 2$ μm for a low temperature and low voltage. The thermal-activated three-dimensional hopping exists at temperatures obove 60 K. The two-dimensional hopping exists for thin films of p-sexiphenyl (Fig. 7) for a low-polarity voltage. The polymer – electrodes contact phenomena play an important role at high voltages. The height of the barriers lowered due to the increase in the electric field shows that the material bulk plays an important role in charge carrier transport between electrodes (Poole-Frenkel effect). The straight-line part of the $\ln I=f(U^{1/2})$ characteristics (Fig. 8) shows that the Poole-Frenkel effect exists in charge carrier transport through the material bulk.

The injection of the charge carrier from Au electrode (Au electronic work function 5.2 eV, Al electronic work function 4.3 eV) [3] into the investigated material proceeds by thermionic emission and field emission. The injection of the charge carrier from electrodes into *p*-sexiphenyl proceeds by field emission for low voltages. In Fig. 10 we can see the influence of the Schotky effect on the charge carrier injection from electrodes into investigated material, which confirms the interface effect. The influence of the photoemission effect is observed at low electric fields (the charge carriers are thermoactivated in the material bulk). The dependence shown in Fig. 9 is called the Fowler-Nordheim curve and shows participation of the influence of the surface and interface polymer – electrode on the current value in the investigated material [4]. The negative gradient of the straight-line part of the characteristic suggests a field emission process from electrode into the *p*-sexiphenyl area.

The number of free-charge carriers injected from electrode into the investigated material and the current-voltage characteristic are dependent on properties of the material bulk in the same region of the electric field. capacity, the electric field inside the material bulk generates a current-voltage dependence described by formula:

$$\frac{I}{d} = \frac{A}{d}\left[\exp Bd^{1/2}\left(\frac{U}{d^2}\right)^{1/2}\right]$$

where *A* and *B* are constants.

The influence of the electrode on the current versus voltage dependence is shown in Fig. 10.

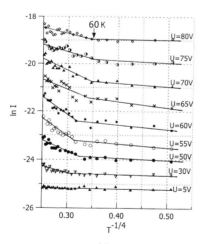 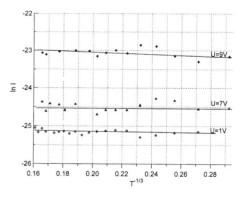

Figure 6. The ln *I*=*f*(*T*^{-1/4}) curves obtained for polycrystalline *p*-sexiphenyl layers 2.033 μm thick, Au (-), Au-Al electrodes, at different polarizing voltages

Figure7. The ln *I*=*f*(*T*^{-1/3}) curves obtained for polycrystalline *p*-sexiphenyl layers 0.73 μm thick, Au (-), Au-Al electrodes, at different polarizing voltages

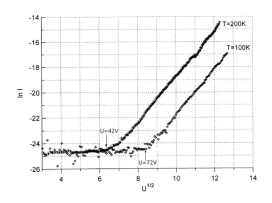

Figure 8. The dependence ln $I=f(U^{1/2})$ for p-sexiphenyl films, sample thickness 2.033 µm at different temperatures; Au-Al electrodes, Au (-)

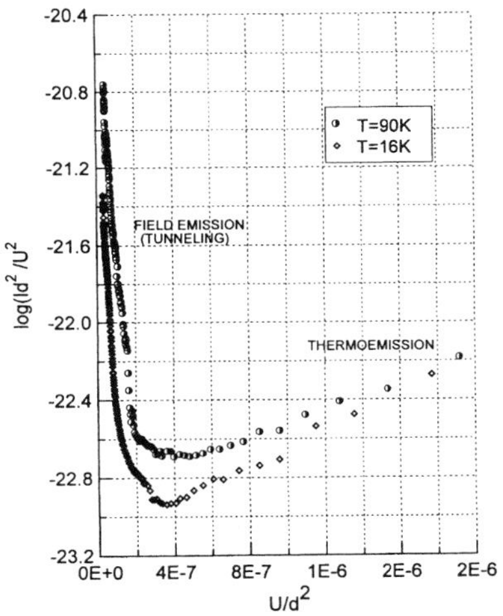

Figure 9. Fowler-Nordheim plot $\log(Id^2/U^2)=f(d/U)$ for p-sexiphenyl layers 0.23 µm thick, Au-Al electrodes, Au(-) at 16 K and 90 K

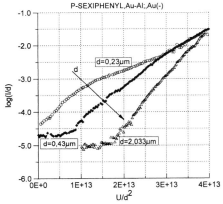

Figure 10. Influence the Schottky effect on space charge limited currents [5]

The current passed through the material is lower than the current which can flow when the electric field increases. A similar effect exists when the interface material-electrode is not ideal injecting. P.N. Murgatroyd [5] has shown that, around the electrode with limited emission.

Conclusions

- The Schottky effect exists on the electrodes (Au, Al) and p-sexiphenyl interface [6].
- The current-voltage characteristics are asymmetric. The reverse current transfer and conduction current transfer exist for the p-sexiphenyl-electrodes interface.
- The injection of the charge current into the p-sexiphenyl area proceeds by field emission[4]
- The charge current transfer proceeds between localized states (traps) in the forbidden energy gap (two- and three-dimensional hopping).
- The Poole-Frenkel phenomenon influences the p-sexiphenyl film conduction process to high electric fields.

[1] N.F.Mott, *Philos. Mag.* **1971**, *24*, 911-934.
[2] V.Voegele, S.Kalbitzer, K.Boringer, *Philos. Mag. B* **1985**, *52*, 153-168.
[3] I.D.Parker, *J. Appl. Phys.* **1994**, *75*, 1656-1666.
[4] S.W.Tkaczyk, *Electron. Technol.* **1996**, *29*, 232-235.
[5] P.N.Murgatroyd, *Thin Solid Films* **1973**, *17*, p.335.
[6] S.W.Tkaczyk, *Synth. Met.* **2000**, *109*, 249-254.

Macromol. Symp. **2004**, *212*, 357-362   357

Photovoltaic Properties of Tetracene and Pentacene Layers

Ryszard Signerski, Grażyna Jarosz,* Jan Godlewski

Gdansk University of Technology, G. Narutowicza 11/12, 80-952 Gdansk, Poland
E-mail: zofia@mif.pg.gda.pl

Summary: Photovoltaic phenomenon in tetracene and pentacene layers evaporated under the same conditions onto a glass substrate and provided with the same couple of electrodes is investigated. Comparison of the results obtained for both organic materials makes it possible to conclude that in spite of differences in mechanisms of charge carrier generation, the values of photovoltaic parameters are very similar.

Keywords: charge transport; photovoltaic effect; tetracene; pentacene; thin films

Introduction

Tetracene and pentacene are materials which have been examined for many years and therefore they can be treated as model organic materials[1,2]. One can find many works on photovoltaic phenomenon in tetracene published hitherto (e.g. [3-6]) and lately also in pentacene [7,8]. The objective of our study is to compare photovoltaic properties of pentacene and tetracene layers evaporated under the same conditions (pressure, substrate temperature and evaporation velocity) and provided with the same couple of electrodes, Au and Al. It is a typical set of electrodes used in the metal-organic layer-metal sandwich systems. The Au electrode (its work function is about 5 eV) effectively injects holes into both materials, while Al (its work function is about 4.2 eV) forms a contact of a barrier with properties similar to Schottky barrier and limits the hole injection.

Experimental

Tetracene and pentacene (Aldrich) powder were purified by vacuum sublimation before they were used in the investigations. Samples were obtained by vacuum evaporation (Auto306 Edwards with a turbomolecular pump, 10^{-3} Pa) of the following layers onto a glass substrate (room temperature): Au, tetracene (Tc) or pentacene (Pn), Al with adequate thicknesses: 10 nm, 400 nm,

     DOI: 10.1002/masy.200450842

15 nm, and with common surface of $0.1 \ cm^2$. The average evaporation velocity of organic layers was about 0.1 nm/s. Samples were made in a series of five cells. We have examined 11 samples with Pn and 9 samples with Tc.

Our experimental research involves spectra of photovoltaic short-circuit currents for illumination through either Al or Au, dependences of short-circuit currents on the light intensity and current-voltage dependences of illuminated or unilluminated samples. All investigations were carried out in air and at room temperature. In spectral measurements, a constant value of flux of photons penetrating into the organic layer was kept for the whole range of wavelengths.

Results and Discussion

Figures 1 and 2 show spectral characteristics of cells with Tc and Pn under illumination through Al or Au and absorption spectra of organic layers. In all cases photovoltaic current flows through organic layers from Al to Au.

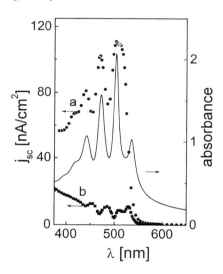
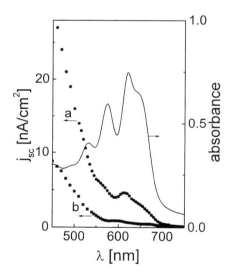

Figure 1. Short-circuit current spectrum for Al/Tc/Au cell illuminated through Al (a) or Au (b) (light intensity $I_o = 10^{13}$ photons/(cm²s)) and absorption spectrum of the Tc layer (solid line).

Figure 2. Short-circuit current spectrum for Al/Pn/Au cell illuminated through Al (a) or Au (b) (light intensity $I_o = 10^{13}$ photons/(cm²s)) and absorption spectrum of the Pn layer (solid line).

For Tc illuminated through Al we observe good correlation between spectra of the photovoltaic current and absorption (we should note that the peak of absorption spectrum at 535 nm is not associated with an excited state but originates from diffraction-interference effects [9]). This relation, called symbatic, results from the dissociation of excitons at the Al electrode and the injection of holes into Tc. If the cell is illuminated through Au, the filtration action of Tc leads to antibatic relation between spectra of the current and absorption (the hole injection by Al which is now a back-electrode) [10]. Similar characteristics can be obtained also if the photogeneration of charge carriers occurs in strong electric field of a barrier formed by a metal of low work function and a p-type semiconductor (Schottky barrier). We cannot entirely exclude this case for the Al/Tc interface. The role of acceptor impurities can be played by oxygen centres [5].

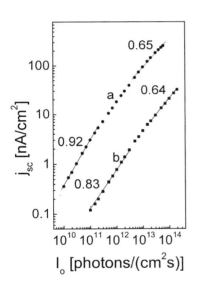

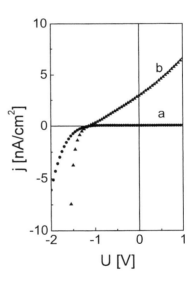

Figure 3. Light intensity characteristics of short-circuit current for Al/Tc/Au cell illuminated through Al by monochromatic light of $\lambda = 505$ nm (a) and for Al/Pn/Au cell illuminated through Al by monochromatic light of $\lambda = 615$ nm (b).

Figure 4. Voltage characteristics of dark current (a) and photocurrent under illumination with monochromatic light ($\lambda = 615$ nm, $I_0 = 10^{13}$ photons/(cm²s)) through Al (b) for Al/Pn/Au cell.

Noticeable smaller differences between spectral characteristics of short-circuit currents obtained for both directions of illumination are observed for the cell with Pn (Fig. 2). Now the structure of current characteristics is determined also by bulk processes of the photogeneration or by the exciton release of charge carriers from traps. The range for $\lambda < 550$ nm corresponds to intrinsic generation. Charge carriers in this range are produced by either dissociation of mixed states of CT and Frenkel or autoionization of molecular excited states [2,11]. At $\lambda > 550$ nm, free charge carriers (holes) can be produced in extrinsic processes such as interactions of excitons with oxygen centres near the Al electrode, trapped charge carriers or with the Al electrode [1,12,13].

Taking into account Figs. 1 and 2, we can calculate the incident-photon-to-current-efficiency: IPCE $= j_{sc}/eI_o$ (where e is the elementary charge), which is 0.7 % for Tc ($\lambda = 505$ nm) and 0.3 % for Pn ($\lambda = 615$ nm). For $\lambda < 550$ nm, IPCE values for Pn greater than for Tc.

Dependences of a short-circuit current on the light intensity for the Al/Tc/Au and Al/Pn/Au cells illuminated through Al are presented in Figure 3. These characteristics are not linear but they can be approached in sections by the relation of the type $j_{sc} \sim I_o^n$ with values of n written over the characteristics. Values of $n < 1$ indicate the existence of trapping processes [14] or the charge carrier recombination [1].

Figure 4 shows voltage dependences of both dark current and photocurrent (corresponding to the photovoltaic range) for illumination through Al obtained for the Pn cell. The sign of the voltage denotes polarization of the Al electode. The obtained open-circuit voltage U_{oc} amounts to - 1.15 V and is the same for illuminated and unilluminated sample. Therefore it is not photovoltage. We observed also a small "battery effect", e.g. the short-circuit current in the dark $j_{scdark} \approx 10^{-12}$ A/cm^2. This effect can result from corrosion of the Al electrode [15]. From the voltage dependence of the photocurrent, we can estimate power conversion efficiency: $\eta = P_{out\,max}/P_{input} \approx$ 0.025 %.

In the case of the Tc cell, the current-voltage dependence is similar and U_{ocdark} amounts to - 1.05 V, but $U_{ocphoto} = -1.25$ V and $\eta = 0.1$ % for $I_o = 10^{13}$ photons/(cm^{2}s) and $\lambda = 505$ nm.

Figure 5 shows current-voltage characteristics of cells with Tc and Pn under illumination through Al with white light of 300 W/m^2 (Xe lamp of 150 W with H_2O and BG 14 filters). η values calculated for these characteristics are almost the same for Tc and Pn, ca. 10^{-4} %. The obtained values are low, because the parameters of our samples (e.g. thickness of layers, the kind of electrodes and their transmission) have not been optimized.

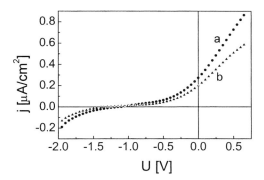

Figure 5. Photocurrent-voltage characteristics for Al/Pn/Au cell (a) and for Al/Tc/Au cell (b) under illumination through Al with white light.

Conclusions

Our investigations were based on a comparison of photovoltaic properties of Al/Tc/Au and Al/Pn/Au cells. We can conclude that the photovoltaic phenomenon for both systems has many features in common but the mechanisms of charge carrier photogeneration can differ. Particularly in the investigated wavelength ranges:

- the photovoltaic short-circuit current in Tc cells results from extrinsic processes of exciton dissociation at the Al electrode or near it,
- the short-circuit current in Pn cells is produced in both intrinsic and extrinsic processes, which occur in a wider region of the sample than in Tc cells,
- the photovoltaic parameter values of Tc and Pn cells obtained under the illumination with white light (of a spectrum similar to solar one) are nearly the same.
-

Acknowledgements

Work was supported by KBN under Program No. 4 T11B 057 22.

[1] M. Pope, Ch. E. Swenberg, *"Electronic Processes in Organic Crystals and Polymers"* Oxford Sci. Publ., New York-Oxford 1999.

[2] E. A. Silinsh, V. Capek, *"Organic Molecular Crystals"*, AIP Press, New York 1994.

[3] L. E. Lyons, O. M. G. Newman, *Aust. J. Chem.* **1971**, *24*, 13.

[4] A. K. Gosh, T. Feng, *J. Appl. Phys.* **1973**, *44*, 2781.

[5] A. J. Twarowski, A. C. Albrecht, *J. Chem. Phys.* **1979**, *70*, 2255.

[6] R. Signerski, J. Kalinowski, I. Koropecky, S. Nespurek, *Thin Solid Films* **1984**, *121*, 175.

[7] Ya. Vertsimakha, A. Verbitsky, *Synth. Met.* **2000**, *109*, 2475.

[8] J. H. Schon, Ch. Kloc, B. Batlogg, *Appl. Phys. Lett.* **2000**, *77*, 2475.

[9] W. Hofberger, *Phys. Status Solidi A* **1975**, *30*, 271.

[10] R. Signerski, J. Kalinowski, *Thin Solid Films* **1981**, *75*, 151.

[11] P. J.Bounds, W. Siebrand, *Chem. Phys. Lett.* **1982**, *85*, 496.

[12] G. Jarosz, R. Signerski, J. Godlewski, *Adv. Mater. Opt. Electron.* **1996**, *6*, 379.

[13] G. Jarosz, R. Signerski, J. Godlewski, *Synth. Met.* **2000**, *109*, 161.

[14] R. Signerski, J. Kalinowski, *Mol. Cryst. Liq. Cryst.* **1993**, *228*, 213.

[15] K. Murata, S. Ito, K. Takahashi, B. M. Hoffman, *Appl. Phys. Lett.* **1997**, *71*, 674.

Macromol. Symp. **2004**, *212*, 363-368

Thermally Stimulated Carrier Transport Due to Stepwise Sample Heating

Władysław Tomaszewicz

Faculty of Technical Physics and Applied Mathematics, Gdansk University of Technology, Narutowicza 11/12, 80-952 Gdansk, Poland
E-mail: wtomasze@sunrise.pg.gda.pl

Summary: The article concerns non-isothermal dispersive carrier transport in an insulating solid with traps. The approximate solutions of transport equations derived previously are extended to the case of stepwise sample heating. The specific features of thermally stimulated currents (TSCs) can be attributed to non-linear dependence of the demarcation level on temperature. In particular, the initial TSC rise has a thermally activated character where activation energy equals to the demarcation energy at the end of previous heating cycle. The accuracy of the formulae describing TSCs is verified by Monte Carlo calculations for Gaussian trap distribution.

Keywords: amorphous solids; dispersive charge transport; Monte Carlo simulation; multiple-trapping model; thermally stimulated currents

Introduction

A characteristic feature of amorphous materials, both inorganic and organic, is a continuous distribution of localised states throughout the energy gap. The excess carrier transport in these materials has frequently dispersive character, due to carrier multiple trapping (MT) or hopping[1]. One of the important tools in determining the energy distribution of localised states in high-resistivity amorphous solids is the measurement of thermally stimulated currents (TSCs). Two kinds of the experiments can be distinguished: (1) The sample with sandwich electrodes is initially excited by strongly absorbed light. TSC is then related to one–sign carrier transport through the sample and to carrier neutralization on the collecting electrode ('TSC drift peak'). (2) The sample having coplanar electrodes is illuminated by weakly absorbed light. TSC is then determined by carrier trapping/detrapping or hopping kinetics as well as by carrier recombination ('TSC recombination peak'). So far, a majority of theoretical analyses of TSCs in amorphous solids, concerning both transport[2-4,6] and recombination peaks[5-8], are based on the MT model. According by, the carriers move in the allowed band, being temporarily captured by the localised states (traps) in the energy gap.

 DOI: 10.1002/masy.200450843

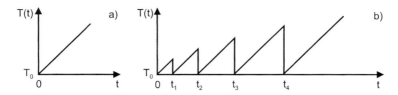

Figure 1. Time dependence of the sample temperature $T(t)$ in the case of linear heating (a) and stepwise heating (b).

One of the advantages of the TSC technique is the possibility of applying various heating schemes. In particular, apart from the most commonly used linear sample heating (Figure 1a), the stepwise heating of the sample (Figure 1b) is also frequently utilised. The TSC measurements using complex heating schemes would provide additional information about gap states and make it possible to test the validity of TSC interpretation. So far, only few theoretical works have appeared concerning the TSC recombination peaks due to stepwise sample heating[5,9]. In the present paper, the results obtained in [3,4], concerning the TSC drift peaks and linear heating scheme, are extended to the stepwise heating regime. The corresponding TSC measurements were performed, e.g., in poly(N-vinylcarbazole)[10].

General formulae

According to refs.[3,4], in the case of strongly dispersive MT transport of carriers, the TSC transport peak is expressed by the approximate formula

$$I(t) = I_0 \frac{d}{dt}\left\{\frac{1-\exp[-\tau_0\Phi(t)]}{\Phi(t)}\right\}, \tag{1}$$

where the function

$$\Phi(t) = C_t \int_{\varepsilon_0(t)}^{\varepsilon_t} N_t(\varepsilon)d\varepsilon, \tag{2}$$

Here, t and ε are the time and energy variables (ε is measured from the edge of allowed band), $\tau_0 = d/\mu_0 E$ is the free-carrier time-of-flight (d – sample thickness, μ_0 – free carrier mobility, E – electric field strength), $I_0 = Q_0/\tau_0$, where Q_0 is the total charge released from the traps, C_t is the carrier capture coefficient, $N_t(\varepsilon)$ is the trap density per energy unit, $\varepsilon_0(t)$ is the demarcation energy defined below and ε_t is the upper limit of trap distribution. The function $\Phi(t)$ determines the probability that the carrier, being free at $t = 0$ is captured in a time unit and stays in the trap until time t. The demarcation level $\varepsilon_0(t)$ is given implicitly by

$$\int_0^t \frac{dt'}{\tau_r[\varepsilon_0(t),t']} \approx 1,$$ (3)

where

$$\tau_r(\varepsilon,t) = \frac{1}{\nu_0}\exp\left[\frac{\varepsilon}{kT(t)}\right]$$ (4)

is the mean carrier dwell-time in the trap (ν_0 – frequency factor, k – Boltzmann constant, $T(t)$ – sample temperature). The level $\varepsilon_0(t)$ separates the shallower states which reached yet the equilibrium occupancy and the deeper states, characterised by non-equilibrium carrier distribution. In the case of linear sample heating,

$$T(t) = T_0 + \beta t$$ (5)

(T_0 – initial temperature, β - heating rate), the demarcation energy varies linearly with temperature,

$$\varepsilon_0(t) \approx k[c^* T(t) - T^*],$$ (6)

with $c^* = 0.967\ln(45.9\ \mathrm{K}\cdot\nu_0/\beta)$ and $T^* = 180$ K (the ratio ν_0/β is expressed in K^{-1}). However, the formulae (1)–(3) are valid for arbitrary time dependence of sample temperature and can be adopted to other heating schemes.

Stepwise heating regime - analytical results

In the case of stepwise sample heating,

$$T(t) = T_0 + \beta(t - t_{i-1}), \quad t_{i-1} \le t < t_i,$$ (7)

Eq. (3) for the $(m+1)$-th heating cycle can be rewritten as:

$$\sum_{i=1}^m \int_{t_{i-1}}^{t_i} \frac{dt'}{\tau_r[\varepsilon_0(t),t']} + \int_{t_m}^t \frac{dt'}{\tau_r[\varepsilon_0(t),t']} \approx 1.$$ (8)

For simplicity, the time of sample cooling after each heating cycle is ignored. Provided that the maximum sample temperature in sequential cycles increases remarkably, only the m-th and $(m+1)$-th terms are of significance in the above sum. Let us denote by $T_m = T(t_m)$ the maximum temperature reached in the m-th cycle. If the time t exceeeds only slightly t_m, such that $T(t) < T_m$, the last term may also be omitted. Then, $\varepsilon_0(t) \approx \varepsilon_m$, where ε_m is the demarcation energy at the end of previous heating cycle, given by

$$\varepsilon_m \approx k\left(c^* T_m - T^*\right).$$ (9)

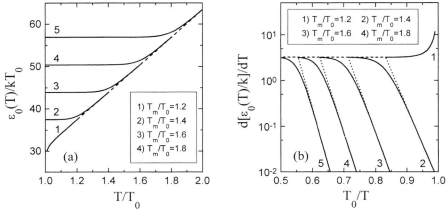

Figure 2. Normalised plots of the demarcation energy $\varepsilon_0(T)$ (a) and its derivative (b) calculated from Eq. (8) for linear (dashed lines) and stepwise (solid lines) sample heating. The derivative of $\varepsilon_0(T)$ computed from Eq. (10) is marked by dotted lines. The numbers on plots refer to sequential heating cycles. The value of $\beta/T_0\nu_0 = 10^{-15}$.

For larger values of t, when $T(t) > T_m$, the last term in Eq. (8) is dominant and the demarcation energy is determined by Eq. (6), independently of the sample heating regime up to the moment t_m (cf. Figure 2a).

We shall consider now the function $d\varepsilon_0(t)/dt$, which determines the 'velocity' of the demarcation level movement in the energy gap. Making use of Eq. (8) and the differentiation rule for implicit function, one gets, after some approximations

$$\frac{d\varepsilon_0(t)}{dt} \approx \frac{\beta\varepsilon_m}{T_m} \exp\left[\frac{\varepsilon_m}{kT_m} - \frac{\varepsilon_m}{kT(t)}\right]. \tag{10}$$

Therefore, the increase in $d\varepsilon_0(t)/dt$ at $t \approx t_m$ has a thermally activated character (see Figure 2b).

On the basis of the above results, the main features of TSCs for the stepwise heating regime can be easily established. From Eqs. (1) and (2) it follows that in the initial time region, $t \approx t_m$, the approximate relationship

$$I(t) \propto \frac{d\varepsilon_0(t)}{dt} \tag{11}$$

holds, which implies that

$$I(t) \propto \exp\left[-\frac{\varepsilon_m}{kT(t)}\right]. \tag{12}$$

Thus, the initial rise of TSC has also a thermally activated character. It should be recalled that, according to Eq. (9), the activation energy ε_m is a linear function of the maximum temperature T_m reached in the former heating cycle. Since the coefficient c^* in Eq. (9) depends on the frequency factor ν_0, the value of ν_0 can be simply determined from the measured dependence of ε_m on T_m. For longer times the demarcation energy and, in consequence, the TSC intensity are independent of the sample heating regime until the time t_m. This implies that the TSC curve, obtained at linear sample heating, constitutes an envelope for the TSC peaks, registered during individual stepwise heating cycles. Such a behaviour was established experimentally for poly(N-vinylcarbazole)[10].

Stepwise heating regime - numerical results

In order to check the accuracy of the derived formulae, we performed the numerical simulation of TSCs corresponding to linear and stepwise heating of the sample by the Monte Carlo method similar to that used in [4]. The calculations were made for the Gaussian distribution of traps,

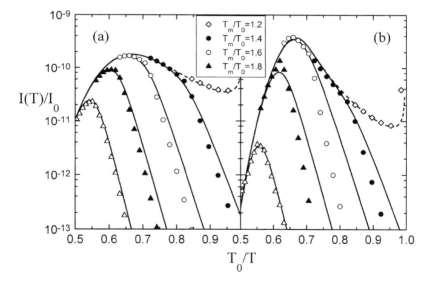

Figure 3. Normalised TSC curves for Gaussian trap distribution of Eq. (13), corresponding to linear (dashed lines) and stepwise (solid lines and points) sample heating. The legend shows maximum temperatures, reached in sequential heating cycles. The calculation parameters: $C_t N_{tot}/\nu_0 = 2.5 \cdot 10^{-4}$ (a), $2.0 \cdot 10^{-3}$ (b); $\tau_0 \nu_0 = 10^5$; $T_c/T_0 = 15$ (a), 10 (b); $\varepsilon_{tm}/kT_0 = 30$; $\beta/T_0\nu_0 = 10^{-15}$.

$$N_t(\varepsilon) = \frac{N_{tot}}{\sqrt{\pi}kT_c} \exp\left[-\left(\frac{\varepsilon - \varepsilon_{tm}}{kT_c}\right)^2\right], \tag{13}$$

and two different values of the characteristic temperature T_c.

In Figure 3 the TSC curves computed from the Eqs. (1), (2) and (8) and obtained numerically (marked respectively by lines and points) are compared. All the TSC features, established in the former section, can be found. In particular, the initial slopes of separate TSC curves increase with the temperature T_m. The initial rise of TSCs computed numerically is somewhat faster than of those calculated from approximate formulae. The discrepancies diminish with increasing heating cycle number and with increasing characteristic temperature T_c. The latter feature follows from the fact that the given formulae show a good accuracy for strongly dispersive transport, i.e. for trap distributions slowly varying with energy.

Conclusion

On the basis of the MT model, the analytical description of TSC drift peaks for stepwise sample heating as well as the corresponding Monte Carlo results have been given. It has been established that the initial TSC rise in the course of individual heating cycle has a thermally activated character. The activation energy approximately equals the demarcation energy for trapped charge carriers at the onset of heating cycle. The subsequent TSC course is independent of the mode of previous sample heating. Analogous TSC features should occur for some other heating modes, e.g. for 'delayed' sample heating, when a significant delay between the carrier photogeneration and the onset of temperature increase exists.

[1] G. Pfister, H. Scher, *Adv. Phys.* **1978**, *27*, 747.
[2] J. Plans, M. Zieliński, M. Kryszewski, *Phys. Rev. B* **1981**, *23*, 6557.
[3] W. Tomaszewicz, B. Jachym, *J. Non-Cryst. Solids* **1984**, *65*, 193.
[4] W. Tomaszewicz, *J. Phys.: Condens. Matter* **1992**, *4*, 3967, 3985.
[5] H. Fritzsche, N. Ibaraki, *Philos. Mag. B* **1985**, *52*, 299.
[6] V. I. Arkhipov, G. J. Adriaenssens, *J. Non-Cryst. Solids* **1995**, *181*, 274.
[7] W. Tomaszewicz, J. Rybicki, P. Grygiel, *J. Non-Cryst. Solids* **1997**, *221*, 84.
[8] W. Tomaszewicz, P. Grygiel, *Synth. Met.* **2000**, *109*, 263.
[9] W. Tomaszewicz, *Proc. VIIth Int. Seminar on Physics and Chemistry of Solids,* Częstochowa, Poland, 2001, p. 67.
[10] J. H. Slovik, *J. Appl. Phys.* **1976**, *47*, 2982.

Photoelectric Properties of Tetracene-Pentacene Heterojunction

Ryszard Signerski, Grażyna Jarosz,* Jan Godlewski

Gdansk University of Technology, G. Narutowicza 11/12, 80-952 Gdansk, Poland
E-mail: zofia@mif.pg.gda.pl

Summary: We investigated vacuum-evaporated sandwich systems formed by two organic layers of pentacene and tetracene. We measured spectral dependences of photovoltaic short-circuit currents and photocurrents as well as current-voltage dependences. In the case of systems equipped with Au and CuI electrodes the structure of spectral characteristics is determined by exciton release of trapped charge carriers either in the bulk of tetracene or near tetracene-pentacene interface and by photogeneration in pentacene layer. For systems with two Au electrodes we can also observe an influence of exciton injection of holes into tetracene on spectral characteristics. The measurements of dark current-voltage characteristics allow to observe the presence of a potential barrier between tetracene and pentacene.

Keywords: charge transport; heterojunction; pentacene; tetracene; thin films

Introduction

Attractive photoelectric and electric properties of organic heterojunction, such as photovoltaic effects, electroluminescence or current rectification are the subject of great interest of scientists. Some features of these phenomena are difficult to explain due to great amount of charge carrier traps and uncontrolled impurities, which lead to complex processes of photogeneration and transport of charge carriers. Heterojunctions formed from organic materials therefore still require intensive investigation.

The object of our paper is to examine photoelectric properties of a heterojunction formed from polycrystalline layers of tetracene (Tc) and pentacene (Pn). Absorption spectra of Tc and Pn are distinctly separated and, as a result, we can analyze the mechanisms of charge carrier photogeneration and find the position of photoactive region in the system [1]. This heterojunction has not been investigated yet although we have already investigated the heterojunction formed from other oligoacenes (anthracene and tetracene) [2].

Experimental

Measurements were performed on sandwich samples evaporated in vacuum with the following materials: Au on a glass substrate, Pn, Tc and Au (Au/Pn/Tc/Au samples) or Au on a glass substrate, Pn, Tc and Cu, which were treated with the vapour of iodine (Au/Pn/Tc/CuI samples). The thicknesses of single organic layers in one junction were the same and amounted to 100 nm or 200 nm. The sample was illuminated with monochromatic light in the range 370 nm - 740 nm, which includes singlet exciton absorptions of both Tc and Pn. Exemplary experimental results, including spectra of absorption, of short-circuit current and of photocurrent as well as dark current-voltage characteristics, are presented in Figs. 1-5. For all the spectra of the currents shown in this work, the flux density of monochromatic light penetrating into organic layers was 10^{13} photons/(cm^{2}s).

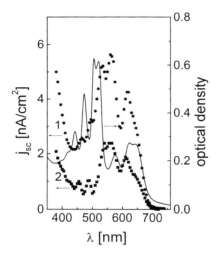
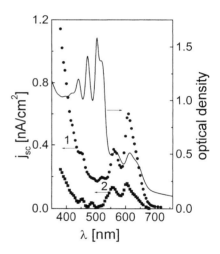

Figure 1. Short-circuit current spectrum of Au/Pn/Tc/CuI cell with 100 nm-thick organic layers illuminated through Au (1) and through CuI (2) and absorption spectrum of Pn/Tc bilayer (solid line).

Figure 2. Short-circuit current spectrum of Au/Pn/Tc/CuI cell with 200 nm-thick organic layers illuminated through Au (1) and through CuI (2) and absorption spectrum of Pn/Tc bilayer (solid line).

The spectra of short-circuit currents of Au/Pn/Tc/CuI cells with 100 nm-thick (Fig. 1) or 200 nm-thick (Fig. 2) organic layers are plotted together with absorption spectra of the respective organic bilayers. Characteristics 1 refer to illumination from the Pn side of the sample while characteristics 2 refer to illumination from the Tc side.

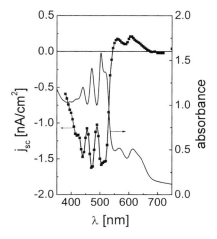 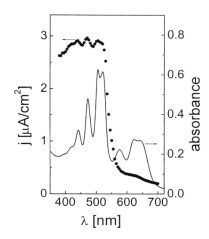

Figure 3. Short-circuit current spectrum of Au/Pn/Tc/Au cell illuminated from the Tc side and absorption spectrum of Pn/Tc bilayer (solid line).

Figure 4. Photocurrent spectrum of Au/Pn/Tc/CuI cell with bias and absorption spectrum of Pn/Tc bilayer (solid line).

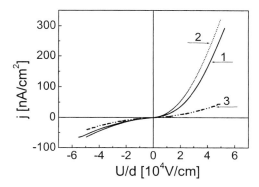

Figure 5. Dark current as a function of average electric field for Au/Pn/Tc/CuI with thicknesses of Tc and Pn 100 nm (1) or 200 nm (2) and for Au/Pn/Tc/Au with 200 nm-thick Tc and Pn layers (3).

Figure 3 shows the spectrum of short-circuit currents of Au/Pn/Tc/Au cell with 200 nm-thick organic layers illuminated from the Tc side. Positive values of short-circuit currents in Figs. 1-3 relate to the current flowing from Pn to Tc.

Figure 4 shows photocurrent spectrum of Au/Pn/Tc/CuI in the presence of bias equal to 0.5 V. The sample was illuminated through the Au electrode, which was negatively polarized. Thicknesses of organic layers were 100 nm.

Dark currents as a function of average electric field ($E=U/d$, where d is the sum of thicknesses of Tc and Pn layers) are presented in Fig. 5. Curves 1 and 2 refer to Au/Pn/Tc/CuI cells with 100 nm-thick (curve 1) or 200 nm-thick (curve 2) organic layers, while curve 3 refers to Au/Pn/Tc/Au cell with organic layers of 200 nm.

Discussion

The investigations of transport of charge carriers in organic systems require the use of appropriate electrodes. To examine the junction of Tc and Pn, between which a potential barrier for holes can be estimated as ca 0.3 eV (the difference between ionization energy of Tc, 5.3 eV, and Pn, 5.0 eV)[3], ohmic electrodes (such as CuI and Au) suitable for these materials should be applied[4]. Then only hole currents can flow through our systems.

Taking into account Figs. 1 and 2 in the range of tetracene absorption, we observe a correlation between maxima of short-circuit currents and minima of absorption spectra for both directions of illumination (antiparallel relations)[5]. Such profiles of short-circuit current spectra indicate that this current is determined by exciton release of charge carriers from traps in the whole bulk of Tc layer or near the Tc/Pn interface. In the case of thinner organic layers (Fig. 1), we do not observe pure antiparallel structure of short-circuit current spectrum for illumination through the Pn layer, which can result from a sequence of exciton detrapping of charge carriers occurring in the whole Tc layer and near the Tc/Pn junction. Comparing Figs. 1 and 2 we also come to the conclusion that the increase in layer thickness leads to a weakening of the spectrum structure in the range of Tc absorption.

On the other hand, in the case of Au/Pn/Tc/Au cell illuminated from the Tc side (Fig. 3) in the same range of absorption, maxima of short-circuit current correspond to maxima of absorption spectrum (parallel relation)[5] and the short-circuit current flows in opposite direction. This relation suggests that for the cell with two Au electrodes the short-circuit current is limited by exciton dissociation and hole injection occurring at the Au/Tc interface, so there is an additional barrier in Tc near Au electrode. The height of this barrier can be estimated as ca 0.3 eV (the difference between ionization energy of Tc (5.3 eV) and work function of Au (5.0 eV)).

Within the region of pentacene absorption ($\lambda > 540$ nm), all spectra of short-circuit currents (Figs. 1-3) correlate reasonably well with the absorption spectra of bilayers. This parallel structure of short-circuit current becomes more noticeable for a thicker layer, where short-circuit current associated with pentacene absorption dominates over the one associated with tetracene absorption. So, in the range of Pn absorption, spectra of short-circuit current are determined by charge carrier generation in the bulk of Pn[3,6]. Simultaneously, the importance of charge carrier injection from electrodes can be excluded because this process does not depend on the sample thickness.

Under the illumination through negatively polarized Au electrode, photocurrent of Au/Pn/Tc/CuI is evidently determined by exciton detrapping of charge carriers in Tc near the Tc/Pn interface (parallel relation in the range of tetracene absorption in Fig. 4) and charge carrier photogeneration in the bulk of Pn can be neglected.

The dependences of dark currents on average electric field are presented in Fig. 5. Curves 1 and 2 obtained from the cells with CuI are similar indicating that the electric properties of these systems are reproducible. Some difference between both curves results partly from the inaccuracy in determining the active surface of the samples and current density. The asymmetry of characteristics 1 and 2 observed for different directions of electric field results from the presence of Pn/Tc barrier, which should be overcome by holes flowing from Pn to Tc (range of negative electric fields). Such a barrier does not exist for the opposite direction of current (range of positive electric fields). On the other hand, in the case of the cell with two Au, a barrier at the Au/Tc interface exists also for positive fields. Therefore characteristic 3 is nearly symmetrical.

Conclusions

Our experimental investigation leads to the following conclusions:

- in the region of Tc absorption, the short-circuit current of Au/Pn/Tc/CuI is determined by charge carrier detrapping by excitons in the whole Tc layer or near the Tc/Pn interface,

- in the region of Pn absorption, the short-circuit current of Au/Pn/Tc/CuI is determined by charge carrier photogeneration in the Pn bulk,

- it is impossible to examine the transport of charge carriers through Pn/Tc heterojunction using Au as electrode for the Tc layer,

- in the presence of bias the photocurrent is determined by exciton detrapping of charge carriers in Tc near the Tc/Pn junction,

- asymmetry of dark current-voltage characteristics of systems with ohmic electrodes (Au and CuI) results from the presence of a potential barrier at the Pn/Tc interface.

Acknowledgements

Work was supported by KBN under Program No. 4 T11B 057 22.

[1]   J. Rostalski, D. Meissner, *Sol. Energy Mater. Solar Cells* **2000**, *63*,37.
[2]   R. Signerski, J. Godlewski, H. Sodolski, *Phys. Status Solidi A* **1995**, *147*, 177.
[3]   E. A. Silinsh, V. Čápek, *"Organic Molecular Crystals"*, AIP Press, New York 1994.
[4]   G. Jarosz, R. Signerski, J. Godlewski, *Synth. Met.* **2000**, *109*, 161.
[5]   M. Pope, Ch. E. Swenberg, *"Electronic Processes in Organic Crystals and Polymers"*, Oxford Sci. Publ., New York-Oxford 1999.
[6]   P. J. Bounds, W. Siebrand, *Chem. Phys. Lett.* **1982**, *85*, 496.

Macromol. Symp. **2004**, *212*, 375-380

Lattice Dynamics and Electron-Phonon Coupling in Pentacene Crystal Structures

Matteo Masino,[1] *Alberto Girlando,*[1] *Aldo Brillante,*[2] *Luca Farina,*[2] *Raffaele Guido Della Valle,*[2] *Elisabetta Venuti*[2]

[1] Dip. Chimica GIAF, University of Parma, Viale delle Scienze 17, 43100 Parma, Italy

[2] Dip. Chimica Fisica e Inorganica, University of Bologna, Viale Risorgimento 4, 40136 Bologna, Italy

Summary: The crystal structures and lattice phonons of pentacene are computed by the quasi harmonic lattice dynamics (QHLD) method. From the eigenvectors of the low frequency phonons we calculate the *e-ph* coupling constants due to the modulation of the transfer integrals. The transfer integrals are computed by the Hückel method and by the INDO/S Hamiltonian for all the nearest neighbor pentacene pairs in the *ab* crystal plane.

Keywords: charge transport; conjugated materials; electron-phonon coupling; lattice dynamics; molecular crystals

Introduction

Pentacene is a well known organic semiconductor, recently object of renewed interest due to a report of very high charge carrier mobilities at low temperature in extremely pure single crystals.[1]

In the present paper we address the lattice phonon dynamics of pentacene, and their interactions with charge carriers, in order to provide a basis for the understanding of the above mentioned phenomena. Indeed, both the temperature dependence of the charge carriers mobilities and the onset of superconductivity have been ascribed to the electron-phonon coupling.[1]

Crystal structure and phonon dynamics

At the light of the five crystallographic analyses so far reported on a single crystal of pentacene,[2-5] the first problem we have to deal with is how many pentacene polymorphic structures exist and

DOI: 10.1002/masy.200450845

how these structures are related to the observed phenomena. On the basis of our recent theoretical[6] and experimental works[7] it is clear that pentacene shows at least two polymorphic structures. We have named the two polymorphs as phase **C** after the structure of Campbell et al.[2] and as phase **H** after the structure which results from the more recent crystallographic investigations.[3-5] The **H** phase is the phase on which all the most recent experiments have been done.[1]

To take into account the temperature effects and to describe vibrational properties, we employ the quasi-harmonic lattice dynamics (QHLD) method.[8,9] In QHLD the vibrational contribution to the Gibbs free energy, $G(p,T)$, is approximated with the Gibbs energy of the harmonic phonons:

$$G(p,T) = \Phi + pV + \sum_{\vec{q},j} \frac{\hbar\omega_{\vec{q},j}}{2} + k_B T \sum_{\vec{q},j} ln\left[1 - exp\left(-\frac{\hbar\omega_{\vec{q},j}}{k_B T}\right)\right]$$

where the intermolecular potential energy of the crystal, Φ, is represented by an atom-atom Buckingham model,[10] with Williams parameters set IV,[11] plus electrostatic interactions described by the *ab initio* atomic charges residing on the atoms. Given an initial lattice structure, one computes Φ and its second derivative with respect to the displacements of the molecular coordinates. The second derivatives form the dynamical matrix, which is numerically diagonalized to obtain the phonon frequencies and the corresponding eigenvectors. We allow for the mixing between lattice and intramolecular vibrations.[9] The equilibrium structure and phonon dynamics as a function of temperature are determined self-consistently by minimizing $G(p,T)$ with respect to the unit cell axes, angles and molecular orientations.

Electronic structure

We have performed electronic structure calculation based on the tight-binding model on the two polymorphic crystal structures of pentacene to study how different packings affect the electronic properties. In the framework of the tight-binding model the band dispersion relations read:

$$\varepsilon_{\pm}(\vec{k}) = (t_{a1} + t_{a2})\cos\vec{k}\vec{a} \pm 4\left(t_{d1}\cos\vec{k}\frac{\vec{a}+\vec{b}}{2} + t_{d2}\cos\vec{k}\frac{\vec{a}-\vec{b}}{2}\right)$$

where the labelling of the *ab* crystal plane intermolecular transfer integral t is the same as in

Ref.[12]. The intermolecular transfer integrals have been estimated using two different quantum chemistry semiempirical methods: The extended Hückel method and the INDO/S Hamiltonian developed by Zerner and coworkers, with Mataga-Nishimoto repulsion potential.[13]

The results as a function of temperature and crystal structures are listed in Tab.1.

Table 1: Calculated VB and CB INDO/S bandwidths as a function of temperature and structure

Extended Hückel bandwidth				
Structure		*0 K*	*150 K*	*300 K*
H	valence band	203 meV	189 meV	169 meV
	conduction band	271 meV	255 meV	229 mev
C	valence band	159 meV	152 meV	135 meV
	conduction band	227 meV	217 meV	193 meV
INDO/S bandwidth				
Structure		*0 K*	*150 K*	*300 K*
H	valence band	632 meV	585 meV	536 meV
	conduction band	672 meV	624 meV	576 mev
C	valence band	548 meV	511 meV	470 meV
	conduction band	588 meV	548 meV	508 meV

The total valence and conduction band of the **H**-polymorph, which has been recognized as the one showing enhanced charge transport properties, are ~15-20% wider than in the **C**-polymorph. Moreover, as already pointed out by Brédas et al.,[12] the INDO/S method is able to reproduce large bandwidth values fully consistent with recent low temperature experimental estimates on the order of 500 meV.[14] These values are then considered more reliable than the simple extended Hückel estimates.

On the other hand, the observed sharp temperature dependence of the transport properties cannot be explained on the basis of bare band theory, since the calculated bandwidth reduction in going from room temperature to 0 K is on the order of 15-20%, to be compared with the experimental estimates of about 90%.[14] It is indeed commonly accepted that in acene crystals band transport is a typical situation at low temperature, but does not usually apply to higher temperature. As temperature increases, low energy vibrational modes start impeding the mobilities of the charge carriers. In this picture, therefore, the coupling between charge carriers and low frequency lattice phonons (*e-lph* coupling) is the key parameter governing charge transport.

E-lph coupling

Whereas extensive theoretical[15,16] and experimental[17] studies have been devoted to the characterization of the local contribution of the electron-phonon coupling, basically the variation of the on-site molecular energy with respect to the intramolecular vibrations, very little is known on the *e-lph* coupling. In molecular crystals, lattice phonons are expected to couple to charge carriers mainly modulating the amplitude of the intermolecular transfer integral. The corresponding linear *e-lph* coupling constants are defined as:

$$g(\vec{q}, j) = \sqrt{\frac{\hbar}{2\omega_{\vec{q},j}}} \left(\frac{\partial t}{\partial Q_{\vec{q},j}} \right)_0$$

where $Q_{\vec{q},j}$ is the normal coordinate for the j-th phonon with wavevector $\vec{q}$. We have assumed that the optical lattice phonons as dispersionless, and have performed calculations for the $\vec{q} = 0$ eigenvectors only. Within this approximation, symmetry arguments show that only the totally symmetric phonons can be coupled to charge carriers. To evaluate the *e-lph* coupling constants we have calculated each intermolecular transfer integral for the equilibrium geometry, as well as for geometries displaced along the QHLD eigenvectors. The various $g(\vec{q}, j)$ are then obtained by numerical differentiation. The *e-lph* coupling constants for the four largest transfer integrals calculated using the INDO/S Hamiltonian are reported in Tab.2.

We can compare the calculated *e-lph* coupling constants with tunneling experiments.[18] Indeed, the conductance derivative spectrum d^2I/dV^2, shown in Fig.1 is proportional to the phonon density of states weighted by an effective electron-phonon coupling function.

Table 2 shows that, although some differences exist, the most strongly coupled modes for both valence and conduction band occur around 25, 19, 12, 8 and 3 meV. The first four modes find correspondence in the experimental tunneling spectrum of Fig.1, where one sees a strong doublet centered at 25 meV, a broad peak at 8 meV and two minor features around 16 and 13 meV. Moreover, our calculations also predict that the lowest energy phonon at 3.4 meV, difficult to access in the tunneling experiment,[18] should be strongly coupled to charge carriers. We associate this phonon mode to the most intense band observed at 4.6 meV in the Raman spectra of **H** pentacene.[7]

Table 2: Low-energy A_g phonons and *e-lph* coupling constants of **H** pentacene calculated at 0 K (me V)

ω	Valence band				Conduction band			
	g_{a1}	g_{a2}	g_{d1}	g_{d2}	g_{a1}	g_{a2}	g_{d1}	g_{d2}
38.4	0.0	0.1	-1.0	0.8	0.0	0.0	1.2	1.1
38.0	0.0	-0.2	-0.3	-0.1	0.0	0.0	0.0	0.0
32.4	-0.1	0.3	-2.4	-1.6	0.0	0.0	-1.4	-1.0
32.1	0.7	0.0	2.7	2.0	-0.3	0.0	1.6	1.5
29.9	0.1	1.4	-1.0	-1.5	0.0	-0.5	0.7	1.3
29.5	1.8	-0.2	1.8	0.3	-0.7	0.0	-0.8	0.0
25.5	0.4	0.0	-1.6	-1.6	-0.9	0.9	1.2	1.8
25.2	-0.4	0.0	3.1	4.5	0.8	0.7	1.2	2.8
20.1	-0.1	1.2	-0.9	-1.7	0.4	-1.4	-0.1	-0.5
19.3	-0.4	0.2	-0.8	-2.8	-1.6	-0.4	-0.7	-3.2
18.4	0.2	0.0	-2.1	-5.0	0.0	0.0	0.0	0.0
16.4	-0.5	-1.0	1.8	0.8	0.5	1.0	-0.4	-0.1
11.9	-0.6	0.0	3.2	1.1	1.0	0.3	-4.3	-2.5
8.7	4.0	-3.0	0.9	-1.4	-2.2	1.7	2.0	1.6
7.6	-1.6	-3.2	-3.6	1.3	1.4	1.2	0.0	2.3
3.4	3.6	4.5	1.5	-1.4	-1.1	-2.5	-3.2	-0.4

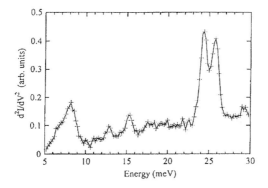

Fig.1: Conductance derivative spectrum of **H** pentacene at 1.4 K. From Ref.[18]

Conclusions

We have calculated the crystal structures, phonon dynamics and electronic structure of both pentacene polymorphs as a function of temperature. Band structure calculations based on tight-binding model can be exploited to reproduce estimated experimental bandwidth of high purity sample at low temperature. The electron-phonon coupling has to be invoked in order to account for the sharp temperature dependence. We, therefore, have fully characterized the interaction between holes/electrons and the low frequency optical phonons, paving the way to understand charge transport properties in pentacene.

[1] J. H. Schön et al., *Phys. Stat. Sol. B* **2001**, 226, 257; *Science* **2000**, 288, 2338; *Nature* **2000**, 406, 702.

[2] R. B. Campbell, J. M. Robertson, J. Trotter, *Acta Cryst.*, **1962**, 15, 289.

[3] D. Holmes, S. Kumaraswamy, A. J. Matzger and K. P. Vollhardt; *Chem. Eur. J.*, **1999**, 5, 3399.

[4] T. Siegrist et al., *Angew. Chem. Int. Ed.*, **2001**, 40, 1732.

[5] C. C. Mattheus, A. B. Dros, J. Baas, A. Meetsma, J. de Boer and T. T. M. Palstra, *Acta Cryst. C*, **2001**, 57, 939.

[6] E. Venuti, R. G. Della Valle, A. Brillante, M. Masino and A. Girlando, *JACS*, **2001**, 124, 2128.

[7] A. Brillante, R. G. Della Valle, L. Farina, A. Girlando, M. Masino, E. Venuti, *Chem. Phys. Lett.*, **2002**, 357, 32.

[8] R. G. Della Valle, E. Venuti and A. Brillante, *Chem. Phys.*, **1996**, 202, 231.

[9] A. Girlando, M. Masino, G. Visentini, R. G. Della Valle, A. Brillante, E. Venuti, *Phys. Rev. B*, **2000**, 62, 14476.

[10] A. J. Pertsin and A. I. Kitaigorodsky, *"The atom-atom potential method"* Springer-Verlag, Berlin, 1987.

[11] D. E. Williams, *J. Chem. Phys.*, **1967**, 47, 4680.

[12] J. Cornil, J. Ph. Calbert, J. L. Brédas, *JACS*, **2001**, 123, 1250.

[13] J. Ridely and M. C. Zerner, *Theor. Chim. Acta*, **1973**, 32, 111.

[14] J. H. Schön, C. Kloc and B. Batllogg, *Phys. Rev. Lett.*, **2001**, 86, 3843.

[15] A. Devos, M. Lannoo, *Phys. Rev. B*, **1998**, 58, 8236.

[16] T. Kato, T. Yamabe, *J. Chem. Phys.*, **2001**, 115, 8592.

[17] N. Gruhn, D.da Silva Filho, T. Bill, M. Malagoli, V. Coropceanu, A. Kahn, J.L. Brédas, *JACS*, **2002**, 124, 7918.

[18] M. Lee, J. H. Schön, Ch. Kloc and B. Batllogg, *Phys. Rev. Lett.*, **2001**, 86, 862.

Effect of a Dipolar Self-Assembly Monolayer Formation on Indium-Tin Oxide on the Performance of Single-Layer Polymer-Based Light-Emitting Diodes

Jorge Morgado,[*1] *Ana Charas,*[1] *Nunzio Barbagallo,*[1] *Luis Alcácer,*[1] *Manuel Matos,*[2] *Franco Cacialli*[3]

[1] Departamento de Engenharia Química, Instituto Superior Técnico, Av. Rovisco Pais, P-1049-001 Lisboa, Portugal

[2] Departamento de Engenharia Química, Instituto Superior de Engenharia de Lisboa, Rua Conselheiro Emídio Navarro – 1, P-1949-001 Lisboa, Portugal

[3] Cavendish Laboratory, Madingley Road, Cambridge, CB3 0HE, UK Department of Physics and Astronomy, University College London, Gower Street, London, WC1E 6BT, UK

Summary: We report on the use of indium-tin oxide surface modification, by grafting of highly polar *p*-disubstituted benzenes, in the fabrication of light-emitting diodes. The polar compounds possess COCl or SO_2Cl grafting groups and CF_3 or NO_2 as highly electronegative groups, leading to the formation of a dipolar monolayer, which brings about an increase in ITO work function, thereby reducing the barrier for hole injection into luminescent polymers. We observe that the effect of this self-assembled monolayer, in terms of light-onset voltage, efficiency and luminance, is at least comparable to the use of a hole injection layer of doped poly[(3,4-ethylenedioxy)thiophene] for LEDs using poly({2-[(2-ethylhexyl)oxy]-5-methoxy-1,4-phenylene}vinylene) (MEH-PPV) and polyfluorene blends as active layers.

Keywords: conjugated polymers; light-emitting diodes (LED); monolayers; polyfluorenes; self-assembly

Introduction

Electroluminescent (EL) polymers offer the prospect for the fabrication of cheap, flexible, large-area displays, among other potential applications. Light-emitting diodes (LEDs) are the key devices behind such an application, and a significant effort has been devoted to the optimization of their performance (efficiency, lifetime, colour stability). The control of charge injection and transport is of paramount importance, as these are fundamental processes in determining both the efficiency and the light-onset voltage of LEDs. More specifically, the

DOI: 10.1002/masy.200450846

improvement of hole injection from the anode side (usually indium-tin oxide, ITO), has involved the insertion of hole injection layers, such as doped polyaniline or poly[(3,4-ethylenedioxy)thiophene] doped with poly(styrenesulfonic acid), PEDOT:PSS[1], or graded hole-injection layers[2]. More recently, the possibility to chemically modify the ITO surface by grafting of polar organic compounds, as a means to increase its work function[3-5] or to increase its hydrophobicity[6], is attracting a significant interest. In particular, it was reported[5] that the work function of ITO could be tuned by grafting p-disubstituted benzene compounds, comprising a grafting group (SO_2Cl, $COCl$, or PO_2Cl_2) and an electron-attracting group, such as Cl or CF_3. We have recently reported on the effect of surface modification of ITO, by formation of such a dipolar self-assembled monolayer (SAM), on optoelectronic properties of LEDs based on poly({2-[(2-ethylhexyl)oxy]-5-methoxy-1,4-phenylene}vinylene) (MEH-PPV) where compounds with a stronger electron-attracting group, NO_2, were also used.[7] We now present the effect of similar ITO surface modifications on the optoelectronic characteristics of LEDs based on polyfluorenes, which are, presently, the most promising luminescent polymers for device applications.[2,8]

Figures 1 and 2 show the molecular structure of the polymers used in this study and of the polar compounds used in the formation of dipolar SAM onto ITO.

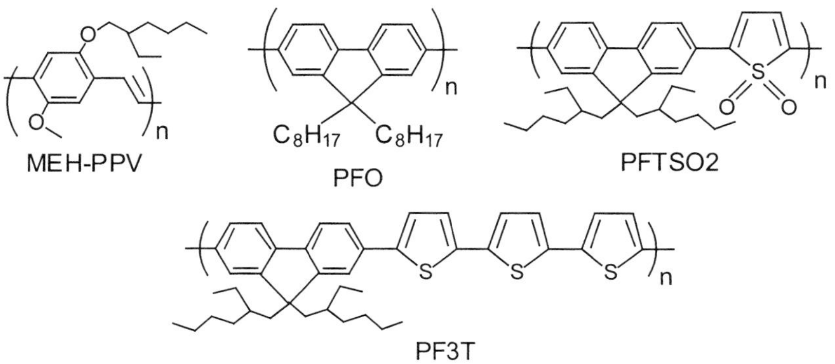

Figure 1. Molecular structure of the luminescent polymers.

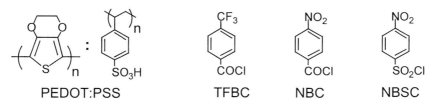

Figure 2. Molecular structure of PEDOT:PSS or PEDOT for short, used as hole-injecting material, and of the polar grafting compounds.

Experimental

MEH-PPV was synthesized by the Gilch route[9] and the copolymers PFTSO2 and PF3T were prepared by Suzuki coupling.[10] PEDOT:PSS was obtained from Bayer and the *p*-disubstituted benzenes were purchased from Aldrich. The SAM-modified ITO substrates were prepared by immersion for 5 min in dichloromethane solutions of the polar grafting compounds (1 mM), then washed with dichloromethane and dried under nitrogen flow. Reference LEDs were prepared using bare ITO substrates and ITO coated with PEDOT:PSS (55 nm thick). The fabrication of LED structures was completed by spin coating with the polymer or polymer blend solutions, followed by thermal evaporation of aluminum electrodes at a pressure of 10^{-3} Pa, defining pixel areas of 4 mm^2. MEH-PPV was used in the neat form, while the polyfluorenes were used as blends, namely, PFO:PF3T and PFO:PFTSO2, where the content of PFO, the host polymer matrix, was 95 % by weight.

Results and Discussion

Figure 3 shows the energetic positions of the relevant frontier levels of the polymers (PFO[11], PF3T[10], PFTSO2[10] and MEH-PPV[7]) and the work function of the electrodes. The ITO work function is considered to lie in the range 4.6-4.8 eV[12,13]; for PEDOT:PSS it is 5.2-5.3 eV.[13]

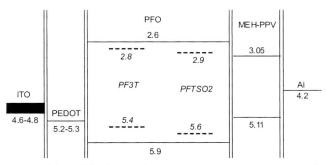

Figure 3. Energetic position of the frontier levels of various components used in fabrication of devices.

TFBC modification of ITO increases its work function by about 0.4 eV[5], which brings it to about 5.0-5.2 eV. Considering that the dipole moment of nitrobenzene is 4.22 D, much higher than the value of 2.86 D for (trifluoromethyl)benzene, we expect the work function of ITO-NBC to be higher than 5.0 eV. Based on the results obtained for chlorobenzene[5], we also expect the work function of ITO-NBSC to be slightly lower than that of ITO-NBC. Figure 4 compares the current and luminance for LEDs based on MEH-PPV. The SAM modification of ITO significantly reduces the light-onset voltage, though the maximum luminance is not significantly improved. Furthermore, the use of NBC or NBSC, which are expected to lead to a higher work function than TFBC, does not further reduce the light-onset voltage from the value of about 2.2 eV closer to 2.13 eV, the energy gap voltage. This result is in agreement with a work function of ca. 5.1 eV for ITO-TFBC, close to the highest occupied molecular orbital energy of MEH-PPV, and therefore corresponding to a negligible or no hole-injection barrier.

Figures 5 and 6 compare the optoelectronic characteristics of LEDs based on PFO:PF3T and PFO:PFTSO2 blends, where ITO was modified with NBSC. In both blends, there is an efficient energy transfer from the host PFO to the guest polymers[14]. The SAM modification of ITO significantly improves both the current density and the luminance of the LEDs in comparison with the use of bare ITO; they are similar or slightly higher than those of the LEDs with PEDOT. The EL efficiencies of the devices with ITO-NBSC anodes are similar to those based on ITO/PEDOT, being 0.05 – 0.09 % of external quantum efficiency.

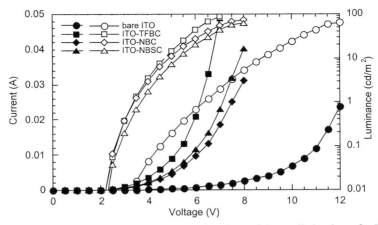

Figure 4. Current (filled symbols) and luminance as functions of the applied voltage for LEDs based on MEH-PPV (140 nm thick).

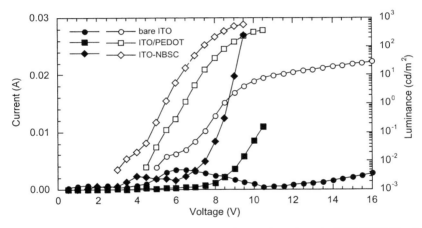

Figure 5. Current (filled symbols) and luminance of LEDs based on the PFO:PF3T blend (95 nm thick).

Conclusions

The SAM modification of ITO with NBSC allows the fabrication of LEDs based on polyfluorene blends with similar or even better performance (light-onset voltage, maximum luminance and EL efficiency) than that of LEDs with PEDOT:PSS-coated ITO.

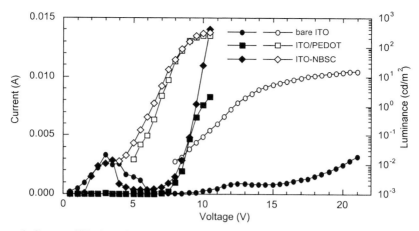

Figure 6. Current (filled symbols) and luminance of LEDs based on the PFO:PFTSO2 blend (80 nm thick).

Acknowledgment

This work benefited from financial support from FCT-Portugal (project POCTI/34668/Fis/2000).

[1] Y. Cao, G. Yu, C. Zhang, R. Menon, A.J. Heeger, *Synth. Met.* **1997**, *87*, 171.
[2] P.K.H. Ho, J.S. Kim, J.H. Burroughes, H. Becker, S.F.Y. Li, T.M. Brown, F. Cacialli, R.H. Friend, *Nature* **2000**, *404*, 481.
[3] F. Nüesch, L. Si-Ahmed, B. François, L. Zuppiroli, *Adv. Mater.* **1997**, *9*, 222.
[4] S.F.J. Appleyard, S.R. Day, R.D. Pickford, M.R. Wills, *J. Mater. Chem.* **2000**, *10*, 169.
[5] C. Ganzorig, K.-J. Kwak, K. Yagi, M. Fujihira, *Appl. Phys. Lett.* **2001**, *79*, 272.
[6] B. Choi, J. Rhee, H.H. Lee, *Appl. Phys. Lett.* **2001**, *79*, 2109.
[7] J. Morgado, A. Charas, N. Barbagallo, *Appl. Phys. Lett.* **2002**, *81*, 933.
[8] M.T. Bernius, M. Inbasekaran, J. O'Brien, W. Wu, *Adv. Mater.* **2000**, *12*, 1737.
[9] H.G. Gilch, W.L. Weelwright, *J. Polym. Sci., Part A-1* **1966**, *4*, 1337.
[10] A. Charas, J. Morgado, J.M.G. Martinho, L. Alcácer, S.F. Lim, R.H. Friend, F. Cacialli, *Polymer*, in press.
[11] A.J. Campbell, D.D.C. Bradley, H. Antoniadis, *J. Appl. Phys.* **2001**, *89*, 3343.
[12] J.S. Kim, B. Lägel, E. Moons, N. Johansson, I.D. Baikie, W.R. Salaneck, R.H. Friend, F. Cacialli, *Synth. Met.* **2000**, *111-112*, 311.
[13] T.M. Brown, J.S. Kim, R.H. Friend, F. Cacialli, R. Daik, W.J. Feast, *Appl. Phys. Lett.* **1999**, *75*, 1679.
[14] A. Charas, J. Morgado, J.M.G. Martinho, A. Fedorov, L. Alcácer, F. Cacialli, *J. Mater. Chem.* **2002**, *12*, 3523.

Current-Voltage Characteristics of Thin Poly(biphenyl-4-ylphthalide) Films

Andrei Bunakov,[1] *Aleksei Lachinov,*[2] *Renat Salikhov* *[1]

[1] Bashkirian State Pedagogical University, Oktyabrskoi revolyutsii 3a, Ufa, 450000 Russia

[2] Institute of Molecular and Crystal Physics, Ufa Research Center, Russian Academy of Sciences, Prospekt Oktyabrya 151, Ufa, 450075 Russia

Summary: The paper considers the features of the charge transport near to the threshold of the transition of polymer films to the high conductive state, induced by a small uniaxial pressure. The problem has not been solved so far, how the energy structure of a wide-band-gap organic dielectric varies near this threshold. The current-voltage characteristics of poly(biphenyl-4-ylphthalide) films at different uniaxial pressures were measured and analyzed. The interpretation of the obtained results is carried out within the framework of the space charge limited conduction model. The estimation of the injection model of transport parameters such as the charge carrier mobility and concentration, trapping state concentration and others are carried out. The analysis of the obtained results allows to make the following preliminary conclusion. Pressure increase promotes formation of a narrow trap band near the quasi-Fermi level resulting from the increase in the injection. This can give rise to a sharp magnification of the charge carrier mobility and even transition to the metallic state.

Keywords: charge carriers mobility; current-voltage characteristics; high conductive state; poly(biphenyl-4-ylphthalide); Space Charge Limited Conduction

Introduction

In thin electroactive polymer films the transition from a dielectric state to a high-conductive state (HCS), induced by such physical influences as electric field[1], small uniaxial pressure[2], temperature[3], is observed. The feature of this phenomenon is that the temperature dependence of the HCS conductivity is of the metal type and the conductivity can be extremely high ($> 10^7$ S/m).[4] The initial dielectric state of the polymer can be described in such parameters as the band gap ~ 4.3 eV, the work function of an electron ~ 4.2 eV, the first potential of ionization ~ 6.2 eV.[5] However, so far there has been the unsolved problem how the energy structure of the wide-band-gap organic dielectric varies near the transition into the HCS. In this

     DOI: 10.1002/masy.200450847

connection, the purpose of the present paper was the study of the laws of the charge transport near the threshold of the transition of polymer films to the HCS, induced by a small uniaxial pressure. For this purpose the current-voltage (*I-V*) characteristics of poly(biphenyl-4-ylphthalide)[6] films of different thickness at different uniaxial pressures were measured and analyzed. The investigations of the *I-V* characteristics of the polymers possessing high conductance are known.[7-11] However, in these papers the data in the vicinity of the transition threshold were not systematic.

Experimental procedure

The choice of poly(diphenylphthalide) (PBP) (Fig. 1) as the object of the investigation was explained by its good film-forming properties on the metal substrates. In paper[12] it was shown that under certain technological conditions the PBP forms continuous homogeneous films of 0.05 - 10 μm thickness. Besides, this polymer has no temperature features in conductivity up to the softenning temperature (360 °C in air). Also PBP is the most investigated polymer from the point of view of HCS inducing.[1,13,14] The quality and homogeneity of the polymer films were checked by optical microscopy methods as described in the paper.[15]

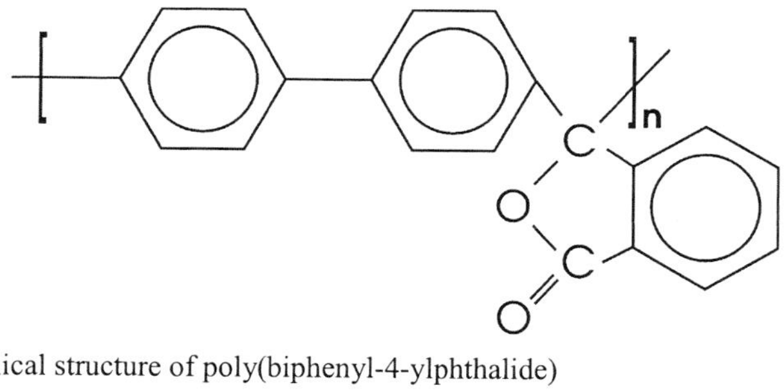

Figure 1. Chemical structure of poly(biphenyl-4-ylphthalide)

The measurements were carried out using the apparatus consisting of a uniaxial pressure device, a pressure sensor and the electrical circuit for the *I-V* characteristics recording. The pressure (*P*) varied up to 900 kPa.

The used samples were of the metal-polymer-metal "sandwich" type. Thin polymer films were prepared by spin coating of the metal electrode surface with a polymer solution. As a solvent, cyclohexanone was used. The metal electrode was deposited on a polished glass plate by the vacuum evaporation method. The other electrode was deposited on the polymer film surface by vacuum evaporation. Various metal films such as Cu, Al and Cr were used for the electrodes. The contacting area was $s \approx 2$ mm^2. The thickness of the films was set by concentration of the solution and checked by the interference method with an MII-4 interferometer; it ranged from 0.8 to 1.5 μm. More than fifty samples were investigated.

Results and discussions

Typical I-V characteristics of the 1-μm thick polymer film near the transition to the metal state are shown in a Fig. 2. These dependences are well fitted by the power function of the $I \approx U^n$ type. Under $P = 0$, the I-V characteristics can be divided into two regions: the linear with $n = 1$ at small voltages and the quadratic with $n = 2$ at the higher voltages. The transition between these regions occurs at a certain voltage U_1.

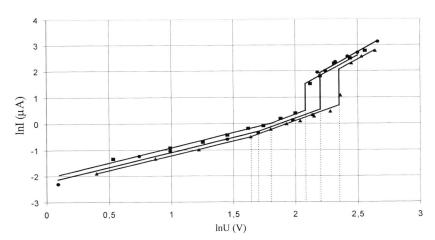

Figure 2. I-V characteristics of the polymer films measured at different pressures (above atmospheric): ■-780 kPa; •-820 kPa; ▲-860 kPa . All points are average of 50 measurements. The measurement error is ca. 5 %

The comparative analysis of the I-V characteristics, introduced in Fig. 2, showed, that increase in pressure leads to the reduction of the voltage U_1. When the voltage exceeds a value of $U_2 >$

U_1, practically vertical growth of the current is observed. The increase in the pressure leads to the shift of U_2 towards higher voltages. After the vertical region at U_2, a quadratic region is observed on the *I-V* characteristics again. A further increase in pressure gives rise to electronic switching of the polymer sample to the HCS.

For interpretation of the obtained *I-V* characteristics the space charge limited conduction (SCLC) model was used[16], which allowed to characterize local states in the polymer band gap and explain the shape of the *I-V* dependence. Before passing to the obtained results, it is necessary to note that within the framework of the problem given in this paper, we use the limit of monochromatic traps. The results of the investigation show that even in this limit we can trace the effect of external influences on the injection model parameters.

The model predicts the Ohmic behavior up to a voltage U_1, at which the concentration of the thermally generated equilibrium charge becomes comparable to that of the injected charge:

$$J \approx e n_0 \mu \frac{U}{L},$$ (1)

where e is the charge of an electron, n_0 the concentration of the equilibrium free charge, μ the electron mobility and L the film thickness.

Further increase in voltage gives rise to the trapping quadratic dependence:

$$J \approx \theta \varepsilon \varepsilon_0 \mu \frac{U^2}{L^3},$$ (2)

where ε is the relative permittivity, ε_0 the permittivity of vacuum and θ the value defining the trap-filling degree.

A practically vertical region follows the quadratic region. It is interpreted as the voltage ($U_2 = U_{TFL}$) of limiting trap-filling. The mode of limiting trap filling is a consequence of the crossing by the quasi-Fermi level of the range of trapping states in the polymer band gap. As a result, the quantity of the injected charge carriers in the polymer sharply increases, which gives rise to a considerable increase in the current. At higher voltages, the *I-V* characteristic is normally described by the square-law (Fig. 2):

$$J \approx \varepsilon \varepsilon_0 \mu \frac{U^2}{L^3}.$$ (3)

The thickness dependence of a current was measured. It was revealed that $I = f(L^{-3})$. Such dependence is in a good correspondence with the equations of injection model.

Table 1 shows the values of voltages U_1, U_{TFL} and basic parameters of the SCLC model: μ - the electron mobility, n_0 – the concentration of the thermally generated equilibrium electrons, $p_{t,0}$ – the concentration of the empty trapping states (this concentration is proportional to their total concentration N_t, which depends on the applied uniaxial pressure). For the estimations, the following relations were used[16,17] :

$$n_0 \approx \frac{\theta \varepsilon \varepsilon_0 U_1}{eL^2} \; ; \; \mu = \frac{JL^3}{\theta \varepsilon \varepsilon_0 U_1^2} \; ; \; p_{t,0} = \frac{\varepsilon \varepsilon_0 U_{TFL}}{eL^2} \tag{4}$$

Table 1. Influence of uniaxial pressure on some parameters of the injection model.

| P | $\mu \cdot 10^{-9}$ | $n_0 \cdot 10^{20}$ | U_1 | U_{TFL} | $p_{t,0} \cdot 10^{21}$ | θ |
kPa	m²/ V·s	m^{-3}	V	V	m^{-3}	
780	1.4	3.8	6.1	8.1	1.4	0,36
820	1.6	2.4	5.5	9.1	1.5	0,26
860	1.8	2.2	5.2	10.6	1.8	0,25

It is necessary to note that the conductivity σ in the linear region of the *I-V* characteristic calculated from the current and voltage values coincides within an order of magnitude with the value σ, calculated from the formula

$$\sigma = en_0 \mu$$

From the data shown in the table follows that the increase in pressure leads to the following changes in the parameters of the charge transport: the concentration of the equilibrium charge carriers diminishes, which apparently results from the change of the quasi-Fermi level. The decrease in U_1 value with increasing pressure confirms our suggestion that the deep trapping states begin to play the basic role and their concentration grows. Growth of the charge carrier mobility is observed, which can be explained most likely by an increase in the concentration of the trapping states and decrease in the trap filling degree.

Conclusions

The analysis of the obtained results allows us to make the following preliminary conclusions. An increase in pressure leads to a change in the charge carrier injection into the polymer film. Probably, this results from the destruction of the relevant surface electronic states.[18] The excess charge in the polymer can create deep trapping states by the mechanism described in the paper.[19] It is possible to assume, that an increase in pressure will lead to a growth of electronic states concentration in a band gap and will cause transition of a localization-delocalization type, for example, by the Mott mechanism. Such state in polymer we term, as HCS. As a result of this process in the polymer band gap near the quasi- Fermi level, a narrow trap band can be formed, which underlies the charge carrier transport and gives rise to an increase in their mobility.

[1] A. N. Lachinov, A. Yu. Zherebov, V. M. Kornilov, *ZETF* **1992**, 102,187.
[2] A. N. Lachinov, A. Yu. Zherebov, V. M. Kornilov, *Pisma ZETF* **1990**, 52, 742
[3] A. N. Lachinov, A. Yu. Zherebov A.Yu., M.G.Zolotukhin, *Synth. Met.* **1993**, 59, 377.
[4] A. N. Ionov, A. N. Lachinov, M. M. Rivkin et al., *Solid State Communs.* **1992**, 82, 609.
[5] B. G. Zykov, V. N. Baydin, Z. Sh. Bayburina et al., *J. Electron Spectrosc. Relat. Phenom.* **1992**, 61, 123.
[6] S. N .Salazkin, S. R. Rafikov, G. A. Tolstikov et al., *Dokl. Akad. Nauk SSSR* **1982**, 262, 355.
[7] A. V. Krayev, S. G. Smirnova, L. N. Grigorov, *Vysokomol. Soedin. Ser. A* **1993**, 35, 1308.
[8] A. M. Elyashevich, A. N. Ionov, V. M. Tuchkevich et al., *Pisma ZETF* **1997**, 23, 8.
[9] S. Tagmouti, A. Oueriagli, A. Outzourhit et al. *Synth. Met.* **1997**, 88, 109.
[10] P. R. Somani, R. Marimuthu , A. B. Mandale *Polymer* **2001**, 42, 2991.
[11] P. R. Somani, D.P. Amalnerkar, S. Radhakrishnan *Synth. Met.* **1998**, 110, 181.
[12] J. R. Rasmusson, Th. Kugler, R. Erlandsson R. et al., *Synth. Met.* **1996**, 76, 195.
[13] A. N. Lachinov, A. Yu. Zherebov A.Yu., M.G.Zolotukhin et al. , *Pisma ZETF* **1986**, 44, 6.
[14] A.Yu. Zherebov, A. N. Lachinov, *Synth. Met.* **1991**, 44, 99.
[15] O.A. Skaldin, A.Yu. Zherebov, V. A. Delev, et al., *Pisma ZETF* **1990**, 51, 141.
[16] M. A. Lampert, P. Mark, *Current Injection in Solids*, Academic Press, New York **1970**.
[17] K. C. Kao, W. Hwang, *Electrical Transport in Solids*, Pergamon Press **1981**.
[18] Yu. A. Berlin, S. I. Beshenko, V. A. Zhorin, et al., *Dokl. Akad. Nauk SSSR* **1981**, 1386.
[19] C. B. Duke, T. J. Fabish, *Phys. Rev. Lett.* **1976**, 37, 1075.

Coherent Electronic Transport through the Single-Walled Carbon Nanotubes

W. Iwo Babiaczyk, Bogdan R. Bułka*

Institute of Molecular Physics, Polish Academy of Science, Smoluchowskiego 17/19, 60-179 Poznań, Poland

Summary: We present the results for coherent electronic transport through the single walled carbon nanotubes. A large value of conductance is obtained for strong coupling to the electrodes, which is close to the ideal transmission of $4e^2/h$ as in experiment. We also consider the system with ferromagnetic electrodes and analyze in detail conductance for separated channels in the coherent regime.

Keywords: carbon nanotubes; coherent transport; quantum conductance

Introduction

Electronic transport through carbon nanotubes (CNT) has recently became an objective of experiment [1-3] due to the technological progress in nanofabrication. It has been known that changing the length of the nanotube, one can change the character of the electronic transport. However, it became evident that the mode (strength) of coupling of the CNT to the electrodes is crucial for the transport properties. When the quality of the contacts is not good, and thus the coupling is weak, the resistances on the junctions are large, much exceeding 13 kΩ, and the transport exhibits an incoherent single-electron tunneling character with the Coulomb blockade effect.[1] If one improves the quality of the contacts, the resistances decrease, which results in higher values of the conductance of the system and one can observe the Kondo resonance, which is due to the exchange interactions between the conducting electrons and an uncompensated localized spin on the CNT.[2] Finally, for very good contacts, one achieves strong coupling to the electrodes and thus the coherent transport character, where the main role is assigned to the multiple scattering on the contacts, which leads to the Fabry-Perot interference.[3] The spin-dependent case measurements have been already performed on multiwalled CNT;[4] however, despite small resistance in that system, it is not possible to observe interference processes and it is very difficult to specify the mechanism of magnetoresistance.

The objective of our work is to determine the coherent transport properties through the system of

DOI: 10.1002/masy.200450848

a single-walled carbon nanotube (SWCNT) connected to metallic electrodes. We calculate the conductance of the systems with Au and Fe (ferromagnetic) electrodes. We show that the main contribution to the conductance spectrum should be ascribed to the superposition of the conducting channels rather than to the interference processes both in the paramagnetic and ferromagnetic systems.

Model description

We consider the system of the SWCNT of the armchair type [5,5], which has 5 carbon rings in the circumference. The dispersion relations of such SWCNT show metallic character with two crossing bands, which can be derived from the bonding (π) and antibonding (π^*) orbitals between neighboring carbon atoms. For the infinite SWCNT the dispersion curves are described by the relation[5]

$$E_m^a(k) = \pm\, t_{C-C}\left\{1 \pm 4\cos\left(\frac{m\pi}{5}\right)\cos\left(\frac{ka}{2}\right) + 4\cos^2\left(\frac{ka}{2}\right)\right\}^{\frac{1}{2}}, \tag{1}$$

where k is the wavevector along the SWCNT, $m = 5$, and a is the lattice constant. The curves (1) have equal slopes of opposite signs. One can say that the transport has an electronic as well as hole character. The Fermi level is assumed as $E_F = 0$ for the neutral SWCNT, and its position can be shifted by applying the gate potential. In the case of the nanotube of length L shorter than the electron-wave coherence length L_{coh}, the standing wave corresponding to $k_n \approx \dfrac{\pi n}{L}$ appears.

To determine the current through the system, we use the tight binding method, within which the Hamiltonian is expressed as

$$H = \sum_{ik,\alpha,\sigma}\varepsilon_{ik\alpha\sigma}c^+_{ik\alpha\sigma}c_{ik\alpha\sigma} + \sum_{ij,\sigma}t_{C-C}\left(c^+_{i\sigma}c_{j\sigma} + h.c.\right) + \sum_{\alpha i,\sigma}t_{\alpha i}\left(c^+_{ki\alpha\sigma}c_{i\sigma} + h.c.\right) \tag{2}$$

where the first term describes the electrons in the electrodes (α = L,R), the second one describes the electrons on the SWCNT, and the last one corresponds to the tunneling between the electrodes and SWCNT. The hopping integrals for C-C bonds are assumed as $t_0 = -2.5$ eV as it is accepted in the literature. The first and the last row of carbon atoms are connected with the electrodes, which are treated as ideal reservoirs. The hopping integral t between the electrodes and the SWCNT is treated as a parameter. Next, we apply the non-equilibrium Green functions in terms of which we express the current as

$$J = \frac{2e}{h} \sum_\sigma \frac{4}{\pi^2 \rho_{L\sigma}\rho_{R\sigma}} \int d\omega \; [f_L(\omega) - f_R(\omega)] \sum_{i,j} \left| G^r_{L\,i,R\,j} \right|^2 , \tag{3}$$

where $G^r_{L\,i,R\,j}$ is the retarded Green function connecting the channels in the electrodes, $\rho_{L\sigma}$ and $\rho_{R\sigma}$ are densities of states for electrons with the spin σ at the Fermi energy in the left ($\alpha = L$) and right ($\alpha = R$) electrode, respectively, and $f_\alpha(\omega)$ is the Fermi distribution function for electrons in the α electrode. The Green functions are obtained by numerical inversion of the matrix $\left| H - \omega \, I \right|$, where the bare Green functions in the electrodes $g^a_{\alpha\sigma} = i\pi\rho_{\alpha\sigma}$. The conductance of the system is given by $G = \left. \dfrac{dJ}{dV} \right|_{V \to 0}$.

Results

Conductance has been calculated numerically for the SWCNT of length $L = 90$ of atomic layers. Figure 1 presents the results for various couplings t, versus the incident electron energy E_F. In the weak coupling regime, G shows sharp resonant peaks, the positions of which correspond to the energies of standing electron waves. The theoretical maximal value of G for one conducting channel (energies lying on one of the branches) is $2e^2/h$. It can occur, however, that the energies on both branches lie close to each other, which causes G peaks to overlap (see the outer peaks on the right and left sides of the plot). With an increase in the coupling t, the width of the peaks grows, and the peaks change their positions. One can see the conductance exceeding the value of $2e^2/h$. In certain cases it can achieve the maximal value $4e^2/h$. This is the case of ideal transmission through the electron and hole channel simultaneously. We have made an effort to match the coupling parameter t to obtain the conductance level similar to the experimental measurements of $G \approx 3.3 \; e^2/h$. We have achieved it for $t = 1.3$ eV, which is shown by the solid line in Fig. 1. The curve resembles the experimental one.[3] One can observe characteristic oscillations corresponding to standing electron waves. In the central part of the plot ($E \approx 0$), the conductance shows a fine structure. Similar features were observed experimentally,[3] and at this point we are in agreement. However, our interpretation of the correspondence of the character of theoretical and experimental curves is different. Liang et al.[3] have assumed the interference effect like in the Fabry-Perot etalon. In our opinion there is a superposition of the conductance for both channels only. Electron waves from both channels are of different symmetry and any matrix element between them is very small, which cannot lead to the interference.

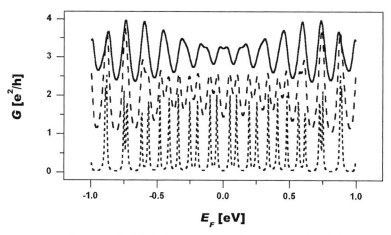

Figure 1. Conductance vs. incident electron energy in the Au-SWCNT-Au system for different couplings values. The solid curve corresponds to the strongest coupling, where the hopping integral is assumed as $t = 1.3$ eV, dashed line $t = 0.9$ eV, and short-dashed line $t = 0.4$ eV. The total density of states for gold at the Fermi level is taken as $\rho_L = \rho_R = 0.294$ states/eV.[6]

The above example has encouraged us further to investigate other systems. The system with ferromagnetic metallic electrodes seems to be interesting. In this case we have different channels for the electrons with spin $\sigma = \uparrow$ and $\sigma = \downarrow$. In such system, the conducting channels for the opposite spin orientations are different. Now, G is a superposition of four components. In Fig. 2, we present the conductance G_P and G_{AP} for the parallel (P) and antiparallel (AP) orientations of polarization, respectively, and $G_{P\uparrow}$ and $G_{P\downarrow}$ corresponding to the electrons with spin $\sigma = \uparrow$ and $\sigma = \downarrow$ in the P configuration. The densities of states of ferromagnetic Fe were determined by the band structure calculations performed using the tight binding version of the linear muffin-tin orbital method in the atomic sphere approximation.[7] The density of states of ferromagnetic Fe for the majority electrons with spin $\sigma = \uparrow$ is $\rho_{Fe\uparrow} = 1.0275$ states/eV, and for the minority electrons $\rho_{Fe\downarrow} = 0.2631$ states/eV. In the parallel configuration, the $G_{P\uparrow}$ curve shows a bit smeared peaks, but still of resonant character of average value about 1 e^2/h, while $G_{P\downarrow}$ shows broad peaks at the level of 2 e^2/h, due to low density of states. In the antiparallel configuration, $G_{AP\uparrow} = G_{AP\downarrow}$ achieves the value slightly lower than 1 e^2/h, i.e. the total conductance G_{AP} takes values below 2 e^2/h. Magnetoresistance, MR = $(G_P - G_{AP})/G_P$, defined as the relative difference of the conductance in the P and AP configuration of spin polarization in the electrodes, shows large oscillations at energies corresponding to the standing waves. The maximal value of MR exceeds 50 %.

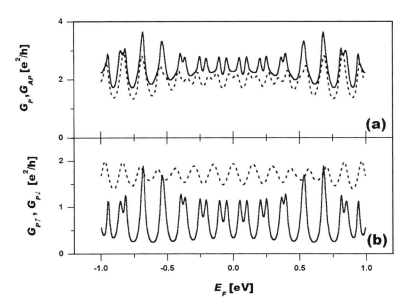

Figure 2. Conductance vs. incident electron energy in the Fe-SWCNT-Fe system. (a) Comparison of G_P (solid line) and G_{AP} (dashed line) for the parallel and antiparallel configuration of the polarization in the electrodes. (b) G_P for separated channels of the electrons with spin $\uparrow$ (solid line) and $\downarrow$ (dashed line). The parameters are as follows: $t = 1.9$ eV, $\rho_{Fe\uparrow} = 1.0275$ states/eV, and $\rho_{Fe\downarrow} = 0.2631$ states/eV.[7]

Conclusions

In our work, calculations of the conductance in the system of SWCNT connected to metallic electrodes have been presented. We have considered the cases of the system with paramagnetic and ferromagnetic electrodes. The obtained results indicate the leading role of the superposition of conductance for different channels and the negligible role of the interference processes between the modes. The method seems credible, as the obtained curves reflect experimental features.[3] We have also calculated magnetoresistance of the system, which shows large oscillations and achieves large values that can also be negative in some points. We believe the experiments on spin-dependent coherent transport through SWCNT can be realized and our theoretical predictions will be verified.

Acknowledgements

We would like to thank Dr. Andrzej Szajek for ab-initio calculations of the density of states for ferromagnetic metals. The work was supported by the Committee for Scientific Research (KBN) under grant No.2 P03B 087 19.

[1] M. Bockrath, D. H. Cobden, P. L. McEuen, N. G. Chopra, A. Zettl, A. Thess and R. E. Smalley, *Science* **1997**, *275*, 1922; S. Tans, M. H. Devoret, R. J. A. Groeneveld and C. Dekker, *Nature* **1998**, *394*, 761.

[2] J. Nygard, D. H. Cobden and P. E. Lindelof, *Nature* **2000**, *408*, 342; W. J. Liang, M. Bockrath and H. Park, *Phys. Rev. Lett.* **2002**, *88*, 126801.

[3] W. J. Liang, M. Bockrath, D. Bozovic, J. H. Hafner, M. Tinkham and H. Park, *Nature* **2001**, *411*, 665.

[4] K. Tsukagoshi, B. W. Alphenhaar, H. Ago, *Nature* **1999**, *401*, 572.

[5] R. Saito, M. Fujita, G. Dresselhaus and M. S. Dresselhaus, *Phys. Rev. B* **1992**, *46*, 1804.

[6] J. M. Seminario, C. E. De La Cruz, P. A. Derosa, *J. Am. Chem. Soc.* **2001**, *123*, 5616.

[7] A. Szajek, unpublished; G. Krier, O. Jepsen, A. Burkhardt, O. K. Andersen, *The TB-LMTO-ASA Program*, source code version 4.7.

Photochromic Liquid Crystalline Structures Containing Azobenzene Moieties

Martin Studenovský,[1] *Zdenka Sedláková,*[1] *Geng Wang,**[1] *Stanislav Nešpůrek,*[1] *Krzysztof Janus,*[2] *Olexandr P. Boiko,*[3] *Francois Kajzar*[4]

[1] Institute of Macromolecular Chemistry, Academy of Sciences of the Czech Republic, Heyrovský Sq. 2, 162 06 Prague, Czech Republic
[2] Institute of Physical and Theoretical Chemistry, Technical University of Wroclaw, Wyb. Wyspianskiego 27, 50-370 Wroclaw, Poland
[3] Institute of Physics, National Academy of Sciences of Ukraine, Prospekt Nauki 46, 03650 Kyiv, Ukraine
[4] Commissariat à l'Energie Atomique, DRT–LIST, DECS/SEMM/LCO, CEA/Saclay, 91191 Gif-sur-Yvette, France

Summary: Novel liquid crystalline photochromic materials of the type 4–R–C$_6$H$_4$–N=N–C$_6$H$_4$–O(CH$_2$)$_n$–N(CH$_2$CH$_2$OH)$_2$, where R is NO$_2$, H, CN, O–*n*-C$_8$H$_{17}$, phenyl, 4–O$_2$NC$_6$H$_4$, were prepared. Some of them are photoconductive. These materials were used for the preparation of light-sensitive polymers in which the photoactive moieties were attached to polyurethane chain. Photochromism of these compounds is based on trans-cis isomerization of azobenzene group. An example of the photochromic activity is presented on solid solution of one material (R = O–*n*-C$_8$H$_{17}$, *n* = 5) in poly(methyl methacrylate) matrix.

Keywords: azobenzene; liquid crystal; photochromism; photoconductivity

Introduction

With the increasing demand for novel devices with optical applications, the search for new materials for data storage becomes a priority. There is a strong interest in this field to find new polymer materials containing photochromic molecules[1], which occupy a prominent position, or functional chromospheres[2]. The azobenzene chromospheres have been widely studied for novel technological applications because their molecules have the ability to photoisomerize[3,4] and the azobenzene moieties can easily undergo photoinduced anisotropy. When exposed to a specific wavelength of polarized light, a reversible anisotropic trans-cis isomerization occurs, which is associated with orientational redistribution of the chromosphere[5]. By irradiating an interference pattern with an appropriate laser light, a surface relief grating can be formed [6] on an azobenzene-

© 2004 International Union of Pure and Applied Chemistry

DOI: 10.1002/masy.200450849

containing thin film. The existence of these effects has stimulated the synthesis of azobenzene copolymers, comonomers of styrene, methyl methacrylate and methyl acrylate, and the preparation of homopolymer blends with azobenzene derivatives[7-9]. Among them, liquid-crystalline structures are expected to show interesting properties.

Experimental

Starting hydroxyazobenzenes **II** (see Scheme 1) were obtained by diazotation and azo coupling of corresponding anilines **I** with phenol. Diols **IV** were obtained by heating corresponding

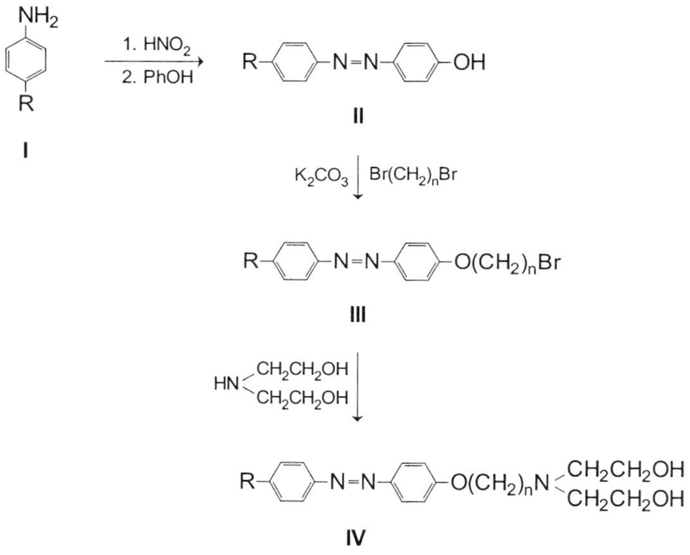

Scheme 1. Synthesis of diols

phenols **II** to 60 °C with five-fold excess of 1,5-dibromopentane or 1,6-dibromohexane in an excess of anhydrous potassium carbonate in acetone. The products were crystallized from acetone and then heated with excess of diethanolamine in isopropyl alcohol at 85 °C. For R = H (L32, L33; Table 1), mesophases do not occur. If R = CN (L46), the nitrile group increases the longitudinal dipole moment of the molecule, which causes higher intermolecular interactions, and a smectic mesophase appears. This mesophase changes to an undefined glassy phase which crystallizes during heating due to a lower viscosity. A longer alkoxy substituent (L41) leads to the expected decrease in the melting point due to a lower dipole moment of the molecule. Introduction of an unsubstituted benzene ring stabilizes the mesophases and a nematic phase appears. The basic characteristics of the synthesized diols **IV** are summarized in Table 1.

The above mentioned diols were used for the preparation of photochromic polymers of the type (e.g., L21 and toluenediisocyanate, TDI).

They were prepared by the reaction of the respective diol **IV** with TDI.

The (L21 – TDI) polymer behaves as an isotropic liquid at 117 °C (during heating) and as an ordered liquid at 120 °C (during cooling). It is photoconductive, the sensitivity parameter S (the ratio of the electrical current of the cell under irradiation (j_{irr}) and in the dark (j_{dark}), $S = j_{irr}/j_{dark}$ was found to be $S = 3.7$.

Absorption spectra of ethanolic solutions of **IV** were measured with a Shimadzu 2401 PC UV-VIS spectrophotometer. Photochromism was studied on thin films of solid solutions of diol IV in poly(methyl methacrylate) (PMMA, M_w 500 000, sample thickness ca. 9 μm) matrix. Samples were irradiated with a xenon discharge lamp 450 W with appropriate filters.

Photoelectrical sensitivity was measured using a Keithley 617 electrometer as a change of the electrical current of the surface photocells with meander-like electrodes (distance 1 mm) at room temperature. The voltage applied to the cell was 200 V, electrode distance 1 mm. The cell was irradiated with a 250-W iodine lamp placed at a distance of 20 cm from the sample.

Results and discussion

The positions of the long-wavelength maxima optical absorption of ethanolic solutions of diols **IV** are given in Table 1 (column 1). All materials absorb light in the region 200 ~ 500 nm. Interestingly, some of the materials show photoconductivity as it follows from column 2. The highest photoconductivity was observed with material L41.

Table 1. Basic characteristics of synthesized diols **IV**

Diol	R	n	1		2		Melting point	Isotropic melt T		Nematic phase T		Smectic phase T
			UV absorption λ (nm)	dark current j_{dark}(A)	sensitivity S		(°C)	(°C)		(°C)		(°C)
L11	O$_2$N	6	373	5.1×10^{-6}	–		102	* 130	*	123	*	68
L21	Ph	6	361	7.6×10^{-5}	–		137	* 146	*	138	*	133
L22	O$_2$N	5	372	8.1×10^{-8}	1.2		90	* 128	*	125	*	69
L32	H	5	345	4.5×10^{-7}	1.1		77	*	–		–	
L33	H	6	345	2.5×10^{-8}	1.2		92	*	–		–	
L41	n-C$_8$H$_{17}$O	5	358	1.7×10^{-7}	1.8		117	* 114	*	52	*	-18
L46	NC	5	360	2.5×10^{-6}	–		103	* 106	–		*	99

Sensitivity S is the ratio of j_{photo}/j_{dark}, where j_{photo} is the current under irradiation.

When the material was irradiated with the light of an appropriate wavelength, cis-trans isomerization was observed. The characteristic spectral changes for diol L41 in PMMA matrix are given in Figure 1. The "dark" spectrum (curve 1) shows a characteristic band with a maximum at 360 nm and a shoulder at 460 nm. Irradiation of the sample with the 360 nm light stimulates the isomerization; the peak at 360 nm decreases and new peaks at 310 and 460 nm appear. The process can be reversed by heat or by irradiation with light of λ = 460 nm (cf. curves 6 - 9 in Figure 1). Full reversibility was obtained by heating to 60 °C (curve 9).

As it follows from Figure 2 the thermal bleaching kinetics taken at different temperatures are not purely exponential. It could be described by a stretched exponential function[10]

$$[M(t)] = [M(0)]\exp\left[-(vt)^\alpha\right],$$ (1)

where $[M(t)]$ and $[M(0)]$ is the concentration of colored species at time t and 0, respectively, v is the decay rate constant, and parameter α ($0 < \alpha < 1$) measures the deviation from the pure exponential behavior. The reason for the dependence (Eq. 1) is that the photochromic reaction is space-demanding process; thus, the kinetics of the photochemical reaction should depend not only on the nature of reacting species but also on their environment. One may, therefore, expect a distribution of the reaction rates.

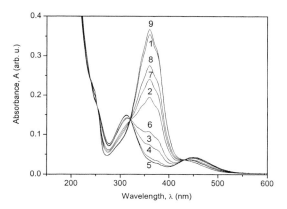

Figure 1. Spectra of L41 in solid PMMA matrix in the dark and after irradiation. Curve 1 – in the dark, curves 2 - 5 – irradiated with light of $\lambda = 360$ nm for 11, 35, 65, and 95 min, respectively; curves 6 - 8 – irradiated with the light of $\lambda = 460$ nm for 10, 40, and 130 min, respectively; curve 9 – after heating to 60 °C.

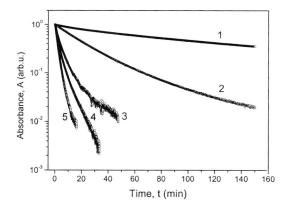

Figure 2. Thermal bleaching kinetics at temperatures 40, 60, 80, 85, and 90 °C (curves 1-5, respectively).

It is not simple to put forward a method allowing to extract information about the rate constant and parameters of its energetic distribution from the shape of the decay curves. A possible method[11], which can be directly employed for measured decays, consists in examining the time dependence of the absorbance × time product: the maximum of the resulting function is directly related to the time constant of the kinetic process, its width to the distribution of the activation energies. The curves replotted from the photochromic decays in Figure 2 are given in Figure 3. The dependences show well-developed maxima. The time constant at 40 °C was $\tau = 2 \times 10^4$ s.

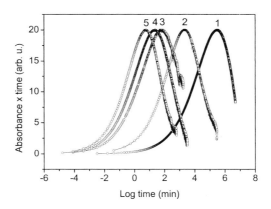

Figure 3. Dependence of the normalized absorbance × time product on logarithm of time.

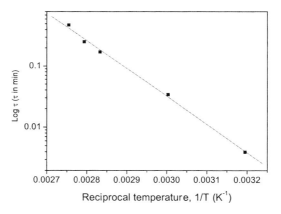

Figure 4. The plot of τ on reciprocal temperature.

The plot of τ vs. reciprocal temperature is linear (cf. Figure 4). The found activation energy of the thermal bleaching process was 87 kJ/mol, the frequency factor $\nu = 2 \times 10^{11}$ s^{-1}.

Conclusion

Novel liquid crystalline photochromic diols of the type $4\text{-R-}C_6H_4\text{-N=N-}C_6H_4\text{-O(CH}_2)_n\text{-}$ $N(CH_2CH_2OH)_2$, where R is NO_2, H, CN, $O\text{-}n\text{-}C_8H_{17}$, phenyl, $4\text{-}O_2NC_6H_4$, were prepared. Their photochromism is based on trans-cis isomerization of the azobenzene group. Typical photochromic parameters for solid solution of the above mentioned material (R = $O\text{-}n\text{-}C_8H_{17}$, n = 5) in poly(methyl methacrylate) were: bleaching time constant at 40 °C, $\tau = 2 \times 10^4$ s, activation energy of the thermal bleaching process, $E_a = 87$ kJ/mol, frequency factor, $v = 2 \times 10^4$ s^{-1}, maximum of UV absorption in the dark, $\lambda = 360$ nm, characteristic absorption bands in the metastable form, $\lambda = 310$ nm and 460 nm. The photochromic process is fully reversible. The above mentioned diols can be used for the preparation of liquid-crystalline, photochromic, and photoconductive polymers by reaction with diisocyanates.

Acknowledgements

Financial support of the Grant Agency of the Academy of Sciences of the Czech Republic (grant AV0Z4050913) and the Ministry of Education, Youth and Sports of the Czech Republic (COST grant OC D14.30) is gratefully appreciated.

[1] S. Hvilsted, F. Andruzzi, C. Kulinna, H. W. Siesler, P. S. Ramanujan, *Macromolecules,* **1995**, 28, 2172.
[2] J. C. Crano and R. J. Guglielmetti (Eds.), Organic Photochromic and Thermochromic Compounds, Plenum, New York, **1999**.
[3] J. Sworakowski, K. Janus, S. Nešpůrek, *IEEE Trans. Dielectr. Electr. Insul.* **2001**, 8, 543.
[4] S. Tripathy, D. Y. Kim, L. Li, J. Kumar, *Pure Appl. Chem.* **1998**, 70, 1267.
[5] N. K. Viswanathan, D. Y. Kim, S. Bian, J. Williams, W. Liu, L. Li, L. Samuelson, J. Kumar, S. Tripathy, *J. Mater. Chem.* **1999**, 9, 1941.
[6] L. Sharma, T. Matsuoka, T. Kimura, H. Matsuda, *Polym. Adv. Technol.* **2002**, 13, 481.
[7] D. Brown, A. Natansohn, P. Rochon, *Macromolecules,* **1995**, 28, 6116.
[8] X. Meng, A. Natansohn, C. Barrett, P. Rochon, *Macromolecules,* **1996**, 29, 946.
[9] L. Sharma, T. Kimura, H. Matsuda, *Polym. Adv. Technol.* **2002**, 13, 450.
[10] R. Richert, H. Bässler, *Chem. Phys. Lett.*, **1985**, 116, 302.
[11] J. Sworakowski, S. Nešpůrek, *Chem. Phys. Lett.*, **1998**, 298, 21.

Macromol. Symp. **2004**, *212*, 407-413

Optical and Dielectric Properties of New Azobenzene Copolyethers Embedded in Polymer Matrices

M. Strat,[*1] *G. Strat,*[2] *V. Scutaru,*[3] *S. Gurlui,*[1] *I. Grecu*[2]

[1] Department of Optics and Spectroscopy, "Al. I. Cuza", University of Iasi, Romania
[2] Department of Physics, "Gh. Asachi" Technical University of Iasi, Romania
[3] Department of Electricity and Magnetism "Al. I. Cuza" University, Romania

Summary: A new group of photochromic azobenzene copolyethers was synthesized. Their photochomic and dielectric properties were studied in poly(vinyl alcohol) and poly(methyl methacrylate) matrices.

Keywords: azobenzene; copolyether; dielectric properties; photochromism

Introduction

In a series of previous works, the synthesis of some polyethers by the phase transfer catalysis technique was presented [1-3]. These polymers were obtained by polycondensation of 3,3-bis(chloromethyl)oxetane (BCMO) and various bisphenols, some of them showing a liquid-crystalline (LC) behavior [4]. The presence of azobenzene units in the polymer structure confers on them important properties with practical applications in optics and optoelectronics. If anisotropic films are made of such a polymer, then the molecules of a nematic liquid crystal (which need not contain azobenzene units) that are in contact with the film can be aligned by photoisomerization of the azobenzene units by the action of the linearly polarized Ar$^+$ laser light [4]. It is assumed that the orientation of the liquid-crystal molecules by azobenzene units is mainly the result of specific molecular interactions between the photocromic units from the polymer layer and the liquid crystal molecules. These specific interactions depend on both the structure of the molecules and the area occupied by the azo units in the polymer surface layer.

In this work, copolyethers containing azobenzene groups in their main chains are studied. Because the copolymer is a thermotropic LC material with thermal transitions situated around 210 °C, it is impossible to study the photoisomerisomerism in the LC state (cis-trans transitions are thermally activated). As a consequence, the copolymer was incorporated in a matrix film.

 DOI: 10.1002/masy.200450850

Thus the photochromic behavior can be studied at room temperature. Therefore, the obtaining of potential surface relief by modification of the polymer matrix and azobenzene concentration could be expected.

Copolymer synthesis

The molecular structure of the studied copolyethers are given in Scheme1.

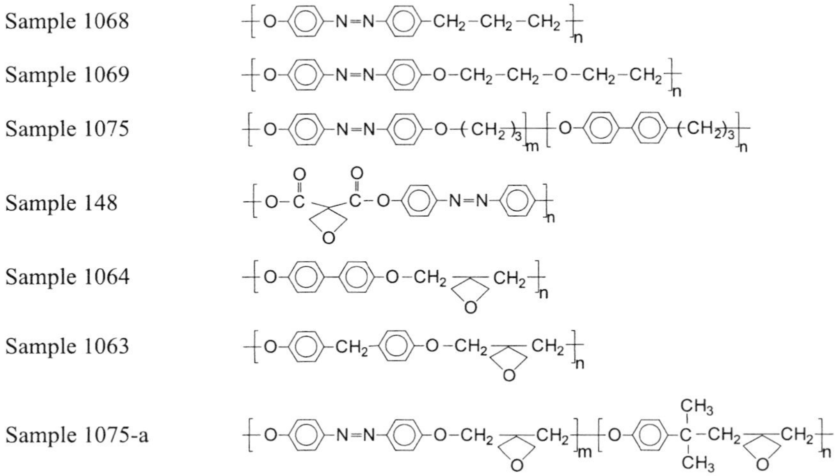

Sample 1068

Sample 1069

Sample 1075

Sample 148

Sample 1064

Sample 1063

Sample 1075-a

Scheme 1

Preparation of PMMA-azo and PVA-azo Films

With the aim to produce homogeneous PVA-based films, the copolymers were dissolved in appropriate quantities of dimethyl sulfoxide and poly(vinyl alcohol) was dissolved in distilled water. In order to obtain homogeneous poly(methyl methacrylate) (PMMA)-based films, the copolymers and PMMA were disolved separately using chloroform as a solvent. The solutions were mixed and homogenized by magnetic stirring. The solutions thus obtained were poured on clean glass surfaces placed horizontally. A special device enables the rotation of the glass substrate at a speed small sufficient to obtain films with quite uniform structure and constant thickness. Chloroform was removed from the films by a special drying procedure. Drawing of the films was performed using a machine at a temperature of about 80 °C (the film was in thermal contact with a metallic cylinder during the stretching process). Films with stretching ratios ranging between 2 and 6 were obtained.

Absorption electronic spectra in natural and polarized light

The behaviour of copolyethers embedded in the PVA matrix was studied using the electronic absorption band at $\bar{\nu}_{max} \approx 27400$ cm^{-1}, a band assigned to benzene rings in azobenzene moieties in trans positions with respect to azo groups.

The spectra in polarized light and the channeled spectrum were recorded with a Specord UV-VIS spectrophotometer (Carl Zeiss Jena). The thickness of the films ranged between 20 and 70 μm. This fact enabled us to use a device, constructed by V. Pop and M. Strat that allowed to obtain polarization and channeled spectra [5,6].

The equipment used consists of a spectrophotometer with two beams and an attachment formed by two optical systems: S_1, placed in the path of the measured beam, and S_2, placed in the path of the reference beam.

The system S_1 is made up of two polarization filters P_1 and P_2 that can be used with their transmission directions perpendicular or parallel. An anisotropic transparent polymer film L (with thickness ≥ 60 μm), for which the birefringence dispersion has to be determined is introduced between P_1 and P_2. In the case of thin polymer films with thickness ranging between ~20 and 60 μm, the attachment also contains an anisotropic transparent plate (stretched polymer or anisotropic crystal), which is placed between P_1 and L [6]. The system S_2 is made up of two polarization filters P_3 and P_4, which compensate the transmission modification introduced by polarizers P_1 and P_2 when they have parallel transmission direction. The parts of the device (Fig. 1) are fixed in settings, which can be rotated around the radiation direction, having the capacity to measure the rotation angle.

The components of the system must be oriented as follows: the polarizers P_1 and P_2 parallel (or crossed) and the film L with its stretching direction at 45° versus the transmission directions of the polarizers. Under these conditions, a channeled spectrum of transmission is obtained. The numbers of maxima and minima in this spectrum are determined by the path length difference between the ordinary and the extraordinary waves introduced by the film L. For obtaining absorption electronic spectra of the film L in polarized light, only polarization filters P_1 and P_3 are needed.

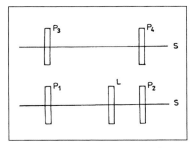

Fig. 1. Scheme of the device for polarization and channeled spectra.

Experimental and Results

i) Photochromic properties

Under the influence of the 27400 cm^{-1} UV radiation, the benzene rings in the azobenzene moieties switched from trans to cis positions. This transition can be rendered evident by spectral methods since the cis photoisomer absorbs at lower wavenumbers ($\tilde{\nu}_{max} \approx 21700\text{-}21300 \ cm^{-1}$).

The assignment and interpretation of the absorption electronic bands has been carried out using existing theories which explain and classify the transitions between molecular electronic states. The absorption bands at around 27400 and 21500 cm^{-1} are associated with the $\pi\text{-}\pi^*$ and $n\text{-}\pi^*$ transitions of the azo aromatic rings, respectively. The results are shown in Figs. 2 and 3.

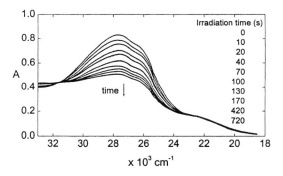

Fig. 2. Electronic absorption spectra of sample 3 in PVA film for different irradiation times.

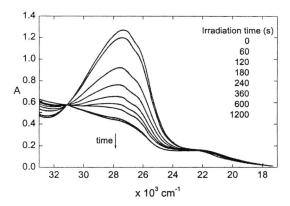

Fig. 3. Electronic absorption spectra of sample 1 in PMMA foil for different irradiation times.

The photochromic effect of these copolyethers embedded in the polymer matrices is reversible. The benzene rings of the azo groups convert from the cis to trans positions under the influence of laser radiation with $\lambda = 514$ nm, produced by an argon laser operating in the continuous regime. The photochromic effect appears as the result of molecular deformation of azobenzene moieties. A general feature of the compounds studied in this work is that the azobenzene moieties are parts of the polymer chain. Under these circumstances, it is more difficult to produce molecular deformation, as compared to other polymer structures in which the azo group is in the side chain.

ii) Polarization spectra

In order to record the spectra in polarized light, an optical device was designed and produced as an auxiliary to a Specord UV-VIS spectrophotometer (Fig. 1). The electronic absorption spectra were recorded by orienting the transmission direction of the polarizer placed in the measured beam both parallel and perpendicular to the sample stretching direction ($P_\parallel$, $P_\perp$). In the path of the reference beam was placed a polarization filter, identical with the one placed in the measured beam, which compensated the transmission modification introduced by the polarizer.

The degrees of polarization were calculated every 50 cm^{-1} all along the existence range of the electronic band, using the relation: $P\,(\%) = \dfrac{A_\parallel - A_\perp}{A_\parallel + A_\perp}100$,

where $A_\parallel$ is the absorbance for $P_\parallel$ and $A_\perp$ - the absorbance for $P_\perp$. By using the values of the polarization degrees, the polarization spectra ($P\,(\%)$, ν) were drawn for all the studied samples.

The birefringence values, Δn, for the stretched samples were determined using a Babinet compensator and the channeled spectra were obtained with the auxiliary device attached to the spectrophotometer (Fig. 1).

Table 1. Characteristic temperatures, order parameter S and birefringence Δn.

Sample	$T_{k\text{-}m}$ (°C)	T_i (°C)	S	Δn
1	210-230	260	0.21	0.015
2	240-250	270	0.15	0.017
3	180	200	0.14	0.016
4	150-190	230	0.22	0.014
5	190-200	212	0.16	0.018

The order parameters, S, of the studied copolyethers were also calculated by using the relation:

$S = \dfrac{A_{\parallel} - A_{\perp}}{A_{\parallel} + 2A_{\perp}}$. The orientation degrees, S, correspond to the substances incorporated in the PVA

matrices. The anisotropy, Δn, results primarily from stretching the matrix. The temperature values (T) at which the mesophase appears in these copolymers are quite high (Table 1), which implies that special attention must be paid to crosslinking, which might occur as the result of breaking oxetane rings at high temperatures (260 – 280 °C) [2].

iii) Dielectric properties. Thermally stimulated polarization current measurements.

The samples investigated from the polarization and depolarization viewpoints were shaped as tablets. The samples were heated at the rate $v = 3 \times 10^{-2}$ K/s in the presence of electrostatic field, $E = 25 \times 10^{5}$ V/m.

The temperature dependence of the polarization current was recorded with a Philips XY recorder. The device also contained a thermostatted furnace, electronic voltmeter, picoampermeter and Al electrodes.

The total current (conductivity and polarization) during the thermal and electric treatment of the sample has one peak at about 330 K for all the studied polymers. The obtained experimental data show that the conductivity current is very low compared with the polarization current. The accuracy of measurements of polarization currents was 10^{-13}A.

The temperature dependence of the polarization current density $J(T)$ is given in Fig. 4. For all the compounds, the dependences are of the same type.

Processing the experimental data using the Gauss function (Eq. 1) has led to a good agreement between the obtained results and the dipolar polarization model using the Langevin function, $L(\beta)$ (Eq. 2) (Fig. 5).

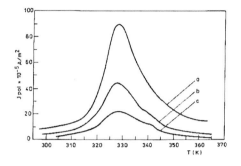

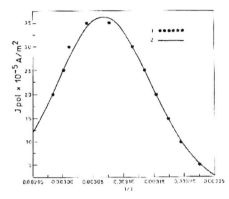

Fig. 4. Thermally stimulated polarization current density for: a 1075 sample, b 1068 sample, c 148 sample.

Fig. 5. Thermally stimulated polarization current density for the 1075 sample (1 experiment, 2 theory).

$$J(1/T) = \frac{A\sqrt{\frac{\pi}{2}}}{w} \exp(\frac{-2(x-x_c)^2}{w^2}) + y_0 \qquad (1)$$

$$L(\beta) = \operatorname{ctgh}\beta - \frac{1}{\beta} \qquad (2)$$

where $\beta = \frac{pE}{kT}$, p is dipole moment, E is electric field. The physical values of the quantities from formula (1) is: $x \sim 1/T$, $x_c = 3 \times 10^{-3}\,\mathrm{K^{-1}}$, $w = 1.568 \times 10^{-4}\,\mathrm{K^{-1}}$; $A = 7 \times 10^{-3}\,\frac{A}{K \cdot m^2}$.

[1] N. Hurduc, V. Bulacovschi and C. I. Simionescu, Eur. Polym. J. 28, 791 (1992).
[2] N. Hurduc, Gh. Surpateanu and V. Bulacovschi, Eur. Polym. J. 28, 1589 (1992).
[3] N. Hurduc, V. Bulacovschi, D. Scutaru, V. Barboiu and C. I. Simionescu, Eur. Polym. J. 29, 1333 (1993).
[4] K. Ikimura, H. Akiyama, N. Ishizuki, Y. Kawanishi, Macromol. Chem. Rapid Commun. 813, (1993).
[5] V. Pop, M. Strat, D. Dorohoi, Romanian Patent 119695 (1985).
[6] V. Pop, D. Dorohoi, E. Cringeanu, J. Macromol. Sci. - Phys. B 33, 373 (1994).

Macromol. Symp. **2004**, *212*, 415-420

Dispersive and Non-Dispersive Hole Transport in Fluorene-Arylamine Copolymers

Dmitry Poplavskyy, Jenny Nelson, Donal D. C. Bradley*

EXSS Group, Blackett Laboratory, Imperial College, Prince Consort Road, London SW7 2BW, United Kingdom

Summary: Hole transport in two fluorene-arylenediamine copolymers is studied by means of time-of-flight (ToF) and space-charge-limited dark-injection (SCL DI) transient current methods. Non-dispersive and dispersive hole transport is observed for a range of sample thicknesses, applied electric fields and temperatures.

Keywords: charge transport; conjugated polymers; polyfluorene; space-charge limited transient current; time-of-flight

Introduction

Poly(9,9-dioctylfluorene) (PFO) and its copolymers are considered to be some of the most promising materials for use in polymeric light-emitting diodes (PLEDs). Recently a new family of PFO copolymers, dioctylfluorene-arylenediamine conjugated polymers, was developed [1]. These materials, which show high hole mobilities reaching $\mu \sim 10^{-3}$ cm^2/Vs combined with low ionisation potentials (~5.0-5.3 eV), have a good potential for applications in PLEDs. Here we present a comparative study of charge transport in two representative polymers (see Fig. 1), poly(9,9-dioctylfluorene-*co-N,N'*-bis(4-butylphenyl)-*N,N'*-diphenyl-1,4-phenylenediamine) (PFB) and poly(9,9-dioctylfluorene-*co-N,N'*-bis(4-butylphenyl)-*N,N'*-diphenylbenzidine) (BFB), which exhibit non-dispersive and dispersive time-of-flight mobilities, respectively [1]. In this work, time-of-flight (ToF) and space-charge-limited dark-injection (SCL DI) transient current methods [2] are used to study the hole mobilities in a range of sample thicknesses, applied electric fields and temperatures. It is shown that the SCL DI method is a useful technique to study charge transport in thin films with thicknesses close to those used in real devices.

Experimental

The samples were prepared by spin-coating a polymer solution in toluene onto an ITO-coated glass substrate (for ToF measurements) or onto poly[(ethylenedioxy)thiophene]/poly(styrenesulfonic acid) (PEDOT:PSS)-pre-coated ITO

                DOI: 10.1002/masy.200450851

substrates (for SCL DI measurements). The Al, Au or Ag top electrodes were thermally evaporated through a shadow mask at a typical pressure of 10^{-6} mbar. A frequency-tripled (λ = 355 nm) Nd:YAG laser provided short light pulses (τ = 6 ns) for the ToF measurements. The samples were housed in an Oxford Instruments cryostat and were kept under helium atmosphere. An HP 8114A pulse generator was used for SCL DI measurements. Both ToF and DI transients were detected with a Tektronix digitising oscilloscope.

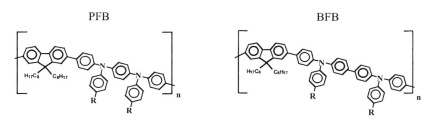

Figure 1. Chemical structures of the copolymers used ($R = n\text{-}C_4H_9$).

Non-Dispersive Hole Transport in PFB

Figure 2 shows typical ToF (a) and SCL DI (b) hole transients in PFB films. It is clear from the ToF transient that the hole transport is non-dispersive (see also [1]), i.e., a distinct current plateau is observed in ToF transients that allows to easily determine the transit time. The DI transient, shown in Fig. 2b, has a characteristic shape with a maximum, as expected for a trap-free insulator with a perfectly injecting (ohmic) contact. The position of the maximum t_{DI} is related to the space-charge-free transit time t_{tr} and, correspondingly, the mobility μ via [2]

$$t_{\mathrm{DI}} = 0.786 \cdot t_{\mathrm{tr}} = 0.786 \cdot d/(\mu E), \tag{1}$$

where d and E are the thickness and applied electric field, respectively.

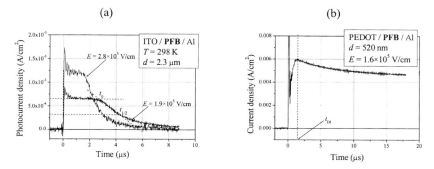

Figure 2. Typical ToF (a) and SCL DI (b) transients in PFB films at room temperature.

The appearance of the SCL maximum also indicates that PEDOT:PSS provides an ohmic contact to PFB. This result is expected from the ionisation potentials of these materials: $I_p =$ 5.2 eV for PEDOT [3] and 5.09 eV for PFB [1].

Figure 3a shows the variation of hole mobilities, determined from the transit time t_0 (as shown in Fig. 2a), as a function of temperature and electric field. We used the Gaussian disorder model (GDM), developed by Bässler and coworkers [4], to analyse the data presented here. The results of this analysis, compared to those for the "building blocks" of PFB, namely homopolymer PFO and small molecule TPD, are shown in Table 1. According to this table, introduction of the arylenediamine blocks into the homopolymer PFO results in a reduced energetic disorder σ, but, on the other hand, the mobility prefactor μ_0 is reduced as well.

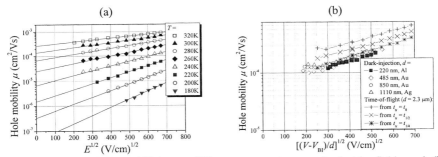

Figure 3. (a) ToF hole mobilities at different temperatures and electric fields and (b) comparison of ToF mobilities with SCL DI mobilities at different sample thicknesses at room temperature in the Poole-Frenkel representation in copolymer PFB.

Table 1. Parameters of the GDM for PFB, compared with those for the homopolymer PFO and triarylamine molecule TPD.

Material	μ_0 (cm^2/Vs)	σ (eV)	Σ	C_0 (cm/V)$^{1/2}$
PFB	0.016	0.085	1.8	$3 \cdot 10^{-4}$
PFO [5]	0.049	0.103	2.7	$2.5 \cdot 10^{-4}$
TPD [6]	0.08	0.077	2.3	$3.7 \cdot 10^{-4}$

Figure 3b presents the comparison of the ToF mobilities with those obtained by the SCL DI method. It is seen that the best agreement is obtained when ToF mobility is calculated using the transit time $t_{1/4}$, i.e. the time at which the photocurrent drops to a quarter of its plateau value (see also Fig. 2a). SCL DI measurements also show that the mobility is independent of sample thickness down to ~200 nm implying that ToF mobilities, measured on thick (~1 μm or more) samples can be used to predict hole mobilities in the thin (~200 nm) samples, which are required for many applications.

Dispersive Hole Transport in BFB

Figure 4a shows a typical ToF hole transient in a 1.3-μm-thick film of BFB coated on an ITO substrate. It is clear that the hole transport in this material is highly dispersive (see also [1]), as the ToF transient exhibits only a weak inflection point, which we assign to the transit time t_{tr} [5,6]. Figure 4b shows a typical dark-injection transient measured on a sample of the same thickness but with an injecting PEDOT electrode. The ionisation potential of BFB ($I_p = 5.26$ eV) well matches that of PEDOT ($I_p = 5.2$ eV), therefore an ohmic contact for hole injection might be expected in such a structure. The observed current transient does not, however, have the characteristic maximum expected from the theoretical predictions for a trap-free insulator [2], and observed in the non-dispersive polymer PFB discussed above. However, a pronounced inflection point in log I - log t presentation is observed. The position of the inflection point scales with the applied voltage and we assign it to the SCL DI transit time t_{DI}, defined in Eq. (1). Hole mobilities calculated from the ToF data agree well with those calculated from these SCL DI data using Eq. (1), measured on samples with $d = 1.3$ μm.

As the thickness of the sample decreases, the experimentally observed SCL DI transients show more distinct features as shown in Fig. 5, where typical transients at the same electric field are compared for samples of different thicknesses. Here the shape of the transient changes from that with an inflection point ($d = 700$ nm and also 1.3 μm; see Fig. 4b) to that with a clear maximum ($d = 420$ and 290 nm). A similar effect is observed with respect to the applied electric field: as the electric field increases, the transit time feature (an inflection point or a maximum) becomes more pronounced.

Such changes of the shape of the SCL DI transients can be well understood assuming the presence of a significant number of traps in the material. As it was shown earlier by Many and Rakavy [7], introduction of traps in a perfect insulator leads to a gradual disappearance of the current maximum in a SCL DI transient as the characteristic trapping time τ becomes comparable with or smaller than the SCL transit time of carriers t_{DI}. As the thickness of the sample decreases, the transit time decreases while the characteristic trapping time remains the same, leading effectively to less trapping. Also, as the electric field grows, the Fermi level rises filling deep traps and reducing the effectiveness of trapping. This results in the same effect, i.e., the transit time feature in the transient becomes more pronounced.

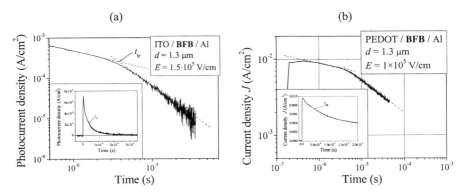

Figure 4. (a) Typical ToF transient and (b) typical dark-injection transient in 1.3 μm thick BFB samples at room temperature.

It has to be noted, however, that the quantitative analysis of the charge mobility is not straightforward in the presence of the strong trapping observed in BFB. Equation (1) that relates the position of the DI peak to the mobility was derived assuming no trapping and therefore should be applied with caution to materials with strong trapping.

Numerical modelling of charge transport in the presence of space-charge in a disordered distribution of energy states with a distribution of trap levels is needed and is currently under way.

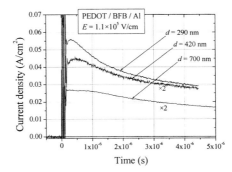

Figure 5. Typical SCL DI transients in BFB samples of different thicknesses at room temperature.

Summary

Hole transport in fluorene-arylenediamine copolymers PFB and BFB was studied by ToF and SCL DI techniques. The Gaussian disorder formalism was succesfully applied to mobility-field-temperature data in PFB yielding hopping transport parameters for this material. It was shown that the hole mobility in PFB measured by the SCL DI method agrees well with that obtained by the ToF method and is independent of sample thickness down to thicknesses of at least 200 nm.

It was shown for BFB that the SCL DI method can be an alternative technique to study charge transport in highly dispersive organic materials. It was demonstrated that the influence of trapping in such experiments becomes less important as the thickness of the sample decreases or the electric field increases.

It is clear that the SCL DI method has advantages over the ToF method in that it can be used in thin films that are representative of real devices; it is also less sensitive to trapping for highly dispersive materials, especially in thin films. The problem could be to find suitable ohmic contacts with which to undertake the experiments.

Acknowledgements

We thank Mark Bernius, Jim O'Brien and Mike Inbasekaran of the Dow Chemical Company for providing the polyfluorenes that we have studied. We also thank the Dow Chemical Company (Polymer Semiconductors) and the UK Engineering and Physical Sciences Research Council for financial support (grants GR/N 34772 and GR/M 45115).

[1] M. Redecker, D. D. C. Bradley, M. Inbasekaran, W. W. Wu, E. P. Woo, *Adv. Mater.* **1999**, 11, 241.
[2] M. A. Lampert, P. Mark, *Current Injection in Solids*, Academic Press, New York 1970, p. 116-121.
[3] T. M. Brown *et al*, *Appl. Phys. Lett.* **1999**, 75, 1679.
[4] H. Bässler, *Phys. Status Solidi (B)* **1993**, 175, 15.
[5] D. Poplavskyy, T. Kreouzis, A. J. Campbell, J. Nelson, D. D. C. Bradley, *Mater. Res. Soc. Symp. Proc.* **2002**, 725, P.1.4.1-11.
[6] M. Stolka, J. F. Yanus, and D. M. Pai, *J. Phys. Chem.* **1984**, 88, 4707; H. H. Fong, K. C. Lun, S. K. Fo, *Chem. Phys. Lett.* **2002**, 353, 407.
[7] A. Many and G. Rakavy, *Phys. Rev.* **1962**, 126, 1980.

Macromol. Symp. **2004**, *212*, 421-426

Studies of Host-Guest Thin Films of Corona-Poled Betaine-Type Polar Molecules by Kelvin Probe Technique and Atomic Force Microscopy

Rorijs Dobulans,[1] *Daiga Cepite,*[1] *Egils Fonavs,*[1] *Inta Muzikante,**[1]
Andrey Tokmakov,[1] *Donats Erts,*[2] *Boris Polakov*[2]

[1] Institute of Physical Energetics, Aizkraukles Str. 21, LV-1006 Riga, Latvia
[2] Institute of Chemical Physics, University of Latvia, Raina Blvd. 19, Riga LV-1586, Latvia

Summary: In this work betaine-type molecules were investigated. As a result of the asymmetry of charge distribution, molecules possess in the ground state a considerable permanent dipole moment. The decay of surface potential of poled polymer films is dependent at least on two relaxation processes. The influence of glass transition of PMMA on thermal dependence of the surface potential is shown. The transition temperature, where no changes of the surface potential appeared, is related to glass transition temperature of the host-guest system. The topography of the film surface was obtained by AFM.

Keywords: AFM; corona poling; host-guest systems; polar dyes; relaxation; surface potential

Introduction

Organic dye-doped materials have received considerable attention because of their large dipole moments and optical nonlinearities. Dipole electrets for photonic applications contain chromophore dipoles, which consist of acceptor and donor groups linked by a bridge of delocalized π-electron system. In this work *N*-(1,3-dioxoindan-2-yl)pyridinium betaine (IPB) molecule was investigated. As a result of the asymmetry of charge distribution, molecules possess in the ground state a considerable permanent dipole moment of 4.5 D[1,2].

In our experiments, host-guest, which have obvious advantages of synthetic simplicity, were chosen. Optically transparent poly(methyl methacrylate) (PMMA) was used as host. The high-electric field corona poling was used to generate dipole orientation of dye molecules in polymer matrices. The topography of the film surface was obtained by

IPB

     DOI: 10.1002/masy.200450852

atomic force microscopy (AFM). In order to investigate the changes of the surface structure, the samples were measured as prepared, after corona poling and after thermal treatment.

Experimental

Isotactic PMMA with molecular weight 440 000 as host was used. IPB was dissolved in PMMA at a concentration of 1 wt %. The glass transition temperature, $T_g = 52 \pm 1$ °C, was obtained by differencial scanning calorimetry. Chloroform was used as solvent to deposit cast films onto gold and ITO electrodes. The thickness of the film was between 0.14 and 1.1 µm.

The high-electric field corona poling was used to generate dipole orientation of IPB molecules in polymer matrix. The corona poling of PMMA and IPB/PMMA films was performed according to Ref.[3]. A high voltage of 4 – 8 kV was applied to a tungsten wire (diameter 25 µm) at the 1 cm distance from the sample surface. During the poling procedure, which continued for 10 min, the corona current was constant, at 1 µA. The poling temperature of IPB/PMMA film was constant in the range 50 – 95 °C. The poling temperature of PMMA film was about 60 °C. The samples were corona-poled through a mask with diameter 1 cm and cooled to room temperature at applied electric field maintaining a constant current.

The changes of the molecular dipole moment of betaine molecules change the surface potential ΔU_S of the host-guest films. The measurements of the surface potential were made by the Kelvin probe technique[4]. The surface potential was measured as the difference between the surface potentials of the grounded bottom electrode and the film. Thermal dependence of the surface potential was measured between 20 and 120 °C. The heating rate was about 8.5 °C/min.

A home-made AFM in the contact mode was used in experiments. The AFM cantilever tips were from standard silicon nitride. The force constant of cantilever, measured using calibrated cantilevers, was around 0.1 and 0.16 N/m for cantilevers with two different dimensions. The typical tip radius was 30 nm.

Topography of polymer films

The topography of the film surface was obtained by AFM. The roughness of the surface was characterized by the amplitude parameter, namely the root-mean-square value S_q. Both

IPB/PMMA and PMMA films exhibited very smooth surface over range 1×1 μm. The evaluated roughness S_q was of ~ 0.4 nm for PMMA and ~ 0.5 nm for IPB/PMMA films (see Table 1). No influence of the substrate on the quality of films was observed.

Table 1. The root-mean-square parameter S_q of as-prepared, poled and annealed PMMA and IPB/PMMA films.

Compound	Substrate	S_q (nm)		
		unpoled	poled	annealed
PMMA	Au	0.4	12.3	0.5
IPB/PMMA	Au	0.5	8	1.8
PMMA	ITO	0.4	3.7	0.6
IPB/PMMA	ITO	0.6	3	1.0

The surface of the corona-poled polymer films drastically changed and the value of S_q increased. In the case of PMMA the structure of the film was with regularly distributed pinholes. The roughness increased several times. In the case of IPB/PMMA films, the increase was lower.[5] As shown in Ref.[6], the corona discharge promoted the rapid diffusion of polar molecules away from the polymer matrix. In our case polar IPB molecules could diffuse to the surface of the film and, due to large permanent dipole moment, the formation of dimers or crystallites of IPB molecules might appear. After annealing above the glass transition temperature, the surface of the films became smooth again.

Surface potential measurements

Surface potential U_s studies provide useful information regarding both structural and electronic properties of oriented films[7,8]. The surface potential of the film depends on both the packing density and orientation of molecules. The film is treated usually as a uniform assembly of molecular dipoles giving rise to polarization of the layer. The surface potential can be related to the dipole moment normal to the plane of the film $\mu_x = \varepsilon \varepsilon_0 A U_s$, where A is the average area occupied by the molecule[8].

The surface potential measurements were performed in the ambient atmosphere mainly on the day after poling. The surface potential of the host-guest film changed going from the unpoled to poled region of the film. In the case of IPB/PMMA films, the surface potential increased from some mV in the unpoled region to several volts in the poled region. The unpoled polymer is isotropic due to random distribution of polar molecules in the polymer matrix. The total dipole moment of the film is close to zero and the value of surface potential is negligible. The increase in the surface potential indicates that orientation of polar molecules is possible in the polymer. For applications, it is important that the highly poled order is maintained for a considerable time. So the efficiency and persistence of alignment and studies of relaxation rates are of considerable interest. As shown in Figure 1, a decay of the surface potential of poled PMMA and IPB/PMMA films has been observed. The decay of the surface potential is well represented by the sum of two exponential terms $U_S = A\exp[-t/\tau_1] + B\exp(-t/\tau_2)$, where τ_1 and τ_2 are the time constants of two relaxation process, a faster and slower one.

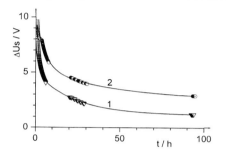

Figure 1. Decay of surface potential of corona-poled PMMA (∇) and IPB/PMMA (o) films fitted with bi-exponential functions.

The initial faster decay of the surface potential in the case of the IPB/PMMA host-guest was $\tau_1 \sim 6$ h. In the case of pure PMMA film, the decay was faster, $\tau_1 \sim 1.4$ h. According to studies of the decay of nonlinear optical properties[9], the initial faster decay of the surface potential may be related to the rotation of other polar molecules in the polymer matrix. At the same time, decay of a surface charge, which arose during corona poling, decreases both the intensity of the SHG signal and the surface potential. The slower process, with $\tau_2 = 43$ h for IPB/PMMA and $\tau_2 = 23$ h for PMMA samples, may be due to thermal relaxation of the polymer matrix. The time constants of the decay of SHG coefficient, d_{33}, was measured in poled host-guest PMMA films. In the case of Disperse Red 1 as guest molecules, the time constants are of the same order $\tau_1 \sim 0.4$ h and $\tau_2 = 10$ h[10].

Temperature dependence of the surface potential of both PMMA and IPB/PMMA poled films was measured in the range from room temperature up to 120 °C. In order to compare the temperature changes of the surface potential, the top vibrating electrode was situated on both films, where the values of surface potential is $\Delta U_S = -1$ V at room temperature. During heating the surface potential $\Delta U_S(T)$ was continuously monitored. As seen from Figure 2, curve 2, the surface potential of IPB/PMMA poled film increased on heating reaching a maximum at ~ 50 ± 5 °C. The maximum around 50 °C is observed also in the $U_S(T)$ dependence of a poled PMMA film (Figure 2, curve 1).

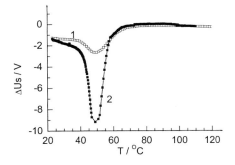

Figure 2. Thermal dependence of the changes of the surface potential of corona poled PMMA (1) and IPB/PMMA (2) films.

The maximum at ~50 °C is associated with glass transition process of the host isotactic PMMA polymer[11,12], where the reorientation of molecular dipoles may take place. In the case of the PMMA film, the changes of the surface potential are comparatively smaller than that of the IPB/PMMA film. After reaching the maximum, the value of the surface potential rapidly decreases with temperature close to zero. In the temperature range 70 – 100 °C, a broad and small peak of the changes of surface potential appeared. At higher temperatures, the signal of surface potential is small and no changes are observed. Above the glass transition temperature the orientation of molecules becomes random and the total dipole moment of the film approaches zero. Similar behavior of the $U_S(T)$ dependence of an IPB/PMMA poled film was observed. In this case, the surface potential approached zero at lower temperatures in comparison with the PMMA film. According to Ref.[12], the dipole moment of PMMA is 1.39 D and interaction with dipoles of IPB molecules may influence the thermal dependence of surface potential of the host-guest film.

As mentioned above, investigation of the glass transition temperature of host-guest films is of great importance both for corona-poling and surface potential measurements. This is especially important in the case of a new host-guest system. Generally, the glass transition temperature of polymer is obtained by differential scanning calorimetry. At the same time, several other experimental methods such as ellipsometry, x-ray reflectivity, second harmonic effect (SHG) characterize the glass transition temperature[6,13,14]. It is shown that the SHG signal delayed and disappeared completely, when the sample is heated above its glass transition temperature[6,14]. As follows from the temperature dependence of the surface potential, the drop-off surface potential signal might be characterized by the glass transition of system.

Conclusions

Corona-poled host-guest films with PMMA as host and N-(1,3-dioxoindan-2-yl)pyridinium betaine as guest show a considerable increase in the surface potential. The increase in surface roughness and formation of grains of the corona-poled films are shown.

An influence of glass transition of PMMA on the thermal dependence of surface potential is observed. The thermal dependence of the surface potential of poled films shows a drop-off of the surface potential at a temperature, which is related to the glass transition temperature of the host-guest system.

Acknowledgments

The authors are grateful to M. Ehlert for the possibility to investigate glass transition temperature of PMMA, B. Stiller and Prof. L. Brehmer for fruitful discussions at Potsdam University. This research was partially supported by project No.19 of the University of Latvia.

[1]   I. Muzikante, E. A. Silinsh, L. Taure, O. Neilands, *Latv. J. Phys. Tech. Sci.* **1998**, 4, 10.
[2]   M. Utinans and O. Neilands, *Proc. SPIE* **1997**, 2968, 13.
[3]   M. A. Mortazavi, A. Knoesen, S. T. Kowel, B. G. Higgins, A. Dienes, *J. Opt. Soc. Am. B* **1989**, 6, 733.
[4]   O. Vilitis, E. Fonavs, I. Muzikante, *Latv. J. Phys. Tech. Sci.* **2001**, 4, 38.
[5]   I. Muzikante, D. Cepite, E. Fonavs, A. Tokmakov, D. Erts, B. Polakov, *11ᵗʰ International Symposium on Electrets ISE-11*, **2002**, Melbourne, Australia, Proceedings, 383.
[6]   M. A. Pauley, H. W. Guan, C. H. Wang, *J. Chem. Phys.* **1996**, 104, 6834.
[7]   D. M. Taylor, *Adv. Colloid Interface Sci.* **2000**, 87, 183.
[8]   O. N. Jr. Oliveira, D. M. Taylor, T. J. Lewis, S. Salvagno, C. J. M. Stirling, *J. Chem. Soc., Faraday Trans. I* **1989**, 85, 1009.
[9]   A. Suzuki, Y. Matsuoka, *J. Appl. Phys.* **1995**, 77, 965.
[10] M. Casalboni, F. Sarcinelli, R. Pizzoferrato, R. D. Amato, A. Furlani, M. V. Russo, *Chem. Phys. Lett.* **2000**, 319, 107.
[11] K. Mazur, *J. Phys. D: Appl. Phys.* **1997**, 30, 1383.
[12] J.-J. Kim, S.-D. Jung and W.-Y. Hwang, *ETRI J.* **1996**, 18, 195.
[13] J. H. Kim, J. Jang, W.-C. Zin, *Langmuir* **2000**, 16, 4064.
[14] W. Shi, C. Fang, Y. Sui, J. Yin, Q. Pan, Q. Gu, D. Xu, H. Wei, H. Hu, J. Yu, *Opt. Commun.* **2000**, 183, 299.

Macromol. Symp. **2004**, *212*, 427-433

Generation of Charge Carrier Pairs in Tetracene Layers

Malgorzata Obarowska,* Ryszard Signerski, Jan Godlewski

Department of Physics of Electronic Phenomena, Gdansk University of Technology, Narutowicza 11/12, 80-952 Gdansk, Poland,
E-mail: mabo@mif.pg.gda/pl

Summary: The mechanism of negative and positive charge carrier generation by light absorption in tetracene layers has been studied. We conclude that there are different processes determining electron and hole production. Positive charge carriers are produced without recombination while the negative charge carrier generation depends strongly on the recombination process. The experimental data for charge carrier generation in tetracene layers are treated theoretically taking into account photogeneration, recombination of charge carriers, trapping and transport processes inside the sample.

Keywords: charge transport; photogeneration; recombination of the charge carriers; tetracene; thin films

Introduction

The charge carriers photogeneration process in polyacene crystals has been studied extensively in many papers, e.g.[1] Particular attention was focused on electron and hole photocurrent production and analysis of the role of trapping in that process was omitted. The study of tetracene has been focused on the analysis of the preliminary charge carrier generation e.g.[2], photoinjection mechanism e.g.[3], the charge-pair separation mechanism, e.g.[4], and description of charge carrier transport phenomena, e.g.[5]

The purpose of this paper is to analyse the electron and hole photogeneration process by studying the photocurrent in tetracene layers (in a sandwich-type arrangement with aluminum electrodes) obtained by ultraviolet light excitation. Particularly, we present experimental results concerning the photocurrent dependence on the wavelength, applied voltage and light intensity. The present study contains theoretical interpretation of the experimental data.

DOI: 10.1002/masy.200450853

Experimental

All measurements were carried out on vacuum-evaporated tetracene layers in a sandwich-type arrangement with Al electrodes. First, tetracene powder was purified by sublimation in a high vacuum. A sample was fabricated by sequential vacuum evaporation of the Al electrode (bottom) - tetracene layer and aluminum as the top electrode onto glass plate. The thickness of the tetracene layer was 1 ± 0.1 μm and the substrate temperature was 300 K. All measurements were performed using an experimental set-up consisting of a monochromator, a xenon lamp producing light from the ultraviolet to visible region and the measurement chamber, where the sample was placed. A picoammeter with internal voltage source was used to measure the current. The entrance slit of the monochromator was adjusted to give the same quantum light intensity at different wavelengths. The thin-layer sample was illuminated through the top Al electrode. Depending on whether the illuminated side had a negative or positive potential, the photocurrents were denoted j^- and j^+, respectively.

Results and discussion

The experimental analysis were based on the measurements of photocurrent as a function of the wavelength, light intensity and applied voltage. The absorption spectrum (A) and the spectral dependences of positive (B) and negative (C) photocurrents are shown in Figure 1. It can be seen that the wavelength dependence of j^+ follows the absorption coefficient (symbatic relationship), in contrast to j^- (antibatic relationship). According to the experimental results presented in Figure 2 (A), the relationship between photocurrent density and the light intensity may be expressed by the formula $j \propto I_0^n$. However, the values of the power of n estimated for j^+ are close to unity whilst are significantly smaller for j^- (Fig. 2 (B)). In Figure 3, the hole and electron photocurrent density as a function of the applied voltage are presented for the values of light wavelength 280, 300, 330 nm and light intensity $I_0 \approx 10^{13}$ photon/cm²s. It is evident that j^- increases more rapidly with the applied voltage than for the hole current j^+.

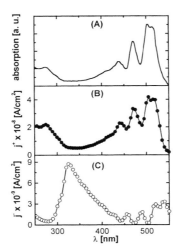

Figure 1 (A) The absorption spectrum of a polycrystalline tetracene film [2]; (B) Spectral dependence of j^+ photocurrent in tetracene layer (positive electrode illuminated); (C) Spectral dependence of j^- photocurrent in tetracene layer (negative electrode illuminated). $I_0 \approx 10^{13}$ photon/cm^{2}s.

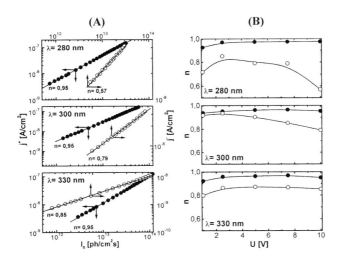

Figure 2. (A) The positive (solid circles) and negative (open circles) photocurrent densities as a function of light intensity (I_0) for tetracene layers at an applied voltage of $U = 10$ V. (B) The values of the power n as a function of applied voltage.

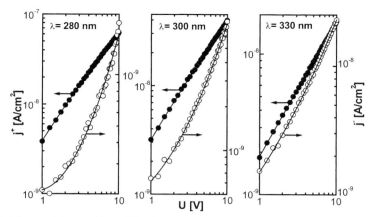

Figure 3. The positive (j^+) (solid circles) and negative (j^-) (open circles) photocurrent densities as a function of applied voltage for a tetracene layer at room temperature and light intensity $I_0 \approx 10^{13}$ photon/cm²s.

On the basis of the above results, it is concluded that the symbatic wavelength dependence of j^+ (Figure 1A) is the result of photogeneration without charge carrier recombination. However, the antibatic wavelength dependence of j^- (Fig. 1 (C)) can be explained as the result of a strong charge carrier recombination at the surface of the electrode. It is interpreted that the negative photocurrent is a result of exciton injection from the bottom electrode (not illuminated) in the long wavelength region ($\lambda > 400$ nm). However, in the short wavelength region ($\lambda < 400$ nm), this process can be hidden as, under the same values of the absorption coefficient, the values of j^- obtained at $\lambda = 330$ nm significantly differ from those obtained for $\lambda = 540$ nm. The negative photocurrent density is generally much smaller than the positive except at 330 nm where the values of density for both currents are similar. This supports the conclusion that the charge carrier generation occurs in the bulk. A theoretical explanation of the obtained results is presented in the next section.

Theoretical model

To describe the voltage and light intensity evolution of the photocurrent, we consider the continuity and the Poisson equations for steady-state condition [6]:

$$\frac{\partial p(x,t)}{\partial t} = G(F,I,x) - \alpha\, n_t(x)p(x) - \frac{1}{e}\frac{dj^+(x)}{dx} = 0 \tag{1}$$

$$\frac{\partial n(x,t)}{\partial t} = G(F,I,x) - \alpha\, p_t(x)n(x) + \frac{1}{e}\frac{dj^-(x)}{dx} = 0 \tag{2}$$

$$j^+(x) = p(x)e\mu_p F(x) \tag{3}$$

$$j^-(x) = n(x)e\mu_n F(x) \tag{4}$$

$$j = j^+ + j^- \tag{5}$$

$$\frac{dF(x)}{dx} = \frac{e}{\varepsilon\varepsilon_0}\left[p_t(x) - n_t(x)\right] \tag{6}$$

$$G(F,x,I) = \Omega(F)I_0\,\kappa\exp(-\kappa x) \tag{7}$$

Here $n(x)$ and $p(x)$ are concentrations of free electrons and holes, respectively, as a function of distance from the electrode x; $n_t(x)$, $p_t(x)$ are concentrations of trapped electrons and holes; $G(F, x, I)$ is the generation term of charge carriers, which can be a function of the electric field (F) and light intensity (I); κ is the absorption coefficient; $\Omega(F)$ is the charge carrier dissociation efficiency; e is elementary charge, α is bimolecular recombination rate of charge carriers; $j^-(x)$ and $j^+(x)$ are electron and hole photocurrent densities, respectively; μ_n and μ_p are electron and hole mobilities; ε_0 and ε are the electric permittivities of vacuum and tetracene crystal. In the above equations we assume that the role of the charge carrier diffusion motion is negligible compared with the drift motion.

The exact solution of the set of equations (1-7) is impossible. To find photocurrent dependences on the light intensity and applied voltage, further analysis is made with the following assumptions:

(i) The recombination between trapped electrons and free holes is not very likely in spite of the recombination between trapped holes and free electrons. In accord with our previous discussion, we assume the recombination process between trapped holes and free electrons as a predominant process in tetracene layers. Hence, we can neglect the second term on the right-hand side of Eq. (1). However, the analogous term in Eq. (2) should be taken into account.

(ii) The trap distribution is exponential, which means that the relation between the concentration of trapped and free holes can be represented by the following formula [7]:

$$p_t(x) = \frac{p(x)}{N_{\text{eff}}} \frac{\pi H}{\ell \sin(\pi/\ell)} \left[\frac{A\kappa I_0}{v} \exp(-\kappa x) \right]^{(1/\ell)-1}. \tag{8}$$

In the above equation, N_{eff} is the effective density of states, H is the concentration of traps, A is the quantity transforming the light flux into free carriers, v is the thermal collision factor, and ℓ is the characteristic trap parameter. The equation can be utilized under the assumption that the optical detrapping is the predominant process.

(iii) There are no space charge carriers. According to the assumption, the relationship between the electric field and applied voltage can be presented as $F = U/d$, where d is the thickness of the tetracene layer.

Based on Eqs. (1) and (7), the following expression can be obtained for j^+:

$$j^+ = e\,\Omega(F)I_0[1 - \exp(-\kappa d)]. \tag{9}$$

It is noteworthy that the relation between the positive photocurrent and the applied voltage is rather complicated and but can be determined in terms of $\Omega(F)$, which can be described in the framework of the Onsager model.

An analogous relationship for the negative photocurrent may be obtained from the set of equations (1-8) taking into account the above assumptions:

$$j^- = \frac{C}{B\,d^2} U^2 I_0^{\left(1-\frac{1}{\ell}\right)} \left\{ 1 - \exp\left[\frac{B}{\kappa d^2} \frac{I_0^{\left(\frac{1}{\ell}\right)}}{U^2} (\exp(-\kappa d) - 1) \right] \right\}, \tag{10}$$

where

$$C = e\,\Omega(F)\kappa, \qquad B = \frac{\alpha\,\pi\,H\,\Omega(F)}{\mu_p\,\mu_n\,N_{\text{eff}}\,\ell\,\sin(\pi/\ell)} \left(\frac{A\kappa}{v} \right)^{\left(\frac{1}{\ell}-1\right)}. \tag{11}$$

From the above equations it follows that the spectral dependence of j^+ is symbatic for $\kappa d < 1$, but the spectral dependence of j^- is antibatic. The relationship between the positive photocurrent density and the applied voltage is determined by the Onsager model. The negative photocurrent density changes more rapidly with voltage than j^+. Both j^+ and j^- photocurrents are functions of the light intensity in the form $j \propto I_0^n$. However, from Eqs. (9) and (10) it follows that the value of the power n for j^+ is equal 1 and for j^-, $n = 1 - (1/\ell)$.

Conclusions

The mechanism of the negative and positive current generation in tetracene layers by light absorption in the ultraviolet region has been studied. From the experimental results it follows that the relation between the current generated by light excitation with a wavelength close to 280 nm and the light intensity is of the type $j \propto I_0^n$ (Fig. 2). For high electric fields the values of coefficient n obtained for the positive current are close to 1 and for the negative current are much smaller and close to 1/2. This remains in a good agreement with predictions of theoretical model (Eqs. 9 and 10) under assumptions that the negative (electron) current is determined by the near-surface recombination and the positive (hole) current avoids that process. The existence of the efficient recombination process between trapped holes and free electrons can be also confirmed by the current dependence on the applied voltage. According to experimental results (Fig. 3), the values of the positive current density are greater than those obtained for the negative current. On the basis of the main experimental results, we can conclude that the positive charge carriers are created directly after light illumination; however, the number of electrons (created inside the sample) depends on the photogeneration yield and the recombination between free electrons and trapped holes.

[1] E. A. Silinsh, H. Inokuchi, *Chem. Phys.* **1991**, 149, 373.
[2] M. Pope, C. E. Swenberg, *Electronic Processes in Organic Crystals and Polymers*, Oxford University Press , New York, 1999.
[3] N. Geacintov, M. Pope, H. Kallmann, *J. Chem. Phys.* **1966**, 45, 2639.
[4] E. A. Silinsh, A. J. Jurgis, *Chem. Phys.* **1985**, 94, 77.
[5] E. A. Silinsh, G. A. Shlihta, A. J. Jurgis, *Chem. Phys.* **1989**, 138, 347.
[6] J. Godlewski, R. Signerski, J. Kalinowski, S. Stizza, *Mol. Cryst. Liq. Cryst.* **1994**, 252, 145.
[7] J. Godlewski, J. Kalinowski, *Phys. Status Solidi (A)*, **1979**, 53, 161.

Macromol. Symp. **2004**, *212*, 435-440

Properties of Optically Addressed Liquid Crystal Spatial Light Modulators Studied by Mach-Zehnder Interferometry

Katarzyna Komorowska,[1] *Andrzej Miniewicz,*[*1] *Janusz Parka*[2]

[1] Institute of Physical and Theoretical Chemistry, Wroclaw University of Technology,
Wybrzeze Wyspianskiego 27, 50-370 Wrocław, Poland

[2] Institute of Technical Physics, Military University of Technology, 00-908 Warszawa, Poland

Summary: In this work optically addressed liquid crystal spatial light modulators are experimentally investigated. Local reorientation of molecules in the modulator and local changes of effective refractive index $n_{eff}(x,y)$ are induced by modulated light intensity $I(x,y)$. We present preliminary results of measurements of phase shifts in optically addressed liquid crystal panels using a Mach-Zehnder interferometer. Experiments were performed for the panels filled with nematic and dye-doped nematic liquid crystals. Optical addressing was realized by Ar+ laser beam ($\lambda = 514.5$ nm) while the reading beam was supplied by He-Ne laser (632.8 nm). The operation voltage was in the range 4 - 20 V. The total phase shift under the influence of addressing light for the studied systems was 2 - 6 π, sensitivity to the addressing light $\sim \mu W/cm^2$ per 2π phase change and speed of response to the light was 20 ms - 30 s with total recovery time 0.5 - 120 s.

Keywords: conducting polymers; liquid-crystalline polymers (LCP)

Introduction

Liquid crystalline materials are attractive for many applications because of their high optical birefringence $\Delta n = n_e - n_o$. Electrically addressed pixelated liquid crystal spatial light modulators (EA LC SLM) are widely used in modern optics [1]. The main drawback of EA LC SLM is that coherent light is diffracted on the well-defined pixelated structure of these modulators limiting optical resolution of system [2]. Optically addressed liquid crystal spatial light modulators (OA LC SLM) were proposed in order to overcome these disadvantages [3]. In such devices phase shift is a function of the externally applied voltage and intensity of addressing light. Spatial controlling and shaping of the wavefront are the main tasks of adaptive optics, for example, a nonlinear Zernike filter wave front sensor based on an OA LC SML has been recently extensively studied [4,5]. The interest in liquid crystalline materials is associated with their suitability for construction of many photonic devices with high speed and high resolution such as optical processors, light amplifiers, etc.

DOI: 10.1002/masy.200450854

The phase shift introduced in OA LC SLM $\delta\phi(x,y)$ under the influence of addressing light can be calculated for extraordinary polarized wave using the formula:

$$\delta\phi\,(x,y) = \frac{1}{L}\int_0^L \frac{2\pi\delta n_e\,(I(x,y),z)}{\lambda}\,\mathrm{d}z, \tag{1}$$

where L - thickness of liquid crystal layer, λ - wavelength of the normally incident light beam, $\delta n_e(I(x,y),\,z)$ - light induced $I(x,y)$ extraordinary refractive index local change: $\delta n_e(z) = n(I(x,y) \neq 0,z) - n(\Gamma(x,y) = 0,z)$. We measure phase shift $\delta\phi(x,y)$ in nematic LC systems employing a Mach-Zehnder interferometer fed with an expanded collimated He-Ne laser beam.

Experimental

The measurements of light induced refractive index changes were performed in two types of planar liquid crystal panels: dye-doped nematic (fourteen-component cyanoester mixture) confined between two transparent ITO/glass electrodes with polyimide layers, and a panel with single photoconducting polymeric layer filled with pure E-7 nematic. Characteristic parameters of the studied LC panels are presented in Table 1.

Table 1. Characteristic parameters of the studied LC panels.

LC Panel	d [μm]	$\Delta\varepsilon$	Δn	Other features
Dye-doped nematic	30	>10	0.34	anthraquinone dye: 1%
Hybrid photoconducting polymer nematic LC structure	10	+13.8	0.225	PVK:TNF 100 nm

d - thickness of LC layer, $\Delta\varepsilon$ - static dielectric anisotropy, Δn - optical birefringence at $\lambda = 589$ nm

The active areas of the presented modulators were 1.5 cm^2. Local phase shifts were monitored at the output of a Mach-Zehnder interferometer (Fig.1).

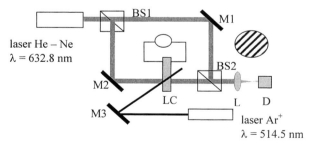

Figure 1. Experimental setup: M1, M2, M3 - mirrors, BS1, BS2 - beamsplitters, LC - liquid crystal modulator placed in one arm of a Mach-Zehnder interferometer, illuminated by addressing beam, D - detector, L - lens.

Optical addressing was realized by monochromatic beam from argon-ion laser (INNOVA 90, coherent, $\lambda = 514.5$ nm) with Gaussian distribution of intensity, extended and filtered to obtain a plane wave. The reading beam was supplied from He-Ne laser (632.8 nm). Operation voltage was in the range 4 - 20 V. All measurements were carried out at room temperature (about 22 °C). Response and recovery times of the samples were measured using a photodiode connected to a digital oscilloscope. Interference patterns showing light-induced phase modulation were captured by a CCD camera connected via a frame grabber to computer.

Results and discussion

The mechanism of light-induced refractive index change in LC modulators is associated with electric-field-driven reorientation of molecules. The internal electric field in LC layer is a superposition of the static field applied externally to the sample and the field induced by the addressing light. In dye-doped LC this external electric field is diminished by a bulk photoconductivity induced by an incoming addressing light [6,7]. In systems with photoconducting polymer layer, the incident spatially modulated light generates the respective surface charge distribution $\rho(x,y)$ in the photoconductor. Then the electric space charge field together with the externally applied static field induces reorientation of liquid crystal molecules [8,9]. The surface space charge field amplifies the externally applied electric field in the bright regions. The dynamic performance of optically addressed LC modulators were measured in a Mach-Zehnder interferometer by monitoring the temporal movement of interference fringes after a sudden opening of the addressing light. The results of recording and decay times for the presented LC modulators are shown in Table 2.

Table 2. Results of recording and decay times for the presented LC modulators.

	Dye-doped nematic	Hybrid photoconducting polymer nematic structure
t_B [s]	3 - 10	30×10^{-3}
t_D [s]	5 - 10	50×10^{-3}

t_R - recording time, t_D - decay time

During measurements we noticed considerable depolarisation of the reading He-Ne laser beam after traversing the LC panel. Especially for externally applied voltage slightly higher than the threshold voltage for the electric Freedericksz transition [10,11], the fringe contrast is decreased. We define the fringe contrast, which depends on external voltage applied to the panel, $C(V)$, by the following formula:

$$C(V) = \frac{I_b(V) - I_d(V)}{I_{max}}$$
(2)

where I_b and I_d - intensities of light in the bright and dark regions of interference pattern respectively, I_{max} - maximum intensity of light in the bright region of interference pattern at $V=0$. In Fig. 2 we present the relation between phase shifts and applied voltages for the described modulators (solid line)

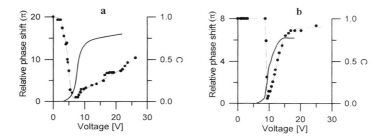

Figure 2. Electrically obtained phase shift (solid line) and contrast C (dashed line) of interference patterns measured for (a) LC modulator filled with dye-doped nematic and (b) hybrid photoconducting polymer nematic LC modulator.

Dashed lines in Fig. 2 represent the changes in interference pattern contrast as function of the applied voltage. In order to determine the usable range of voltages (for $C > 0.8$), we performed our experiments without the addressing light. The interference pattern amplitude $I_b(V) - I_d(V)$ as function of voltage and the phase shift were measured in the same experimental run. Just above the threshold voltage, even a small change of the applied field causes large changes of the effective refractive index (large phase shift in the reading beam). At the same time, the contrast of interference pattern significantly decreases. With a further increase in the voltage, one can observe how the LC undergoes transition to a more ordered state, and the contrast of interference pattern again rises. The spatial light modulator should work within the voltage range where the contrast is high (i.e. $C > 0.8$), so the usable voltage range is limited to several volts. Under this criterion the maximum phase shift under the influence of addressing light for the dye-doped nematic phase modulator is 0.4 π and for the panel with photoconducting polymer layer it is about 1.5 π. The panel filled with dye-doped liquid crystal has a too small interference pattern contrast (less than 0.9) in almost the whole range of voltages, so it is useless as a spatial light modulator. In Fig. 3 we present the dependence of light-induced phase shift on its intensity (λ = 514 nm) in high-contrast voltage regions for the nematic with photoconducting layer.

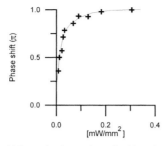

Figure 3. Dependence of phase shift on the intensity of addressing laser beam 514 nm for hybrid photoconducting polymer nematic LC modulator (15 V).

The presence of photoconducting layer allowed to obtain 1π phase shift in reading beam at writing light intensities 0.4 mW/mm^2. In order to demonstrate the phase change abilities of LC spatial phase modulators, we present photographs (Fig. 4) of interference patterns obtained for addressing beam with Gaussian distribution (a,b) of intensity and uniform distribution of intensity (c-f).

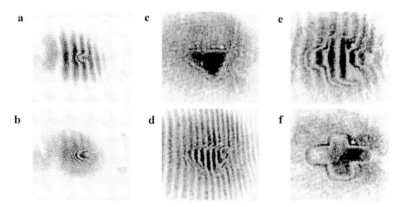

Figure 4. Local changes of phase obtained in nematic liquid crystal modulator with photoconducting layer for different distribution of intensity: (a,b) for Gaussian distribution, (c-f) for plane wave cut by a cross (c,d) and a triangle (e,f) mask.

The experiments were carried out with different values of light intensity and applied voltage. In the first experiments we did not use expanded laser beam (Fig. 4a, b). The phase shift seen by the reading beam can be larger than 2π (a), with the contrast in the illuminated region a little higher than outside it. In Fig. 5a we applied such a value of voltage, for which the contrast was very weak. In the illuminated region the interference pattern is better visible because of a higher value of electric field induced by the addressing light. In Fig. 4 c-f we present results of SLM addressing with expanded laser beam, passing through amplitude masks, a cross and a triangle. Fringe shifts were observed for different values of voltage applied to the LC panel as well as for different conditions of observation in Mach – Zehnder interferometers. The results clearly indicate how the light intensity changes locally the refractive index of LC and the $\Delta n \cdot d$ – changes optical path for the reading plane wave. From the practical point of view, it is necessary to study spatial resolution of the presented systems using different shapes and sizes of amplitude masks.

Conclusions

In this paper we presented preliminary experimental results for optically addressed liquid crystal phase modulators (OA LC SML), which seem to be capable of overcoming some of the limitations of electrically driven phase modulators or conventional phase filters. OA LC SLM offer interesting potentials in this field. No pixelation and high sensitivity allow to create an effective device able to modulate (or correct) the phase of laser beam or even filter it in the Fourier plane. The presented results show a potential for optimization of these systems with respect to their performance as phase modulators.

[1] J. Zmija, S.J. Klosowicz, J. Kedzierski, E. Nowinowski - Kruszelnicki, J. Zielinski, Z. Raszewski, A. Walczak, J. Parka, Optoelectron. Rev. 5 (1997) 93.
[2] G. Moddel, in: U. Efron (Ed.), Spatial Light Modulator Technology: Materials, Devices and Applications, Chap. 6, Marcel Dekker, New York 1994, 287.
[3] K. Komorowska, A. Miniewicz, J. Parka, Synth. Met. 109 (2000) 189.
[4] Gary W. Carhart, Mikhail A. Vorontsov, Eric W. Justh, SPIE 4124 (2000), 138.
[5] Mikhail A. Vorontsov, Eric W. Justh, A. Beresnev, SPIE 4124 (2000), 138.
[6] S. Bartkiewicz, A. Miniewicz, Adv. Mater. Opt. Electron. 6 (1996), 219.
[7] A. Miniewicz, K. Komorowska, O.V. Koval'chuk, J. Vanhanen, J. Sworakowski, M.V. Kurik, Adv. Mater. Opt. Electron. 10 (2000), 55.
[8] G. Pawlik, A. C. Mitus, A. Miniewicz, Opt. Commun., 182 (2000), 249.
[9] N.V. Tabiryan and C. Umeton, J. Opt. Soc. Am. B, 15 (1998), 1912.
[10] I.C. Khoo, Liquid Crystal, Wiley, New York 1995.
[11] L.M. Blinov, V.G. Chigrinov, Electrooptic Effects in Liquid Crystal Materials, Springer, New York 1995.

Polarized Electroluminescence from Bilayer Structures

Geng Wang,[1] *Stanislav Nešpůrek,*[1,2] *Jan Rakušan,*[3] *Marie Karásková,*[3] *František Schauer,*[2] *Ludwig Brehmer*[4]

[1] Institute of Macromolecular Chemistry, Academy of Sciences of the Czech Republic, Heyrovsky Sq. 2, 162 06 Prague 6, Czech Republic
[2] Faculty of Chemistry, Technical University of Brno, Purkyňova 118, 612 00 Brno, Czech Republic
[3] Research Institute of Organic Syntheses 532 18 Pardubice-Rybitví, Czech Republic
[4] Department of Condensed Matter Physics, Institute of Physics, Potsdam University, Am Neuen Palais 10, 14469 Potsdam, Germany

Summary: The photoalignment ability of poly[methyl(phenyl)silylene] (PMPSi) films makes it possible to use them as hole-transporting substrates for the preparation of organic oriented films. A PMPSi layer prepared by spin coating was irradiated, after drying, with linearly polarized UV light. Then, water-soluble hydroxyaluminium phthalocyaninesulfonate $[Al(OH)Pc(SO_3Na)_{1-2}]$ was deposited by casting. The cell $ITO/PMPSi/Al(OH)Pc(SO_3Na)_{1-2}/Al$ showed non-linear current-voltage characteristics. For applied voltages higher than 10 V, polarized electroluminescence was observed. Its spectral characteristic consisted of two peaks with maxima at about 320 and 700 nm; their polarized anisotropies $R_{EL} = \Phi_{||} / \Phi_{\perp}$ were ca. 15 and 0.5, respectively.

Keywords: hydroxyaluminium phthalocyaninedisulfonate; photoalignment; polarized electroluminescence; poly[methyl(phenyl)silylene]

Introduction

Organic light emitting diodes (OLEDs), as one of the most promising candidates for the next generation of flat panel displays, have been extensively studied. Polarized OLEDs can be used in many electro-optical applications. Their fabrication needs an oriented surface for the deposition of an electroluminescent (EL) polymer or a special treatment (orientation) of the EL material. Several attempts have been made to prepare polarized OLEDs devices[1-4]. The development of new rubbing-less technique has been of great interest in the last years. Some of these techniques are based on photosensitive polymers. Their photoaligning ability is determined by the anisotropy of properties induced by light.

Poly[methyl(phenyl)silylene] (PMPSi) with significant electron delocalization along the chain[5] seems to be a suitable substrate material for the preparation of oriented organic films. If irradiated with linearly polarized light, it shows an angular-dependent photoinduced cleavage of Si-Si bonds, formation of polysiloxane structures, and quasi-stable photogenerated ion-pairs (dipoles), preferentially within segments oriented along the light polarization[6]. The organic material deposited in this way on the treated PMPSi surface is oriented[7]. Thus, we can simply prepare the ITO glass/hole-transporting PMPSi film/oriented EL material/ Al electrode cell, which can produce polarized electroluminescence. In this paper we report the fabrication of polarized OLEDs based on bilayer structure, consisting of PMPSi and a water-soluble phthalocyanine.

Experimental

Poly[methyl(phenyl)silylene] (PMPSi), M_w = 4×10^4, was prepared by the Wurtz-coupling polymerization. The low-molecular-weight fraction was extracted with boiling diethyl ether. Hydroxyaluminium phthalocyaninesulfonate [Al(OH)Pc(SO$_3$Na)$_{1-2}$] sodium salt was synthesized by the following method. Hydroxyaluminium phthalocyanine was dissolved in 10 % fuming sulfuric acid and subsequently heated at 85 °C for 6 h. The reaction mixture was poured into a mixture of water and ice. The solid was filtered off, washed, dispersed in water and pH of the dispersion was adjusted with NaOH to ca. 11. Hydroxyaluminium phthalocyaninesulfonate changed into the dark, blue water-soluble sodium salt, which was isolated by evaporation of water on a water bath. The product, analyzed by HPLC, contained 47.3 % monosulfonated HOAlPC, 45.9 % disulfonated HOAlPc, 6.2 % trisulfonated HOAlPC and 0.7 % tetrasulfonated HOAlPc.

Thin (120 nm) PMPSi layers were prepared by spin-coating ITO glass with a toluene solution of PMPSi. Before deposition, PMPSi was centrifuged (12 000 rpm, 15 min). After deposition, the films were dried at 0.1 Pa and 330 K for at least 4 h. Then, the layer was irradiated with polarized UV light for 90 min using a mercury discharge lamp (HBO 100 W). A film of Al(OH)Pc(SO$_3$Na)$_2$ was deposited on the PMPSi layer by casting its water solution. The top Al electrode was vacuum-evaporated. Current-voltage ($j\sim U$) characteristics were measured using a Keithley 6175A electrometer. The emission intensity was measured with a photomultiplier system R928 (Hamamatsu Photonics Co.). The measurements were performed at room temperature. The polarized EL was measured using a polarizer placed between the sample and light detector.

Result and discussion

From the measurements of $j \sim U$ characteristics of the ITO/PMPSi/Al cell (Figure 1) follows that current increases with increasing forward bias voltage, but the reverse bias current remains small; the cell exhibits rectifying characteristics. The emission intensity increases monotonically with increasing current injection. The electroluminescent efficiency was 3×10^{-5} %. The overall EL intensity of the photodegraded PMPSi film was about twice lower than that of the nondegraded film.

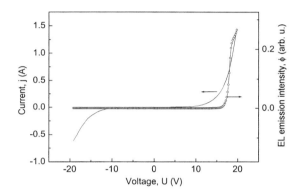

Figure 1. Current-voltage characteristics and dependence of EL emission intensity of the ITO/PMPSi/Al cell on applied voltage at room temperature, (PMPSi layer thickness 110 nm).

EL spectrum (Figure 2), measured at 77 K, nearly independent of the applied voltage, shows peaks at λ_{max} = 354 nm and λ_{max} = 470 nm; the peaks almost coincide with those in the photoluminescence spectrum. The sharp emission band, with its full width at half-maximum, FWHM = 0.15 eV, is narrower than that of the PL spectrum (0.24 eV). The narrow UV band is of excitonic nature associated with the polymer Si backbone.

The visible luminescence is related to the presence of branching points. The fluorescence quantum efficiency of PMPSi in solution was quite high, about 0.15 at 360 nm. The Stokes shift was ca. 19 nm in solution and ca. 26 nm in the solid state.

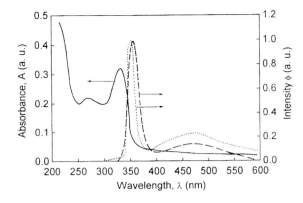

Figure 2. Absorption (solid line), photoluminescence (dashed line) and electroluminescence (dotted line) spectra of PMPSi film.

Irradiation of PMPSi with UV light (~ 350 nm) leads to Si-Si bond scission. The absorption at 340 nm (σ-σ^* transitions) decreases and the maximum is shifted to short wavelengths (Figure 3; degradation was induced by linearly polarized light from an HBO 100 W mercury discharge lamp).

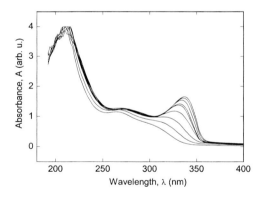

Figure 3. Absorption spectra of PMPSi film after UV photodegradation. The top curve is the spectrum of the non-degraded film; other curves are spectra of gradually photodegraded films (bottom curve – degradation 50 min).

Bond scission preferentially oriented along the light polarization resulted in anisotropic UV absorption and dichroism was also detected[7]. At the same time, polysiloxane structures were detected by IR spectroscopy, and cation radicals by flash photolysis[5]. In the presence of the cleaved Si-Si bonds, which form hole traps ca. 0.45 eV deep[8], quite a long life-time of ion-pairs was detected[9] (up to 2000 s at 160 K).

The anisotropic distribution of species, generated by angular-dependent photoselection, allowed for the preparation of oriented hydroxyaluminium phthalocyaninesulfonate layers. The feasibility of the preparation of oriented structures was preliminarily tested using liquid crystals; a twisted direction distribution was observed. The direction of the easy axis was parallel to the UV light polarization. The $j \sim U$ characteristics of the electroluminescent ITO/PMPSi (photodegraded)/Al(OH)Pc(SO$_3$Na)$_2$/Al cell (Figure 4), shows a linear dependence for the applied voltages $U < 6.4$ V. For higher voltages, the current increases superlinearly. At $U = 10$ V, a strong increase in the current was observed with applied voltage. For the voltages $U > 11.5$ V, a trap-free current was observed ($j \sim U^2$) and polarized electroluminescence was detected.

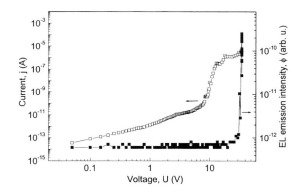

Figure 4. Current-voltage characteristic and electroluminescence emission intensity of the ITO/PMPSi/AlOHPc(SO$_3$Na)$_2$/Al cell at room temperature.

Polarized EL spectra are given in Figure 5. The spectrum is typical of a phthalocyanine layer [10]. The spectral characteristic consists of two peaks with maxima at ca. 320 and 700 nm; their polarizing anisotropy $R_{EL} = \Phi_{||} / \Phi_{\perp}$ is about 15 and 0.5, respectively.

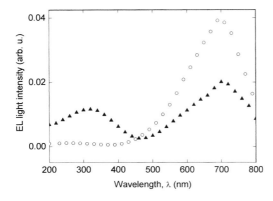

Figure 5. Polarized EL spectra measured at 80 V. The polarizing directions are parallel (solid triangle) or perpendicular (open circle) to the electric vector of the photoaligning light.

Conclusion

Polarized OLEDs of the ITO glass/photodegraded PMPSi/EL material/Al type can be prepared using PMPSi films treated with polarized light. The EL film deposited on the treated PMPSi surface shows an oriented structure, which produces polarized electroluminescence. It seems that the method can be generalized: EL films deposited on the treated PMPSi surface, both from solution and by vacuum evaporation, show polarized EL characteristics. Because PMPSi is a good charge-transporting material, the injection of holes into the diode is not limited.

Acknowledgements

Financial support by grants No. A1050901 (Grant Agency of the Academy of Sciences of the Czech Republic), No. 202/01/0518 (Grant Agency of the Czech Republic), and No. ME440 (Ministry of Education, Youth and Sports of the Czech Republic; KONTAKT) are gratefully appreciated.

[1] P. Dyreklev, M. Berggren, O. Inganas, M. R. Andersson, O. Wennerstrom, T. Hjertberg, *Adv. Mater.* **1995**, 7, 43.
[2] M. Hamaguchi, K. Yoshino, *Appl. Phys. Lett.* **1995**, 67, 3381.
[3] A. Bolognesi, G. Bajo. J. Paloheimo, T. Ostergard, H. Stubb, *Adv. Mater.* **1997**, 9, 121.
[4] M. Grell, W. Knoll, D. Lupo, A. Meisel, T. Miteva, D. Neher, H. G. Nothofer, U. Scherf, A. Yasuda, *Adv. Mater.* **1999**, 11, 671.
[5] S. Nešpůrek, A. Eckhardt, *Polym. Adv. Technol.* **2001**, 12, 427.
[6] A. Dyadyusha, S. Nešpůrek, Yu. Reznikov, A. Kadashchuk, J. Stumpe, B. Sapich, *Mol. Cryst. Liq. Cryst.,* **2001**, 359, 67.
[7] G. Wang, S. Nešpůrek, J. Pospíšil, Y. Kaminorz, J. Stumpe, B. Sapich, *European Conference on Organic Electronics and Related Phenomena* 2001, Potsdam Germany.
[8] A. Kadashchuk, N. Ostapenko, V. Zaika, S. Nešpůrek, *Chem. Phys.* **1998**, 234, 285.
[9] S. Nešpůrek, A. Kadashchuk, Y. Skryshevski, A. Fujii, K. Yoshino, *J. Jpn. Soc. Electr. Mater. Eng.* **2000**, 119, 39.
[10] A. Fujii, M. Yoshida, Y. Ohmori, K. Yoshino, *Jpn. J. Appl. Phys.* **1996**, 35, 37.

Macromol. Symp. **2004**, *212*, 447-454

Temperature- and Humidity-Related Degradation of Conducting Polyaniline Films

Eva Tobolková,[1] *Jan Prokeš,*[1] *Ivo Křivka,*[1] *Miroslava Trchová,*[2] *Jaroslav Stejskal*[2]

[1] Faculty of Mathematics and Physics, Charles University, Ke Karlovu 5, 121 16 Praha 2, Czech Republic
[2] Institute of Macromolecular Chemistry, Academy of Sciences of the Czech Republic, Heyrovského nám. 2, 162 06 Praha 6, Czech Republic

Summary: The time dependence of dc conductivity of conducting polyaniline films was measured in relation to temperature and relative humidity of the environment. Optical and structural properties of the samples were checked using Fourier transform infrared (FTIR) spectroscopy.

Keywords: ageing; conducting polymers; humidity; polyaniline; temperature

Introduction

Polyaniline (PANI) has been reported as a material having one of the best environmental stabilities among conducting polymers[1,2]. However, this is true mainly for its non-conducting form, emeraldine base, while the more interesting conducting form, protonated emeraldine salt, is much less stable. The decay of the electrical conductivity with time under environmental conditions is an undesirable phenomenon, which sets fundamental restrictions to possible technical applications, and, so far, it has not been satisfactorily solved. The variations of conductivity of PANI can be related to both intrinsic and external factors. The former are related to polymer structure through the perfection of quinoid and benzenoid unit repetition, the chain ordering, and the dopant selection. The latter are linked to the conditions of the ageing tests through the effect of the atmosphere and temperature. The decline of the conductivity of protonated PANI appears as a result of combination of these factors, which can interact through different mechanisms (loss of conjugation, chain degradation, deprotonation, oxidation, and crosslinking)[3]. Therefore, a simplification becomes crucial. In this work, our interest is reduced to the effect of the two most common factors, temperature and humidity, on ageing of direct-

DOI: 10.1002/masy.200450856

current conductivity (dc) of PANI.HCl films. An attempt to describe the observed behaviour as a sum of two parallel processes is presented.

Experimental

Preparation of samples. PANI hydrochloride was prepared by the oxidation of aniline hydrochloride (0.20 M) with ammonium peroxodisulfate (0.25 M) in dilute hydrochloric acid[4]. The films were polymerized in situ on substrates immersed in the reaction mixture. Disc-shaped glass supports, 13 mm in diameter, with deposited gold electrodes, were used for dc measurement, while doubleside-polished crystalline silicon substrates were coated for FTIR spectroscopy study.

Properties. The average conductivity of the films polymerized on glass supports was of the order 10 S cm^{-1}. The average film thickness was assumed to be 130 nm, which was determined by Stejskal et al.[4] from the correlation between interferometric and absorption measurements.

Equipment. *The dc conductivity* (σ) was measured by the four-point van der Pauw method[5]. The sample holder was placed in a humidity chamber Heraeus-Vötsch VLK 07/35 operating in temperature range from +5 °C to +90 °C and at relative humidity (RH) from 30 % to 90 %. *Thickness* of several samples was checked on a Surfometer SF200 (Planer Products Ltd.) or Taly Step (Taylor Hobson Taly Step Surface Profiler S/N) and the values obtained were in good agreement with the previous result[4]. *Infrared measurements* were performed on a Nicolet Impact 400 FTIR spectrometer. The transmission spectra in the range $400 - 4000 \text{ cm}^{-1}$ with 2 cm^{-1} spectral resolution were obtained on films deposited on silicon substrates.

Regime. Samples are put into a chamber to following conditions:

1. start-up conditions (25 °C, 50 % RH)   - for 2 h
2. at 25 °C the RH is set to 30 %, 45 %, 60 %, 70 %, 75 %, or 90 %   - for 17 h
3. RH is held at the same value, the temperature is elevated to 90 °C.   - for 85 h

Results and Discussion

Time variations of dc conductivity at constant increased temperature and various relative humidities. Dc conductivity was measured as a function of time at increased temperature, relative humidity was set to $30 - 90$ % according to the regime described above. In this work, the data of the 90 °C series are presented. Time dependence of conductivity of PANI.HCl films aged at 90 °C at four humidities is shown in Figure 1 (the time scale begins just after the temperature 90 °C was stabilized; see Regime/step 3 above). The shape of curves depends remarkably on RH. Initially, a simple decrease in conductivity with time, $\sigma(t)$, is a response to the previous temperature increase. Subsequently, the shape of $\sigma(t)$ strongly depends on RH. It reaches a constant value for lower RH (30 % < RH < 60 %), while for higher RH (70 % < RH < 90 %) it is growing.

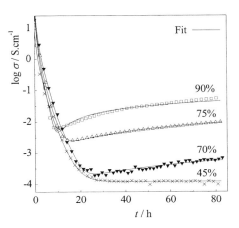

Figure 1. Time dependence of dc conductivity, σ, of PANI.HCl at 90 °C and various relative humidity; the fit (4) to experimental curves $\sigma = \sigma_D \exp\left[-(t/t_D)^{1/2}\right] - \sigma_R \exp\left(-t/t_R\right) + \sigma_\infty$

There are three possible reasons for the initial steep decrease in conductivity mentioned in literature most often[6–9]. These are: (1) loss of moisture, as most of the ageing experiments have been carried out in ambient atmosphere and only the effect of elevated temperature was followed, (2) oxidation due to diffusion of oxygen, (3) deprotonation. As to the morphology, PANI may be considered as a heterogeneous structure consisting of conducting grains (polaronic clusters) embedded in insulating, less ordered regions[10]. In this case the decrease in conductivity can be correlated with a decrease in grain size[7,11,12].

Long-term evolution of σ was studied[7] for another conducting polymer, polypyrrole, which shows similar properties to PANI in many respects. Polypyrrole pellets or coatings on various substrates were annealed at elevated temperature in ambient atmosphere. For $t > 40$ h, the fitting of $\sigma_1(t)$ according to stretched exponential of the ageing time t was successful

$$\sigma_1(t) = \sigma_D \exp\left[-(t/t_D)^{1/2}\right] \qquad (1)$$

where σ_D is an initial dc conductivity value and t_D is the characteristic time of the degradation process. (1) gave satisfactory results from the very beginning of the ageing process and the t_D parameter was used to characterize the stability of the materials under study[7,11,12].

The use of Eq. (1) to our samples brings good results for ageing at low RH, while the subsequent increase in conductivity at higher RH requires introduction of another term, closely related to the presence of water in the material, which would act against the conductivity loss. If we assume that the measured changes of conductivity are attributed to development of the conducting grain size, we can refer to the "variable-size conducting-island" model proposed by Kahol[8]. It is based on the concept that the absorbed water increases the effective size of conducting grains. Assuming in the first approximation that degradation and recovery processes develop independently, we can treat the total conductivity σ (t) of a film as a sum of the final conductivity σ_∞ of the aged film at time tending to infinity and two time-dependent terms

$$\sigma(t) = \sigma_1(t) + \sigma_2(t) + \sigma_\infty \qquad (2)$$

where $\sigma_2(t)$ is the dc conductivity associated with the recovery process occurring under given conditions.

For establishing the recovery component $\sigma_2(t)$, let us consider the processes that are likely to occur. There is the diffusion of water molecules to the interior of the sample. Rannou[3] and Travers[13] estimated the diffusion coefficient $D \sim 10^{-9}$–10^{-10} cm^2 s^{-1} for their samples. Taking into account the average thickness of our samples ($\sim$100 nm) and that range for D, we can say that our films reach the equilibrium with environment within seconds, which is negligible compared with ageing times of 5×10^5 s. Having been absorbed in the polymer, the water molecules enhance (probably through the gradually increasing grain size[8]) the electric charge transfer. Let us assume for the recovery component

$$\sigma_2(t) = -\sigma_R \exp(-t/t_R) \qquad (3)$$

where σ_R and t_R are characteristic parameters of the recovery process. According to Eqs. (1), (3) and the initial assumption (2), the final conductivity can be expressed as

$$\sigma(t) = \sigma_D \exp\left[-(t/t_D)^{1/2}\right] - \sigma_R \exp(-t/t_R) + \sigma_\infty \qquad (4)$$

The fit of Eq.(4) to experimental curves is shown in Figure 1. The data from long-term ageing were analysed according to the classic model Eq. (1) and the model with the recovery process given by Eq. (3). For low RH, the fit parameters calculated according to both models give close values, while at RH > 60 % the fit (1) is not satisfactory. Following figures illustrate the dependence of the mean fit parameters on RH. Figure 2 offers information about the parameters related to the degradation process. The time t_D is interpreted as a main measure of the resistance to ageing. It is not very sensitive to RH; it is of the order of 10^2 s for each of the tested samples.

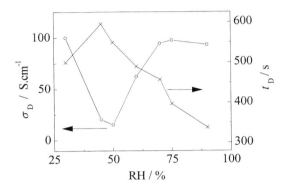

Figure 2. Dependence of the fit parameters σ_D and t_D on RH

However, while the stability at both low and high RH decreases, it reaches the maximum at RH 45 %. This may follow from the fact that the samples, which were prepared in air, are close to equilibrium conditions at RH ca. 45 %. The conductivity σ_D is associated with the volume fraction of the polymer that can be degraded while ageing. It shows the minimum at RH 50 %, where, according to t_D, the highest stability can be expected. The RH dependence of parameters t_R and σ_R on RH, related to the conductivity recovery process, is shown in Figure 3. Similarly to the degradation, t_R is a measure of the recovery process and σ_R relates to the conductivity of the volume fraction that is able, at given RH, to react with the absorbed water, which results in an increase in conductivity. Obviously, the decreasing t_R parameter and increasing σ_R reflect a direct correlation with the fact, that the recovery process is induced by the presence of water.

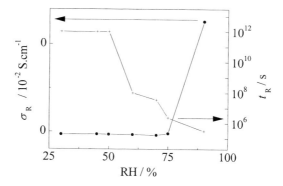

Figure 3. Dependence of the fit parameters σ_R and t_R on RH

FTIR spectroscopic study. Films were annealed in the humidity chamber together with those tested for dc conductivity. Infrared spectra were measured after the ageing (Regime/step 3) was finished. The FTIR spectrum of PANI.HCl film (curve 1) is shown in Figure 4.

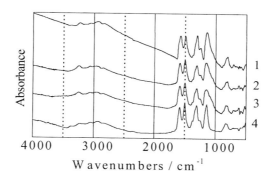

Figure 4. The FTIR spectra of PANI hydrochloride film

The spectra of samples annealed for 85 h at 90 °C and two values of RH are shown in the same figure (curve 2 – 30 %, curve 3 – 90 %). The spectrum of sample annealed at 120 °C for 30 min under ambient conditions (curve 4) is presented for comparison. The most significant feature of the annealed films is the disappearance of a broad band at wavenumbers higher than 2000 cm^{-1}, which is characteristic of the conducting form of PANI. It is more pronounced for the film annealed for 30 min at 120 °C. It is in good correlation with the decrease in conductivity of the samples, thus confirming the deprotonation of PANI.HCl film. Comparing the spectrum of the fresh film with the spectra of the annealed samples, we observe many changes which indicate deprotonation. The main peaks observed at 1578 cm^{-1} and 1495 cm^{-1} in the PANI.HCl film are blue-shifted to 1584 cm^{-1} and 1502 cm^{-1} in the spectra of films annealed at 90 °C for 85 h and exposed to different humidities (curves 3 and 4). The absorption band at 1308 cm^{-1}, the band observed at about 1240 cm^{-1}, characteristic of the conducting protonated form, and the 1140 cm^{-1} band decreased in the spectra of films annealed at 90 °C and RH 90 % for 85 h (curve 3) and annealed at 120 °C under ambient conditions (curve 4). This confirms that deprotonation increases with humidity and temperature.

Conclusion

Analysis of the time dependence of dc conductivity with respect to the two processes involved in ageing at elevated temperature and at various relative humidities has been presented. The recovery observed in dc conductivity is probably associated with the presence of water in the film. It is more pronounced for films annealed at higher relative humidity. Both the degradation and recovery processes are compatible with the concept of heterogeneous metallic structure[10] and can be correlated with changes in the conducting grain size[7,8,11]. Further analysis of the parameters obtained from fitting is envisaged. FTIR study brings evidence of deprotonation of the films related to the conditions of ageing. The degree of deprotonation increases with increasing both temperature and humidity. The structures in the presence of water are more pronounced in the samples annealed at higher RH. No evidence of the oxidation process expected in the paper by Sixou et al.[7] is apparent.

Acknowledgements. This work is a part of the research programs MSM113200001–2, which are financed by the Ministry of Education, Youth and Sports of the Czech Republic. We are also grateful for support of the Grant Agency of the Czech Republic (202/02/0698), the Academy of Sciences of the Czech Republic (A 4050313) and the Grant Agency of Charles University (186/2001/B).

454

[1]   A. Yue, A.J. Epstein, Z. Zhong, P.K. Gallagher, A.G. MacDiarmid, *Synth. Met.*, **1991**, *41–43*, 765.
[2]   M. Angelopoulos, A. Ray, A.G. MacDiarmid, A.J. Epstein, *Synth. Met.*, **1987**, *21*, 21.
[3]   P. Rannou, M. Nechtschein, *Synth. Met.*, **1997**, *84*, 755.
[4]   J. Stejskal, I. Sapurina, J. Prokeš, J. Zemek, *Synth. Met.*, **1999**, *105*, 195.
[5]   L.J. van der Pauw, *Philips Res. Rep.*, **1958**, *13*, 1.
[6]   A. Wolter, P. Rannou, J.P. Travers, B. Gilles, D. Djurado, *Phys. Rev. B*, **1998**, *58*, 7637.
[7]   B. Sixou, N. Mermilliod, J.P. Travers, *Phys. Rev. B*, **1996**, *53*, 4509.
[8]   P.K. Kahol, A.J. Dyakonov, B.J. McCormick, *Synth. Met.*, **1997**, *84*, 691.
[9]   J. Prokeš, I. Křivka, T. Sulimenko, J. Stejskal, *Synth. Met.*, **2001**, *119*, 479.
[10] L. Zuppiroli, M.N. Bussac, S. Paschen, O. Chauvet, L. Forro, *Synth. Met.*, **1994**, *50*, 5196.
[11] S. Sakkopoulos, E. Vitoratos, E. Dalas, *Synth. Met.*, **1998**, *92*, 63.
[12] E. Dalas, S. Sakkopoulos, E. Vitoratos, *Synth. Met.*, **2000**, *114*, 365.
[13] J.P. Travers, M. Nechtschein, *Synth. Met.*, **1987**, *21*, 135.

Charge Transport in Polyaniline Doped with 3-Nitro-1,2,4-triazol-5(4H)-one

Oleksiy Starykov,[1] *Jan Prokeš, ***[2] *Ivo Křivka,*[1] *Jaroslav Stejskal*[2]

[1] Faculty of Mathematics and Physics, Charles University, Ke Karlovu 3, 121 16 Prague, Czech Republic
[2] Institute of Macromolecular Chemistry, Academy of Sciences of the Czech Republic, Heyrovsky Sq. 2, 162 06, Prague, Czech Republic

Summary: Polyaniline (PANI) base was protonated in aqueous solutions of an organic acid, 3-nitro-1,2,4-triazol-5(4H)-one (NTO). The temperature dependence of DC conductivity of PANI-NTO seems to correspond to the theory of variable range hopping (VRH) in three dimensions. The frequency dependence of AC conductivity also reflects the hopping nature of mobile charges. The activation energy for the polymers with protonation degree above 0.12 remains constant with increasing dopant concentration and DC conductivity. The value of this constant may correspond to the energy needed for the ionization of dopant counterion. The fit of the electric relaxation function to the stretched exponential function $\varphi(t) = \exp[-(t/\tau)^{\beta}]$ gives the stretch parameter β about 0.35, which shows that the distribution of relaxation times is broad and indicates a high inhomogeneity in the distribution of a dopant.

Keywords: charge transfer; charge transport; dielectric properties; disorder; polyaniline

Introduction

The electric properties of conducting polymers have been the topic of intensive study since the intrinsic conductivity in these materials was discovered[1]. Doping of PANI with organic or inorganic acids is used to create free or localized charges in a polymer medium. In this procedure, the conductivity of PANI increases by several orders[2], from ca. 10^{-10} to 10^{1} S cm^{-1}. PANI-NTO is a new type of material combining the properties of a conducting polymer, PANI, and the features of an explosive, NTO. Several mechanisms of charge transport were found in conducting polymers[1]. Despite high doping of a polymer, the charge carriers seem to be localized, so they should move in the material by hopping over or tunneling through the potential barriers between the localized states. Variable-range hopping[3] assumes that carriers tunnel through the energy barriers between the localized states by means of phonon thermal energy. 3D-VRH occurs in the disordered polymers with structures similar to amorphous semiconductors, where charges can choose an easiest hop in any direction. Quasi-1D VRH occurs in the polymers where charges move along the polymer chain, with sudden interchain hops[4]. In some cases, polymerization and protonation lead to creation of small conducting islands distributed in a non-conductive matrix. Then charges tunnel between these regions according to the charged-energy-limited tunneling (CELT) mechanism[5].

Experimental

PANI base was protonated in solutions of various NTO concentrations[6]. The protonation degree is defined as the molar ratio of NTO and PANI. Samples were dried and pressed into pellets 13 mm in diameter. DC and AC parameters were measured after vacuum deposition of gold electrodes (10 mm in diameter) on both sides of pelletized samples. The AC properties were measured in temperature range 100–320 K in the frequency range 20 Hz–5 MHz using precision LCR meters HP4284A and HP4285A. The temperature dependence of DC conductivity was measured in Kelvin configuration using a Keithley 2001 multimeter in temperature range 100–320 K.

Results and discussion

The temperature dependence of DC conductivity for the samples with different protonation levels is shown in Figure 1. Fitting the temperature dependence of DC conductivity showed the best agreement for the model of 3D VRH[3]

$$\sigma_{DC} \propto \exp\left[-\left(T_0/T\right)^\gamma\right] \tag{1}$$

with $\gamma = 1/4$. The frequency dependence of conductivity, $\sigma(\omega)$, measured at 215 K for the polymers with protonation degree from 0.02 to 0.23 is shown in Figure 2. The low-frequency limit of $\sigma(\omega)$ increases with increasing protonation degree with a simultaneous increase in characteristic or onset frequency ω_c. Below this frequency, the conductivity is almost constant, but beyond it increases with frequency. The approximate power law of AC conductivity $\sigma(\omega) \sim \omega^{s}$ [3] holds above ω_c with the exponent s ca. 0.65.

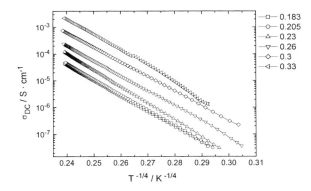

Figure 1. Temperature dependence of DC conductivity for polyanilines with different protonation degrees.

Fitting the frequency dependence of AC conductivity to the equations

$$\sigma(\omega) = \sigma_{AC}(\omega) + \sigma_{DC}(\omega),$$ (2)

representing the total conductivity of a material, where σ_{DC} is the DC conductivity and

$$\sigma_{AC}(\omega) \propto \omega \ln^{d+1}\left(\frac{\nu_{ph}}{\omega}\right) \sim \omega^s$$ (3)

is the AC conductivity, we obtained values of $\nu_{ph} = 10^{11} \text{ s}^{-1}$ for 3D-VRH ($d = 3$) and $\nu_{ph} = 10^8 \text{ s}^{-1}$ for quasi-1D VRH ($d = 1$). These values are independent of temperature. The latter value, corresponding to quasi–1D hopping, is too low in comparison with that known from literature ($\sim 10^{13} \text{ s}^{-1}$)[3]. This indicates that the 3D VRH takes place in these materials.

Due to the high DC conductivity of the materials, representation of AC measurements in terms of dielectric permittivity is impossible, because the peak in ε'' is masked by the σ_{DC} contribution[7]. It was found that, in the case of conducting materials, it is convenient to present the results of AC measurements in terms of complex conductivity or complex modulus, which is the quantity reciprocal to complex permittivity $M^*(\omega) = 1/\varepsilon^*$. This quantity represents a Fourier transform[8]

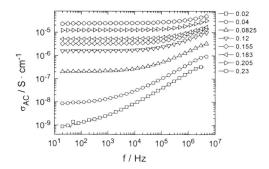

Figure 2. Frequency dependence of AC conductivity measured at 215 K for polyanilines with different protonation degree.

$$M^*(\omega) = M_\infty \left[1 - \int_0^\infty \left(-\frac{d\varphi(t)}{dt} \right) \exp(-i\omega t) dt \right] \qquad (4)$$

of the decay function of electric field $\varphi(t)$, which in most cases can be presented in the form of Kolhrausch-Williams-Watts (KWW) function[8]

$$\varphi(t) = exp\left[-(t/\tau)^\beta \right] . \qquad (5)$$

Figure 3 shows the frequency dependence of the imaginary part of electric modulus (points). The broad asymmetric curves indicate high disorder and broad distribution of relaxation times.

Fitting of the experimental plots to Equation (4) was made (Figure 3, full lines). Parameter β was found to have a constant value of 0.35 for all protonation degrees. This value correlates with the power exponent s characterizing this frequency behavior of AC conductivity $\beta + s \sim 1$ [9]. Low β indicates broad distribution of relaxation times and high inhomogeneity of the distribution of dopant counterions.

Figure 4 shows the dependences of DC conductivity and activation energy on protonation degree. DC conductivity was measured at room temperature. The activation energy W was calculated from the temperature dependence of relaxation time τ, given by $\tau = \tau_0 \exp(W/kT)$, where τ_0 is the inverse phonon attempt frequency $\tau_0 = \nu_{ph}^{-1}$.

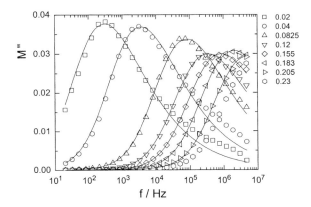

Figure 3. Frequency dependence of the imaginary part of complex electric modulus at 215 K for the polyanilines with protonation degree from 0.02 to 0.23. Full lines – fit to the KWW equation.

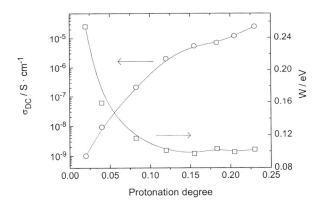

Figure 4. Dependence of activation energy W (circles) and DC conductivity σ_{DC} (squares) of polyanilines on the protonation degree (full lines – guide to eye).

At low concentrations of NTO, the activation energy and DC conductivity rapidly change. For the protonation degrees higher than 0.12, the activation energy of charge hopping remains constant, ~0.1 eV. In this region, DC conductivity increases slower than at low dopant concentrations. At low degrees of protonation, the decreasing activation energy is given by the increasing number of localized energy states and decrease in energy barriers between them. At high doping levels, the activation energy is defined by the energy required for the ionization of dopant[10], which is independent of the protonation degree. DC conductivity slowly increases due to increasing charge concentration.

Conclusions

Changing protonation degree is an effective method of controlling the electric properties of PANI. Conductance in PANI doped with NTO takes place *via* variable-range hopping among localized energy states, created during doping. The frequency dependence of AC conductivity indicates VRH in three dimensions; the onset frequency and activation energy increase with growing degree of protonation. Broadened and asymmetric $M''(\omega)$ plots indicate a highly disordered material. The low value of stretch parameter in the KWW function characterizes the broad distribution of relaxation times and high inhomogeneity of dopant distribution. The constant activation energy for samples with high degrees of protonation may correspond to the energy needed for the ionization of dopant counterion.

Acknowledgements

This work is a part of the research program MSM113200002 financed by the Ministry of Education of the Czech Republic. Financial support of the Grant Agency of the Charles University (186/2001/B) and the Academy of Sciences of the Czech Republic (K4050111) is also gratefully acknowledged.

[1] *Advances in Synthetic Metals. Twenty Years of Progress in Science and Technology*, P. Bernier, S. Lefrant, G. Bidan, Eds., Elsevier, Amsterdam, 1999.
[2] S. Roth, *One-Dimensional Metals*, VCH, Weinheim, 1995.
[3] N. Mott, A. Davis. *Electronic Processes in Non-crystalline Solids,* Clarendon, Oxford, 1979.
[4] Z. H. Wang, A. Ray, A. G. MacDiarmid, A. J. Epstein, *Phys. Rev. B,* **1991**, 43, 4373.
[5] F. M. Zuo, M. Angeloupoulos, A. G. MacDiarmid, A. J. Epstein, *Phys. Rev. B,* **1987**, 36, 3475.
[6] J. Stejskal, I. Sapurina, M. Trchová, J. Prokeš, *Chem. Mater.*, **2002**, *in press.*
[7] A. K. Jonscher, *Dielectric Relaxation in Solids*, Chelsea Dielectric Press, London, 1983.
[8] C. T. Moynihan, L. P. Boesch, N. L. Laberge, *Phys. Chem. Glasses*, **1973**, 14, 122.
[9] K. L. Ngai, R. W. Rendell, *Phys. Rev. B,* **2000**, 61, 9393.
[10] V. Čapek, private communication.

Macromol. Symp. **2004**, *212*, 461-466

Optical Spectroscopy of Polysilylene Films

Yevhen V. Demchenko,[1] *Josef Klimovič,*[*1] *Stanislav Nešpůrek*[1,2]

[1] Faculty of Mathematics and Physics, Charles University, V Holešovičkách 2,
180 00 Prague 8, Czech Republic
E-mail: klimovic@kmf.troja.mff.cuni.cz; Fax: (+420) 221 912 351
[2] Institute of Macromolecular Chemistry, Academy of Sciences of the Czech
Republic, Heyrovský Sq. 2, 162 06 Prague 6, Czech Republic

Summary: Luminescence of poly[cyclohexyl(methyl)silylene] films was
studied by time-resolved emission spectroscopy at temperatures from 15 K to
300 K. The main short-wavelength peak, shifted with rising temperature to
lower energies, showed time features typical of an energy donor. Three weak
emission centres were found in the visible spectral region. They decayed more
slowly having the time features of an energy acceptor. The rise in temperature
made both the energy transfer and extension of the effective σ-conjugation
length more efficient. Model calculations of the chain conformation energy
showed the close-to-all-trans and gauche helix minima for iso- and syndiotactic
sequences and barriers between them.

Keywords: conjugated polymers; luminescence; molecular modeling;
polysilylenes; time-resolved spectroscopy

Introduction

Polysilylenes are a linear arrangement of covalently bonded silicon atoms with interesting
electronic structure and photophysical properties[1-3]. Optical spectra of polysilylenes with
alkyl side groups reflect mainly the transitions between electronic states of the σ-conjugated
system of the silicon main-chain atoms. If the side groups contain a π-electron system, the
energy states contain a contribution of a mixture of π- and σ-orbitals. Experiments with
oligomers shorter than 20 silicon atoms in solutions showed[2] that with rising number of
chain atoms, both the absorption and luminescence spectral maxima related to the lowest
electron transitions were shifted to the red and the luminescence lifetime shortened.

An important condition for an effective π-conjugation is the stiffness of the backbone of
the macromolecule. An array of single-bonded silicon chain atoms in polysilylenes is

DOI: 10.1002/masy.200450858

different. Its conformation can change due to the possibility of rotation around the chain single bonds. Thus, the extent of σ-conjugation must be a function of both the length and the conformation state of the oligomer or polymer chain segments. This fact complicates the explanation of a number of experimental facts. For example, thermochromism of polysilylene solutions often observed during cooling was explained by temperature changes in the conformational preferences of the chains[1]. By conformation changes was also explained the considerable spectral red shift observed in stress-oriented polysilylene samples[1].

We report here the changes in luminescence spectra of poly[cyclohexyl(methyl)silylene], p(cHxMeSi), films caused by temperature changes in the region from 15 to 300 K. p(cHxMeSi) is structurally similar to the often studied poly[methyl(phenyl)silylene]. On the other hand, the electronic structure of p(cHxMeSi) is simpler because the alkyl side groups do not change the nature of the electronic states of the σ-conjugated system of silicon atoms. In the used temperature region, the polymer under study was deep in its glassy state and, thus, only small conformation changes could occur. The aim of this work is to find the influence of temperature on the conjugated system of p(cHxMeSi) under the condition that conformation changes are strongly limited.

Experimental

Poly[cyclohexyl(methyl)silylene], p(cHxMeSi), T_g = 399 K, was prepared by Wurtz coupling polymerization. Films were prepared from a toluene (UV spectroscopy grade) solution by casting on polished Cu substrates or UV quartz glasses, dried under toluene vapor and then in vacuum. Their thickness was around 10 μm for luminescence and less than 1 μm for absorption measurements. Samples were placed inside an optical refrigerator (Air Products Ltd.). Absorption was measured using a Hitachi 3000 spectrophotometer.

In emission studies, the samples were excited by the 311 nm nitrogen line of a pulse nanosecond lamp (Applied Photophysics Ltd.) selected with an absorption UV filter and interference filter (Hg313, Melles Griot), at 45° to the film surface. Luminescence was measured at 90° to the sample surface, analysed with a grating monochromator (Jobin-Yvon Model H.20 VIS), and detected with a cooled photomultiplier (Philips Model XP2254B) in the time-correlated single-photon counting regime. At every wavelength (usual bandpass 8 nm), the whole time dependence of the luminescence response was detected. By integration (computer summation) in the time period of the whole response curves, the total spectrum (equivalent of the stationary spectrum) was obtained. The spectrum was corrected for the spectral dependence of the sensitivity of the detection part of the spectrometer. The emission decay function was determined from the time-response curve of luminescence at a chosen wavelength and from the time shape of the scattered excitation pulse by computer deconvolution (Levenberg-Marquardt procedure).

Results

The conformation of the p(cHxMeSi) chain was modeled using commercial program *Cerius2* (Biosym Technologies, Inc.), with the force field based on the quantum mechanical calculations of Sun[4]. Segments of Si chains with Si atoms were constructed in the zig-zag (all-trans) conformation, the side groups being positioned in both the isotactic and syndiotactic sequence. All torsion angles around the chain Si-Si bonds were fixed at a certain value (of a set of angle values) and, by the minimization procedure, all the other space parameters were determined for the state of minimal intramolecular interaction energy of the segment at a chosen torsion angle. The results are seen in Figure 1.

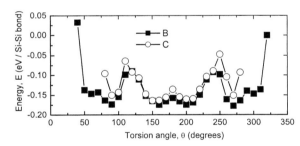

Figure 1. The conformation energy of isotactic and syndiotactic chain segments of p(cHxMeSi) as a function of the torsion angle of the Si-Si bonds.

Due to the bulky side-groups, the all-trans conformation is not exactly zig-zag but forms an open helix (torsion angle near 20°). The torsion angle dependence of conformation energy does not differ substantially between iso- and syndiotactic chains. The position and the energy of trans and gauche minima differ only little, the barrier between them is ca. 0.1 eV high. These results show that the intramolecular conformation energy determines the space arrangement of the chains only partly; a substantial role in this respect must play the intermolecular interactions.

UV absorption and photoluminescence parts of spectra of p(cHxMeSi) in toluene solution at room temperature are shown in Figure 2. The absorption spectrum corresponds well to the literature data: a strong narrow absorption peak with a maximum at 320 nm. The absorption spectrum of a p(cHxMeSi) film is almost the same.

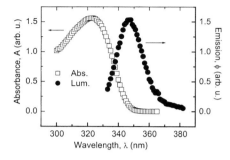

Figure 2. UV absorption and emission parts of the spectra of p(cHxMeSi) in a toluene solution at room temperature.

The photoluminescence of p(cHxMeSi) films consists of a strong narrow peak at 348 nm observed also in solution and of a broad very weak emission in the range from 380 to 550 nm. The intensity difference of the two emissions was so high that it was necessary to measure and analyze them separately.

The shapes of the short-wavelength emission peaks at different temperatures are plotted in Figure 3. It can be seen that the position of the peak maximum is influenced by temperature; it is shifted from approx. 3.6 eV at 45 K to 3.5 eV at 270 K. In the time-resolved spectra, no important differences in the shapes were detected. The luminescence decay measurements on films confirmed a rapid decay. For achieving satisfactory deconvolution, a sum of four exponentials was necessary to express the decay. The shape of the decay curves was typical of a donor of excitation energy in the presence of acceptors (the Foerster-type decay curve, see Ref.[5] for details). Within the frame of the Foerster model, the lifetime of the donor chromophores (conjugated parts of chains) without the presence of energy acceptors is around 0.6 ns (in the spectral band from 340 to 348 nm) at 15 K. By rising the temperature, the energy transfer becomes faster and the lifetime a little shorter. In a toluene solution at room temperature, this lifetime was close to 0.5 ns.

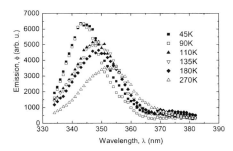

Figure 3. The total spectra of the short-wavelength strong emission peak of a p(cHxMeSi) film at different temperatures.

In Figure 4 is plotted the emission spectrum of p(cHxMeSi) film measured in the spectral region from 380 to 550 nm at 15 K. At this temperature, the intensity of the emission is sufficiently high and the shape of possible components is well expressed (raising the temperature the total intensity falls down).

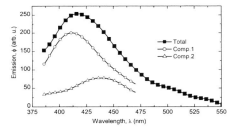

Figure 4. The total long-wavelength emission spectrum of p(cHxMeSi) films at 15 K and its components.

The time-resolved spectra showed a shift of the emission maximum towards lower energies with time. The analysis leads to the conclusion that the total spectrum consists of two major overlapping components with maxima at around 412 and 438 nm (which are also shown in Figure 4) and a weak additional band at around 500 nm. The long-wavelength luminescence decays much more slowly than the UV main band emission. The decay curves (measured in the 8-nm-broad region around 430 nm) were again complicated (satisfactory was deconvolution using a sum of several exponential functions was satisfactory). The shape of the decay curve was typical of an acceptor of excitation energy including a short initial period of rising intensity (cf. Ref.[5] for details). The lifetime of the excited state of acceptor(s) was found to be 5.9 ns at 15 K. On raising temperature, the intensity of this emission is strongly quenched. In toluene solution at room temperature, the intensity was too low to allow a detailed analysis.

Discussion and conclusions

The absorbing chromophores of the alkyl-substituted polysilylene films are mainly the segments of the Si chain regular in tacticity and conformation (very probably the gauche helices). Due to the proximity of other chain segments, the excitation can rapidly migrate through the film. The distribution of emitting conjugated segments can thus differ from that of absorbing species (this is a part of the observed Stokes shift). Excitation is also transferred to other centres emitting at lower energies. The rate of the energy transfer added to the rate of nonradiative and of fluorescence processes causes the observed rapid decay of the fluorescence. The observed red shift of the emission on heating the film above 90 K (Figure 3) can hardly be ascribed to the conformation changes (glassy state). It can be the result of the changes in the relaxation properties of the chain parts surrounding the emitting segment or of the lowering of the barriers between conjugated segments related to local irregularities in the conformation caused by rising chaotic thermal motion.

The time-resolved measurements of the weak broad visible emission of p(cHxMeSi) show that it consists of three components and that their emission decays much more slowly than that of the main short-wavelength emission. For the explanation of the sources of similar emission observed in other polysilylenes, a number of suggestions can be found in the literature including impurities, local chemical defects, triplet or excimer states, branching points or weakened Si–Si bonds. Our above-described experiments are not sufficient for their identification.

Acknowledgement

The authors wish to express their thanks to Mr. S. Zubarev for help with the model calculations and to the Computer Center of the Czech Technical University for access to the facilities. The financial support by grants No. 12/96/K of the Grant Agency of the Academy of Sciences of the Czech Republic, No. 202/01/0518 of the Grant Agency of the Czech Republic, and No. ME 440 of the Ministry of Education, Youth and Sports of the Czech Republic is gratefully appreciated.

[1]   L.A. Harrah, J.M. Zeigler, *Macromolecules* **1987**, *20*, 601.
[2]   Y. Kanemitsu, in: *"Photonic Polymer Systems"*, D.L. Wise, G.E. Wnek, D.J. Trantolo, T.M. Cooper, J.D. Gresser, Eds., M. Dekker, Inc., NewYork-Basel-Hong Kong 1998, pp.1-32.
[3]   S. Nešpůrek, A. Eckhardt, *Polym. Adv. Technol.* **2001**, *12*, 427.
[4]   H. Sun, *Macromolecules* **1995**, *28*, 701.
[5]   J.B. Birks: *"Photophysics of Aromatic Molecules"*, Wiley-Interscience, London-N.Y.-Sydney-Toronto 1970, Chapter 11.

Macromol. Symp. **2004**, *212*, 467-472

The Role of Micelles of Multiblock Copolymers of Ethylene Oxide and Propylene Oxide in Ion Transport during Acid Bright Copper Electrodeposition

Nadya Tabakova,[1] *Ivan Pojarlieff,*[1] *Vera Mircheva,*[2] *Jaroslav Stejskal**[3]

[1] Institute of Organic Chemistry, The Bulgarian Academy of Sciences, Sofia, Bulgaria
[2] Institute of Engineering Chemistry, The Bulgarian Academy of Sciences, Sofia, Bulgaria
[3] Institute of Macromolecular Chemistry, Academy of Sciences of the Czech Republic, Prague, Czech Republic

Summary: The role of micelle formation in the levelling activity of electrolytes for copper electrodeposition was investigated. It was established that micelle formation in the electrolytes caused a dramatic increase in the degree of levelling. The effect takes places *via* the solubilization of the disodium bis(sulfopropyl) disulfide brightener in copolymer micelles and changes in the distribution of the brightener on cathode surface. The positive effect of a conducting-polymer additive, a colloidal polyaniline dispersion, on the brightness of electrodeposited copper coatings is discussed.

Keywords: block copolymer; conducting polymer; copper deposition micelles; polyaniline

Introduction

Surface-active agents and nitrogen- and sulfur-containing organic compounds are essential additives to non-cyanide electrolytes for copper electrodeposition[1]. These compounds control the structure, the physico-mechanical properties and anticorrosion resistance of electrodeposited metal coatings, and influence certain technological parameters of the cathode deposition of metals. Usually, copper coatings are used as a supportive underlayer for the multilayer copper–nickel–chromium coatings. They have to possess specific properties, the most important being the high degree of levelling. During the last few years, we have designed non-cyanide electrolytes for acid bright copper deposition with a high levelling activity[2] – above 90 %. This means that the surface structure becomes smooth and achieves a mirror-like appearance. The application of these

 DOI: 10.1002/masy.200450859

electrolytes brings both ecological and economic benefits, because the high levelling activity allows elimination of all manual operations, including polishing, and provides the possibility of fully-automated processing. The surfactants used in these electrolytes were multiblock copolymers of ethylene oxide (EO) and propylene oxide (PO)[2] with a specific hydrophilic/hydrophobic balance, a low degree of crystallinity, and high solubilization performance. In aqueous solutions, they form micelles with a hydrophilic shell comprising poly(ethylene oxide) (PEO) segments and a hydrophobic core composed of poly(propylene oxide) (PPO) segments.

Unlike the currently-used wetting agents in electrolytes for copper electrodeposition – PEO and statistical copolymers of EO and PO – the multiblock copolymers are characterized by micelle formation, solubilization efficiency, and specific adsorption behaviour. The objective of this study is to investigate the role of copolymer micelles in the ion transport and electrochemical levelling of copper coatings during acid bright copper electrodeposition. The role of a conducting polymer colloid, a polyaniline (PANI) dispersion used as an additive, has also been studied.

Experimental

All materials and methods have been described in a previous paper[2]. The main electrolyte contained 220 g dm^{-3} copper sulfate pentahydrate, 50 g sulfuric acid (98 %) and 0.09 g dm^{-3} sodium chloride. A symmetrical pentablock copolymer of ethylene oxide and propylene oxide, $(EO)_8(PO)_{15}(EO)_{10}(PO)_{15}(EO)_8$ (EPEPE), was used as a wetting agent, disodium bis(sulfopropyl) disulfide (DS) as a brightener. Colloidal PANI particles were prepared by dispersion polymerization of aniline hydrochloride[3] using poly(N-vinylpyrrolidone) as stabilizer[4].

Results and discussion

Block copolymer micelles

The influence of the concentration of EPEPE on the degree of levelling of the copper coating at DS concentrations of 7 mg dm^{-3} and 17 mg dm^{-3}, and at a PANI concentration of 4.5 mg dm^{-3} has been studied (Fig. 1). It was established that, in the absence of EPEPE or at concentrations lower than the critical micellar concentration (cmc), the copper coatings were matt and rough. X-ray photoelectron analysis of these coatings revealed that, regardless of the DS concentration, sulfur was present both in the concavities and on the peaks of the uneven cathode surface. At an EPEPE concentration of 1.6 mg dm^{-3}, which corresponds to the cmc, sulfur was detectable only on the surface peaks. This change in sulfur distribution was parallelled by an abrupt increase in the degree of levelling. At EPEPE concentrations of 2.5–5 mg dm^{-3}, depending on the DS concentration, a plateau was reached and a further increase in the EPEPE concentration (even to 35 mg dm^{-3}) did not lead to a significant increase in the degree of levelling.

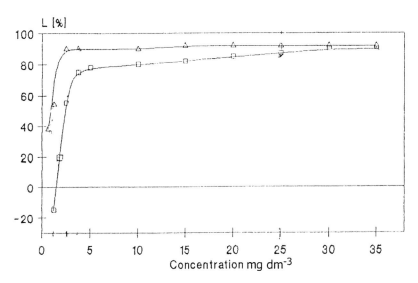

Figure 1. Influence of the EPEPE concentration on the degree of levelling L estimated by profilometry[2] at DS concentrations of 7 mg dm^{-3} (□) and 17 mg dm^{-3} (Δ).

For electrolytes containing PEO instead of EPEPE, no dramatic increase in the degree of levelling, nor changes in the sulfur distribution were observed[5]. In our opinion, this is due to the fact that poly(ethylene oxide) does not form micelles. The DS concentration has an influence on the degree of levelling only in the lower EPEPE concentration range (up to *ca* 10 mg dm^{-3}).

Effect of colloidal polyaniline particles

We have also studied the adsorption of dispersed PANI particles. X-ray photoelectron analyses showed that nitrogen from PANI was present only at the peaks of the cathode surface[5], independently of the presence of EPEPE and DS. The dominant adsorption of PANI particles, which had a hydrodynamic radius of $R_h = 185$ nm, was observed by scanning electron microscopy (Fig. 2). In acid electrolytes, protonated PANI is a polycation and therefore PANI particles are attracted to the cathode; they adsorb at the peaks of the cathode surface and block copper deposition at those sites. Because PANI is electrically conducting, this process does not result in the passivation of the electrode. During PANI adsorption from copper electrolytes containing EPEPE and DS, the dominant adsorption of colloidal PANI on the peaks persists but the adsorption layer formed is denser and continuous due to the DS-loaded EPEPE micelles chemisorbed on the peaks[2].

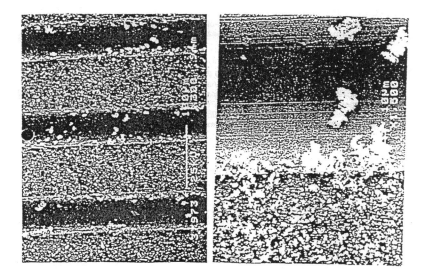

Figure 2. A micrograph of cathode surfaces with PANI particles adsorbed on the surface peaks (two different magnifications).

Thus, there are three types of positively charged particles, which travel during electrolysis to the cathode at different speeds depending on their size and their relative charge density (Fig. 3). The micelles of EPEPE containing adsorbed copper ions are the smallest objects and have the highest relative charge density. Therefore, they travel to the cathode rapidly. Having lost their copper ions, they are desorbed and redirected back into the electrolyte, where they are again positively charged by interaction with copper ions. Larger micelles, with a lower relative charge density, containing in addition to copper ions also the solubilized DS, reach the cathode more slowly and adsorb predominantly at the peaks of the cathode surface. After losing their copper ions and positive charge, they remain attached by chemisorption of DS to the metal surface and block copper deposition at those sites. The higher the DS content, the larger is the size of the micelles and, hence, the larger the area of cathode where copper deposition at peaks is prevented and surface brightness increases. This may explain the influence of the DS concentration on levelling (Fig. 1). Colloidal PANI particles are the largest and they also adsorb at the cathode peaks. Therefore, a dense non-interrupted layer – with a specific composition and structure – is formed at the cathode; this hampers electron transport and the free access of copper ions, and affects the nucleation of the metal phase.

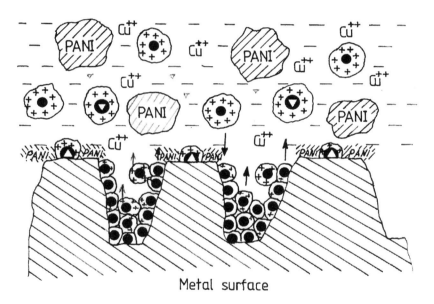

Figure 3. A proposed model of ion transport by EPEPE micelles during copper electrodeposition: DS brightener molecules (open triangles), the hydrophobic cores of EPEPE micelles (full circles), the copper ions in the micellar shell (crosses), and colloidal polyaniline particles (PANI) participate in the copper deposition.

Conclusions

The formation of a dense continuous adsorbed layer of a brightener and colloidal PANI particles at the peaks of cathode surface impedes electron transport and the free passage of copper ions to the cathode surface during electrodeposition of copper. This results in the formation of highly levelled copper layers. The proposed model is in good agreement with all the experimental findings and serves to explain the importance of micelle-forming surfactants in obtaining mirror-bright copper layers and its role in ion transport. The EPEPE, in addition to its micelle-forming properties in aqueous media, is also non-toxic, it does not foam, and is commercially available. These properties make it a suitable surfactant in electrolytes for copper electrodeposition. Moreover, the introduction of a conducting-polymer colloid, a PANI dispersion, can substantially improve the levelling of electrodeposited copper coatings and thus their brightness.

[1] M. Loshkaryov, J. Lochkaryov, *Surf. Technol.* **1978**, *6*, 397.
[2] N. Tabakova, N. Petkova, J. Stejskal, *J. Appl. Electrochem.* **1998**, *28*, 1083.
[3] J. Stejskal, *J. Polym. Mater.* **2001**, *18*, 225.
[4] J. Stejskal, P. Kratochvíl, M. Helmstedt, *Langmuir* **1996**, *12*, 3389.
[5] S. Rashkov, DSc. Thesis, Sofia 1983, p. 38–54.

Macromol. Symp. **2004**, *212*, 473-478

Pyrazoloquinolines – Alternative Chromophores for Organic LED Fabrication

Jacek Niziol,[*1,4] *Andrzej Danel,*[2] *Ewa Gondek,*[3] *Pawel Armatys,*[1] *Jerzy Sanetra,*[3] *Gisele Boiteux*[4]

[1] Faculty of Physics & Nuclear Techniques, University of Mining & Metallurgy, al. Mickiewicza 30, 30-059 Kraków, Poland
[2] Department of Chemistry, Hugon Kollataj Agricultural University, al. Mickiewicza 24/28, 30-059 Kraków, Poland
[3] Institute of Physics, Cracow University of Technology, ul. Podchorążych 1, 30-84 Kraków, Poland
[4] Laboratoire des Materiaux Polymères et Biomateriaux, Université Lyon 1, 43, Bd. du 11 Novembre 1918, 69622 Villeurbanne, France

Summary: In this paper the synthesis of some new chromophores which could be used in polymer/organic LED fabrication are presented. All of them are pyrazoloquinoline (PAQ) derivatives. Their emission properties were tuned by side group substitution. They were characterized by absorption and photoemission spectroscopy. Some were used, dispersed in poly(*N*-vinylcarbazole) (PVK) matrix, as the emissive layer in LED structures.

Keywords: dyes; electroluminescence; fluorescence; light emitting diodes (LED); pyrazoloquinolines

Introduction

Polymer and/or organic light-emitting diodes (LEDs) have the potential to radically change the display technology by enabling cheap, colour and flat displays with viewing properties similar to traditional cathode ray tubes. Other interesting applications like large surface lighting or back-lighting are also possible. Although organic and/or polymer LEDs have been the subject of extensive research since the beginning of the nineties and many compounds emitting in the green region are already known, there is still important demand for materials having their intense emission in the blue and red regions to produce full-colour RGB displays. In this paper new 1*H*-pyrazolo[3,4-*b*]quinoline (PAQ) derivatives with properties appropriate for these applications are reported. Such compounds exhibit strong fluorescence in solution [1-3] and also in the solid state [1] - a desirable property for high-performance electroluminescent devices. Recently, some other 1*H*-pyrazolo[3,4-*b*]quinoline derivatives have been used in light-emitting devices in the form of molecular layers [4,5].

 DOI: 10.1002/masy.200450860

Synthesis

The general structure of synthesized PAQ derivatives is in Fig. 1.

Compound	R^1	R^2	R^3	R^4
PAQ1	Ph	Me	4-MeOC$_6$H$_4$	Me$_3$C
PAQ2	Ph	Me	4-MeOC$_6$H$_4$	F
PAQ3	Ph	Me	4-MeOC$_6$H$_4$	MeO
PAQ4	Ph	Me	Ph	-
PAQ5	Ph	Ph	Ph	-
PAQ6	Me	Me	Me	-
PAQ7	Me	Ph	Ph	-

Figure 1. General structure of 1H-pyrazolo[3,4-b]quinoline derivatives

Depending on the substituents, one of the three methods described below was used for the synthesis of a particular PAQ derivative. The most important synthetic method includes the Friedländer condensation of 1,3-disubstituted 4,5-dihydro-1H-pyrazol-5-ones with anthranilaldehyde (Fig. 2). The reaction is performed usually in the melt or in diethylene glycol. The other reactions include cyclisations of benzoic acid derivatives (Fig. 3). Since anthranilaldehydes and 2-aminobenzophenones and acetophenones are not readily available, the reaction of 5-chloro-4-formylpyrazoles or 4-aroyl-5-chloropyrazoles with substituted anilines [6] (Fig. 4) was used for the synthesis of 1H-pyrazolo[3,4-b]quinolines. Due to these reactions the modifications of the carbocyclic ring is very easy.

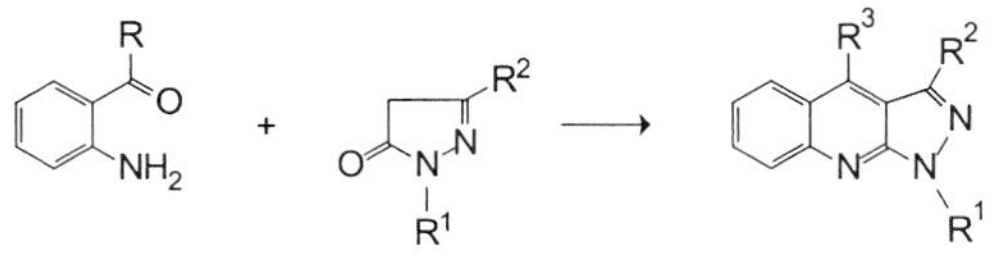

Figure 2. Friedländer condensation. R = H, Me, Ph ; R^1, R^2 = H, Me , Ph

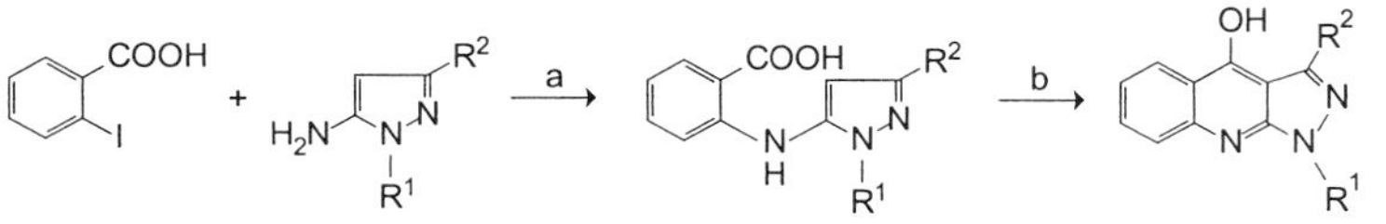

Figure 3. Cyclisations of benzoic acid derivatives a K$_2$CO$_3$ / CuO /H$_2$O; b polyphosphoric acid, 100 °C

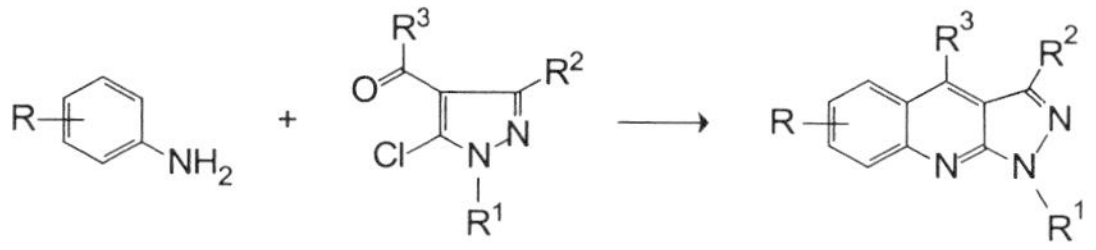

Figure 4. Reaction of 5-chloro-4-formylpyrazoles or 4-aroyl-5-chloropyrazoles with substituted anilines (R = H, MeO, NEt$_2$, Cl, Br; R^1, R^2 = Me, Ar; R^3 = H, Ar)

These methods suffer from certain limitations and disadvantages. Their application to the preparation of 4-aryl derivatives is restricted by the availability of 2-aminobenzophenones and aroylpyrazoles. Mostly, multistep syntheses are required for their preparation. Recently, an improved synthesis of 4-aryl-1H-pyrazolo[3,4-b]quinolines from aromatic amines and 4-benzylidene-4,5-dihydro-1H-pyrazol-5-ones [7] (Fig. 5), and also their one-pot synthesis [8] were described (Fig. 6). In the latter case it is very easy to modify aromatic ring in position 4 and to introduce the heterocycling ring.

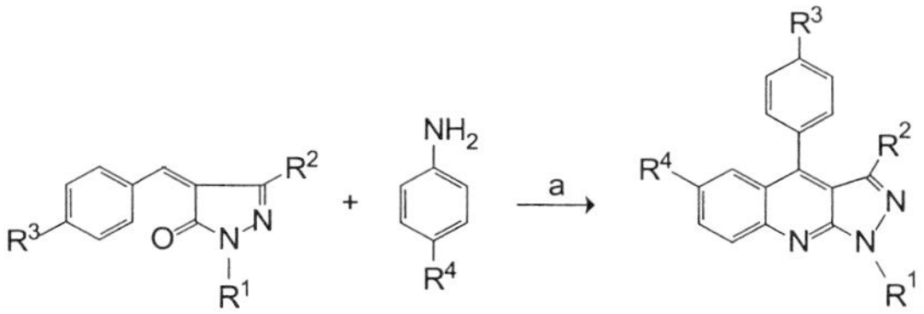

Figure 5. Improved synthesis of 4-aryl-1H-pyrazolo[3,4-b]quinolines (R = H, MeO, NEt$_2$, Cl, Br; R^1, R^2 = Me, Ar; R^3 = H, Ar) (under microwave irradiation)

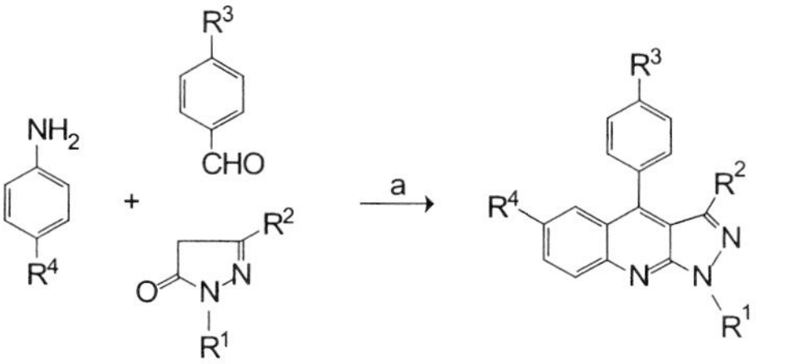

Figure 6. One-pot synthesis of 4-aryl-1H-pyrazolo[3,4-b]quinolines (in ethylene glycol, 190 °C, 1-2 h)

Experimental

Absorption and photoluminescence spectra of the synthesized compounds were measured in THF solution (Table 1). No simple correlation between positions of photoluminescence maxima and structures of the molecules was observed.

Table 1. Absorption (HOMO-LUMO transition), photoluminescence and electroluminescence maxima (nm).

Compound	Absorption	Photoluminescence	Electroluminescence
PAQ1	394	448	453
PAQ2	399	465	460
PAQ3	412	457	456
PAQ4	394	466	456
PAQ5	397	474	459
PAQ6	396	438	434
PAQ7	398	429	429

It was found in one of previous publications [1,4,5] that almost all PAQ derivatives rapidly crystallize from solutions when evaporated, which normally shifts electroluminescence spectra towards the red end of visible region. For this reason, the electroluminescence properties were measured in a simple sandwich structure ITO/PVK-PAQ/Ca/Al, where the active layer of PVK polymer was obtained from a solution of PVK and PAQ in THF by spin-casting. The most advantageous concentration of this dopant, at which the emitted light was of the highest intensity, was 1 wt. % As one can see in Table 1 and in Fig. 7, the maximum of electroluminescence spectra ranged between 430 and 460 nm. The shortest maxima wavelengths were observed for compounds PAQ6 and PAQ7. These spectra were less broad (ca. 50 nm) compared with the other (70 – 80 nm).

No effort was made to improve the geometry of the tested LEDs, because the main interest was to measure the electroluminescence spectra. Therefore the "turn-on" voltages were typically rather high, on average 12 V. Despite this inconvenience, it was still possible to estimate the relative quantum efficiency of the studied devices, because all of them had equal thickness. In Fig. 8, where luminescence/current dependences are shown, one can see that the strongest radiation combined with the least power consumption were found for LEDs with active layers doped with either PAQ4 or again PAQ7. In this last case, an estimate of the external quantum efficiency was ca. 1.5 %.

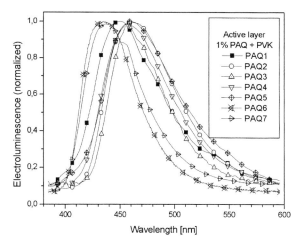

Figure 7. Normalized elektroluminescence spectra of LEDs with a PVK active layer doped with different pyrazoloquinolines

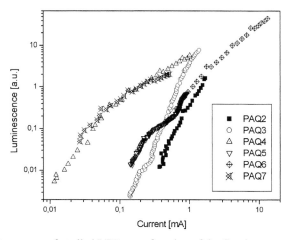

Figure 8. Luminescence of studied LEDs as a function of the flowing current

Conclusions

In this work seven new derivatives of 1*H*-pyrazolo[3,4-*b*]quinolines were synthesized and their photo- and electroluminescence properties were measured. 1-Methyl-3,4-diphenyl-1*H*-pyrazolo[3,4-*b*]quinoline was the most efficient dopant in the tested LEDs regarding its emission spectrum and quantum efficiency.

Acknowledgements

This work was partly supported through the KBN grant No. 8-T11-B-075-18. J. N. acknowledges the aid of the Marie Curie Fellowship Program No. HPMCFT-2000-00580.

[1] J. Niziol, A. Danel, G. Boiteux, J. Davenas, B. Jarosz, A. Wisla, G. Seytre, *Synth. Met.* **2002**, *127,* 175.
[2] K. Rechthaler; K. Rotkiewicz, A. Danel, P. Tomasik, K. Khatchatryan, G. Köhler, *J. Fluoresc*ence **1997**, *7*, 301.
[3] A.B.J. Parusel, K. Rechthaler, D. Piorun, A. Danel, K. Khatchatryan, K. Rotkiewicz, G. Köhler, *J. Fluoresc*ence **1998**, *8,* 375.
[4] Y.T. Tao, E. Balasubramaniam, A. Danel, A. Wisła, P. Tomasik, *J. Mater. Chem.***2001**, *11*, 768.
[5] Y.T. Tao, E. Balasubramaniam, A. Danel, B. Jarosz, P. Tomasik, *Chem. Mater.* **2001**, *13*, 1207.
[6] L. Hennig, T. Müller, M. Grosche, *J. Prakt. Chem.* **1990**, *332*, 693.
[7] G. Chaczatrian, K. Chaczatrian, A. Danel, P. Tomasik, *ARKIVOC,* **2001,** *2*(6), 1.
[8] A. Danel, K. Chaczatrian, P. Tomasik, *ARKIVOC*, **2000**, *2*(6),63.

Macromol. Symp. **2004**, *212*, 479-484

Influence of Morphological and Conformational Changes on the Molecular Mobility in Poly(ethylene naphthalene-2,6-dicarboxylate)

Didier Colombini,[1] *Noureddine Zouzou,*[2] *Juan Jorge Martinez-Vega**[2]

[1] Department of Polymer Science & Engineering, Lund University, Lund Institute of Technology, Box 124, SE – 22100 Lund, Sweden
[2] Laboratoire de Génie Electrique de Toulouse (UMR-CNRS 5003) Université Paul Sabatier, 118 route de Narbonne, 31062 Toulouse Cedex, France
E-mail: juan.martinez@lget.ups-tlse.fr

Summary: Dynamic electric (DEA) and mechanical (DMA) analyses were combined to explore the relaxational processes in amorphous and semicrystalline poly(ethylene naphthalene-2,6-dicarboxylate) (PEN) samples. Differential scanning calorimetry measurements were carried out to investigate the crystallinity of the samples following isothermal annealing treatment at 443 K. The two secondary relaxations β and β^*, the main α relaxation, as well as the ρ-relaxational process, were revealed by both electric and mechanical viscoelastic responses of the PEN samples. DMA results clearly identified the above T_α loss factor peak, ρ, as a probe of the cold crystallization. However, the association of both DMA and DEA investigations pointed out that electric and non-electric aspects might govern the ρ-process.

Keywords: cold crystallization; dynamic electric analysis; dynamic mechanical analysis; PEN; ρ-relaxation

Introduction

Poly(ethylene naphthalene-2,6-dicarboxylate) (PEN) is of particular interest in electrical applications, since its aromaticity leads to advantageous mechanical properties and to chemical and thermal stability. Recently, the microstructure/property relationship in amorphous and various semicrystalline PEN samples has been investigated[1-4], and four distinct relaxational processes (β, β^*, α, and ρ, with increasing temperature) were reported. The present work deals with both dynamic mechanical and electric analyses to explore the influence of the crystallinity on the molecular mobility in PEN.

 DOI: 10.1002/masy.200450861

Experimental

Amorphous PEN supplied by DuPont de Nemours (Luxembourg) was used. As previously detailed [3,5,6], as-received amorphous PEN samples were maintained at 443 K for 5, 15, 30, 45, 60, 120, and 180 min in order to reach various crystallinities, which were then estimated by differential scanning calorimetry DSC 2010, TA Instruments). Thus, following their respective isothermal annealing treatment (performed at 443 K under nitrogen atmosphere), the samples were cooled down to 303 K (10 K/min), and the DSC traces were recorded from 303 K to 573 K (10 K/min).

The DMA 2980 (TA Instruments) was used operating in tensile mode under isochronal conditions (10 Hz, from 173 K to 513 K with a heating rate of 2 K/min) to measure the temperature dependence of mechanical viscoelastic properties of PEN samples.

The dynamic electric analysis was carried out by using the DEA 2970 (TA Instruments) under isochronal conditions (10 Hz, from 173 K to 453 K with a heating rate of 2 K/min) to measure the temperature dependence of complex permittivity ε^*.

In order to achieve same controlled conditions for all the samples, the isothermal annealing treatment (at 443 K) was systematically performed in the furnace of the three instruments.

Results and Discussion

The DSC traces of as-received and annealed PEN samples are given in Fig. 1. First, the glass transition of the amorphous phase can be seen (around 398 K): obviously, the change of ΔC_p baseline becomes less pronounced when the isothermal annealing time increases. A cold crystallization exothermic peak is then clearly visible in curves a to e in Fig. 1.

Following the increase in the amount of the crystalline phase resulting from the increase of the annealing exposure, the area of this exothermic contribution significantly decreases going from curves a to e. After 45 min at 443 K (curves f - h), the cold crystallisation peak cannot be detected any longer and a small pre-melting peak is observed instead. As expected, at higher temperatures (around 443 K), the melting peak of PEN does not appear to be affected by the duration of the isothermal annealing treatment. In addition, on the basis of melting enthalpy results, it equals 103.4 J/g [7] for a 100 % crystalline PEN sample. The crystallinities of the annealed samples were estimated and are given in the table in Fig. 1.

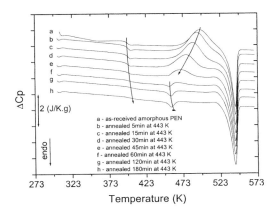

Sample	Isothermal annealing At 443 K (min)	Crystallinity (%)
a	0	0.3
b	5	1.1
c	15	5.1
d	30	14.2
e	45	29.8
f	60	39.4
g	120	41.5
h	180	42.3

Figure 1. DSC traces of the as-received amorphous and annealed PEN samples.

Figure 2 reports the temperature dependence of the storage modulus (E') and the loss factor (tan δ) for all amorphous and annealed PEN samples. When the temperature increases, three relaxation processes – β (around 223 K), β^* (around 358 K), and α (around 413 K) relaxations – are evidenced. Their molecular assignments have been discussed in detail previously[6]. Insofar as the α relaxation is associated with the anelastic manifestation of the glass transition of PEN, the corresponding peaks of tan δ are accompanied by an important decrease in E' for all the samples. After passing through the glass transition, the amorphous PEN phase begins to crystallize. As a result of the cold crystallization (occurring in amorphous and short-time annealed samples), a rapid increase in the storage modulus as well as the occurrence of a shoulder in tan δ related to the ρ-process can be observed above 453 K.

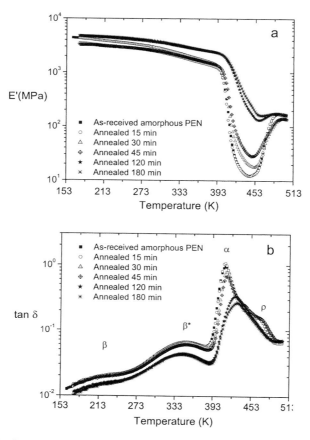

Figure 2. Mechanical viscoelastic properties of the as-received amorphous and annealed PEN samples: (a) storage modulus E' and (b) loss factor tan δ *versus* temperature (at 10 Hz).

Similarly to mechanical thermograms, the temperature dependence of permittivity (ε') and the loss factor (tan δ) also reveal the three β, β*, and α relaxations of PEN, the corresponding peaks of tan δ being associated with an increase in ε' (Fig. 3). It was then deemed to be of interest to combine both mechanical and electric approaches to further investigate the molecular mobility in the amorphous and annealed PEN samples, especially for the high-temperature processes. Independently of the technique, the maxima of either the mechanical or the electric loss peaks associated with the α relaxation are shifted toward higher temperatures, when the crystallinity increases. Both decreasing values of the dielectric permittivity and increasing mechanical storage moduli are obtained, when the cold crystallization process takes place. However, the ratio of the magnitude of both α- and ρ-relaxation peaks appears to be much more important in the case of dielectric measurements than when a mechanical stress is applied to the samples. As a result, the so-called ρ-relaxation can be considered as governed by both electrical and non-electrical aspects.

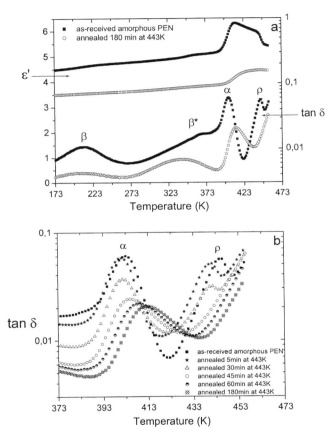

Figure 3. Dielectric viscoelastic properties of the as-received amorphous and annealed PEN samples: permittivity ε' and loss factor tan δ *versus* temperature (at 10 Hz).

Similarly to mechanical thermograms, the temperature dependence of permittivity (ε') and the loss factor (tan δ) also reveal the three β, β*, and α relaxations of PEN, the corresponding peaks of tan δ being associated with an increase in ε' (Fig. 3). It was then deemed to be of interest to combine both mechanical and electric approaches to further investigate the molecular mobility in the amorphous and annealed PEN samples, especially for the high-temperature processes. Independently of the technique, the maxima of either the mechanical or the electric loss peaks associated with the α relaxation are shifted toward higher temperatures, when the crystallinity increases. Both decreasing values of the dielectric permittivity and increasing mechanical storage moduli are obtained, when the cold crystallization process takes place. However, the ratio of the magnitude of both α- and ρ-relaxation peaks appears to be much more important in the case of dielectric measurements than when a mechanical stress is applied to the samples. As a result, the so-called ρ-relaxation can be considered as governed by both electric and non-electric aspects.

Conclusion

In this paper, both dynamic electric and mechanical analyses were successfully associated to explore the relaxational processes in amorphous and semicrystalline PEN samples. First, differential scanning calorimetry measurements were performed to estimate the consequence of more or less long-time isothermal annealing at 443 K for the crystallinity of PEN samples. Then, the crystallinity dependence of the viscoelastic properties of the samples was investigated. In particular, DMA results clearly identified the above T_g loss factor peak as a probe of the cold crystallization (observed only for the amorphous and the short-time annealed samples). Although the influence of the cold crystallization also occurs in the dielectric response of the samples, DEA results showed that the high-temperature loss peak (associated with the ρ-relaxation) is simultaneously governed by electric and non-electric aspects. The ambiguity of this relaxation in the literature could therefore find its source in such a bimodal origin.

[1] J.C. Cañadas, J.A. Diego, M. Mudarra, J. Belana, R. Diaz-Calleja, M.J. Sanchis, C. Jaimés, *Polymer*, **1999**, *40*, 1181.
[2] A. Nogales, Z. Denchev, I. Sics, T.A. Ezquerra, *Macromolecules*, **2000**, *33*, 9367.
[3] J.J. Martinez-Vega, N. Zouzou, L. Boudou, J. Guastavino, *IEEE Trans. Diel. Electr. Insul.*, **2001**, *8*, 776.
[4] L. Hardy, I. Stevenson, G. Boiteux, G. Seytre and A. Schönhals, *Polymer*, **2001**, *42*, 5679.
[5] N. Zouzou, C. Mayoux, J.J. Martinez-Vega, *Rev. Int. Génie Electr.*, **2000**, *3*, 513.
[6] N. Zouzou, D. Colombini, J.J. Martinez-Vega, *Polymer*, **2002**, submitted.
[7] S. Z. Cheng, B. Wunderlich, *Macromolecules*, **1988**, *21*, 789.

Influence of Morphology on Electric Conductance in Poly(ethylene naphthalene-2,6-dicarboxylate)

Jose Fidel Chavez,* Juan Jorge Martinez-Vega, Noureddine Zouzou,
Jean Guastavino

Laboratoire de Génie Electrique de Toulouse (UMR-CNRS 5003), Université Paul Sabatier, 118 route de Narbonne, 31062 Toulouse Cedex, France.

Summary: Polymers find an important field of application in the electrical industry because of their utilization as insulators and dielectrics. The challenge over the last several years has been to make electrical compounds smaller in size. In that way the material morphology is of strategic importance in the performance of new components in relation to their geometry. Therefore, it is important to know the influence of different types of stress on electric properties of the materials.

We have studied the influence of the crystallinity on electric conductance of poly(ethylenenaphthalene-2,6-dicarboxylate) (PEN), which is increasingly replacing some dielectrics (polypropylene and PET) in certain electric applications.

Starting from an amorphous PEN, samples with various crystallinities were obtained and analyzed using DSC in order to determine their thermal characteristics.

Moreover, measurements of charge and discharge currents were carried out on PEN samples to characterize their electric properties. The steady state current observed after 120 min, in response to the electric field, shows an increase in the insulator resistivity with the crystallinity. The increase in the crystallinity favors a decrease in the free volume, so the movement of charges is restricted especially by the species of large sizes.

Keywords: crystallization; electric conductance; insulator resistivity; morphology; PEN films

Introduction

Poly (ethylenenaphthalene-2,6-dicarboxylate) (PEN) is a polymer obtained by polycondensation of naphthalene-2,6-dicarboxylic acid and ethylene glycol.[1]

PEN competes with PET (poly(ethylene terephthalate)) in engineering applications that require high performance in terms of dielectric properties, thermal stability and physical and chemical properties.[2] The rigid aromatic ring system of naphthalene explains a better performance of this polyester. Indeed, the naphthalene moiety in PEN provides stiffness of the linear backbone, leading to improved characteristics fitting its applications, such as thermal resistance, excellent mechanical properties like tensile properties and dimensional stability, and outstanding gas barrier characteristics.[3] Due to all these advantages, the interest of the market in PEN growing also due to the study of its electric and mechanical behavior, and of other prominent electronic characteristics (such as an increase in the product of capacitance and voltage per unit of volume and its application in the technology of surface mounted technology (SMT) insulator).[4]

The electric conductance across the surfaces of insulating polymers is an aspect of considerable technical and commercial interest but poorly understood (as regards the mechanisms involved), despite the effort made by several authors of theoretical end experimental studies. [5] With PEN, the problem is even greater due to its recent appearance on the market and obviously due to little experimental and theoretical study, especially at high electric fields. In particular, the clear and transparent viewpoint of the charge injection and carrier migration processes will be essential in the future of these materials. Our paper contributes to explanation of the behavior of PEN, in electric fields from 7 MV/m to 120 MV/m using amorphous and semicrystalline PEN with crystallinity of 40 % .

Experimental

The PEN films used in this work were kindly provided by DuPont de Nemours Luxembourg. The films were obtained by biaxial orientation. The samples contain additives, which allow the rolling up of films. Two types of PEN were characterized: films of 70 μm thickness of very low crystallinity (amorphous films) and films of the same thickness with a partly crystalline structure (40 %).

The stages of sample preparation were metallization and annealing followed by secondary vacuum processing. The sample used was metallized on its two faces, yielding a metal-insulator-metal structure. Thus two zones were formed: the first, the interface between the material and metal, and the other was the bulk. With the aim to study the influence of morphology and of the mode of preparation of the samples on their electric behavior, three studies were carried out:

(a) Initially, we studied an amorphous sample of metallized PEN.

(b) Next, a sample of amorphous PEN was metallized and then annealed (PEN-SC1)

(c) A sample of amorphous PEN was annealed and then metallized (PEN-SC2)

In the cases b and c, the samples were placed in vacuum ($\approx 10^{-6}$ mbar) for 8 h before experiment. The objective was to remove traces of moisture and air bubbles formed sometimes in the sample.

In Fig. 1, we show thermal characteristics and crystallinities of the samples. Crystallinity was calculated using the relation:

$$\chi\,(\%) = 100 \times \frac{\Delta H_f - \Delta H_c}{\Delta H_\infty}$$

Where ΔH_f is the enthalpy of fusion, ΔH_c the enthalpy of crystallization and $\Delta H_\infty = 103.4$ J/g the enthalpy of fusion of fully crystalline PEN. [6]

With PEN-SC1 and PEN-SC2, we did not find any important difference with respect to semicrystallinity and other thermal characteristics like glass transition.

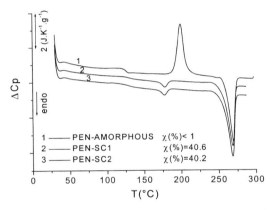

Figure 1. Comparison of thermal characteristics. ΔC_P is the specific heat and $\chi\,(\%)$ the crystallinity.

The measurements proceed in vacuum ($\approx 10^{-6}$ mbar) to avoid the problems of contour discharges. The characteristics current/voltage, I/E, were obtained at ambient temperature by application of a continuous voltage.

The samples were metallized with evaporated gold under vacuum. This technique makes it possible to obtain gold electrodes of 20 mm diameter and 30 nm thickness. The process of metallization is carried out in vacuum (1 mbar) to keep the purity of gold.

On annealing, the material was heated from room temperature to 160 °C at a heating rate of 10

°C/min. The temperature remained constant for 180 min to obtain the desired semicrystalline morphology. The sample was then cooled down to room temperature at the same rate. The heat treatment was carried out in a Mettler FP82HT device; the temperature control was performed with a regulator of Eurotherm-2404 type.

With regard to the instrumentation, we used a controlled feeding, Fug Electronik GmbH HCN 35-20000. Its use eliminated current disturbances due to the oscillations of excitation. Its operation was 0-20 kV with a maximum current of 5 mA. We used a vibrating read electrometer (Cary 401), which allows a reliable measurement of current from 5×10^{-15} A to 10^{-11} A. The principle of operation is the collection of charges by a vibrating capacity. The voltage which appears at the terminals of the capacitor is converted to an alternating signal by the vibrating electrode. By combining a synchronous detection with this alternating signal, the current is obtained. In order to perform acquisition of the data by this analog device, we used a numerical multimeter and a numerical and controllable (bus IEEE 488) picoammeter Keithley 617.[7]

Results and Discussion

We can approach the discussion of the influence of the morphology of the samples on their electric behavior.

In Fig. 2, we traced I/E characteristics of the three studied samples. In particular one of them shows that the insulator resistivity (the ratio of the direct voltage to the total current which crosses the electrodes and the sample for 120 min) depends at the same time on the bulk resistivity and the sheet resistivity of the sample (IEC standard resistivity 167).[8] The resistivities of the samples were found as: amorphous PEN is $\rho = 2.93 \times 10^{16}$ Ωm, for the PEN-SC1 $\rho = 5.73 \times 10^{16}$ Ωm, and finally for the PEN-SC2 $\rho = 3.76 \times 10^{16}$ Ωm .

In the region of weak fields, the rise in voltage induces an increase in current; for this linear behavior the resistivity is independent of the applied electric field.[9] The displacement of the charge carriers between the two electrodes is the source of conductivity in the material.

An increase in the crystallinity causes a decrease in the conductivity of PEN, since an increase in crystallinity favors a decrease in free volume making thus difficult the charge movement. The difference in the conduction levels of the two morphologies of PEN, i.e., amorphous and partly crystalline, depends on the quality of this course. Indeed, assuming a constant number of carriers, we can observe a strong reduction in mobility associated with the semicrystalline samples. We point out that this semicrystalline structure is heterogeneous because it contains amorphous parts, amorphous parts under constraint (thus more rigid) and crystalline parts. Probably, this heterogeneous nature of the structure causes the reduction in the conduction level.

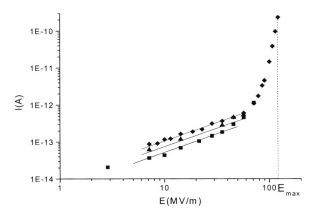

Figure 2. Comparison of I/E characteristics: ♦ PENam (amorphous PEN); ■ PEN-SC1. ▲ PEN-SC2 . (E_{max} is the maximum field applied to amorphous PEN, 120 MV/m).

The ionic conductance phenomenon is very sensitive to a reduction in the free volume associated with the formation of new intermolecular connections of the van-der-Waals type as well as to the proportion of the amorphous region more susceptible to lodge this type of transport.[10] However, the presence of the ionic conduction phenomenon due to the presence of impurities remains hypothetical.

Even though the difference in resistivity between the amorphous PEN and semicrystalline PEN is obvious, that which differentiates the PEN-SC1 and PEN-SC2 is less clear from the electric point of view. The measurement is qualitative in terms of the proportion of the ordered phases but not quantitative in terms of the nature, size and concentration of crystallites. On the other hand, a relation between the adhesion strength of the metal and the film resistance, particularly the sheet resistance, may exist. The latter is due to a combination of surface roughness and interfacial reactivity, which is very low for evaporated gold films; so a contribution to sheet resistance can be entirely attributed to surface roughness.[11] Another aspect of I/E characteristic is the value of the threshold field from which the number of the injected carriers is significant compared with that of intrinsic carriers.

A phenomenon of demetallization was found with the semicrystalline samples, PEN-SC1 and PEN-SC2, when getting into the region of 70 MV/m fields. This phenomenon forced us to stop the experiment due to great variation of the measurements. However, it is important to remark that it is not a predisruptive phenomenon. In amorphous samples, such phenomena do not occur.

Conclusion

The objective of this paper and its originality consists in making a comparison of the electric response of two different structures of PEN. The first was characterized by a homogeneous amorphous phase and the other by a heterogeneous morphology, corresponding to a crystallinity of 40 %. The heterogeneous PEN was prepared by two methods, which had an influence on the its properties.

The experimental results revealed an increase in the insulator resistivity with increasing crystallinity. The increase in crystallinity supports the reduction in free volume; thus, the displacements of the charge carriers are limited in particular to the species whose size is significant. The reduction in mobility is significant also for crystalline morphology characterized by a more compact and heterogeneous structure.

[1] M.Vesly, Z. Zamorsky, *Plaste Kautsch*. **1963**, *10*, 146.
[2] E. Krause, *PhD Thesis*, Paul Sabatier University, Toulouse, **1996**.
[3] J.C. Cañadas, J.A. Diego, M. Mudarra, J. Belana, R. Diaz-Calleja, M.J. Sanchis, C. Jaimés, *Polymer*, **1999**, *40*, 1181.
[4] J.J.Martinez-Vega, N. Zouzou, L. Boudou, J. Guastavino, *IEEE Trans. Diel. Electr. Insul.*, **2001**, *8*, 776.
[5] H.J. Wintle, *Conduction Processes in Polymers*, ASTM Publication, Philadelphia PA, **1983**, pp. 239-354.
[6] N. Zouzou, C. Mayoux, J. J. Martinez-Vega, *Rev. Int. Génie Electr.*, **2000**, *3*, 513.
[7] E. Duhayon, *PhD Thesis*, Paul Sabatier University, **2001**.
[8] C. Menguy, *Traité Génie Electr.* **1997**, *D2310*, 1.
[9] R. Coelho, B. Aladenize, *Les diélectriques : propriétés diélectriques des matériaux isolants*, Hermès, Paris, **1993**.
[10] P. J. Phillips, *Morphology and Molecular Structure of Polymers and Their Dielectric Behavior*, ASTM Publication, Philadelphia Pa, **1983**, pp. 119-238.
[11] A.O Ibidunni, R.J Brunner, *Metal/polymer adhesion: effect of ion bombardment on polymer interfacial reactivity*, in *Metallized Plastics*, K.L. Mittal, Ed., New York, **1997**, pp. 281-289.

Macromol. Symp. **2004**, *212*, 491-496

Influence of Carbazol-9-yl Substitution in Polysilanes on Charge Carrier Trapping and Recombination

Ireneusz Glowacki, * *Ewa Dobruchowska, Beata Luszczynska, Jacek Ulanski*

Department of Molecular Physics, Technical University of Lodz, Zeromskiego Str. 116, 90-924 Lodz, Poland

Summary: A series of poly{[3-(carbazol-9-yl)propyl]silane-*co*-methylphenyl-silane}s were investigated by optical absorption, photoluminescence and thermoluminescence measurements. It was found that the optical absorption bands of the carbazolyl side groups superimpose on those of the Si backbones in the ultraviolet range. This feature reduces photodegradability of Si-Si bonds during UV irradiation. The TL spectra recorded in the 15 – 325 K temperature range after photoexcitation at 15 K show that the carbazolyl side groups act as trapping sites in polysilanes. Increasing density of carbazolyl groups results in increasing population of deeper (*ca.* 150 meV) traps. Spectral analyses of the thermoluminescence at different temperatures are discussed and compared with analogous results for poly(9-vinylcarbazole) (PVK). It is concluded that the monomeric mechanism of luminescence dominates at low temperatures while the excimeric mechanism prevails at higher temperatures, similarly to PVK.

Keywords: carbazol-9-yl; luminescence; polysilanes; radiative recombination; thermoluminescence; trapping

Introduction

Silicon-backbone linear oligomers and polymers – polysilanes (also named polysilylenes) are functional polymer materials showing interesting photophysical properties.[1,2] The unique photoelectric properties of polysilanes, high-hole drift mobility (of the order of 10^{-4} cm^2/V s) and strong ultraviolet optical absorption and photoluminescence (PL) are associated with σ conjugation along the Si backbone.[3,4] The carbazol-9-yl-substituted polysilanes are systems of special interest (in spite of their low degree of polymerization), because these materials possess both charge generation and charge transport sites.[5-7]

          DOI: 10.1002/masy.200450863

Experimental

The copolymers and homopolymers of the carbazol-9-yl-substituted silane were prepared and purified according to the methods described by Tabei *et al.*[8] The polymer compositions, molecular weight distributions and glass transitions temperatures (T_g) are listed in Table 1.

The films of carbazol-9-yl-substituted polymers were obtained by casting from dichloromethane/dichloroethane (1:1) solution on quartz glass substrates for optical absorption and PL measurements (thickness > than 1 μm) and on steel plates for thermoluminescence (TL) measurements (7 – 10 μm). The poly(methylphenylsilane) (PMPSi) films were cast from tetrahydrofuran solution. For TL experiments, the polymer films were placed between a thermostatted stage and sapphire plate and squeezed together with a brazen frame anchored thermally to the stage in vacuum chamber (closed-cycle cryogenic system APD Cryogenics, type Displex); the heating rate was 7 K/min. The measurements were carried out in the 15 K - 325 K range after photoexcitation at 15 K using nitrogen pulsed laser. The luminescence was monitored with a photomultiplier Thorn EMI 9789 QB. Spectral analyses of the thermoluminescence were carried out using a grating monochromator with sweeping time 40-60 s in the range from 350 nm to 600 nm.

Table 1. Characteristics of investigated polymers.

Polymer	Monomers		Copolymer composition [1]	T_g [K]	Molecular weight [2]	
	MPSi	CzSi			M_n	M_w
PMPSi	10	0	–	298	5220	8890
PSiK (10:1)	10	1	10:0.5	299	3380	5830
PSiK (10:3)	10	3	10:2.8	300	8460	16400
PSiK (10:5)	10	5	10:3.3	297	5020	8150
PSiK (10:10)	10	10	10:6.0	295	850	1940
PSiK	0	10	–	296	1580	2440

MPSi – methylphenyldichlorosilane, CzSi – [3-(carbazol-9-yl)propyl] dichloro(methyl)silane
[1] By ^{29}Si NMR
[2] By GPC in dichloromethane, polystyrene calibration

Results and discussion

Two characteristic bands at 275 and 335 nm are seen in the absorption spectrum of PMPSi. Introduction of the carbazole groups causes an appearance of three new maxima at 264, 295 and 345 nm (see Fig. 1). The presence of two isosbestic points (at 270 and 300 nm) in the absorption spectra of solutions (not shown here) for all substituted polysilanes proves the existence of two independent chromophoric groups.

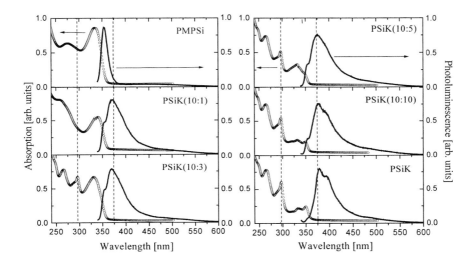

Figure 1. Optical absorption spectra (circles) and photoluminescence spectra under 337 nm excitation (solid lines) of polysilane films. The dashed lines indicate the position of the most characteristic absorption and PL maxima of carbazol-9-yl group.

The photoluminescence spectra of PMPSi, obtained after excitation with 337 nm wavelength, show a single emission band at about 360 nm. In the long-wavelength range (over 400 nm), one can observe also another, very weak broad band reported also by others.[4] The intensity of the 360 nm band associated with the σ^*-σ transition of silane chains, decreases upon increasing the carbazole group concentration due to the screening effect. We cannot exclude that for the systems with low carbazole contents also monomeric emission of carbazole groups may appear in this wavelength range (355 nm). At the same time the excimer emission of the carbazole groups (at ca. 370 and 420 nm) increases gradually. Similar photoluminescence spectra for such systems were reported earlier.[4,6] In the case of PSiK (10:5), PSiK (10:10) and PSiK, one can see the long-wavelength shoulder probably associated with the aggregates or extended excimer of carbazole rings, which may be formed in the systems with higher concentration of carbazol-9-yl groups. The TL spectra for different polysilane samples photoexcited at 15 K are presented in Fig. 2a. The TL spectrum of PMPSi

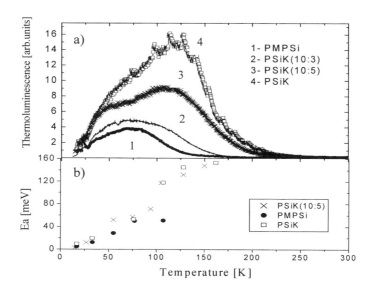

Figure 2. Thermoluminescence of silane homo- and copolymers (a), activation energy diagram (b), estimated from the partial heating experiments.

is the weakest, with a maximum at *ca.* 75 K and with a low-temperature (35-45 K) shoulder. This indicates that only shallow traps for charge carriers are present in PMPSi, which remains in good accordance with the former results obtained by Kadashchuk *et al.*[9] From the partial heating experiments, one can estimate that the traps with depth about 50 meV play a dominant role (see Fig. 2b). Introduction of carbazol-9-yl side groups results in a progressive increase in the thermoluminescence intensity and in the change of the TL spectra shape. In the TL spectrum for PSiK (10:5), one can clearly distinguish two maxima: the high-temperature maximum at around 120 K and the low-temperature one at around 45 K. This spectrum is similar to that previously obtained for poly(vinylcarbazole) (PVK), but in the latter the two maxima are better separated.[10] The occurrence of two TL maxima indicates that upon heating, the charge carriers are subsequently liberated from two different trapping levels. From the partial heating experiments, the depths of these two levels can be estimated as 40 meV and 120-130 meV, respectively. The TL spectrum of PSiK has the main maximum around 140 K and a shoulder on the low-temperature side, indicating the presence of another maximum. Partial heating results show that traps with depth 150 meV predominate in the PSiK (see Fig. 2b).

To identify the origin of radiative recombination processes giving rise to the TL maxima, the spectral analyses of the light emitted during the TL runs (at temperatures close to the TL maxima) have been performed (see Fig. 3a, b). For PMPSi, one band in the range

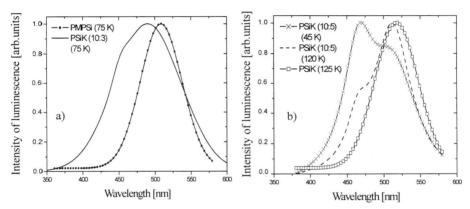

Figure 3. The normalized spectra of the light emitted in the course of the TL at temperatures close to the TL maxima for the investigated polysilanes.

440 - 540 nm is observed, corresponding to the weak, long-wavelength band in the photoluminescence spectra. Spectral analysis of the TL for all copolymers with carbazol-9-yl groups (with the exception of PSiK (10:1)) showed two bands. For the PSiK (10:5), the spectral analysis was performed at the first TL maximum at *ca.* 45 K and at the second, stronger TL maximum at *ca.* 120 K – in the same TL run. In both cases the spectra show two distinct bands at *ca.* 470 nm and *ca.* 520 nm (Fig. 3b). The spectrum of isothermal phosphorescence at 15 K, although very weak (not shown here), also reveals these two bands. These bands are associated with triplet states of carbazole groups: monomeric at about 470

nm and excimeric at about 520 nm. The ratio of the excimeric to monomeric emission intensity is higher for the high-temperature TL maximum. For PSiK, the TL emission is dominated by excimer sites (see Fig. 3b). Similar spectral evolution of the light emitted in the course of TL with increasing temperature was found for PVK.[9] Comparing the spectra of the PSiK (10:5) and PSiK samples shown in Figs. 2 and 3, one can conclude that full substitution of polysilane with carbazol-9-yl groups causes an increase in the population of deeper traps and of excimeric states contributing to thermoluminescence.

Conclusion

Introduction of carbazol-9-yl side-groups into polysilanes chain influences absorption, photoluminescence and thermoluminescence spectra. The carbazol-9-yl groups in polysilanes are chromophores separated from the main chain, protecting the Si-Si bonds from photodegradation by UV light. Photoluminescence spectra indicate the presence of aggregates (or other extended excimer structures) at high carbazole group concentrations. Spectral analysis of the emitted light indicates that the triplet monomeric mechanism of luminescence dominates at low temperatures and the triplet excimeric mechanism prevails at higher temperatures. These results are similar to those found for PVK. The carbazol-9-yl side-groups introduced to polysilanes form new photogeneration/recombination centres and act also as trapping sites. Increasing density of carbazol-9-yl groups results in increasing population of deep traps.

[1] R. D. Miller, J. Michl, *Chem Rev.* **1989**, *89*, 1359.
[2] Y. Kanemitsu, in: *Photonic Polymer Systems, Fundamentals, Methods, and Applications,* D.L. Wise *et al.*, Eds., Marcel Dekker, New York, 1998, p. 1 - 32.
[3] S. Nespurek, F. Schauer, A. Kadashchuk, *Chem. Monthly* **2001**, *132*, 159.
[4] A. Watanabe, Y. Tsutsumi, M. Matsuda, *Synth. Met.* **1995**, *74*, 191.
[5] I. Glowacki, J. Jung, J. Ulanski, *Synth. Met.* **2000**, *109*, 143.
[6] S. Mimura, H. Naito, T. Dohmaru, Y. Kanemitsu, M. Aramata, *Appl. Phys. Lett.* **2000**, *77*, 2198.
[7] I. Glowacki, E. Dobruchowska, J. Ulanski, *Synth. Met.* **2000**, *109*, 139.
[8] E. Tabei, M. Fukushima, S. Mori, *Synth. Met.* **1995**, *73*, 113.
[9] A. Kadashchuk, N. Ostapenko, V. Zaika, S. Nešpůrek, *Chem. Phys.* **1998**, *234*, 285.
[10] I. Głowacki, J. Ulański and B. Kozankiewicz , in: *Space Charge in Solid Dielectrics,* J. C. Fothergill and L.A. Dissado (Eds.), 1998, p. 251.

Phase Transitions in Poly(heptane-1,7-diyl biphenyl-4,4'-dicarboxylate). SAXS, WAXS and DSC Study

Ginka Todorova,[*][1] *Ernesto Pérez,*[2] *Mónica M. Marugán,*[2] *Manya Kresteva*[1]

[1] Sofia University, Faculty of Physics, James Bourchier 5 Blvd., 1126 Sofia, Bulgaria

[2] Instituto de Ciencia y Tecnología de Polímeros, Juan de la Cierva 3, 28006 Madrid, Spain

Summary: The main-chain thermotropic liquid-crystalline poly(heptane-1,7-diyl biphenyl-4,4'-dicarboxylate) (P7MB) was investigated by time-resolved small-angle X-ray scattering (SAXS), wide-angle X-ray scattering (WAXS), and differential scanning calorimerty (DSC). Nonisothermal crystallisation with different rates of cooling and heating was used. On cooling, two phase transitions are observed, isotropic melt - smectic (I-Sm) and Sm- three-dimensional crystalline structure (Sm-Cr), whereas on heating only one transition is observed, Cr-I transition. The transition enthalpies were calculated. Temperature dependences of d-spacings of all crystalline peaks and of the peak observed at high values of scattering vector in the SAXS region were derived. The temperature dependence of the degree of crystallinity was established, based on the integrated intensities of the crystalline peaks and amorphous halo in WAXS.

Keywords: DSC; liquid crystalline polymers; phase transitions; polyesters; WAXS

Introduction

In the last decades, liquid crystals (LC) attract great attention because of their broad spectrum of applications. A deep understanding of their structure formation and processes under different conditions as well as the nature of phase transitions will show the way to creating materials with predictable and desired properties.

P7MB belongs to the class of main-chain liquid-crystalline polyesters. It consists of biphenyl group as mesogen, and seven methylene groups as flexible spacer. So far, several publications have been devoted to investigations of polymeric dibenzoate series with all-methylene[1-4] and oxymethylene spacers[5], as well as blends[6] and statistic copolymers[7], composed of monomers with spacers of different lengths.

Since P7MB is known to be thermotropic LC[1-4], we focus our attention on the investigation of phase transitions, observed during nonisothermal processes.

Experimental

The chemical structure of P7MB is:

© 2004 International Union of Pure and Applied Chemistry

DOI: 10.1002/masy.200450864

P7MB was synthesized by melt transesterification of the diethyl ester of biphenyl-4,4'-dicarboxylic acid and the corresponding diol, using isopropyl titanate as catalyst. The details of the preparation have been described elsewhere[3]. It inherent viscosity is 1.03 dL/g, measured in chloroform at 25 °C.

Time-resolved SAXS and WAXS measurement were carried out at synchrotron DESY, Hamburg, Germany. Cooling (v_c) and heating (v_h) rates were 2 °C/min and 12 °C/min, respectively. Data treatment was performed using Bragg's law ($d = \lambda/2\sin\theta = 1/s$), where λ is the X-ray beam wavelength, θ is the scattering angle, and s is the scattering vector. Degree of crystallinity was calculated from the equation: $X_{cr} = \dfrac{\Sigma I_i}{I_{tot}}$, where ΣI_i is the sum of integrated intensity of crystalline reflections and I_{tot} is the total integrated intensity under the WAXS pattern.

Differential scanning calorimetry was applied to identify the number and temperature intervals of the observed phase transitions. Measurements were carried out on a Perkin-Elmer DSC7 apparatus under nitrogen atmosphere. A heating rate of 10 °C/min cooling rates were used, 2, 4, 6, 10, 20 °C/min.

Results and discussion

On cooling, P7MB undergoes two phase transitions, whereas during heating, only one transition is observed (Fig. 1a). The first exotherm is single for all cooling rates, whereas the second transition is split into two exotherms.

The high-temperature transition is attributed to I-Sm phase transformation[1,2]. Upon cooling from isotropic state (Fig. 1a, curve at 167.5 °C), the transition to smectic phase manifests itself by the growth of a diffraction peak at $s \approx 0.587$ nm^{-1} (1.705 nm), originating from the smectic layer periodicity, d_s, and a liquid-like diffuse scattering in WAXS pattern, centered at $s \approx 2.164$ nm^{-1} (0.462 nm), which is characteristic of interchain packing correlation (Fig. 2a, curve at 138.5 °C). The smectic layers are formed by the mesogens and one and the same macromolecule passes through many layers. Watanabe and Hayashi[1,2] elucidated the smectic type as S_{CA}, where the average macromolecular axis is normal to the smectic layer surface and the mesogen axes are tilted with respect to the layer normal in such a way that the consecutive mesogens form an angle of about 60° - i.e. the mesogens in every neighboring layers are tilted with respect to the layer normal at the same angle but in the opposite directions.

The maximum of the transition exotherm shows a linear dependence on v_c (see Fig.1b - upper plot). The shift of the transition with v_c is due to the undercooling effect. The enthalpy for all the used cooling rates is constant, $\Delta H_{I-S} \approx 15.8$ J.g^{-1}, suggesting that the quantity of the material, which is transformed into a smectic phase, is independent of v_c. The temperature interval, in which Sm phase is stable, varies from 5.22 °C (at $v_c = 4$ °C/min) to 14.79 °C (at $v_c = 20$ °C/min). The general tendency is expansion of the interval with increasing cooling rate, which could be explained by the conditions of crystallization onset. At high cooling rates, the crystallization is hindered and it starts at lower temperatures.

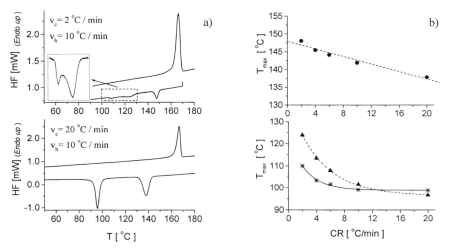

Figure 1. a) DSC curves of P7MB, starting with cooling from isotropic melt a subsequent heating: cooling rate 2 °C/min (upper plot) and 20 °C/min (bottom plot), heating rate 10 °C/min. HF is the heat flow; b) Dependence of transition temperatures on cooling rate, presented by the exotherm maxima: I-Sm transition (upper plot) Sm-Cr transition with double exotherm (lower plot).

Second transition interval of the DSC cooling runs corresponds to transformation to the Cr phase. When v_c is low (2 °C/min), two different exotherms are well seen (Fig.1a - upper curve), while for 10 °C/min the peaks merge and only the peak shape asymmetry is a hint for the existence of two exotherms. Once the high cooling rate is used, the shape is even almost symmetric (Fig.1a - lower curve). The results for the cooling rate dependence of the two exotherm maxima are plotted in Figure 1b (lower graph). They follow an exponential law. T_{max} of the first DSC peak component decreases faster than the second one. The enthalpies of the exotherms vary in a different way (see Table I). $\Delta H_{Sm-1-Cr}$ decreases with v_c, whereas

$\Delta H_{Sm-2-cr}$ increases. It is worth noting that the sum of both enthalpies is constant ($H_{Sm-Cr} \approx$ 19.4 J.g^{-1}). The fact could be explained as a competition between two kinetics dependent processes. At low v_c, the prevailing mechanism is associated with the first exotherm component, whereas at high v_c the mechanism connected with the second exotherm component is dominant. It could be speculated that there exists polymorphism, similar to the case discussed by Bae et al.[8], and the structure transformation goes through different metastable phases under different crystallization conditions, but since we do not have any unambiguous proof, we will not concern the problem here.

Table I. Transition enthalpies, calculated from DSC curves at different cooling rates and a heating rate of 10 °C/min. $\Delta H_{Sm-1-Cr}$ and $\Delta H_{Sm-2-Cr}$ are enthalpies for first and second component of the double exotherm, observed at Sm-Cr transition and ΔH_{Cr-I} is the enthalpy of fusion.

		Cooling rate (°C/min)				
		2	4	6	10	20
$\Delta H_{Sm-1-Cr}$	(J.g^{-1})	13.96	13.15	10.91	7.60	4.59
$\Delta H_{Sm-2-Cr}$	(J.g^{-1})	5.54	6.49	8.50	12.01	14.48
		Subsequent heating at 10 °C/min				
ΔH_{Cr-I}	(J.g^{-1})	31.2	31.2	31.9	32.1	32.6

In the case of v_c=2 °C/min, at the beginning of this interval, six crystalline peaks overlapping the amorphous halo are seen on WAXS patterns (see Fig.2a, the curve at 67.5 °C and the inset). Their intensities gradually increase with decreasing temperature. It seems that the smectic layer spacing, d_s, gradually decreases from the beginning to the end of the transitional area, and the peak is preserved in the crystalline phase (Fig. 2b). Its value in Cr phase is $d_s \approx$ 1.672 nm. In fact, the smectic peak probably disappears and a crystalline peak, with d-spacing very close to that of the smectic periodicity, grows[3,9] but, because of the difficulties originating from the deconvolution procedure, we were not able to distinguish them.

In Fig. 2c, index of crystallinity, X_{Cr}, deduced from WAXS, is shown. Since some of the crystalline peaks are very weak and, after the deconvolution procedure, they were estimated with comparatively high levels of errors, we used only the integrated areas under the highest two peaks (p.1 and p.3 in the inset of Fig 2a) in calculations of crystallinity. We believe that the degree of crystallinity is proportional to these values since we assume only one crystalline structure on the basis of some previous results. On cooling, X_{Cr} grows stepwise during transformation from the Sm into Cr phase.

The final crystalline structure is stable in the low-temperature region with an upper limit of ca. 112 °C (derived from X-ray experiments) and the peaks depend very slightly on temperature. The particular values for the temperature expansion coefficients (β_T) of all Cr peaks are shown in Table II.

On heating, the crystalline structure obtained on cooling is preserved up to the isotropisation temperature. This fact is supported by the presence of all crystalline peaks. The temperature dependences are the same as those for cooling at temperatures below 112 °C (see Table II). The degree of crystallinity gradually decreases (from values of ca. 20 % to 10 %) up to the melting interval, in which it reaches zero.

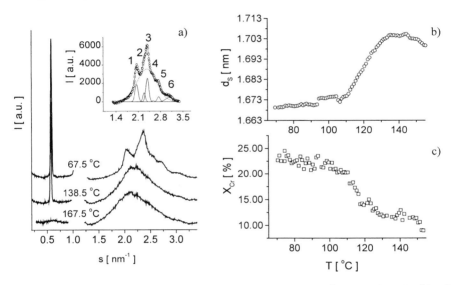

Figure 2. a) X-ray diffractograms at different temperatures: 167.5 °C - I melt, 138.5 °C - Sm LC phase, 67.5 °C - Cr structure. The inset shows assignment of the crystalline peaks (for details, see the text); b) smectic or crystalline d-spacing, d_s, versus temperature; c) index of crystallinity, X_{Cr}, versus temperature.

The transition temperature during heating is constant (≈ 166.7 °C), showing that the final structure, formed in the solid state, is the same for all cooling rates. One could make a conclusion that despite of possible different ways of reaching the final structure, it is the same and independent of the cooling rate. The enthalpies of fusion are shown in Table I (last row). Their values slightly increase with the rising cooling rate and constant heating rate.

Unfortunately, because of the low degree of crystallinity, we were not able to calculate the parameters of crystal unit cell and the type of symmetry. We only assumed monoclinic or triclinic system of symmetry, similarly to the conclusion made by Osada et al.[10] for P5MB. The direction of the d-spacing between consecutive mesogens is collinear with crystallographic c-axis; because of the zig-zag mesogen arrangement, the length of the unit cell along (001) is around 3.344 nm - i.e., twice d_S.

Table II. d-spacings of crystalline peaks in the Cr phase (at 41 °C) and their temperature dependences, described by temperature expansion coefficients, β_T. Cooling rate $- 2$ °C/min and heating rate $- 12$ °C/min. The cooling values are valid below 112°C.

Peak	1	2	3	4	5	6
d^{cool} (nm)	0.489	0.438	0.419	0.390	0.365	0.326
$\beta_T^{cool} = \delta d/\delta T$ (nm/°C)10^{-5}	4.2 ± 0.2	8.7 ± 0.6	11.8 ± 2	80 ± 4	3.0 ± 0.2	8.2 ± 0.5
d^{heat} (nm)	0.489	0.437	0.419	0.389	0.364	0.326
$\beta_T^{heat} = \delta d/\delta T$ (nm/°C)10^{-5}	4.6 ± 0.4	8 ± 0.9	11.5 ± 0.7	93 ± 8	3.25 ± 0.04	9.2 ± 0.9

Conclusion

P7MB shows enantiotropic behavior on cooling and monotropic on heating for all the used cooling and heating rates. Smectic structure is formed for all cooling rates on cooling and the temperature interval of its stability increases with rising cooling rate. The quantity of the material, which is transformed into smectic phase, is independent of the cooling rate. The smectic layer spacing slightly increases with decreasing temperature. The crystalline structure, formed at the end of structure transformation during cooling, is the same for all the used cooling rates. The structure is stable below 112 °C on cooling and up to the melting interval on heating. Cr structure has low symmetry, with plausible monoclinic or triclinic symmetry of the unit cell.

[1] J. Watanabe, M. Hayashi, *Macromolecules* **1988**, *21*, 278.
[2] J. Watanabe and M. Hayashi, *Macromolecules* **1989**, *22*, 4083.
[3] E. Pérez, A. Bello, M. M. Marugán, J. Pereña, *Polym. Commun.* **1990**, *31*, 386.
[4] E. Pérez, J. Pereña, R. Benavente, A. Bello, *Handbook of Engineering Polymeric Materials,* New York-Basel-Hong Kong **1997**, 383.
[5] E. Pérez, M. M. Marugán, D. L. VanderHart, *Macromolecules* **1993**, *26*, 5852.
[6] J. Watanabe, M. Hayashi, Y. Nakata, T. Niori, M. Tokita, *Prog. Polym. Sci.* **1997**, *22*, 1053.
[7] A. Nakai, T. Shiwaku, W. Wang, H. Hasegawa, T. Hashimoto, *Polymer* **1996**, *37*, 2259.
[8] H. Bae, J. Watanabe, Y. Maeda, *Macromolecules* **1998**, *31*, 5947.
[9] Unpublished results concerning isothermal crystallization.
[10] K. Osada, H. Niwano, M. Tokita, S. Kawauchi, J. Watanabe, *Macromolecules* **2000**, *33*, 7420.

Macromol. Symp. **2004**, *212,* 503-508

Polydiacetylene Nanowires in Ultrathin Films

Marina Alloisio,[1] *Dina Cavallo,*[1] *Carlo Dell'Erba,*[1] *Carla Cuniberti,*[1] *Giovanna Dellepiane,*[1] *Emilia Giorgetti,*[2] *Giancarlo Margheri,*[2] *Stefano Sottini*[2]

[1] INFM, INSTM, Dipartimento di Chimica e Chimica Industriale, Università di Genova, Via Dodecaneso, 31, I-16146 Genova, Italy
[2] IFAC-CNR, Via Panciatichi, 64, 50127 Firenze, Italy

Summary: In this paper the results of studies carried out on thin films of new poly[bis(carbazol-9-ylmethyl)diacetylene]s (PCDAs) are reported. The preparation of the films has required clever synthesis to make processable the conjugated polymers without degrading their optoelectronic properties. To this end, the parent poly(diacetylene), (polyDCHD), has been modified by introducing long alkyl or acyl chains in the 3 and 6 positions of the carbazole rings. Electronic absorption spectra and linear and nonlinear optical characterization of three types of PCDAs are reported and compared.

Keywords: electronic spectra; NLO properties; polycarbazolyldiacetylenes; thin films

Introduction

The formation of ultrathin films of conjugated polymers with nanometer control over thickness is an area of intense study, as these films have desirable electronic, linear and nonlinear optical properties. In recent years we have been involved in the preparation of symmetrical derivatives of poly(carbazolyldiacetylene)s ($-CCz=CCz-C\equiv C-$)$_n$ by topochemical polymerization of alkyl- and acyl-substituted diacetylenes homologous to 1,6-di(carbazol-9-yl)hexa-2,4-diyne (DCHD). Because of its outstanding properties[1], polyDCHD is one of the most studied polydiacetylenes (PDAs). Among the poly[bis(carbazol-9-ylmethyl)diacetylene]s (PCDAs) prepared in our laboratory, polyDCHD-HS and polyDPCHD have shown a more interesting behavior. PolyDCHD-HS, with long hexadecyl chains in the 3 and 6 positions of carbazolyl substituents, in microcrystalline powder of the red form, is self-assembled into cylindrical shapes producing a two-dimensional columnar structure with hexagonal symmetry Col$_{ho,}$ which undergoes a transition to a less ordered one (Col$_{hd}$) around 85 °C[2]. This structure differs from the ordered one by a reduced correlation between the repeat units inside the columns. The alkyl chains in a liquid-

DOI: 10.1002/masy.200450865

like state fill the space between the columns and are responsible for a large separation of polymer molecules (3.8 nm). The quasi-one-dimensional behavior and good processability of this polymer, which allows the preparation of homogeneous films by standard spin coating technique, has stimulated full investigation of its linear and nonlinear optical properties and so far very interesting results have been obtained[3]. Although the third-order susceptibility $\chi^{(3)}$ of polyDCHD-HS, measured in the telecommunication window either in the solid state by third harmonic generation (THG) technique[4] or in solution by Z-scan[5] is of the same order as that measured for other PDAs (10^{-11} esu), the low absorption in the same spectral region together with the polymer sub-picosecond response time provides a rather nice figure of merit, which makes it promising for applications. On the other hand, the measurement of $\chi^{(3)}$ performed by surface plasmon spectroscopy (SPS) with ps pulses on quasi-monomolecular layers of the polymer spun on silver-coated plates has given a surprisingly large off-resonance nonlinearity $|\chi^{(3)}| > 10^{-9}$ esu at 1064 nm. The giant nonlinearity observed has been related to the nanostructured surface of the silver film that provides an electromagnetic mechanism for the enhancement of the nonlinearity through local field effects[6], and makes these hybrid structures interesting for applications in nanostructural devices having photonic bandgap properties[7].

PolyDPCHD has palmitoyl chains in the 3 and 6 positions of the carbazoles designed to investigate the effects of the decreased electrondonor properties of the carbazole rings on optical and electronic properties of the polymer. Powder X-ray diffraction studies have revealed also in this polymer the formation at room temperature of a Col$_{hd}$ supramolecular structure characterized by an intercolumnar distance of 3.9 nm, equal to that found for the high-temperature mesophase of polyDCHD-HS. PolyDPCHD is poorly soluble at room temperature in common organic solvents, but red films of good optical quality can be obtained by spin-coating or casting of monomer solutions followed by thermal polymerization. Only preliminary data on the characterization of the polymer and its films have so far been reported[8].

In order to exploit the effect of asymmetrical substitution of the carbazole rings on the polymer optoelectronic properties, we have synthesized a novel diacetylene (a-DCHD) carrying the hexadecyl substituents only on one of the rings. The resulting polymer, named polya-DCHD (a = asymmetric) has been obtained by thermal topochemical polymerization and carefully purified according to the procedure reported in Ref. 2.

Results and Discussion

The chemical structures of the repeat units of the polymers described in the present paper are in Figure 1.

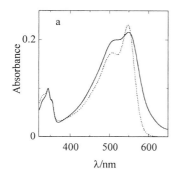

X=Y= n-C$_{16}$H$_{33}$ polyDCHD-HS

X=Y= n-C$_{15}$H$_{31}$CO   polyDPCHD

X= n-C$_{16}$H$_{33}$, Y=H   poly$a$-DCHD

Figure 1. Chemical structures of the PCDAs.

Poly$a$-DCHD and polyDCHD-HS are soluble in common organic solvents so that thin films of optical quality can be prepared by standard spin-coating techniques. The room-temperature absorption spectra of thin films spun on glass substrates from toluene solutions of both polymers are reported in Figure 2 and compared with the solution spectra.

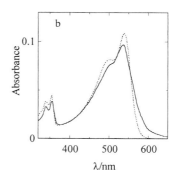

Figure 2. Absorption spectra of the films on glass (full line) and toluene solutions (dotted line) of (a) poly$a$-DCHD and (b) polyDCHD-HS.

The correspondence of the solution and film spectra shows that also in the solid state the polymer backbones are isolated. At this point it is interesting to notice that the absorption spectrum of a thin film of polyDPCHD (Figure 3), cast on glass as reported in Ref. 8, shows a strong and extremely narrow exciton band similar to those observed in PDA single crystals, and a well resolved fluorescence emission. These properties appear quite different from those reported in Figure 2 for the other polymers, which, as films, do not give appreciable emission at room temperature. However, both polya-DCHD and polyDCHD-HS in benzene solution show absorption and emission profiles practically identical to those of Figure 3[9,10]. These results indicate that isolated red PDA chains with ordered conformations show measurable fluorescence. This finding is in agreement with the observation of a very high fluorescence yield for red chains of poly4BCMU isolated in the monomer matrix at very low temperature[11]. Unfortunately, at present, we cannot quantify the fluorescence yield for our film.

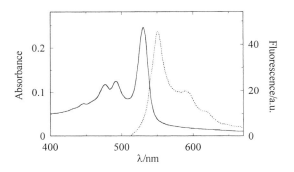

Figure 3. Absorption (full line) and emission (dotted line) spectra of a polyDPCHD film on glass. Excitation wavelength 480 nm, $T = 20\ °C$.

Since these polymers have been designed for application in photonic devices, their thermal and photochemical stability have been checked. PolyDPCHD films are the most stable because they do not undergo changes up to 180 °C and after six hours of UV irradiation with 35 W lamp. All the polymers, however, are completely stable at a conventional temperature of 75 °C.

The optical linear characterization so far performed on these polymers has provided the refractive index values reported in Table 1. For the first two polymers, measurements have been carried out with thicker guided films, which have permitted the determination of both transverse electric (TE) and transverse magnetic (TM) refractive indices (n_{TE} and n_{TM}). It is interesting to notice negligible film birefrigence evidenced by these values. We believe that this insensitivity to polarization may also hold for the nonlinearity, thus making these polymers very interesting for telecommunications. The only slightly lower value of n_{TM} found for the thinner film of polyDCHD-HS spun on a silver layer indicates that different substrates and film thickness do not substantially affect the refractive index. The refractive index of polyDPCHD has been measured by ellipsometry[12] with unpolarized light. The higher values found for this polymer are probably due to a different chemical nature of the substituents in the carbazole rings.

Table 1. Linear optical properties of polya-DCHD, polyDCHD-HS and polyDPCHD.

PCDA	λ	Thickness	n_{TM}[1]	n_{TE}[1]	n[2]
	nm	μm			
Polya-DCHD	632.8	2.0	1.583	1.589	-
PolyDCHD-HS	1321	2.4	1.567	1.548	-
	849	2.4	1.585	1.563	-
	849	0.02[3]	1.550	-	-
PolyDPCHD	1321	0.032	-	-	1.648
	849	0.032	-	-	1.663
	632.8	0.032	-	-	1.702

[1] From m-line spectroscopy measurements, Ref. 3
[2] From ellipsometry measurements, Ref. 12
[3] Film spun on silver-coated plate

So far the third-order nonlinear polarizability coefficient $\chi^{(3)}$ has been measured on thin films of polyDCHD-HS and polyDPCHD by employing two different techniques. THG technique has been applied[4,12] to study films spun on glass. Surface plasmon spectroscopy (SPS) has been used to study thin films of polyDCHD-HS spun on silver-coated plates[6].

Table 2. Nonlinear optical properties of polyDCHD-HS and polyDPCHD.

PCDA	λ	Thickness	$\chi^{(3)}$ [1]	$\chi^{(3)}$ [2]
	nm	nm	esu	esu
PolyDCHD-HS	1500	140	4.0×10^{-11}	-
	1800	140	2.9×10^{-11}	-
	1064	157	-	$\sim 10^{-9}$
	1064	43	-	$\sim 10^{-8}$
PolyDPCHD	1600	32	4.5×10^{-11}	

[1] From THG measurements, Refs. 4,12
[2] From SPS measurements, Ref. 6

Table 2 collects all the obtained data. As already noted, very different values have been obtained by the two methods for polyDCHD-HS, while similar data are observed for the two polymers by using the THG technique. However, we would like to stress that the THG dispersion curves of $\chi^{(3)}$ for the two polymers are quite different as a result of different contributions of multiphotonic processes[4,12].

Conclusions

We have shown that in thin red films the conjugated backbones of our PCDAs are isolated and of polyDPCHD also in ordered conformations. The presence of ordered chains in organized nanostructured particles enhances the fluorescence emission. The $\chi^{(3)}$ values from THG are of the same order of magnitude at fixed wavelength and seem to be unaffected by detailed chemical structure of substituents. However, the $\chi^{(3)}$ dispersion evidences different contributions of multiphoton processes[4,12], which deserves further investigations.

Acknowledgments

Financial support from the Italian Ministry of University and Scientific and Technological Research and from the PF MSTA II of the National Research Council of Italy is acknowledged.

[1]   H. Matsuda, H. Nakanishi, N. Minami, M. Kato, *Mol. Cryst. Liq. Cryst.* **1988**, *160*, 241.
[2]   B. Gallot, A. Cravino, I. Moggio, D. Comoretto, C. Cuniberti, C. Dell'Erba, G. Dellepiane, *Liq. Cryst.* **1999**, *26*, 1437.
[3]   E. Giorgetti, G. Margheri, F. Gelli, S. Sottini, D. Comoretto, A. Cravino, C. Cuniberti, C. Dell'Erba, I. Moggio, G. Dellepiane, *Synth. Met.* **2001**, *116*, 129.
[4]   F. D'Amore, A. Zappettini, G. Facchini, S.M. Pietralunga, M. Martinelli, C. Dell'Erba, C. Cuniberti, D. Comoretto, G. Dellepiane, *Synth. Met.* **2002**, *127*, 143.
[5]   E. Giorgetti, G. Toci, M. Vannini, D. Cavallo, G. Dellepiane, *Synth. Met.* **2002**, *127*, 139.
[6]   E. Giorgetti, G. Margheri, S. Sottini, G. Toci, M. Muniz-Miranda, L. Moroni, G. Dellepiane, *Phys. Chem. Chem. Phys.*, **2002**, *4*, 2762.
[7]   S.I. Bozhevolnyi, J. Ereland, K. Leosson, P.M.W. Skovgaard, J.M. Hvam, *Phys. Rev. Lett.*, **2001**, *86*, 3008.
[8]   D. Cavallo, M. Alloisio, C. Dell'Erba, C. Cuniberti, D. Comoretto, G. Dellepiane, *Synth. Met.* **2002**, *127*, 71.
[9]   M. Alloisio, A. Cravino, I. Moggio, D. Comoretto, S. Bernocco, C. Cuniberti, C. Dell'Erba, G. Dellepiane, *J. Chem. Soc., Perkin Trans 2*, **2001**, 146.
[10] M. Alloisio, S. Sottini, D. Cavallo, C. Dell'Erba, C. Cuniberti, G. Dellepiane, B. Gallot, paper presented at the E-MRS 2002 meeting, Strasbourg, 18-21 June 2002.
[11] R. Lécuiller, J. Berréhar, C. Lapersonne-Meyer, M. Schott, J.-D. Ganière, *Chem. Phys. Lett.*, **1999**, *314*, 255.
[12] F. D'Amore, S.M. Pietralunga, M. Martinelli, M. Alloisio, D. Cavallo, C. Cuniberti, C. Dell'Erba, G. Dellepiane, paper presented at the E-MRS 2002 meeting, Strasbourg, 18-21 June 2002.

Macromol. Symp. **2004**, *212*, 509-514

Organic Electroluminescent Devices Containing Phosphorescent Molecules in Molecularly Doped Hole Transporting Layer

Waldemar Stampor,[*1] *Jakub Mężyk,*[1] *Jan Kalinowski,*[1] *Massimo Cocchi,*[2] *Dalia Virgili,*[2] *Valeria Fattori,*[2] *Piergiulio Di Marco*[2]

[1] Department of Molecular Physics, Technical University of Gdańsk, ul. Narutowicza 11/12, 80-952 Gdańsk, Poland
[2] Institute of Organic Synthesis and Photoreactivity, National Research Council of Italy, via P. Gobetti 101, 40129 Bologna, Italy

Summary: We demonstrate high-efficient simple electrophosphorescent devices comprised of tris{4-[*N*-(3-methylphenyl)anilino]phenyl}amine (m-MTDATA) dispersed in a polycarbonate (PC) matrix as a hole-transporting layer (HTL), and 2-(biphenyl-4-yl)-5-(4-*tert*-butylphenyl)-1,3,4-oxadiazole (PBD) as an electron-transporting layer (ETL). The HTL doped with a complex phosphor *fac*-tris(2-phenylpyridine)iridium, [Ir(ppy)$_3$], and/or 5,6,11,12-tetraphenyltetracene (rubrene) fluorescent dye is shown to act as an emitter. Devices containing [Ir(ppy)$_3$] as a single HTL dopant show the highest external quantum efficiency (QE) reaching 9 % (photon/electron) due to direct electron-hole recombination on phosphorescent [Ir(ppy)$_3$]. A decrease in QE of one order of magnitude at high current densities is observed in all devices. Addition of rubrene to [Ir(ppy)$_3$]-doped devices shifts the maximum QE towards larger current densities.

Keywords: charge transport; host-guest systems; luminescence; organic light-emitting diodes (LEDs); thin films

Introduction

The electron-hole recombination in organic light-emitting diodes (LEDs) results in the production of singlet- and triplet-emitting states [1]. Due to spin statistics, three times more triplets than singlets are usually created. However, in common electrofluorescent LEDs only the singlets contribute directly to the emission and triplets relaxing nonradiatively are lost for light generation. Using phosphorescent materials with radiatively active triplets provides a straight way to a significant increase in quantum efficiency (QE) of organic LEDs. Indeed, the successful application of phosphorescent molecules such as porphyrinplatinum (PtOEP) [2-5] and *fac*-tris(2-phenylpyridine)iridium, [Ir(ppy)$_3$] [4,6-11], and some other Ir(III) complexes [12-14] in

DOI: 10.1002/masy.200450866

generation of organic electroluminescence has enabled the fabrication of organic LEDs with an external QE up to 19 % ph/carrier [12]. The optimizing device performance led to the design of state-of-the art multilayer LEDs having the structure ITO/HTL/EML/HBL/ETL/Mg with four organic layers – hole-transporting layer (HTL), charge carrier combination and emission layer (EML), electron-transport/hole blocking layer (HBL) and electron-transporting layer (ETL). Typically, the EML consists of organic phosphor embedded in electron-transporting host like (CBP) [4,6-8,10,13]. So far, very few organic LEDs with simpler structures and high QE have been demonstrated [10,11,14]. Using poly(9-vinylcarbazole) (PVCz) as a host material for [Ir(ppy)₃] guest emitter and an oxadiazole with dimeric structure of PBD, named OXD-7, as ETL, double layer (DL) devices ITO/PVCz:Ir(ppy)₃/OXD-7/Mg with the maximum external QE of 7.5 % (ph/e) were fabricated [11]. In turn, the DL device comprising m-MTDATA HTL and CBP host doped with [Ir(ppy)₃] phosphor exhibited the peak external QE of 12 % (ph/e) [10]. In the present work we demonstrate another simple high-efficiency devices with DL structure. In our LEDs, [Ir(ppy)₃] dispersed in m-MTDATA-doped polycarbonate (PC) matrix form HTL and oxadiazole PBD is used as ETL. This configuration allows to obtain the external QE of 9 % (ph/carrier). We also show that the QE efficiency decrease at high current densities, commonly observed in all organic LEDs, can be partly reduced by incorporating additionally in HTL a fluorescent dye (rubrene).

Results and discussion

The molecular structures, EL device configuration and energy level diagram of the materials used in this study are shown in Fig. 1. The DL LEDs were prepared by spin-coating of HTL and vacuum evaporation of ETL. The (m-MTDATA:PC:dopant) HTL was spin-cast onto ITO (indium tin oxide) glass substrates. Then, a 100 % PBD layer (ETL) was deposited by vacuum evaporation. The sandwich devices were completed by a vacuum-deposited Ca covered with a protecting layer of Ag. The weight content of m-MTDATA in HTL was 69 – 75 %. As HTL dopants, phosphorescent [Ir(ppy)₃] (designated hereafter as Ir) and/or fluorescent rubrene (Rb) were incorporated. The ionization potential of m-MTDATA (5.0-5.1 eV [10,15]) makes the ITO anode a good hole-injecting contact to the HTL. In turn, the electron affinity of PBD (2.6 eV [16]) enables the Ca cathode to be an efficient injector of electrons to the ETL. From the HOMO and LUMO positions of Ir (5.4 eV and 3.0 eV [10,17], respectively) and Rb (5.4 eV and

3.2 eV [15]) it is apparent that the molecules of Ir and Rb in the (m-MTDATA:PC) matrix form effective electron-trapping sites. The method of external EL QE measurement and other experimental details are described elsewhere [17,18].

(a)

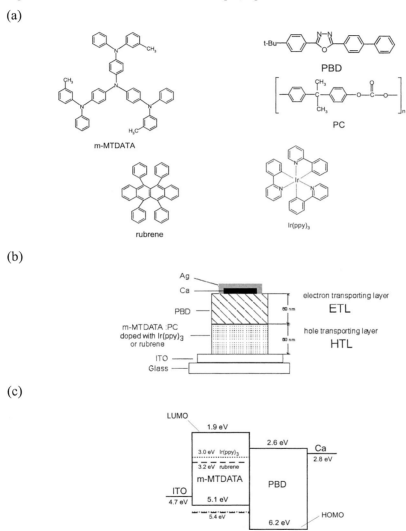

(b)

(c)

Figure 1. The molecular structures (a), EL device configuration (b) and energy-level diagram (c) of the materials used in this study

512

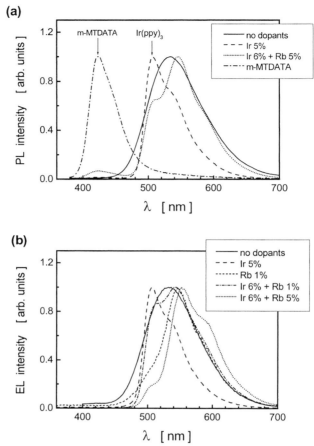

Figure 2. Normalized PL and EL spectra of devices with different HTLs: [Ir(ppy)₃] (Ir) and rubrene (Rb) in (m-MTDADA:PC)

Figure 2 shows the PL and EL spectra of the DL structures from Fig. 1. The spectra of undoped devices [18] are shown for comparison. The emission spectra of all devices are characteristic of the dopants (either phosphorescence of Ir or fluorescence of Rb) when doped separately, and their combination if both dispersed in the HTL. The devices exhibit the EL external QE between 0.1 % (ph/carrier) up to 9 % (ph/carrier), dependent on the type and concentration of dopant and the driving current (Fig. 3). Generally, the LEDs with Rb-doped HTLs show efficient emission at higher voltages compared with Ir-doped or undoped devices. The devices containing a small

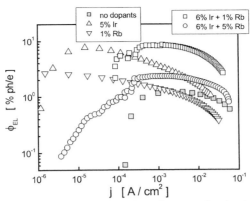

Figure 3. External quantum efficiency as a function of current density j for doped and undoped devices

admixture of Rb (1 %) as a single fluorescent HTL dopant show the QE values close to those of undoped devices. Since electroluminescence in dopant-free devices originates from exciplexes (m-MTDATA/PBD) [18], an efficient energy transfer from these exciplexes to Rb molecules is assumed to occur most probably in Rb-doped devices. A direct charge-trapping on Rb molecules can also contribute to this emission, similar to that for Ir-doped devices (see below). Devices containing [Ir(ppy)₃] as a single HTL dopant show the highest quantum efficiency (9 % ph/carrier). Considering that holes are blocked at the HTL/ETL interface and electrons are likely to be directly injected from PBD to Ir and then readily trapped on Ir molecules, direct exciton formation on Ir should be considered as a significant pathway to the efficient EL emission. Alternatively, charge recombination and exciton formation may occur first on m-MTDATA and then excitons may be subsequently transferred to Ir through the Dexter or Förster process. However, a rather high barrier for electron injection from PBD to HTL and the absence of any m-MTDATA emission in the EL spectra of undoped devices suggest that the latter process is less probable. Adding Rb to Ir-doped devices shifts the maximum QE towards larger current densities (Fig. 3). Whereas a small amount of Rb does not change the maximum value of QE, a significant decrease in QE is observed at larger concentrations of Rb. These phenomena are most probably associated with modification of electron-transporting properties of the (m-MTDATA:PC) system by Rb molecules like in rubrene-doped TPD HTL layer [19] in electrofluorescent LEDs. The common feature of all devices is a decrease in QE of nearly one order of magnitude at high current densities, independent of emission type (fluorescence or phosphorescence) of dopants. The following three mechanisms can contribute to the observed effect: exciton-exciton annihilation, exciton-charge carrier interaction and electric field-induced quenching of fluorescence or phosphorescence [4,17]. Currently we are unable to determine which process is dominant in the EL devices studied in the present work. Details of the quenching mechanism are under investigation.

Conclusions

$[Ir(ppy)_3]$ and rubrene doped into HTL (m-MTDATA:PC) act as the emitters in DL devices with PBD as an ETL. Devices containing $[Ir(ppy)_3]$ as a single HTL dopant show the highest QE (9% ph/carrier) due to direct electron-hole recombination on phosphorescent $[Ir(ppy)_3]$ molecules. Adding rubrene to $[Ir(ppy)_3]$-doped devices shifts the maximum QE towards large current densities. The one order of magnitude decrease in QE at high current densities is observed in all devices, independent of emission type (fluorescence or phosphorescence) of dopants. Three mechanisms possibly contributing to the observed effect (exciton-exciton annihilation, electric field-induced quenching of photoluminescence, and exciton-charge carrier interaction) should be considered in further study.

[1]   J. Kalinowski, *J. Phys. D: Appl. Phys.* **1999**, 32, R179.
[2]   M. A. Baldo, D. F. O'Brien, Y. You, A. Shoustikov, S. Sibley, M. E. Thompson and S. R. Forrest, *Nature* **1998**, 395, 151.
[3]   V. Cleave, G. Yahioglu , P. Le Barny, R. Friend and N. Tessler, *Adv. Mater.* **1999**, 11, 285.
[4]   M. A. Baldo, C. Adachi and S. R. Forrest, *Phys. Rev. B* **2000**, 62, 10967.
[5]   G.E. Jabbour, J.-F. Wang and N. Peyghambarian, *Appl. Phys. Lett.* **2002**, 80, 2026.
[6]   M. A. Baldo, S. Lamansky, P. E. Burrows, M. E. Thompson and S. R. Forrest, *Appl. Phys. Lett.* **1999**, 75, 4.
[7]   T. Tsutsui, M.-J. Yang, M. Yahiro, K. Nakamura, T. Watanabe, T. Tsuji, Y. Fukuda, T. Wakimoto and S. Miyaguchi, *Jpn. J. Appl. Phys.* **1999**, 38, L1502.
[8]   M.A. Baldo, M.E. Thompson and S.R. Forrest, *Nature* **2000**, 403, 750.
[9]   M. Ikai, S. Tokito, Y. Sakamoto, T. Suzuki and Y. Taga, *Appl. Phys. Lett.* **2001**, 79, 156.
[10] C. Adachi, R. Kwong and S.R. Forrest, *Org. Electron.* **2001**, 2, 37.
[11] M.-J. Yang and T. Tsutsui, *Jpn. J. Appl. Phys.* **2000**, 39, L828.
[12] C. Adachi, M.A. Baldo, M. E. Thompson and S. R. Forrest, *J. Appl. Phys.* **2001**, 90, 5048.
[13] H.Z. Xie, M.W. Liu, O.Y. Wang, X.H. Zhang, C.S. Lee, L.S. Hung, S.T. Lee, P.F. Teng, H.L. Kwong, H. Zheng and C.M. Che, *Adv. Mater.* **2001**, 13, 1245.
[14] X. Gong, M.R. Robinson, J.C. Ostrowski, D. Moses, G.C. Bazan and A.J. Heeger, *Adv. Mater.* **2002**, 14, 581.
[15] H. Fuji, T. Sano, Y. Nishio, Y. Hamada and K. Shibata, *Macromol. Symp.* **1997**, 125, 77.
[16] Z.L. Zhang, X.Y. Jiang, S.H. Xu and T. Nagamoto, in *Organic Electroluminescent Devices*, S. Miyata and H.S. Nalva Eds., Gordon & Breach, Amsterdam 1997, p. 228.
[17] J. Kalinowski, W. Stampor, J. Mężyk, M. Cocchi, D. Virgili, V. Fattori and P. Di Marco, *Phys Rev. B* **2002**, 66, 235321.
[18] M. Cocchi, D. Virgili, G. Giro, V. Fattori, P. Di Marco, J. Kalinowski and Y. Shirota, *Appl. Phys. Lett.* **2002**, 80, 2401.
[19] H. Murata, C.D. Merrit and Z.H. Kafafi, *IEEE J. Select. Top. Quantum Electron.* **1998**, 4, 119.

Density-of-States Function in Metal-Free Phthalocyanine: Effect of Exposure to Chlorodifluoromethane

I. Zhivkov,*[1,2] *G. Danev,*[1] *S. Nešpůrek,*[2] *J. Sworakowski*[3]

[1] Central Laboratory of Photoprocesses, Bulgarian Academy of Sciences, G. Bonchev Str., bl. 109, 1113 Sofia, Bulgaria
[2] Institute of Macromolecular Chemistry, Academy of Sciences of the Czech Republic, Heyrovský Sq. 2, 162 06 Prague 6, Czech Republic
[3] Institute of Physical and Theoretical Chemistry, Technical University of Wroclaw, Wyb. Wyspianskiego 27, 50-370 Wroclaw, Poland

Summary: Changes in the energy structure of local states during the exposure of metal-free phthalocyanine films to chlorodifluoromethane were studied using the technique of thermomodulated space-charge-limited currents. The broadening of the energy distribution of the states is explained on the basis of charge-dipole interactions.

Keywords: chlorodifluoromethane; electron-dipole interaction; gas sensing; local states; phthalocyanine

Introduction

Phthalocyanines have been known for many years as organic semiconductors[1]. Among many possibilities of their applications as organic metals, photoconductors, solar cells, non-linear optical media, light limitators, electroluminescent diodes, optical data storage devices, and photosensitizers[2] is also gas sensing. Due to formation of charge-transfer species[3,4], phthalocyanines change their electrical conductivities over several orders of magnitude upon exposure to BF_3, BCl_2, F_2, Cl_2, I_2, NO, NO_2, O_2, and other gases[5,6]. However, the changes in capacitance, optical absorption, and optical refractivity can also be used for gas detection. These changes usually reflect the variations in the energy structure of the sensing material. It was found that polar dopants broaden the distribution of the transport hopping states due to electron-dipole interaction[7,8]. One can also expect a similar influence of some gases on deeper local states.

In this paper we describe changes in the energy structure of local states upon exposure of metal-free phthalocyanine (H_2Pc) films to polar chlorodifluoromethane ($CHClF_2$). Two methods were used in the study: temperature dependence of the DC electrical conductivity and thermomodulated space-charge-limited current (TM-SCLC) spectroscopy.

Experimental

H$_2$Pc was purified by sublimation in a temperature gradient. Gold coplanar electrodes, separated by 0.7 mm and 13 μm gaps for the measurements of the tempearture dependence of electrical conductivity and TM-SCLC, respectively, were deposited on Corning 7059 glass substrates. Thin (100 ÷ 150 nm) H$_2$Pc films were deposited on the electrode system by vacuum evaporation; the deposition rate was 15 nm min^{-1}. The presence of the α-form of H$_2$Pc after the evaporation was checked by UV-VIS and IR spectroscopies[9,10].

After mounting a sample in the measuring chamber, the system was evacuated to a pressure of 7 Pa and then filled with nitrogen (purity 99.996 %) to atmospheric pressure. This procedure was repeated three times. CHClF$_2$ was applied into the chamber through a rubber membrane by injection. In this way N$_2$ – CHClF$_2$ mixtures were prepared, with the concentration of the latter ranging between 60 and 500 ppm.

Electrical conductivity was measured using a Keithley 617 electrometer. The samples for the TM-SCLC measurements were first heated in N$_2$ atmosphere to 140 °C with the aim to desorb oxygen. Then TM-SCLC characteristics, i.e. current (I) vs. voltage (U) and activation energy of the current (E_a) vs. U dependences were measured at room temperature (temperature modulation $\Delta T = 6$ °C). After the CHClF$_2$ exposure, the sample was heated again to 140 °C and then slowly cooled (8 h in the dark) to room temperature. The energy spectrum of density-of-states (DOS) function $h(E)$ was constructed from the TM-SCLC characteristics using the procedure described in Ref.[11].

Outline of the TM-SCLC theory

For any position of the energy of the Fermi level E_F, the equation is fulfilled

$$E_F(L) = -kT \ln \frac{I}{U} \chi_1 + C \qquad (1)$$

where $C = kT \ln(e\mu_0 N_b/L)$, L is the sample thickness (the charge injection is realized by a contact at coordinate $x = 0$), k is the Boltzmann constant, T is temperature, e is the elementary charge, μ_0 is the charge carier mobility, N_b is the effective density of states, and χ_1 is parameter of the order of unity whose exact relation to experimental variables is $\chi_1 = 1/(2 - \gamma)$, where $\gamma = d(\ln U)/d(\ln j)$ is the slope of the $U\text{-}I$ characteristic.

The total concentration of the charge carriers (n_s) (in most cases practically equal to the concentration of the trapped carriers) is given by the equation

$$n_s = \int_E h(E) f(E - E_f) \, dE = \frac{\varepsilon \varepsilon_0}{e L^2} (2 - \gamma)(1 - \gamma)(1 + B)U \qquad (2)$$

where $h(E)$ is the density-of-states (DOS) function to be determined, f is the Fermi function, $\varepsilon\varepsilon_0$ is the electric permittivity and B is a term containing higher-order corrections.

Changing the voltage V, E_F moves from its "thermodynamic" value towards the extended states scanning the local states by changing their occupancy

$$\frac{d n_s}{d E_f} = \frac{d}{d E_f} \left(\int_E n_s(E) \, dE \right) = \int_E h(E) \frac{d f(E - E_f)}{d(E - E_f)} \, dE \qquad (3)$$

The increment of the space charge due to the shift of the Fermi level dn/dE_F is connected to the experimentally measured data via Eqs. 1 and 2. The DOS function $h(E)$ is obtained after deconvolution of the integral in (3) as

$$h(E) = \frac{d n_s}{d E_f} \left[1 - M_2 \left(a^2 + a' \right) \right] \qquad (4)$$

where

$$M_2 = \int_{-\infty}^{+\infty} (E_f - E)^2 f(E - E_f)[1 - f(E - E_f)] \, dE = 3.2898 (kT)^2 \qquad (5)$$

and $a = d(\ln dn_s/dE_f)/dE_f$ and $a' = da/dE_f$.

The energy scale can be determined using a dominant energy level E_d. Mathematically, E_d represents the first momentum of the energy distribution of trapped carriers

$$E_d = \frac{\int_E h(E)(E_v - E) f(1 - f) \, dE}{\int_E h(E) f(1 - f) \, dE} \qquad (6)$$

E_d can be determined from the experimental data as

$$E_d = E_a + \frac{(3\gamma - 4)\gamma}{(2 - \gamma)(1 - \gamma)} \frac{dE_a}{d(\ln I)} + \frac{1}{1 + B} \frac{dB}{d(1/kT)} \qquad (7)$$

where E_a is the experimentally measured activation energy of the SCL current determined from the relation $I = I_0 \exp(-E_a/kT)$.

Results and discussion

Temperature dependences of the electrical conductivity

Temperature dependences of the dc electrical conductivity (S) are plotted in Figure 1. Curve 1 was measured in $CHClF_2$ atmosphere. The log S vs. $1/T$ dependence can be approximated with a stright line, with the activation energy $E_a = 0.66$ eV. Curve 2 was measured on increasing the temperature after replacing $CHClF_2$ with N_2. The non-Arrhenius part of the curve at elevated temperatures can be associated with the desorption of $CHClF_2$. Curve 3 was measured in N_2 atmosphere, $E_a = 0.91$ eV.

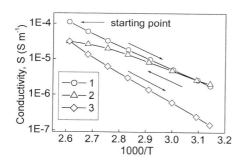

Figure 1. Temperature dependences of the dark conductivity. Curve 1 – in $CHClF_2$ atmosphere, T decreasing, 2 – in N_2, T increasing, 3 – in N_2, T decreasing

Thermomodulated SCLC spectroscopy

The TM-SCLC characteristics, i.e. I vs. U (a) and E_a vs. U (b) are presented in Figure 2. The low-voltage part of the I-U characteristic is quadratic followed by a superquadratic one at voltages exceeding 5 V. As follows from the E_a - U characteristics, the shift of the Fermi level during a voltage run in N_2 atmosphere is expected to be from about 0.8 eV to 0.2 eV.

Generally, the slopes of the I - U characteristics in the voltage range 5 – 50 V decrease with increasing CHClF$_2$ concentration, as does the low-voltage activation energy. The set of curves presented in Figure 2 allows to construct the DOS function (see Figure 3). Four peaks, with maxima at 0.26, 0.32, 0.40, and 0.52 eV and low-energy tail states were detetected. The trap 0.52 eV deep is often attributed to oxygen Coulombic centres for holes[6,12]. Traps in the energy interval (0.3 - 0.4) eV have been reported in the literature[13,14]; they are usually characterized as structural ones. Their concentration and energy depend on, e.g., the rate of the film deposition[13,14,15]. Our experiments reveal the presence of two peaks in this energy range; at present we cannot decide whether the doubling is a real phenomenon or an artifact associated with the calculation procedure we employed. The trap 0.25 eV deep was also found from steady-state photoconductivity measurements.

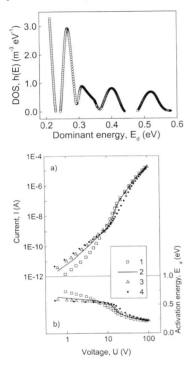

Figure 2. Energy distribution of DOS of H$_2$Pc film in N$_2$ atmosphere

Figure 3. (a) *I-U*, (b) E_a-*U* characteristics as a function of CHClF$_2$ concentration. 1 - N$_2$ only, 2 - 120 ppm, 3 - 235 ppm, 4 - 470 ppm

Because of the polarity of the CHClF$_2$ molecule (dipole moment $\mu = 1.56$ D), one can expect a broadening of the trap distribution due to charge-dipole interactions[7,8]. The effect was indeed found in our experiments. The evolution of the peak at 0.26 eV with CHClF₂ concentration is given in Figure 4.

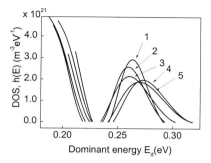

Figure 4. Broadening of the peak at 0.26 eV with increasing CHClF₂ concentration: Curve 1 - N₂ atmosphere, other curves N₂ – CHClF₂. Concentration: curve 2 - 60 ppm, curve 3 - 120 ppm, curve 4 - 175 ppm, curve 5 - 235 ppm

Two important features were observed with increasing concentration of CHClF₂: (i) the half-width of the peak increased, (ii) the overall concentration of local states was constant. The FWHM of the peak as a function of the CHClF₂ concentration is plotted in Figure 5. The broadening of the DOS function with increasing concentration of CHClF₂ can be explained on the basis of charge-dipole interactions[7] and it is in a good agreement with theoretical predictions[8]. Similar broadening was also observed for the deeper local states.

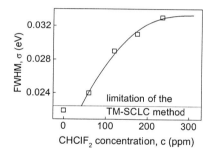

Figure 5. Dependence of FWHM of the 0.26 eV peak on the CHClF₂ concentration

Conclusion

The broadening of the energy distribution of local states for charge carriers was observed during the exposure of metal-free phthalocyanine films to chlorodifluoromethane. The effect is explained on the basis of charge-dipole interactions.

Acknowledgement

The research was supported by grants H 806 from the Bulgarian Ministry of Education and Science, A1050901 and AV0Z4050913 from the Grant Agency of the Academy of Sciences of the Czech Republic, ME 270 from the Ministry of Education, Youth and Sports of the Czech Republic and from the Technical University of Wroclaw.

[1] D. D. Eley, *Nature* **1948**, 162, 819; A. T. Vartanyan, *Zh. Fiz. Khim.* **1948**, 22, 769.
[2] M. Hanack, M. Lang, *Adv. Mater.* **1994**, 6, 819.
[3] R. Rella, A Serra, P. Siciliano, A. Tepore, L. Vally and A. Zocco, *Langmuir* **1997**, 13, 6563.
[4] I. Zhivkov, S. Nešpůrek, G. Danev, *Proc. ISCMP'2000*, Varna, Bulgaria.
[5] M. S. Nieuwenhuizen, A. W. Barendsz, *Sensors Actuators* **1987**, 11, 45.
[6] A. Chyla, J. Sworakowski, A. Szczurek, E. Brynda, S. Nešpůrek, *Mol. Cryst. Liq. Cryst.* **1993**, 230, 1.
[7] S. Nešpůrek, H. Valerián, A. Eckhardt, V. Herden, W. Schnabel, *Polym. Adv. Technol.* **2001**, 12, 306.
[8] H. Valerián, E. Brynda, S. Nešpůrek, *J. Appl. Phys.* **1995**, 78, 6071,
[9] B. R. Holebone and M. J. Stillman, *J. Chem. Soc., Faraday Trans. 2*, **1978**, 74, 284.
[10] J. H. Sharp, *J. Phys. Chem.* **1968**, 72, 3230.
[11] F. Schauer, S. Nešpůrek, O. Zmeškal, *J. Phys. C, Solid State Phys.* **1986**, 19, 7231.
[12] C. Hamann, *Phys. Status Solidi* **1968**, 26, 311.
[13] I. Zhivkov, S. Nešpůrek, J. Sworakowski, *Acta Phys. Pol., A* **2000**, 100, 215.
[14] D. F. Barbe, C. R. Westgate, *Solid State Commun.* **1969**, 7, 563.
[15] K. Yoshino, K. Kento, K. Tatsuno, Y. Inuishi, *Technol. Rep. Osaka Univ.* **1972**, 22, 585.

Macromol. Symp. **2004**, *212*, 523-528

Modification of Poly(styrene-*alt*-maleic anhydride) with 1,3,4-Oxadiazole Units for Electroluminescent Devices

Gabriela Aldea,[1] *Drahomír Výprachtický,*[*2] *Věra Cimrová*[2]

[1] "Petru Poni" Institute of Macromolecular Chemistry, Romanian Academy, Aleea Grigore Ghica Voda 41A, 6600 Iasi, Romania
E-mail: aldeagro@netscape.net
[2] Institute of Macromolecular Chemistry, Academy of Sciences of the Czech Republic, Heyrovský Sq. 2, 162 06 Prague 6, Czech Republic
E-mail: vyprach@imc.cas.cz , cimrova@imc.cas.cz

Summary: New copolymers of poly(styrene-*alt*-maleic anhydride) (PSMA) modified with 2-(4-aminophenyl)-5-(biphenyl-4-yl)-1,3,4-oxadiazole and hexylamine were prepared. The copolymers, characterized by UV-vis and FT IR spectroscopy, reached 1.22 mol % of the oxadiazole units relative to anhydride groups at the maximum (PSMA-4). Electric and optical properties of the copolymers were studied. The currents obtained depend strongly on the content of oxadiazole units in the copolymers. Currents measured in PSMA-4 were more than two orders of magnitude higher than those measured in the copolymers without oxadiazole. Using polymer blends made of poly(9,9-dihexadecylfluorene-2,7-diyl) and PSMA-4, blue light-emitting devices were fabricated and their photoluminescence and electroluminescence spectra were measured.

Keywords: electroluminescence; electron-transporting properties; oxadiazole; poly(styrene-*alt*-maleic anhydride)

Introduction

Recently, oxadiazole-containing polymers have received a great deal of interest in the field of polymeric light-emitting devices (LEDs). Low-molecular-weight oxadiazoles like 2-(biphenyl-4-yl)-5-(*tert*-butylphenyl)-1,3,4-oxadiazole (PBD) have been used as electron-transporting moieties in LEDs.[1-3] To overcome the diffusion and crystallization of low-molecular-weight oxadiazoles, polymers with attached oxadiazole moieties have been synthesized.[4,5] Recently, amphiphilic

DOI: 10.1002/masy.200450868

copolymers containing the oxadiazole rings as pendant groups were prepared by aminolysis of poly(styrene-*alt*-maleic anhydride) (PSMA) with 2-(4-aminophenyl)-5-(4-nitrophenyl)-1,3,4-oxadiazole and alkylamines (C_{10}, C_{18}).[6] The content of oxadiazole units in copolymers was found to be low (0.5 mol % relative to anhydride groups). Here we describe the modification of PSMA with 2-(4-aminophenyl)-5-(biphenyl-4-yl)-1,3,4-oxadiazole affording higher contents of the oxadiazole units in material. Electric properties of thin films were studied as well.

Experimental

Synthesis. Poly(styrene–*alt*–maleic anhydride), PSMA, (50 mol % of styrene units, M.w. 1600), was purchased from Polysciences, Inc. and used as received. 2-(4-Aminophenyl)-5-(biphenyl-4-yl)-1,3,4-oxadiazole (Oxad-NH_2) was synthesized from 4-nitrobenzohydrazide (Aldrich) and biphenyl-4-carbonyl chloride (Fluka) according to the recently published procedure.[7]

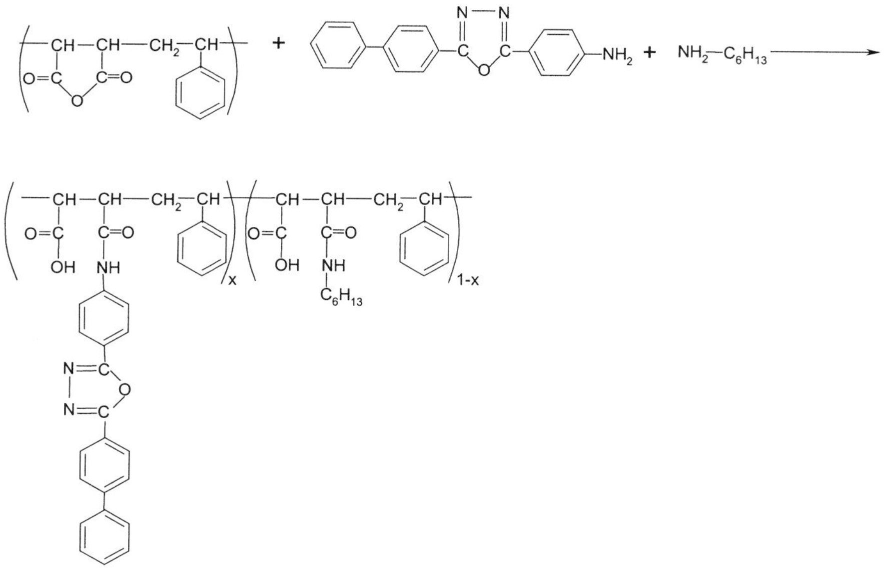

Figure 1: Modification of PSMA with Oxad-NH_2 and hexylamine

The polymer modification (Figure 1) was carried out as follows: A solution of 0.31 g (0.001 mol) of Oxad-NH$_2$ in N-methylpyrrolidone (20 ml) was added slowly (1 h) to a solution of 4.0 g of PSMA (0.02 mol of anhydride groups) in N-methylpyrrolidone (40 ml). The mixture was heated with stirring (Table 1, PSMA-1 – PSMA-4), then cooled to room temperature and 2.7 ml (0.02 mol) of hexylamine was added. The solution was heated to 100 °C for 5 h. After cooling, the reaction mixture was poured into ethyl acetate and the precipitate was filtered off and dried. The modified polymer was reprecipitated from methanol into ethyl acetate. PSMA-0 was prepared reacting 4.0 g of PSMA with 3.0 ml (0.022 mol) of hexylamine in N-methylpyrrolidone (40 ml) at 100 °C for 5 h.

Sample preparation. Thin copolymer films were prepared by spin coating from chloroform solutions, for optical studies on fused silica substrates. Polymer LEDs with a hole-injecting indium-tin-oxide (ITO) electrode and an electron-injecting aluminium electrode were fabricated. In this case the polymer layers were spin-coated on ITO substrates covered with a thin layer of poly[3,4-(ethylenedioxy)thiophene]/poly(styrenesulfonate) (PEDT:PSS). Finally, 60-80 nm thick top aluminium electrodes were vacuum-evaporated on the top of polymer films to form LEDs with typical active areas of 4 mm^2.

Electroluminescence and photoluminescence measurements. Electroluminescence (EL) and photoluminescence (PL) spectra were measured using a home-made spectrofluorometer with single photon-counting detection (SPEX, RCA C31034 photomultiplier). LEDs were supplied from a Keithley 237 source measure unit, which served for the simultaneous recording of the current flowing through the sample. PL spectra were taken perpendicular to the sample surface and the incidence angle for the excitation beam was 30° relative to the sample surface normal to minimise reabsorption. 300 W Xe lamp (Oriel) was used as the excitation source. The measured EL and PL emission spectra were corrected for the spectral response of the detection system. Current-voltage and luminance-voltage characteristics were recorded simultaneously using the Keithley 237 source measure unit and a silicon photodiode with integrated amplifier (EG&G HUV-4000B) for the detection of total light output. A voltage signal from the photodiode was recorded with a Hewlett Packard 34401A multimeter. The LED characteristics were measured in a vacuum chamber to prevent any electrode degradation. PL and absorption measurements were performed under ambient laboratory conditions.

Results and Discussion

The amidation reaction of the anhydride groups in PSMA with Oxad-NH$_2$ is slow. In first experiments under mild conditions, we reached very low contents of oxadiazole-modified copolymer (0.10 mol % relative to anhydride groups). Increasing the reaction time and temperature (Table 1), we reached 1.22 mol % of oxadiazole units in copolymer. This means that approximately 25 % of the feeded (5 mol %) amount of Oxad-NH$_2$ reacted with the anhydride copolymer. The oxadiazole content in copolymers was determined by UV-vis spectrometry in methanol (Figure 2), assuming that the molar absorption coefficient of oxadiazole structure unit in copolymer is the same as that of Oxad-NH$_2$ (ε_{328} = 36 500 L mol^{-1} cm^{-1}). On the other hand, the reaction of hexylamine with anhydride groups was quantitative. Analyses of FT IR spectra of PSMA-1 – PSMA-4 revealed that all anhydride group signals (1858, asym C=O; 1778, sym C=O; 1224, 922, C-O-C stretching; cm^{-1}) disappeared giving rise to those of COOH (1700, C=O stretching; 1400, O-H bending; cm^{-1}) and amide (3061, N-H stretching; 1640, C=O stretching amide I; 1563, N-H bending amide II; cm^{-1}) groups.

Table 1. Reaction of PSMA with Oxad-NH$_2$ and hexylamine (100 °C, 5 h)

Copolymer	Temperature	Reaction time	Oxadiazole content [a]
	°C	h	mol %
PSMA-0 [b]	-	-	0.00
PSMA-1	100	2	0.10
PSMA-2	135	24	0.84
PSMA-3	135	48	1.04
PSMA-4	135	72	1.22

[a] Calculated relative to anhydride groups.
[b] The copolymer modified with hexylamine only.

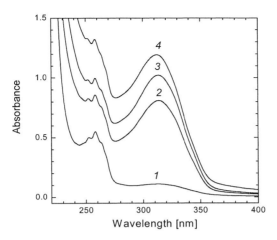

Figure 2: UV-vis spectra of PSMA copolymers modified with Oxad-NH$_2$ in methanol (c = 0.80 g/l, 1 cm): *1* PSMA-1, *2* PSMA-2, *3* PSMA-3, *4* PSMA-4

Electric properties of single-layer ITO/polymer/Al samples were studied. Typical dependences of the current on the applied electric field are shown in Figure 3a. It is evident that the current depends strongly on the content of oxadiazole units in the modified polymer. Significant changes in the current density were already observed in the PSMA with a low content of oxadiazole units. In comparison with PSMA-0 devices, in PSMA-1 devices (0.10 mol % of oxadiazole units) one order of magnitude higher currents and in PSMA-4 devices (1.22 mol % of oxadiazole units) more than two orders of magnitude higher currents were measured. Shapes of the electric characteristics indicate that both transport and injection properties are modified.

Recently, we have shown that the EL efficiency and time stability of polyfluorene-based LEDs were improved in blend LEDs made of poly(9,9-dihexadecylfluorene-2,7-diyl) (PFC16) and polysilane.[8] The increase in the EL efficiency was attributed to modification of the charge transport and recombination in the blend. As shown above, changes in the oxadiazole content in modified PSMA changed dramatically transport properties; therefore we tested new polymers in blends with luminescent PFC16. We succeeded in fabrication of blue light-emitting devices using a blend of PSMA-4 and PFC16. An example of electroluminescence spectrum of the ITO/PEDT:PSS/PFC16+PSMA-4/Al is shown in Figure 3b. Compared with the devices fabricated from neat PFC16, an improvement of the EL efficiency and stability was achieved and

528

lower onset voltages were observed in the blend devices.

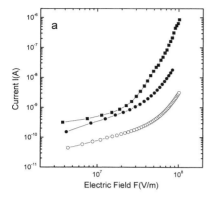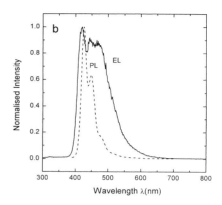

Figure 3: (a) Current-electric field characteristics for the ITO/PSMA-0/Al (open circles), ITO/PSMA-1/Al (solid circles) and ITO/PSMA-4/Al (solid squares) samples. (b) Photo-luminescence and electroluminescence in the blend made of copolymer PSMA-4 and luminescent SPFC16 LED: ITO/PEDT:PSS/PFC16+PSMA-4/Al

Acknowledgment

We thank the Grant Agency of the Czech Republic (grant No. 203/01/0512) and the Grant Agency of the Academy of Sciences of the Czech Republic (grant No. AV0Z 4050913) for support.

[1] J. Morgado, E. Moons, R. Friend, F. Gacialli, *Adv. Mater.* **2001**, *13*, 810.
[2] Y. Liang, Q. Lin, H. Zhang, Y. Zheng, *Synth. Met.* **2001**, *123*, 377.
[3] M. Zheng, L. Ding, E. E. Gurel, F. E. Karasz, *J. Polym. Sci., Part A: Polym. Chem.* **2002**, *40*, 235.
[4] L. Boiteau, M. Moroni, A. Hilberer, M. Werts, B. Boer, G. Hadziioannou, *Macromolecules* **2002**, *35*, 1543.
[5] S.-Y. Song, T. Ahn, H.-K. Shim, I.-S. Song, W.-H. Kim, *Polymer* **2001**, *42*, 4803.
[6] B. Schultz, B. Dietzel, Y. Kaminorz, *Macromol. Symp.* **2001**, *164*, 457.
[7] R. Kotva, V. Cimrová, D. Výprachtický, *Polym. Bull.*, accepted.
[8] V. Cimrová, D. Výprachtický, *Appl. Phys. Lett.* **2003**, *82*, 642.

Correlation between Large Polarons in Molecular Chains

Larissa Brizhik, Alexander Eremko*

Bogolyubov Institute for Theoretical Physics, Ukrainian National Academy of Sciences, Metrologichna Str., 14-b, 03150 Kyiv, Ukraine

Summary: We studied few extra electrons in a molecular chain with respect of electron-phonon coupling in the adiabatic approximation. It is shown that the lowest state of two extra electrons in a chain corresponds to the singlet bisoliton state with one deformational potential well. Two electrons with parallel spins form a localised triplet state, which corresponds to the two-hump charge distribution function. Three extra electrons form an almost independent nonlinear superposition of a soliton and bisoliton states. In the case of four electrons, the two almost independent bisolitons are formed. These two states tend to separate in the chain at the maximal distance due to the Fermi repulsion, accounted for in the zero-order adiabatic approximation. This repulsion is partly compensated by the attraction between the solitons due to their exchange with virtual phonons, described by the non-adiabatic part of the Hamiltonian. The formation of solitons is characterised by the appearance of the bound soliton and bisoliton levels in the forbidden energy band. This constitutes the qualitative difference of the large polaron (soliton) states from the almost free electron states and small polaron states.

Keywords: collective state; conducting polymers; electron-phonon interaction; polaron; soliton

Introduction

The concept of a 'soliton' [1,2] or large polaron [3] in low-dimensional molecular systems has been intensively used to explain various phenomena in biological physics and condensed matter physics [2-4]. A large polaron state is formed due to the electron-phonon interaction at moderate values of the interaction constant and is theoretically described in adiabatic approximation. There are clear experimental pieces of evidence of the existence of large polarons or solitons in conducting polymers [3,5]. In this respect, a question arises about the interaction between large polarons. It was shown in [6] that, due to the interaction with local deformation of the chain, two electrosolitons with opposite spins bind in a localized bound singlet spin state called 'bisoliton'.

© 2004 International Union of Pure and Applied Chemistry   DOI: 10.1002/masy.200450869

Here we consider the localized states of few extra electrons in the conducting band of a molecular chain in the adiabatic approximation.

Model Hamiltonian

The state of electrons in an isolated conduction band accounting for the electron-phonon interaction and neglecting direct electron-electron interaction, is described by Fröhlich Hamiltonian $\hat{H} = \hat{H}_q + \hat{H}_{e-ph} + \hat{H}_{ph}$. This Hamiltonian is in the site or quasimomentum representation:

$$\hat{H} = \sum_{n,\sigma} E_0 a_{n,\sigma}^+ a_{n,\sigma} + \sum_{n \neq m,\sigma} J_{n,m} a_{n,\sigma}^+ a_{m,\sigma} + \frac{1}{\sqrt{N}} \sum_{n,\sigma,q} \chi(q) e^{iqan} a_{n,\sigma}^+ a_{n,\sigma} (b_q + b_{-q}^+) + \sum_q \hbar \omega_q b_q^+ b_q \quad (1a)$$

$$\hat{H} = \sum_{k,\sigma} E(k) \ a_{k,\sigma}^+ a_{k,\sigma} + \frac{1}{\sqrt{N}} \sum_{k,\sigma,q} \chi(q) a_{k,\sigma}^+ a_{k-q,\sigma} (b_q + b_{-q}^+) + \sum_q \hbar \omega_q b_q^+ b_q, \quad (1b)$$

where creation and annihilation operators of an electron of the site n with the spin σ are associated with the creation and annihilation operators of an electron with the wavevector k by the unitary transformation $a_{k,\sigma}^+ = \frac{1}{\sqrt{N}} \sum_n e^{ikan} a_{n,\sigma}^+$. The electron energy dispersion in the conduction band, $E(k)$, is related to the matrix elements of the exchange interaction, $J_{n,m}$, by the Fourier transformation, and the function $\chi(q)$ determines the short-range interaction of electrons with phonons of the frequency ω_q.

The ground electron state in the conduction band is determined from the Schrödinger equation:

$$\hat{H} |\Psi\rangle = E |\Psi\rangle, \quad (2)$$

Depending on the value of the electron-phonon coupling, the three different types of the ground electron states can be realised in the system: (1) almost free electrons, (2) large polaron states, (3) small polaron states. Physical properties and mathematical tools of description of these three principal states of charge carriers are different. The realisation of one of the three regimes is determined by the relations between the three main parameters of the system [7]: (1) the electron band width, which is determined by matrix elements, (2) characteristic (or maximal) phonon frequency, ω_0, (3) electron-phonon interaction energy, which can be represented in the form:

$$E_b = \frac{1}{N} \sum_q \frac{|\chi(q)|^2}{\hbar\omega_q} .$$ (3)

Here we discuss the state of few extra electrons in a molecular chain under adiabatic condition, which is necessary for the existence of large polarons. In this case the electron correlation is qualitatively different from the one in the two other cases. This is clear from the following analysis of the three limiting approximations.

Since the Hamiltonian (1) conserves the number of electrons, the state-vector of N_e electrons depends on the N_e creation operators. The solution of the Schrödinger equation can be written as $|\Psi\rangle = \hat{U}|\Psi_0\rangle$, where $\hat{U}(\{a^+\},\{b^+\}) = \exp(\hat{S})$ is a unitary operator. Then Eq. (2) can be transformed into the following one:

$$e^{-\hat{S}}\hat{H}e^{\hat{S}}|\Psi_0\rangle \equiv \hat{\tilde{H}}|\Psi_0\rangle = E|\Psi_0\rangle .$$ (4)

1. In the **weak interaction** limit, the electron-phonon interaction is assumed to be small, i.e., it is proportional to a small parameter ε. In this case the operator $\hat{S}$ in (4) is also proportional to the small parameter. The operator expansion can be used: $\hat{\tilde{H}} = \hat{H} + [\hat{H},\hat{S}] + \frac{1}{2}[[\hat{H},\hat{S}],\hat{S}] + ...$ In the first order of the perturbation theory the operator $\hat{S}$ can be found from the condition $\hat{H}_{e-ph} + [\hat{H}_e + \hat{H}_{ph}, \hat{S}] = 0$, which gives

$$\hat{S} = \sum_{k,q,\sigma} f(k,q)a_{k,\sigma}^+ a_{k-q,\sigma} b_q^+ - h..c., \quad f(k,q) = \frac{1}{\sqrt{N}} \frac{\chi^*(q)}{\hbar\omega_q + E(k+q) - E(k)} .$$ (5)

This transforms Hamiltonian (4) into the next one

$$\hat{\tilde{H}} = \sum_{k,\sigma}[E(k) + \Delta E(k)]a_{k,\sigma}^+ a_{k,\sigma} + \sum_{k,k',q,\sigma,\sigma'}V(k,k',q)a_{k+q,\sigma}^+ a_{k'-q,\sigma'}^+ a_{k',\sigma'}a_{k,\sigma} + O(\varepsilon^3) .$$ (6)

Here $\Delta E(k)$ determines the standard renormalisation of the electron energy and effective mass in the second order of the perturbation theory [8]. The new term in (6) accounts for the direct electron-electron interaction, induced by phonons, $V(k,k',q)\sim|\chi(q)|^2$. This interaction is of the attraction type, it does not influence the state of an extra isolated electron, but it leads to the binding of the two isolated electrons into a Cooper pair [9]. In the case of a large number of electrons at the finite density, the ground electron state at zero temperature is the superconducting state.

2. In the **small polaron approximation** the matrix elements of the exchange interaction, $J_{n,m}$, are assumed to be small in the Hamiltonian. The unitary transformation is chosen in the form [10]:

$$\hat{S} = \sum_{n,q,\sigma} f(n,q) a_{n,\sigma}^{+} a_{n,\sigma} b_{q}^{+} - h.c., \qquad f(k,q) = \frac{1}{\sqrt{N}} \frac{\chi^{*}(q)}{\hbar \omega_q}. \tag{7}$$

In this case the transformed Hamiltonian in the nearest-neighbour approximation reads

$$\widetilde{\widetilde{H}} = \sum_{n,\sigma} [(E_n - E_b) a_{n,\sigma}^{+} a_{n,\sigma} - J e^{-G} (a_{n,\sigma}^{+} a_{n+1,\sigma} e^{-B_n^{+}} e^{B_n} + h.c.)] -$$

$$- \sum_{n,m,\sigma,\sigma'} V(n-m) a_{n,\sigma}^{+} a_{m,\sigma'}^{+} a_{m,\sigma'} a_{n,\sigma} + \sum_{q} \hbar \omega_q b_q^{+} b_q. \tag{8}$$

Here the term proportional to the exponential operators describes jumps of a polaron from a site to a site with the creation of a certain number of phonons: $B_n^{+} \sim b_q^{+}$. In view of the small value of this term at strong coupling, $G \sim |\chi(q)|^2 \gg 1$, it can be considered as a perturbation, so that the zero-order term of the Hamiltonian accounts for the processes with the conservation of the total number of phonons only. Then, in the wavevector presentation, one extra electron in a chain is described by a small polaron state in a narrow conduction band, $\widetilde{J} = J\exp(-G)$. Moreover, the direct attraction type polaron-polaron interaction via phonons appears in Hamiltonian (8), which is given by the term $V(n-m)$. Therefore, the two extra electrons in a chain form a bipolaron [11]. The adiabatic limit will be considered in more details in the next section.

Adiabatic limit

In this case the kinetic energy of atom vibrations, which is a part of phonon Hamiltonian, $\hat{H}_{ph}$, is considered as a small term of the total Hamiltonian. Due to this, the vector state can be represented as the product of electron and phonon wavefunctions, which corresponds to the Born-Oppenheimer approximation. Accordingly, let us represent the unitary transformation, $\hat{U}$, as a product of $\hat{U}_e$ and $\hat{U}_{ph} = \exp(\bar{S})$.

$$\bar{S} = \frac{1}{\sqrt{N}} \sum_{q} \beta_q b_q^{+} - h.c., \qquad a_{k,\sigma} = \sum_{\lambda} \psi_{\lambda}(k) A_{\lambda,\sigma} \tag{9}$$

where the coefficients of the unitary operators are chosen in such a form that the electron part of the Hamiltonian $\widetilde{\widetilde{H}}$ is diagonal and, thus, they satisfy the equations:

$$E(k)\psi_\lambda(k) + \frac{1}{N}\sum_q \chi(q)(\beta_q + \beta_{-q}^*)\psi_\lambda(k-q) = E_\lambda\psi_\lambda(k). \tag{10}$$

These two transformations are not independent, electron state is determined by the lattice configuration, which, in turn, depends on the electron state. The transformed Hamiltonian now takes the form:

$$\widetilde{H} = W + \sum_{\lambda,\sigma} E_\lambda A_{\lambda,\sigma}^+ A_{\lambda,\sigma} + \sum_q \hbar\omega_q b_q^+ b_q + \sum_q \{[\beta_q + \frac{1}{\sqrt{N}}\sum_{\lambda,\sigma}\varphi_{\lambda,\lambda}(q)A_{\lambda,\sigma}^+ A_{\lambda,\sigma}]b_q^+ + \text{h.c.}\} +$$
$$+ \sum_{\lambda \neq \lambda',\sigma}\varphi_{\lambda,\lambda'}(q)A_{\lambda,\sigma}^+ A_{\lambda',\sigma}(b_q + b_{-q}^+). \tag{11}$$

Here $\varphi_{\lambda,\lambda'}(q) = \chi(q)\sum_k \Psi_\lambda^*(k)\Psi_{\lambda'}(k-q)$. The last term in (11) describes the transitions between the adiabatic terms with the absorption and radiation of phonons and is the nonadiabaticity operator. Provided the condition of the adiabatic approximation is fulfilled, it can be neglected and the problem can be studied in the zero-order adiabatic approximation. Unlike in the considered cases of almost free electrons and small polarons, the direct electron-electron interaction is absent in the zero-order Hamiltonian. Nevertheless, as mentioned above, electrons are not independent and move in the common self-consistent potential. In this case the ground electron state can be represented as the product of N_e electron creation operators and of the vacuum state

$$|\Psi_0\rangle = \prod_{\lambda,\sigma} A_{\lambda,\sigma}^+|0\rangle \tag{12}$$

which is the eigenstate with the energy $E = W + \sum_\lambda n_\lambda E_\lambda$. Here $n_\lambda = 0,1,2$ are the occupation numbers of the adiabatic level and $\sum_\lambda n_\lambda = N_e$. The coefficients β_q ought to satisfy the relation

$$\beta_q + \frac{1}{\sqrt{N}}\sum_\lambda n_\lambda\varphi_{\lambda\lambda}^*(q) = 0. \tag{13}$$

Using the longwave (continuum) approximation and switching to spatial representation, we can give Eq. (10) in the form:

$$-\frac{d^2\Psi_\lambda}{dx^2} + U(x)\Psi_\lambda = \frac{2ma^2}{\hbar^2}E_\lambda\Psi_\lambda, \tag{14}$$

where the deformational potential, accounting for the expression (13), has the form

$$U(x) = -2g\sum_{\lambda} n_{\lambda} |\Psi_{\lambda}|^2 \tag{15}$$

where m is the effective electron mass and g is the electron-phonon interaction constant. Because the deformational potential (15) is determined by the occupied adiabatic levels, the problem reduces to solving the system of many-component nonlinear Schrödinger equations for the occupied states [12]. Unoccupied excited states are determined by Eq. (15) with the given potential.

One extra electron is described by the conventional nonlinear Schrödinger equation for a single adiabatic level, $n = 1$, whose solution corresponds to the soliton state with the energy $E_1 = -\hbar^2\mu^2/(2ma^2)$, where $\mu = g/2$. The ground state of the two extra electrons in a chain also corresponds to one adiabatic level occupied by the two electrons with opposite spins, $n = 2$. The corresponding energy is $E_1 = -\hbar^2\mu^2/(2ma^2)$, where $\mu = g$.

In the case of three extra electrons, the ground state corresponds to the two occupied adiabatic levels with the occupation numbers $n_1 = 2$, $n_2 = 1$, which are determined by the two-component nonlinear Schrödinger equation which admits the solution[12]

$$\Psi_1 = \sqrt{\frac{\mu_1^2 - \mu_2^2}{g}} \frac{\mu_1\cosh(\phi_2)}{D}, \quad \Psi_2 = \sqrt{\frac{\mu_1^2 - \mu_2^2}{g}} \frac{\mu_2\sinh(\phi_1)}{D}, \tag{16}$$

where

$$D = \mu_1\cosh(\phi_1)\cosh(\phi_2) - \mu_2\sinh(\phi_1)\sinh(\phi_2), \quad \phi_1 = \mu_1(x+l), \quad \phi_2 = \mu_2(x-l). \tag{17}$$

The constant l in (17) can take arbitrary values. These two states correspond to energies $E_j = -\hbar^2\mu_j^2/(2ma^2)$, with $\mu_1 = g.$, $\mu_2 = g/2$. Therefore, the three electrons form a bisoliton and a soliton, which are described by essentially changed wavefunctions as compared with an isolated soliton function, and which have the total energy in the form of the sum of soliton and bisoliton energies. The constant l in this case determines the distance between soliton and bisoliton center of mass coordinates. This gives us an example of a nonlinear superposition of quasiparticles, when the energy is independent of the distance between them, although the total wavefunction is not a superposition of the functions. The corresponding wavefunctions (16) are shown in Fig. 1.

In the case of four extra electrons, the two adiabatic levels are occupied, with the occupation numbers $n_1 = n_2 = 2$. Similar situation takes place for the two electrons with polarised spins, so that $n_1 = n_2 = 1$. In these latter cases $\mu_1, \mu_2 \rightarrow \mu_0$, where $\mu_0 = g$ for two bisolitons, and $\mu_0 = g/2$ for the triplet state. The wavefunctions $\Psi_{1,2}$ describe the two bisolitons (two solitons) separated by the large distance

$$L = \frac{1}{2\mu_0}\ln\frac{2\mu_0\cosh(2\mu_0 l)}{\Delta} \to \infty \quad \text{at} \quad \Delta = \mu_1 - \mu_2 \to 0. \tag{18}$$

The total energy

$$E = -\frac{n^2}{6}Jg^2 + 2Jg^2\cosh^2(gl)\exp(-2ngL), \tag{19}$$

where $n = 1$ for the triplet state, and $n = 2$ for the two bisolitons. The corresponding wavefunctions of four electrons are shown in Fig. 2. In this case the constant l characterises the level of collectivisation of the two separated potential wells.

The energy expression includes the term which depends on the distance between the centers of mass coordinates, L, and describes the repulsion due to the Pauli principle.

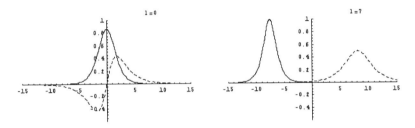

Fig.1. The wavefunctions of three extra electrons in a chain: (a) $l = 0$, (b) $l = 7$.

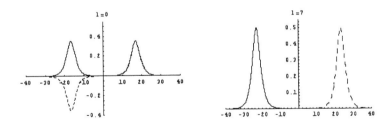

Fig.2. The wavefunctions of four extra electrons in a chain: (a) $l = 0$, (b) $l = 7$.

Non-adiabatic corrections

The nonadiabatic term in the Hamiltonian (11) results in the additional direct interaction between the two solitons or bisolitons via the phonon field. This additional interaction partly compensates the Fermi repulsion and stabilizes the two-bisoliton and triplet state solutions. Considering the nonadiabatic term of the Hamiltonian as a perturbation, one can write the first-order energy correction in the form

$$\Delta E = \left\langle \Psi_0 \left| \hat{H}_{na} \frac{1}{E - H_0} \hat{H}_{na} \right| \Psi_0 \right\rangle. \tag{20}$$

Here the wavefunctions correspond to the zero-order adiabatic approximation $\hat{H}_0$, and $\hat{H}_{na}$ is the non-adiabatic part of the Hamiltonian, i.e., the last term in (11). To calculate the energy of (20), it is necessary to find the total energy spectrum of the system. This includes the localised (bound) levels and delocalised (continuum) spectrum. The localised levels are always occupied, and unoccupied levels belong to the continuum spectrum. The corresponding calculations show that the energy correction term (20) contains two terms. One of these terms is independent of the distance between centres of mass coordinates and determines the renormalisation of their energy, while the second term depends on the distance. The latter term accounts for the additional attraction, and results in the stabilisation of the state. The total energy with the account of the correction term (20) has the form [13]:

$$E = -\frac{n^2}{6} Jg^2 \left(1 + \frac{3\gamma}{8\pi^6 g}\right) + 2Jg^2 \cosh^2(gl)\exp(-2ngL) - \frac{3Jg\gamma}{8\pi^3 L^2 \cosh(\lg)}. \tag{21}$$

Here γ is the so-called non-adiabaticity parameter determined as $\gamma = \hbar\omega_{max}/J$.

Because of the competition between the repulsion force due to the Pauli principle and the attraction force due to the phonon field, the two bisolitons or two solitons with parallel spins are separated by the equilibrium distance, which is determined from the condition $dE/dL = 0$, where E is determined in (21).

Conclusion

We have shown that the lowest state of two extra electrons in a chain corresponds to the singlet bisoliton. Three extra electrons form almost independent nonlinear superposition of a soliton and bisoliton states. In the case of four electrons or two electrons with parallel spins, the two independent bisolitons (solitons) are formed. These two states tend to separate in the chain at the maximal distance due to the Fermi repulsion. This repulsion is partly compensated by the attraction between the solitons due to their exchange by the virtual phonons. The formation of large polarons is characterised by the appearance of the bound soliton and bisoliton levels in the forbidden energy band. This constitutes a qualitative difference of the large polaron (soliton) states from the almost free electron states and small polaron states. In the case of a finite density of extra electrons in a chain at the zero temperature, the charge density wave is formed in a system [14]. The repulsion between bisolitons leads to the formation of a periodic lattice of bisolitons with the distribution period $l = N/N_{bs}$, $N_{bs} = N_e/2$, which corresponds to the many-electron solution of the Peierls-Froehlich problem at zero temperature [15].

It is worth noting that the self-consistent deformation potential of the lattice is a reflectionless one. In the case of one extra electron or a singlet state of two electrons that occupy the same level (bisoliton), this potential has a single bound state. In the case of a triplet state of two electrons (two bisolitons), when Fermi statistics forbids one-level occupation, lattice deformation forms a reflectionless potential with two bound levels. In the case of an arbitrary number of electrons, the corresponding self-consistent deformation potential is a reflectionless single-band potential with one gap, which separates the occupied sublevels from the vacant ones [15].

Acknowledgement

This work was partly supported by a grant of the Programme of Fundamental Research of the Ukrainian National Academy of Sciences.

[1] A.S. Davydov, N.I. Kislukha, *Phys. Status Solidi*, (b) **1973**, *59*, 465.
[2] A.S. Davydov, *"Solitons in Molecular Systems"*, Reidel, Dordrecht, 1985.
[3] A.J. Heeger, et al., *Rev. Mod. Phys.,* **1988**, *60*, 781.
[4] A.C. Scott, *Phys. Rep.,* **1992**, *217*, 1.
[5] I. Gontia, et al., *Phys. Rev. Lett.,* **1999**, *82*, 4058.
[6] L.S. Brizhik, A.S. Davydov, *Fiz. Nizk. Temp.,* **1984**, *10*, 748.
[7] L.S. Brizhik, A.A. Eremko, *Synth. Met.,* **2000**, *109*, 117.
[8] H. Haken, *Quantenfeldtheorie des Festkörpers*, B.G. Teubner, Stuttgart, 1973.
[9] L.N. Cooper, *Phys. Rev.,* **1956**, *104*, 1189.
[10] Yu.A. Firsov (Ed.), *"Polarons"*, Nauka, Moscow, 1975.
[11] A.S. Alexandrov, J. Ranninger, *Phys. Rev. B,* **1981**, *23*, 1794; *24*, 1164.
[12] L.S. Brizhik, A.A. Eremko, *Physica D,* **1995**, *81*, 295.
[13] L.S. Brizhik, A.A. Eremko, *Ukr. J. Phys.,* **1999**, *44*, 1022.
[14] G. Gruner, *Rev. Mod. Phys.,* **1988**, *60*, 1129.
[15] A.A. Eremko, *Phys. Rev. B,* **1992**, *46*, 3721.

Macromol. Symp. **2004**, *212*, 539-548

Photosensitive Layers of TiO₂/Polythiophene Composites Prepared by Electrodeposition

J. Pfleger,[*1] M. Pavlík,[1] N. Hebestreit,[2] W. Plieth[2]*

[1] Institute of Macromolecular Chemistry, Academy of Sciences of the Czech Republic, Heyrovský Sq. 2, 162 06 Prague 6, Czech Republic
[2] Institute of Physical Chemistry and Electrochemistry, Dresden Technical University, Mommsenstr. 13, 01062 Dresden, Germany

Summary: We adopted an electrophoretic deposition method for the preparation of thin layers of insoluble composite nanoparticles composed of TiO₂ core and about 2 nm thick shell of polythiophene, prepared by oxidative polymerization of thiophene. The reduced form of TiO₂-polythiophene composite material was deposited on the conductive surface from an ultrasonically generated microdispersion. Varying the dispersion media, applied voltage and the electrode arrangement made it possible to control the quality and morphology of the films. Compact semitransparent films deposited on ITO electrodes, suitable for photoelectrical measurements, were obtained within short deposition times.

Keywords: electrodeposition; photoconductivity; polythiophene; thin films; titanium dioxide

Introduction

Since 1991, when O'Regan and Grätzel reported on a photoelectrochemical solar cell composed of a ruthenium dye-sensitized nanoporous titanium dioxide, achieving a photoelectrical conversion efficiency of about 10 % [1], many researchers have focused on the utilization of the heterojunction formed between a wide-band-gap inorganic semiconductor and organic visible light absorbing material for solar cell applications. Particularly, there is an increasing effort to overcome the limitations originated in utilization of the liquid electrolyte used in the original Grätzel-type solar cells by substituting it either with a solid electrolyte [2] or an amorphous organic hole-transporting material [3]. Combinations of ultrathin films of polythiophene (PT) and titanium dioxide (prepared chemically [4], electrochemically or by vapour deposition [5,6]) were also described, in which a p/n junction, formed between the

DOI: 10.1002/masy.200450870

TiO$_2$ film (n-type semiconductor) and the polythiophene film in the reduced state (p-type semiconductor), contributed to the creation and effective separation of charges upon illumination [7]. A nebulized vapour deposition method was reported recently [8], which enables the formation of nanoparticles of metal oxides and a simultaneous build-up of the charge transporting polymer network. The synthesis of hybrid organic/inorganic materials based on metal oxides and semiconducting polymers was also realized by the sol – gel process [9], and many further data about synthesis, properties and applications of nanocomposites, based on conductive or semiconductive polymers and various inorganic materials were reported [10].

Electrodeposition has evolved recently to a simple and effective method for fabrication of homogenous thin layers of insoluble materials, such as ceramics, with many possible applications. The electrodeposition, covering the conductive surface by a functional material in an applied electric field, brings, in addition to experimental simplicity, several remarkable advantages: good layer-forming control and fine regulation via optimization of many parameters (varying dispersion media, applied voltage, electrode type, etc.), guaranteed contact, perfect coverage of the electrode, and the possibility of preparation of very thin layers. Limitations in the use of the electrodeposition method are evident: only electrophoretically active, relatively stable dispersion-forming materials can be deposited and the substrates must be conductive.

Here we report on the application of the cathodic electrophoretic method as a possible method of the preparation of the solid-state photoelectrically sensitive cell based on the heterojunction between wide-band-gap titanium dioxide and polythiophene in the core-shell nanocomposite.

Experimental

Preparation of PT-coated TiO$_2$ nanoparticles: As a core we used a TiO$_2$ powder P25 (Degussa), with average particle size 21 nm and surface area 50 m^2/g, The polythiophene was prepared from fresh distilled thiophene (Aldrich – Chemistry, 99+ %) and anhydrous iron(III) chloride (Fluka Chemie, purum) [4,11].

An aliquot of 25 g TiO$_2$ (dried at 60 – 100 °C for 2 - 4 h) was dispersed together with 0.03 – 0.05 M thiophene in ca. 200 ml of chloroform (anhydrous, max. 0.03 % water, Merck) by stirring for a few minutes [4]. By this procedure, the oxide particles with thiophene adsorbed on the surface were obtained. To this dispersion, 50 ml of a saturated solution of 0.2 – 0.3 M iron(III) chloride in chloroform was added. After stirring for 1 - 2 h, the color of the mixture changed from gray to black and hydrogen chloride was produced. These changes were used for monitoring the reaction progress. At the end, the dispersion was filtered off and the dark gray product was dried. As a result, the oxide particles covered with the oxidized polythiophene shell were obtained. To obtain the reduced form of the polythiophene, iron chloride was extracted from the product by washing with methanol (2 - 4 h). During this procedure, the color changed from gray (oxidized state) to red (reduced state). Finally, the powder was dried at 60-80 °C for 2 - 3 h.

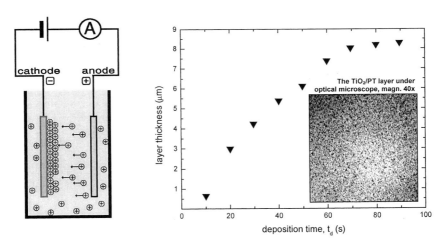

Figure 1. Schematic drawing of the typical anodic electrophoretic deposition setup.

Figure 2. Dependence of the layer thickness on deposition time for TiO$_2$/PT particles (applied voltage 200 V, current limited to 0.1 mA).

The electrophoretic deposition was performed in the experimental setup shown schematically in Figure 1, consisting of the power supply in serial connection with an amperemeter and two electrodes inserted in the dispersion of particles in organic medium. The used power supply made it possible to work in the potentiostatic or galvanostatic mode. The dispersions were ultrasonicated before and stirred during the deposition.

The optical absorption spectra were obtained as diffusion reflection measured on pressed pellets prepared from the powder material or on films (thickness of several μm) electrodeposited on ITO substrates, using a Perkin Elmer 340 spectrophotometer equipped with an integration sphere.

The upper electrodes on the top surface of the deposited films were prepared by vacuum deposition. DC electrical conductivity was measured on pressed pellets using the van der Pauw method. The photoelectrical characteristics were measured under steady state illumination conditions using a Xe lamp (XBO 75) and a Jobin Yvon H25 monochromator, with a photodiode (EG&G HUV-1100B) in the reference beam, in the serial connection of the sample, a power supply (Keithley 230) and electrometer (Keithley 617). For the voltammetric measurements during the electrodeposition studies, the same electric devices were used.

Results and Discussion

Material characterization

The optical absorption spectrum of TiO_2/PT nanoparticles is given in Figure 4A as the Kubelka-Munk function. It consists of the optical absorption of TiO_2 nanoparticles with the absorption band $E_g = 3.3$ eV and the optical absorption of PT ($E_g = 2.1$ eV). The optical absorption of pure TiO_2 nanoparticles is shown for comparison.

The DC electrical conductivity obtained from measurements on pressed pellets in ambient atmosphere was $\sigma = 1.5 \times 10^{-6}$ S m^{-1}. Recalculated to the PT content in the composite (ca. 7 % by weight), this value corresponds to the conductivity of an undoped polythiophene.

Electrophoretic deposition

The dispersion systems containing TiO_2 nanoparticles, PT nanoparticles and TiO_2/PT composite nanoparticles in ethanol or butyl acetate, were tested for the possible preparation of thin layers on conductive substrates by the electrodeposition method in the experimental setup given in Figure 1. These dispersion systems differed in the electrodeposition ability as summarized in Table 1. The symmetric arrangement with both Ni electrodes was used in the electrodeposition studies. Generally, the nanoparticles based on TiO_2 were capable of electrodeposition in ethanol medium on the negative electrode only; no electrodeposition was observed on either cathode or anode when butyl acetate was used as a supporting medium, even at high electric potentials. The presence of the PT shell covering the TiO_2 nanoparticles surface caused an increase in the electrodeposition current by about one order of magnitude. On the other hand, no electrodeposition of the composite nanoparticles was observed in butyl acetate. The dispersion of neat PT behaved in opposite way. The polymer particles did not deposit in ethanol medium but the deposition in butyl acetate was observed, although at a high voltage exceeding 300 V and at a much lower deposition current, yielding only layers of poor quality. In this system, the dispersion stability was low, the dispersion tends to sedimentation, which suggests a smaller zeta potential of the nanoparticles.

Table 1. Summary of electrodeposition in various dispersion systems.

Material	Electrodeposition media	
	Ethanol	Butyl acetate (+1% EtOH)
TiO_2/PT particles	DEPOSITION (cathodic) $U > 30$ V, $I \sim 10^{-4}$ A	NO DEPOSITION $U \sim 300$ V, $I \sim 10^{-8}$ A
TiO_2	DEPOSITION (cathodic) $U > 30$ V, $I > 10^{-5}$ A	NO DEPOSITION $U \sim 300$ V, $I \sim 10^{-8}$ A
Polythiophene	NO DEPOSITION $U \sim 30$ V, $I \sim 10^{-4}$ A	DEPOSITION (cathodic) $U \sim 300$ V, $I \sim 10^{-6}$ A

The stability of the dispersion, electrophoretic mobility and the capability of coagulation and adhesion to the substrate have to be considered to discuss the electrophoretic deposition process. To achieve a stable dispersion of nanoparticles, their particulate surface charge is required. In nonaqueous dispersions, the dissociation/ionization of surface groups (electrostatic stabilization) and adsorption of ionic surfactants (electrosteric stabilization) are the only possible mechanisms for dispersion stabilization [12]. Oxide surfaces have a large concentration of surface amphoteric hydroxyl groups, which can react to yield both positively or negatively charged particles depending on pH of the medium [13]:

$$(M\text{-}OH)^0 + H^+ \rightarrow (M\text{-}OH_2)^+ \quad \text{at low pH}$$
$$(M\text{-}OH)^0 + OH^- \rightarrow (M\text{-}O)^- \quad \text{at high pH}$$

For the La, Zr or La oxides dispersed in ethanol, the three-step process of the surface charge stabilization was proposed [14]: (i) adsorption of ethanol molecules in undissociated form on the basic surface sites of the oxide particles, (ii) dissociation of ethanol molecules by H^+

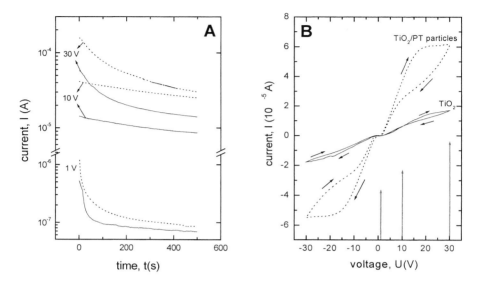

Figure 3. A. Relaxation of the electrodeposition current for a TiO_2/PT dispersion (dottted lines) and for TiO_2 particles (full lines) in ethanol after applying various voltages. **B.** Cyclovoltammetry of a TiO_2/PT dispersion (dotted line) and of neat TiO_2 (full line) in ethanol.

transfer to the basic surface sites, and (iii) desorption of EtO^- anions. As a result, the nanoparticles are positively charged. In SiO_2 particles, the second step of the mechanism proposed above proceeds via the charge transfer from acid surface sites to ethanol, which yields the negatively charged particles. As can be seen from the results summarized in Table 1, the former mechanism explains the situation for TiO_2 as well as for PT-covered TiO_2 particles in ethanol medium.

The layer formation by electrodeposition from the dispersion of TiO_2/PT nanoparticles in ethanol, under the constant current conditions ($I = 0.1$ mA, electrode distance 1 cm and area 0.5 cm^2) is shown in Figure 2. The deposition rate was decreasing with time, the thickness of the layer reached its saturation value at 8 μm within 80 s. The structure of the resulting layer is shown in the inset of Figure 2.

The time dependences of the electrodeposition of TiO_2/PT composite nanoparticles under the constant voltage conditions are shown in Figure 3A for the applied voltage 1, 10 and 30 V. The electrodeposition kinetics of neat TiO_2 nanoparticles is also shown for comparison. The starting concentration of nanoparticles in the dispersion was 14 mg/ml with the total amount of 500 mg in the volume, which allows to consider the deposition to take place under the constant concentration conditions.

The dependences of the electrodeposition current on the applied voltage for the same system as described above are presented in Figure 3B. It should be pointed out that the curves do not show the pure voltammetric behavior of the system, but reflect also a contribution of the growing layer to the total resistance in the electrical circuit. The symmetric shape of the voltammetric curves could be used as a proof of the electrophoretic mechanism of the deposition process: the nanoparticles preserve charge when deposited and the reversed electric field results in the stripping-off of the electrode and in the deposition on the opposite one. The UV-VIS optical absorption spectra of the electrodeposited composite nanoparticles did not show any observable optical absorption in the near-infrared region suggesting that the polymer did not undergo any oxidation during the electrodeposition process.

The observed time evolution of the electrodeposition current and its dependence on the applied voltage could be explained as follows. The deposition current is given by the product of nanoparticle concentration, electric field and reciprocal viscosity of the solvent. The electric field in the vicinity of the coated electrode has to exceed a certain threshold value to make the coagulation of the nanoparticles and their adherence to the electrode surface possible. In our case this threshold was ca. 5 V when the distance between the electrodes was 1 cm. During the deposition process, the growing layer forms an increasing resistance in the circuit on which the voltage drop occurs, which diminishes the driving force for the electrophoretic movement as well as for the particle coagulation on the surface of the growing layer. As a result, saturation of the layer thickness has been achieved.

Photoelectrical characteristics

We made an attempt to exploit the electrodeposition method for the fabrication of a photoelectrically sensitive hybrid polymer/inorganic solid cell based on the concept published recently [15]. In this type of the device, the photoelectrical conversion takes place on the interface between the TiO_2 core and PT shell of the composite particles, the bulk of the sample being sandwiched between two metal electrodes.

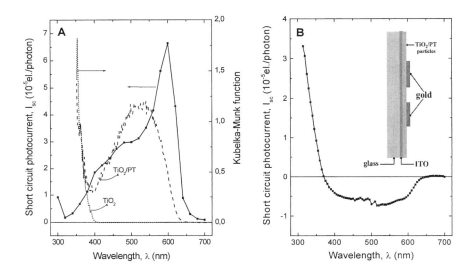

Figure 4. - A. Spectral dependence of the short-circuit photocurrent of TiO_2/PT samples measured in vacuum (10^{-3} Pa) - full line, UV-VIS absorption spectra of TiO_2/PT particles - dashed line, and neat TiO_2 - dotted lines. **B.** Spectral dependence of the short-circuit photocurrent of TiO_2/PT samples measured in air. Inset: sandwich structure of the sample under study.

The spectrum of the short-circuit photocurrent (I_{sc}) obtained on the samples of thickness 2 μm (for sample structure, see the inset of Figure 4B) at 10^{-3} Pa and room temperature is shown in Figure 4A, normalized to the incident photon flux. The sample was illuminated through the ITO electrode, the sign of the photocurrent refers to the top Au electrode. Compared with the optical absorption, the I_{sc} spectrum exhibits an antibatic behavior. It shows that the region near the ITO/polythiophene interface only weakly contributes to the free charge carrier photogeneration. This is in accordance with the mechanism in which the photogeneration takes place effectively in the bulk of the composite on the distributed p/n junction between TiO_2 and PT. The photogeneration in the visible region proceeds via the absorption of photons in PT shell. Because the LUMO level of the polymer is well above the TiO_2 conductance band, the photoexcited electron is transferred into the TiO_2 core leaving the hole in the polymer, through which it can move towards the top Au electrode, which forms with the polymer an Ohmic contact. The polymer thus works both as a photosensitizer and as a charge-transporting medium. This process is in agreement with the observed polarity of the Au electrode, which is positive upon illumination. In the UV spectral region, the photons are preferably absorbed by the TiO_2 core. Because the transfer of an electron from TiO_2 into PT is not energetically favorable, the charges recombine and do not contribute to the photocurrent.

The photoelectrical characteristics of the cell were found to be strongly dependent on ambient conditions. When the samples were exposed to the open atmosphere, the shape of the short-circuit photocurrent has changed dramatically as shown in Figure 4B. First, in the visible spectral region corresponding to the PT optical absorption, the direction of the photocurrent was reversed compared with the experimental conditions in vacuum, reaching a higher absolute value. Second, the photocurrent peak with opposite sign was observed for the wavelengths shorter than 380 nm, corresponding to the optical absorption of TiO_2. Also the charging/discharging effects were observed during the current-voltage characteristics measurements in the dark and under illumination. Similar behavior was found previously with photoelectrochemical cells composed of titanium electrode coated with the TiO_2/PT particles [16] or TiO_2/poly(bithiophene) bilayers [17] in $LiClO_4$/acetonitrile solution. The photocurrent was ascribed to the redox behavior of the PT (cathodic photocurrent) and TiO_2 (anodic peak). This suggests that the solid-state nature of the photoelectrical behavior of the cell under study measured in vacuum changed to the photoelectrochemical behavior upon the influence of ambient humidity.

Conclusions

The electrophoretic deposition was found to be a feasible way of preparing layers of insoluble polythiophene coated TiO_2 composite nanoparticles. Although the system was not optimized to achieve the best performance (e.g., neither adhesion promoter nor the binder was used in the electrodeposition), the quality of the layers was sufficient for the preparation of sandwich structures for photoelectrically active devices.

Acknowledgements

Financial support of the Grant Agency of the Czech Republic (grant No. 202/00/1152) is gratefully acknowledged. M. P. was partly supported by the European Graduate School EGK 720/1.

[1]   B. O'Regan, M. Grätzel, *Nature*, 1991, *353*, 737.
[2]   F. Cao, G. Oskam, P.C. Searson, *J. Phys. Chem*, 1995, *99(47)*, 17071.
[3]   U. Bach, D. Lupo, P. Comte, J. E. Moser, F. Weissortel, J. Salbeck, H. Spreitzer, M. Grätzel, *Nature*, 1998, *395*, 583.
[4]   W. Plieth, N. Hebestreit, German patent announcement no. 19919261.8.
[5]   L. Torsi, P. G. Zambonin, A. R. Hillman, D. C. Loveday, *J. Chem. Soc., Faraday Trans.*, 1993, *9*, 3941.
[6]   K. Idla, O. Inganäs, M. Strandberg, *Electrochim. Acta*, 2000, *45*, 2121.
[7]   S. Marchant, P. J. S. Foot, *J. Mater. Sci.: Mater. Electron.*, 1995, *6*, 144.
[8]   C. L. Huisma, A. Goossens, J. Schoonman, Proceedings of the 17th Workshop on Quantum Solar Energy Conversion (QUANTASOL 2002), March 17-23, 2002, Rauris, Salzburg, Austria.
[9]   J. Wen, G. L. Wilkes, *Chem. Mater.*, 1996, *8*, 1667.
[10]   R. Gangopadhyay, A. De, *Chem. Mater.*, 2000, *12*, 608.
[11]   J. Hofmann, M.S. Thesis, TU Dresden, 2000.
[12]   J. Lyklema, *Adv. Colloid Interface Sci.*, 1968, *2*, 65.
[13]   P. Sarkar, P.S. Nicholson, *J. Am. Ceram. Soc.*, 1996, *79*, 1987.
[14]   G. Wang, P. Sarkar, P.S. Nicholson, *J. Am. Ceram. Soc.*, 1997, *80*, 965.
[15]   D. Gebeyehu, C.J. Brabec, N.S. Sariciftci, D. Vaneneugden, R. Kiebooms, D. Vanderzande, F. Kienberger, H. Schindler, *Synth. Met.*, 2002, *125*, 279.
[16]   N. Hebestreit, J. Hofmann, U. Rammelt, W. Plieth, *Electrochim. Acta*, 2002, accepted.
[17]   U. Rammelt, N. Hebestreit, A. Fikus, W. Plieth, *Electrochim. Acta*, 2001, *46*, 2363.

Macromol. Symp. **2004**, *212*, 549-554

Vibration-Induced Energy Relaxation in Two-Level System

Miroslav Menšík, * *Stanislav Nešpůrek*

Institute of Macromolecular Chemistry, Academy of Sciences of the Czech Republic, Heyrovsky Sq. 2, 162 06 Prague 6, Czech Republic
E-mail: mensik@imc.cas.cz

Summary: Energy decay in a general two-level electronic system coupled to a vibrational harmonic mode interacting with a thermal bath is studied theoretically. The model assumes a general form of the off-diagonal elements (in the electronic basis) of the vibrational-electronic interaction. The cases for constant, linear, and quadratic dependence with respect to the vibrational displacement are investigated. For short-time regime, fast oscillations corresponding to a coherent energy exchange between electrons and vibrations appear. Their frequency Ω corresponds to the energy difference ($\hbar\Omega$) between electronic levels. Additionally, for the case of linear or quadratic coupling, the amplitude modulation of the oscillations with the frequency ω equal to that of vibrational motion is found.

Keywords: electron-vibrational interaction; energy transfer; excited state decay

Introduction

The process of energy relaxation in a system consisting of two interacting harmonic-potential energy surfaces (PES) is frequently studied (cf., e.g., [1-6] and references therein). Usually, the coupling between two PESs is assumed to be independent of a vibrational coordinate [1,2], but sometimes a linear [3,4,6] or quadratic dependence [5] is taken into account. In this article we present the studies of all these types of dependences. The Hamiltonian of the electron-vibrational system is taken in the following form (see also Fig. 1)

$$H^S = \hbar\omega[(\varepsilon + DQ + \frac{Q^2}{2} + \frac{P^2}{2})|1\rangle\langle1| + (-\varepsilon + \frac{Q^2}{2} + \frac{P^2}{2})|2\rangle\langle2| + W(Q)(|1\rangle\langle2| + |2\rangle\langle1|)] \quad (1)$$

The first (second) round bracket in the Hamiltonian (1) describes PES of, e.g., excited (ground electronic) state (see also Fig. 1). $2\varepsilon\hbar\omega$ is the vertical energy gap. The last bracket denotes the vibration-dependent coupling of both PESs. We assume three cases: (a) $W(Q) = V$ (b) $W(Q) =$

 DOI: 10.1002/masy.200450871

ΔQ, and (c) $W(Q) = \alpha Q^2$ (where V, Δ and α are interaction constants). Q (P) is a vibrational coordinate (momentum), $\hbar\omega$ is a vibrational energy quantum, and D describes the shift of the position of the minimum of the excited state with respect to the ground state. The electron-vibrational system is also attached to a thermal bath. This interaction is assumed to be linear in the vibrational displacement. Its impact on the system of interest is described by a standard master equation and the resulting tensor L^{REL} of a vibrational relaxation is of zero-th order in the vibration-induced coupling of the respective PES. Its explicit form can be found in, e.g., Ref. [1]. For simplicity, we assume that it is characterized by one rate constant k of the vibrational relaxation. The density matrix operator $\rho^S(t)$ of the relevant system then satisfies the relation

$$\frac{\partial}{\partial t}\rho^S(t) = -\mathrm{i}[H^S, \rho^S(t)]/\hbar + L^{REL}\rho^S(t) \qquad (2)$$

For the initial condition, we assume a fast excitation from the ground to excited state by vertical optical transitions without any change of the vibrational distribution, which is thermalized in the minimum of the ground state (see Fig. 1).

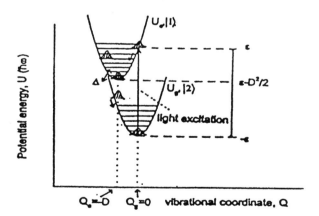

Figure 1. Potential energy surfaces of the excited and ground electronic states. The arrows indicate the process of vibrational relaxation.

The variable of interest, the excited state population $p(t)$, defined as

$$p(t) \equiv \mathrm{Tr}_{vib}\langle 1|\rho^S(t)|1\rangle \qquad (3)$$

is calculated in the time (t) interval ($0.50/\omega$). For the value of parameters introduced below, the edge of the interval is 0.13 ps. Here Tr_{vib} means a trace over vibrational mode.

Results

In the numerical simulation, we use the following values of the model parameters: temperature $T = 300$ K, vibrational energy quantum $\hbar\omega = 0.25$ eV, vertical excited state energy $2\varepsilon\hbar\omega = 2.5$ eV, $D = 0$, 0.5 and 1, rate of the vibrational relaxation $k = 0.3\omega$, 0.1ω and 0.03ω. The interaction energy terms are taken to be comparable with $\hbar\omega$, so V, Δ, and $\alpha = 0.5$, 1 and 2. The calculated dependences of $p(t)$ for short-time scale are given in Figs. 2 - 4. In general, very fast oscillations of $p(t)$ are superimposed on decay curves. They correspond to a coherent energy exchange between the electronic states and vibrational mode. In Fig. 2 the case of constant interaction is shown. The frequency Ω of these oscillations is associated with the energy difference $\hbar\Omega = 2\varepsilon\hbar\omega$ between the excited and ground electronic states.

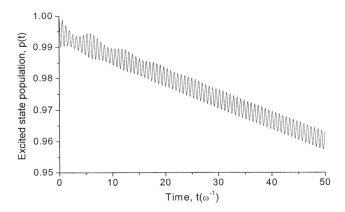

Figure 2. Time dependences of the excited state population for the constant coupling. $V = 0.5$, $k = 0.1\omega$ and $D = 1$.

A slight modulation of the oscillations with the frequency equal to ω were observed. In the case of linear or quadratic dependence (see Figs. 3 and 4), the situation changes.

552

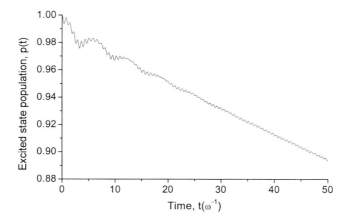

Figure 3. Time dependences of the excited state population for the linear coupling. $\Delta = 0.5$, $k = 0.1\omega$ and $D = 1$.

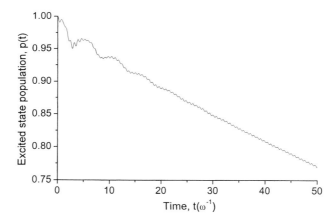

Figure 4. Time dependences of the excited state population for the quadratic coupling. $\alpha = 0.5$, $k = 0.1\omega$ and $D = 1$.

In general, time duration of the fast oscillations is equal to k^{-1}. The oscillation modulation with the frequency ω is more effective and even dominates if the Stokes shift is larger. As a result, strong oscillations with frequency equal to ω are found. The duration times of these oscillations are equal to k^{-1}. The dependence of the amplitude, frequency, durability of the oscillations and their modulations on the interaction parameters are given in Table 1.

Table 1. Characteristics of coherent oscillations for the constant and linear cases

Frequency Ω	Amplitude A	Modulation frequency φ	Oscillation durability τ^1	Modulation durability τ^2
Constant case: $W(Q) = V$				
$(2\varepsilon(1+\frac{V^2}{\varepsilon^2}) - \frac{D^2}{2})\omega$	$\frac{V^2}{2(\varepsilon^2+V^2)}$	ω	$> k^{-1}$	k^{-1}
Linear case: $W(Q) = \Delta Q \sim V_{\text{eff}}$				
$(2\varepsilon(1+\frac{V_{\text{eff}}^2}{\varepsilon^2}) - \frac{D^2}{2})\omega - \omega$	$\frac{V_{\text{eff}}^2}{2(\varepsilon^2+V_{\text{eff}}^2)}$	ω	k^{-1}	k^{-1}
Quadratic case: $W(Q) = aQ \sim V_{\text{eff}}$				
$(2\varepsilon(1+\frac{V_{\text{eff}}^2}{\varepsilon^2}) - \frac{D^2}{2})\omega$	$\frac{V_{\text{eff}}^2}{2(\varepsilon^2+V_{\text{eff}}^2)}$	ω	$> k^{-1}$	k^{-1}
		2ω (for strong coupling)		

The dependences of the frequencies and amplitudes of coherent oscillations are very similar to that of simple two-level model without inclusion of vibrational states. It is also interesting to see that the frequency of the fast oscillations for the linear case is lower by ω compared to other cases, because the interaction Hamiltonian for linear case causes the vibrational energy exchange $\hbar\omega$. The introduction of the effective value V_{eff} of the interaction for the linear and quadratic case should be understood as a measure of the influence of interaction constants Δ and α on the frequency Ω and amplitude A. For the values of vertical and vibrational energies used above, the period of fast oscillations is about 1.65 fs, while that of the modulation is 16.5 fs. As far as the frequency Ω of fast oscillations equals to the frequency of emitted light (if the radiative processes were taken into account), these oscillations cannot be directly observed optically. On the other hand, the period of the oscillation modulation is, for a real system, at least one order of magnitude slower. Thus, if in the case of a non-radiative excited state decay the ω-dependent modulation exists, then it must be visible in the optical emission decay curves. Such effect does not occur very often in molecular system, because the dynamics of the relevant non-radiative system is only rarely reduced to a single vibrational mode [7]. In the case of several vibrational modes, a destructive interference of vibrational modes will blur the vibrational modulation.

For a long time range, the oscillations disappear and the occupation probability $p(t)$ decays almost exponentially. In Ref. [6] it was shown that in the case of linear dependence of the off-diagonal interaction on the vibrational displacement Q and for small values of the Stokes shift, the rate constant Γ of the occupation probability decay satisfies the following relation

$$\Gamma \sim \Delta^2 k,$$ (4)

which reminds of the Fermi golden rule, where the delta function is replaced by the rate constant k of vibrational relaxation. This formula illustrates well the energy decay process. The energy of electronic excitation is transferred coherently to the vibrational mode (this process is represented by the constant Δ of the electron-vibrational interaction) and after that it is dissipated to a bath.

Conclusion

The calculated dependences of the excited state population show that the curve of non-radiative decay contains fast oscillations modulated with the frequency ω, which is associated with a vibrational motion. The modulation is strongly increased in the case of a large Stokes shift and, particularly, if the off-diagonal (in the electronic basis) matrix elements of the electron-vibrational interaction are explicitly dependent on vibrational displacement.

Acknowledgement

The financial supports by grants AV0Z4050913 and A1050901 of the Grant Agency of the Academy of Sciences of the Czech Republic and ME 270 from the Ministry of Education, Youth and Sports of the Czech Republic are gratefully appreciated.

[1] O. Kühn, V. May, M. Schreiber, *J. Chem. Phys.* **1994**, *101*, 10404
[2] U. Kleinekatöfer, I. Kondov, M. Schreiber, *Chem. Phys.* **2001**, *268*, 121
[3] M. Menšík, *J. Lumin.* **1996**, *65*, 349
[4] M. Menšík, *J. Phys.: Condens. Matter* **1995**, *7*, 7349
[5] M. Menšík, S. Nešpůrek, *Synth. Met.* **2000**, *109*, 207
[6] M. Menšík, S. Nešpůrek, *Czech. J. Phys.* **2002,** *52*, 945
[7] I. V. Rubtsov, K. Yoshihara, *J. Phys. Chem.* **1999**, *103*, 10202

End-Functionalized π-Conjugated Oligomeric Materials for Photoelectrical Applications

Martin Pavlík,[*1,2] *Jiří Pfleger,*[1] *Christo Jossifov,*[3] *Jiří Vohlídal*[2]

[1]Institute of Macromolecular Chemistry, Academy of Sciences of the Czech Republic, Heyrovský Sq. 2, 162 06 Prague 6, Czech Republic
[2]Department of Physical and Macromolecular Chemistry, Laboratory of Specialty Polymers, Charles University, Albertov 2030, 128 40 Prague 2, Czech Republic
[3]Institute of Polymers, Bulgarian Academy of Sciences, 1113 Sofia, Bulgaria

Summary: Soluble poly(diphenylacetylene)s (PDA) capped with PhCO– groups (PDA-C) and $Ph_2C=$ groups (PDA-P) were prepared via McMurry reductive coupling of benzil and carbonyl-olefin exchange reaction of tetraphenylethene and benzil, respectively. Fluoren-9-ylidene groups have been introduced into PDA by the McMurry coupling of PDA-C with fluoren-9-one and via copolymerization of benzil and fluoren-9-one. The oligomers prepared are stable in air, soluble in a variety of solvents. They can be processed by casting to form good-quality thin films suitable for measurements of electrical and optoelectrical properties.

Keywords: conjugated polymers; oligomers

Introduction

Poly(vinylene)s derived from monosubstituted acetylenes are the first large group of π–conjugated organic materials which showed impressive electrical and optoelectrical properties. However, these polymers have insufficient stability under electrical and optical loading as well as low stability in air, which is the main drawback preventing them from practical use [1-8]. On the other hand, fully substituted polyvinylenes, such as poly(diphenylacetylene)s (PDA) introduced by Masuda and coworkers [4,9], are sufficiently stable in air; however, unsubstituted PDAs prepared by their procedures are insoluble in any solvent, and ring-substituted PDAs, which are soluble, do not show desired electrical and optoelectrical properties. The procedures of Masuda et al. consist in the chain polymerization of diphenylacetylene with $TaCl_5$- and $NbCl_5$-based catalysts, which is not controlled as to the polymer molecular weight. It seems that a too high molecular weight of the so obtained PDAs is the main reason for their insolubility. Since polymers of low-to-medium molecular weights are usually soluble and, in addition, it is well known that various oligomeric conjugated oligomers show functional properties that are good enough for their applications as materials in construction of electronic and optoelectronic devices [10], it seems useful to find a synthetic path to soluble and processable low-molecular-weight PDAs.

We have tried to accomplish such synthesis by using Masuda's catalyst systems at low diphenylacetylene/catalyst ratios and various temperatures; however, we always obtained high-molecular-weight, insoluble PDA. Also attempts to prepare processable PDAs by using tungsten dinuclear complexes containing unbridged quadruple bonds [11] were unsuccessful. Thus, it seems that formation of growing species is too slow compared with the propagation in these systems, which does not allow to synthesize low-molecular-weight PDAs in that way. More than ten years ago, Jossifov et al. [12-15] used the carbonyl-olefin exchange reaction, which reminds of the McMurry reaction [16], for preparation of various conjugated polymers. They prepared PDA from diphenylethane-1,2-dione (benzil) and 1,1,2,2-tetraphenylethene and from 1,2,3,3-tetraphenylprop-2-en-1-one using WCl$_6$, MoCl$_5$ and WOCl$_4$ in combination with AlCl$_3$ as catalysts, which, however, must be used at high monomer/catalyst mole ratios ranging from 0.1 to 1 to achieve a good yield of polymer. According to the results obtained and the mechanism proposed by Jossifov [14,15], the carbonyl-olefin exchange reactions give soluble PDAs with benzoyl (Ph-CO) and diphenylmethylidene (Ph$_2$C=) end groups. In principle, the carbonyl-olefin exchange reaction can also be used for additional end-group modification of PDAs. In the present contribution we report the preparation of soluble PDAs capped with Ph-CO, Ph$_2$C=, and fluoren-9-ylidene end groups by combining the carbonyl-olefin exchange reaction and McMurry reaction.

Polymerization via carbonyl-olefin exchange reaction

McMurry reaction

Experimental

General. Benzil (diphenylethane-1,2-dione), tetraphenylethene, tungsten hexachloride (WCl$_6$), aluminium trichloride (AlCl$_3$), chlorobenzene dried over molecular sieve, 1,4-dioxane for fluorescence measurements (all Aldrich), fluoren-9-one (Fluka) and methanol and sodium hydroxide (both Lachema, Czech Republic) were used as supplied. Toluene (Lachema, Czech Republic) was distilled from P_2O_5 and kept over molecular sieve. Size exclusion chromatography (SEC) analyses of polymers were made on a TSP (Thermo Separation Products, Florida, USA) chromatograph equipped with a UV detector operating at 254 nm and a series of two PL-gel columns (Mixed B and Mixed-C, Polymer Laboratories Bristol, UK, THF eluent, flow rate 0.7 mL min^{-1}); molecular weight values based on polystyrene standards are reported. IR spectra were measured on a Nicolet Magna IR-760 spectrophotometer using the diffusion reflectance method. UV-VIS absorption spectra were recorded on a double beam Perkin Elmer 340 spectrophotometer and steady-state photoluminescence spectra on a SPEX Fluorolog 3-11 instrument using polymer solutions in 1,4-dioxane.

Polymerization and copolymerization reactions were carried out under dry argon atmosphere in chlorobenzene or toluene using the following concentrations of reactants: $[WCl_6]_0$ = $[AlCl_3]_0$ = 0.5 M; $[monomer]_0$ = 0.5 M or 1 M. The monomer or a mixture of monomers was dissolved in particular solvent (25 mL) at 60 °C, then thermostatted at a desired temperature, weighed amounts of powdered $AlCl_3$ and WCl_6 were added under stirring and the resulting mixture was allowed to react for a given time 1 – 6 h. The reaction was terminated by adding aqueous NaOH (50 mL, 40 % w/w) under shaking, the organic phase was separated and washed with distilled water till the neutral reaction. The product was precipitated by adding methanol (100 mL), separated by centrifugation, four times reprecipitated from toluene/methanol and dried in vacuum to constant weight. PDA with terminal =CPh_2 groups (hereinafter denoted as PDA-P) was prepared by reacting a mixture of benzil and 1,1,2,2-tetraphenylethene, PDA with terminal benzoyl groups (PDA-C) by reacting only benzil as monomer. A copolymer of benzil and fluoren-9-one prepared using the latter procedure is denoted PDA-F.

End-capping of PDA-C with fluoren-9-one. PDA containing fluoren-9-ylidene end groups, PDA-FM, was prepared by a modification of PDA-C using the reaction of the oligomer with fluoren-9-one. Solid WCl_6 (0.1 g, 0.253 mmol) and solid $AlCl_3$ (0.05 g, 0.375 mmol) were added under argon to a solution of PDA-C (100 mg) of M_w = 1 290 and fluoren-9-one (50 mg, 0.277 mmol) in chlorobenzene (13 ml), the suspension was heated up and stirred for 3 h at 100 °C. Then the mixture was cooled down to room temperature and worked up as described above (M_w = 1 410).

Results and Discussion

Polymerization data summarized in Table 1 clearly show that the studied reaction is rather complex; nevertheless, some qualitative conclusions can be drawn. As can be seen, a longer reaction time and mainly a higher reaction temperature and higher monomer concentration have generally a positive effect on the yield and M_w of the benzil homopolymer and copolymer. Also, the influence of the solvent used for the polymerization is evident. In general, higher yields as well as higher molecular weights are obtained in chlorobenzene rather than in toluene.

Table 1. Polymerization data for poly(diphenylacetylene)s

No.	Monomer (mole ratio)	$[WCl_6]/$ [monomer(s)]	T, °C	t, h	Solvent	Yield %	M_w
1	B	1	60	1	PhCl	nd[b]	690
2	B	1	60	2	PhCl	nd[b]	740
3	B	1	60	3	PhCl	48	1 200
4	B	1	60	6	PhCl	53	1 300
5	B	1	100	3	PhCl	49	1 400
6	B	0.5	100	1	PhCl	91	2 500
7	B	0.5	100	3	PhCl	93	3 600
8	B	1	131	3	PhCl	~100	7 200 [a]
9	B	1	60	3	toluene	57	350
10	B	0.5	60	3	toluene	81	1 400
11	B + T (1:1)	1	25	6	PhCl	33	690
12	B + T (1:1)	1	25	48	PhCl	93	2 100
13	B + T (1:1)	1	60	6	PhCl	25	650
14	B + T (1:1)	1	80	6	PhCl	38	830
15	B + T (1:1)	1	100	6	PhCl	91	2 200
16	B + T (2:1)	1	60	3	toluene	51	410
17	B + T (1:1)	1	60	3	toluene	30	460
18	B + T (1:2)	1	60	3	toluene	22	510
19	B + T (1:0.8)	0.7	60	1	toluene	60	1 100
20	B + F (1:1)	1	80	3	PhCl	23	1 100
21	B + F (1:1)	0.5	80	3	PhCl	48	7 900
22	B + F (1:1)	0.5	80	3	toluene	52	1600
23	F	1	80	6	PhCl	12	2 900

[a] Molecular weight of a soluble fraction of the polymer (ca. 8 % of the isolated polymer).
[b] not determined
B – benzil, T – tetraphenylethene, F – fluoren-9-one.

Using the observed effects of reaction conditions, the character of the prepared PDAs can be tuned from short-chain oligomers to almost insoluble polymers formed under reflux of the solvent (Table 1, entry No. 8).

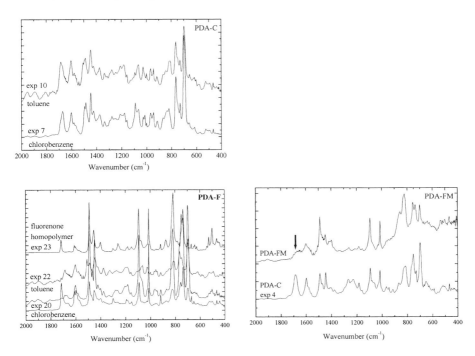

Figure 1. - IR spectra of selected PDA oligomers acquired by diffusion reflectance method.

Copolymerization of benzil and fluoren-9-one provided interesting results. We assumed that fluoren-9-one will act as a precursor of terminal units only, introduced to both ends of PDA via the McMurry reaction. However, the observed results surprisingly indicate that fluoren-9-one also acts as a bifunctional monomer, the units of which are also incorporated in the polymer backbones. This suspicion was confirmed by a successful homopolymerization of this monomer (experiment 23). On the other hand, functionalization of the prepared PDA-C gave a polymer, in which terminal carbonyl groups are almost absent being replaced with fluoren-9-ylidene groups (PDA-FM). However, nor in this case can we exclude participation of homopolymerization of fluoren-9-one in the overall reaction.

IR spectra of selected oligomers of all four classes are compared in Figure 1. As can be seen, the band at 1686 cm^{-1}, which is typical of carbonyl groups, is absent in IR spectra of PDA with =CPh$_2$ end groups as expected according to the results of Jossifov [13-15]. However, this band is present in IR spectrum of PDA-F sample prepared in toluene, which proves incomplete end-capping of this polymer with fluoren-9-ylidene units, most probably owing to prevailing incorporation of fluorene units into the polymer backbones. On the other hand, the CO band is practically absent in the IR spectrum of PDA-F prepared in chlorobenzene, which corresponds with the above discussed easier course of the coupling reaction in this solvent. We cannot provide more detailed data on the structure of these copolymers because only poorly resolved NMR spectra were obtained.

In the case of copolymerization of benzil with 1,1,2,2-tetraphenylethene (Table 1, experiments 11-19), we found that an increasing amount of tetraphenylethene in the starting mixture leads to a decrease in the yield of methanol-insoluble polymer (experiments 16-18), which can be ascribed to the preferred formation of non-functionalized benzil homopolymer. According to IR spectra, all copolymers of benzil and tetraphenylethene contain some carbonyl groups, which indicates that, under the used reaction conditions, the McMurry reaction is favored compared with the carbonyl-olefin exchange reaction.

UV-vis absorption and fluorescence spectra of PDA-P oligomers in dioxane solution are presented in Figure 2. PDA prepared in toluene was probably partly contaminated with a very stable complex of the oligomer with the catalyst residues [17], which absorbs in the visible region and, as can be seen from Figure 2, quenches the polymer photoluminescence. Compared with PDA-P, the optical absorption of PDA-C is red-shifted possessing much higher absorbance and broader distribution thus indicating a higher extent of effective π-conjugation in PDA-C. The bathochromic shift of absorption spectra of π-conjugated polymers caused by carbonyl groups was discussed recently [18] as an increase in the effective conjugation length. In addition to the conjugative effect, carbonyl groups introduce n→π* electronic transition originating in the nonbonding π-orbitals on the oxygen atom, which is not observed in the spectrum due to much lower absorbance. In absorption spectra of PDA-F, optical transitions centered at 370 nm, typical of fluorene groups, as well as vibronic peaks in the fluorescence emission spectrum centered at 420 nm are clearly visible.

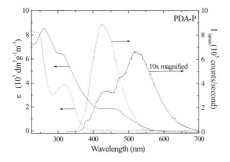

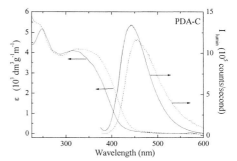

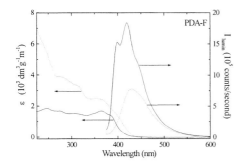

Figure 2. UV-vis absorption spectra and photoluminescence emission spectra of dioxane solutions of the PDAs prepared in chlorobenzene (dashed lines) and toluene (solid lines). Excitation wavelength 320 nm for PDA-P prepared in chlorobenzene, 366 nm in all other cases.

Conclusions

Although the application of the McMurry reaction to preparation of special polymers was already described [7,8], we extended this procedure to synthesis of telechelic soluble oligomers with various functional end-groups. Molecular weight of these oligomers can be controlled by an appropriate choice of reaction conditions. The described method is very simple and provides end-functionalized soluble oligomers in a single step. The resulting materials are film-forming, well processable and they show high stability in air. These properties meet demands of practice for organic materials for optoelectrical measurements and applications. Other end-groups can also be introduced into PDA chains by using these procedures and suitable precursors containing carbonyl and/or diphenylmethylidene end-groups of appropriate reactivity. Furthermore, we have found that fluoren-9-one homopolymerizes in the presence of WCl_6 and $AlCl_3$; however, we did not yet obtain evidence that would enable proposal of a mechanism of this reaction. This interesting new reaction providing a novel material is to be studied more in detail.

Acknowledgements

Financial support of the Grant Agency of the Czech Republic (project 202/00/1152), Grant Agency of the Academy of Sciences of the Czech Republic (project 12/96/K), Grant Agency of the Charles University (project 241/2002/B-CH/PrF) and Ministry of Education, Youth and Sports of the Czech Republic (MSM 1131000001) is gratefully acknowledged.

[1] H. S. Nalva, ed., Handbook of Organic Conductive Molecules, Wiley, New York 1996.
[2] J. Sedláček, J. Vohlídal, Z. Grubišič-Gallot, *Makromol. Chem. Rapid Commun.* 1993, *14*, 51.
[3] J. Vohlídal, D. Rédrová, M. Pacovská, J. Sedláček, *Collect. Czech. Chem. Commun.* 1993, *58*, 2651.
[4] H. Shirakawa, T. Masuda, K. Takeda in: The Chemistry of Triple-bonded Fumctional Groups (ed. S. Patai), Wiley, New York 1994, Chapter 17, pp. 945-1016.
[5] J. Vohlídal, Z. Kabátek, M. Pacovská, J. Sedláček, Z. Grubišic-Gallot, *Collect. Czech. Chem. Commun.* 1996, *61*, 120.
[6] Z. Kabátek, B. Gaš, J. Vohlídal, *J. Chromatogr. A* 1997, *786*, 209.
[7] S. M. A. Karim, R. Nomura, T. Masuda, *Polym. Bull.* 1999, *43*, 305.
[8] J. Vohlídal, J. Sedláček in: Chromatography of Polymers: Hyphenated and Multidimensional Techniques (ed T. Provder); *ACS Symp. Ser.* 1999, 731, 263.
[9] A. Niki, T. Masuda, T. Higashimura, *J. Polym. Sci.* 1987, *25*, 1553.
[10] J. Jortner, M. Ratner, Molecular Electronics; Blackwell Science, Oxford 1997.
[11] K. Mertis, S. Arbilias, D. Argyris, N. Psaroudakis, J. Vohlídal, O. Lavastre, P.H. Dixneuf, *Collect. Czech. Chem. Commun.* 2003, *68*, 1094.
[12] I. Schopov, C. Jossifov, *Makromol. Chem., Rapid Commun.* 1983, *4*, 659.
[13] C. Jossifov, I. Schopov, *Makromol. Chem.* 1991, *192*, 857, 863.
[14] C. Jossifov, *Eur. Polym. J.* 1993, *29*, 9.
[15] C. Jossifov, *Eur. Polym. J.* 1998, *34*, 883.
[16] J. McMurry, *Chem. Rev.* 1989, *89*, 1513.
[17] I. Schopov, L. Mladenova, *Makromol. Chem. Rapid Commun.* 1985, *6*, 659.
[18] M. Belletete, J.F. Morin, S. Beaupre, M. Ranger, M. Leclerc, G. Durocher, *Macromolecules* 2001, 34, 2288.

Macromol. Symp. **2004**, *212*, 563-569

Comparative Study of Photodegradation and Metastability in Solution-Processed and Plasmatic Polysilylenes

F. Schauer,[*1,2] N. Dokoupil,[2] P. Horváth,[2] I. Kuřitka,[2] S. Nešpůrek,[2,3] J. Pospíšil[3]*

[1] Centre for Polymer Materials, Faculty of Technology, T. Bata University, T.G. Masaryk Sq. 272, 762 72 Zlín, Czech Republic
[2] Faculty of Chemistry, University of Technology, Purkyňova 118, 612 00 Brno, Czech Republic
[3] Institute of Macromolecular Chemistry, Academy of Sciences of the Czech Republic, Heyrovsky Sq. 2, 162 06 Prague, Czech Republic

Summary: This study deals with photodegradation and metastability of polysilylenes. Physical properties of chemically prepared poly[methyl(phenyl)silylene] and polysilylenes prepared in radio-frequency and microwave discharges are compared. The position and width of the photoluminescence band due to the main chain excitations for both materials preserve but its intensity strongly depends on the preparation conditions and changes considerably during the photodegradation. The difference in photodegradation rate of both materials is explained by their different susceptibility to silyl radical formation and reverse recombination due to the chain and side groups rigidity.

Keywords: metastability; photodegradation; photoluminesecence; plasma; polymerization; polysilylenes

Introduction

Photodegradation and metastability is a widely studied phenomenon and hitherto unsolved problem in silicon (sp^3)-based and σ-bonded materials with the dimensionality changing from one dimensional (1D) polysilylenes[1] to three-dimensional (3D) systems like amorphous hydrogenated silicon (a-Si:H)[2], where 3D covalently bonded network with hydrogen as broken bond terminator and stress relieving agent is formed. The complexity of the photodegradability and/or metastability rests in many intermediate steps or chemical reactions, e.g. bond scission, crosslinking, oxidation, and bond breaking and their dynamics together with the hydrogen subsystem behaviour[3]. As real systems are disordered, a decisive role play also the topological and energetical disorder parameters, leading, e.g., to the concepts of defect pool[4]. Polysilylenes seem to be interesting materials for the production of near-UV electroluminescence devices (LEDs)[5]. Their main advantage is good processability and the possibility to modify their luminescence spectra by the chemical structure of side groups. However, the liability of polysilylenes to energetical radiation resulting in their photodegradation and metastability is one of the problems limiting their applications. Recently in[6] we have demonstrated the feasibility of the UV LED production from plasma polysilanes. The purpose of the present paper is to compare the photodegradation and metastability of solid films of poly[methyl(phenyl)silylene] (PMPSi) prepared from solution and by plasma polymerization.

Experiment

The "standard" PMPSi (chemically synthesized) was prepared as described earlier[7]. Films were prepared from toluene solution by spin-coating on quartz (absorption) or on Si wafer (luminescence, IR absorption). For the preparation of "plasma" PMPSi, we used radio-frequency (RF) and microwave (MW) plasma techniques. Both techniques are different in many aspects[8], namely: (1) the excitation of the monomer and presence of the energy-carrier feed gas, (2) the electron temperature and plasma density, and (3) the ion bombardment. The CVD RF plasma reactor was a 13.56 MHz radiofrequency capacitively coupled discharge unit with maximum power 1.5 W cm^{-2}. For the film deposition, we used a mixture of hydrogen (0-60Pa) and monomer (5-30 Pa). The substrate temperature was 80 °C. In microwave plasma the feed gas (helium and/or hydrogen) was ionised and excited by microwave power in an excitation discharge chamber with the frequency of 2.45 GHz under the conditions of electron cyclotron resonance (ECR) and fed to the monomer. As input monomer material methyl(phenyl)silane (Fluka, MPS) was used.

For UV-VIS absorption, IR absorption and luminescence measurements, a Hitachi U300, a Nicolet 400 FTIR, and a Perkin-Elmer LS 55 fluorimeter, respectively, were used. For the degradation experiments, the 266 nm and 355 nm light of NdYAG laser (Continuum Minilite II) was used.

Optical absorption and photoluminescence (PL) emission spectra of both standard- and plasma-PMPSi films are given in Fig. 1. Optical absorption of standard-PMPSi consists of three bands with maxima at 338 nm (σ-σ* transitions), 276 nm (π-π* transitions in phenyl groups), and 195 nm (mainly π-π* transitions as follows from quantum chemical calculations). The luminescence spectrum shows the excitonic deactivation at about 355 nm. The fluorescence quantum efficiency is quite high, about 0.15. The excitonic PL (band with maximum at 325 nm) was observed for RF plasma-PMPSi samples prepared under the partial pressure of the monomer higher than P_{mono} = 20 Pa; the maximum of the band was detected at

$\lambda = 310 \div 360$ nm depending on P_{mono}. Maxima of other bands was observed at about 260 and 225 nm; also a long-wavelength absorption tail, typical of disordered materials, was detected.

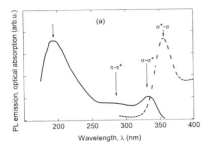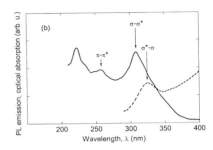

Figure 1. Absorption (full line) and emission PL ($\lambda_{exc} = 280$ nm, dotted line) spectra of (a) standard PMPSi, (b) RF plasma PMPSi ($P_{mono} = 20$ Pa, $P_{H2} = 60$ Pa).

In Fig. 2a there are the emission spectra of standard PMPSi for different excitation wavelengths. The differences indicate that the PL spectrum consists of several components. It has been revealed that the structural defects of PMPSi, which consist not only of the branching points[9] but also of defects produced inevitably during the polymerization and material ageing, act as radiative recombination centres for the luminescence in the visible region. In Fig. 2b this effect is depicted for the plasma PMPSi. The PL spectrum measured with $\lambda_{exc} = 280$ nm consists of both the σ^*-σ exciton emission with maximum at $\lambda = 370$ nm and a broad defect luminescence centred at $\lambda \sim 480$ nm. The PL emission spectrum obtained with $\lambda_{exc} = 210$ nm differs considerably: the σ^*-σ emission at $\lambda = 370$ nm is very strong and the defect luminescence is hardly observable.

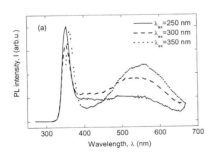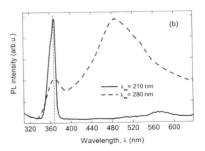

Figure 2. Emission PL spectra of (a) standard PMPSi and (b) RF plasma PMPSi; parameter is the excitation wavelength.

Degradation and stability

In Fig. 3 are compared IR spectra of standard and plasma PMPSi before and after degradation by UV radiation of 266 nm for 3600 s. There is hardly any change in bonding of phenyls in both materials after the degradation (Si-C_{ar} at 1120 and 1427 cm^{-1}), but there are changes in bonding of aliphatic carbon Si–C_{ali} (1046 and 1253 cm^{-1}). The most expressed change is the increase in absorption of the more stable Si–CH_2–Si units (1039 - 1050 cm^{-1}) and subsequent formation of Si–C units near 800 cm^{-1} in both materials.

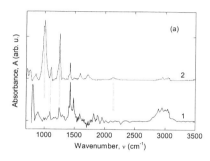
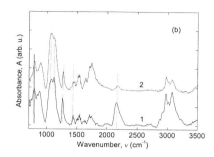

Figure 3. IR spectra of (a) standard and (b) RF plasma PMPSi before and after degradation with 266 nm light.

The scission of the Si–Si bonds of the main chain was confirmed by luminescence quenching. In Fig. 4a there is a spectral dependence of the luminescence emission of as-prepared and degraded (1 h, λ_{deg} = 260 nm) standard samples. The intensity exciton peak (A)

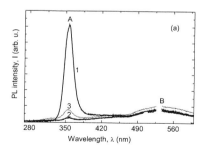
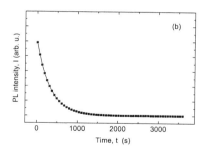

Figure 4. The quenching of the σ^*-σ excitonic PL of standard PMPSi film: (a) spectral dependence, λ_{ex} = 266 nm, curve 1 – virgin sample, curve 2 – 1 h degradation with 266 nm light radiation, curve 3 – after 2 days of RT annealing; (b) time dependence of peak A.

strongly decreases with the degradation, whereas the peak of defect luminescence (B) is hardly changed. It is also remarkable that the σ*-σ peak position in luminescence (355 nm) is not changed during the degradation. In Fig. 4b is the corresponding degradation time dependence of the luminescence quenching.

We also observed remarkable changes in the dependence of the excitation spectra on degradation time (λ_{deg} = 200 nm). In Fig. 5 are excitation PL spectra detected at 360 nm for three degradation times, 0, 5 and 15 min. The most remarkable feature is a tail of the long-wavelength part of the excitation PL peak, depicting the increase of material disorder expressed by the changing slope on progressed UV irradiation. As it follows from Fig. 5b the created disorder is also wavelength-dependent. An interesting and not-explained feature of the excitation spectra is the peak near 220 nm, which decreases after UV degradation. This feature was observed also in plasma PMPSi[10]. An additional information of this excitation anomaly comes from plasma PMPSi subjected to physical degradation by thermal treatment.

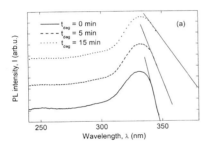 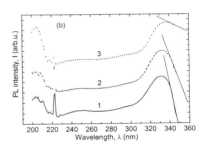

Figure 5. Excitation spectra of PL of standard-PMPSi, λ_{em} = 360 nm: (a) detail of the disorder effect, λ_{deg} = 200 nm, curve 1 – virgin sample, curve 2 – t_{deg} = 5 min, curve 3 – t_{deg} = 15 min; (b) complete excitation spectrum, curve 1 – non-degraded sample, curve 2 – t_{deg} = 5 min, λ_{deg} = 200 nm, curve 3 – t_{deg} = 5 min, λ_{deg} = 325 nm.

In effusion diagrams of plasma PMPSi, three species were observed – phenyl, CH_3 and H with corresponding temperature ranges at 400 - 500 °C, 370 - 800 °C (with maxima at 400 and 720 °C), and 500 - 800 °C, respectively. The radiative exciton photoluminescence at about 330 nm (λ_{exc} = 220 nm) strongly decreases with phenyl effusion.

Discussion

From the comparison of solution-processed and plasmatic PMPSi it may be concluded that the plasma PMPSi produced under optimised conditions is a nanostructural material with controlled 1D islands of polysilylene embedded in 3D amorphous network producing a cage effect[6]. Its optical properties change in a broad interval, as they strongly depend on the plasma parameters. Similarity of plasma and standard PMPSi with respect to PL is well documented in Figs. 1 and 2. The PL due to σ^*-σ deexcitation preserves roughly the peak position. Similar are also absorption spectra, but the values of absorption coefficients are different, cf. $\alpha \approx 1 \times 10^4$ cm^{-1} vs. 1×10^5 cm^{-1}.

Photodegradation of PMPSi films is a consequence mainly of Si–Si bond scission. From transport and charge-collection experiments follows that the photogeneration efficiency decreases with degradation[11,12]. It is argued that the weak bond formed after irradiation and susceptible to annealing, represents the defect in conformation resulting in the recombination state and thus creating the non-radiative path for exciton trapping and deactivation. The hole transport during the photodegradation is influenced by the increase of the disorder[13] (cf. Fig. 5.) However, the picture of the Si–Si backbone photophysics in the solid state is more complicated and it is a matter of debate. A controversy is about the absorption and scission of Si–Si bonds[14]. There are calculations supporting various stages of deformations of Si–Si bonds. Alan et al.[13] reported, on the basis of theoretical calculations, that the σ^*-σ transition tends to the scission of Si–Si bonds but, usually, they do not break directly because of interaction with the surrounding medium forming mainly weak Si–Si bonds [15]. Takeda et al.[16] found, by light-induced ESR and model calculations, two types of photocreated metastable states, one for lower photoexcitation energy (3.5 eV) due to the Si skeleton stretching and the other for higher-energy photoexcitations (4.8 eV). The weak bond concept was also used in the model of thermally activated scission with the distribution of activation energies[17]. This model is close to the more general defect-pool model[2] with the distributed formation energies put forward for amorphous silicon and its alloys.

Acknowledgements

This work was supported by grants from the Grant Agency of the Czech Republic (202/01/0518) and from the Ministry of Education, Youth and Sports of the Czech Republic (OC D14.30).

[1]   R.D. Miller and J. Michl, *Chem. Rev.* **1989**, 89, 1359.
[2]   R.A. Street, Hydrogenated Amorphous Silicon, Cambridge University Press, Cambridge, 1991.
[3]   N. Matsumoto, Jpn. J. Appl. Phys. **1998**, 37, 5425.
[4]   Z.E. Smith and S. Wagner, in Advances in Disordered Semiconductors: Amorphous Silicon and Related Materials, ed. H. Fritzsche, World Scientific, Singapore, N. Jersey, London, Hong Kong, 1989, pp. 409.
[5]   H. Suzuki in Organic Electroluminescent Materials and Devices, ed. S. Miyata and H.S. Nalwa, Gordon and Breach Publishers, 1997, p. 231.
[6]   F. Schauer, I. Kuřitka, N. Dokoupil and P. Horváth, *Physica* E 14 **2002**, 272.
[7]   F. Schauer, R. Handlir and S. Nespurek, *Adv. Mater. Opt.Electron.* **1997**, 7, 61.
[8]   N. Dokoupil, I . Kuřitka and F. Schauer, Proc. 6th Seminary on Physics and Chemistry of Molecular Systems, December 2000, Brno.
[9]   O. Itoh, M. Terazima, T. Azumi, N. Matsumoto, K. Takeda and M. Fujino, *Macromolecules* **1989**, 22, 10718.
[10] P. Horváth, F. Schauer, O. Salyk, I. Kuřitka, S. Nešpůrek, J. Zemek, and V. Fidler, *J. Non-Cryst. Solids* **2000**, 266-269, 989.
[11] R. Handlíř, F. Schauer, S. Nešpůrek, I. Kuřitka, M. Weiter and P. Schauer, *J. Non-Cryst. Solids* **1998**, 227-230, 669.
[12] O. Salyk, A. Poruba and F. Schauer, *Chem. Pap.* **1996**, 50, 177.
[13] G. Allan, C. Delerue and M. Lannoo, *Phys. Rev.* B 48, **1993**, 7951.
[14] H. Naito, S. Kodama, Q.Z. Kang and M. Okuda, *J. Non-Cryst. Solids* **1996**, 198-200, 653.
[15] Y. Nakayama, T. Kurando, H. Hayashi, K. Oka and T. Dohmaru, *J. Non-Cryst. Solids* **1996**, 198-200, 657.
[16] K. Takeda, K. Shiraishi, M. Fujiki, M. Kondo, K. Morigaki, *Phys. Rev. B* **1994**, 50, 5171.
[17] P. Trefonas, R. West, R.D. Miller, *J. Am.Chem. Soc.* **1985**, 107, 2737.

Frequency Analysis of Diffusion in 1D Systems with Energy and Spatial Disorder

Angeles Pitarch,* Germà Garcia-Belmonte, Juan Bisquert

Departament de Ciències Experimentals, Universitat Jaume I, 12080 Castelló, Spain
E-mail: pitarchm@guest.uji.es

Summary: We have analyzed the frequency-dependent features of diffusive transport of mobile carriers by hopping in a disordered environment composed of varying energy barriers. We developed a simple approach based on the solution of the master equation for a gradient of concentration (chemical potential) under a sinusoidal perturbation. This method extends the well-known result of the steady state case, $D = \{1/\Gamma\}^{-1}$, to the frequency domain. The results of our calculations are in agreement with the approximate analytical solution of the CTRW formalism. We are able to determine the onset of frequency-dependent diffusivity in terms of probabilities of highest energy barriers.

Keywords: ac conductivity; activation energy; diffusion; dispersive transport; hopping

Introduction

Electronically conducting polymers possess a unique combination of physical and chemical properties making them prospective materials for numerous applications such as in batteries and supercapacitors, sensors and ion release systems. In these applications the organic conductor is embedded in a secondary, ionically conducting phase that shields the electric fields beyond the nanoscopic scale, so that diffusion of electrons becomes the dominant transport mechanism. It is well known that, in the presence of a severe structural disorder, the transport by hopping between localized states introduces dispersion- or frequency-dependent mechanisms. Recently it has been shown that electronically conducting polymers permeated with electrolyte display those effects at moderately high frequencies [1,2]. Therefore, for a realistic description of devices, it is required to extend Fickean diffusion models to account for the dependence of the local diffusivity on factors such as concentration and frequency. Here we describe a first step towards that goal,

© 2004 International Union of Pure and Applied Chemistry   DOI: 10.1002/masy.200450874

providing a simple method to determine the local diffusivity on a random energy-barrier model system.

Method

We consider a finite chain of N sites with randomly distributed energy barriers between them. A particle probability current dependent on time flows into site 1 and is extracted at site N. The rate equations governing the kinetics of hopping have the form:

$$\frac{\mathrm{d}}{\mathrm{d}t} P_i(t) = \Gamma_{i+1,i} P_{i+1}(t) + \Gamma_{i-1,i} P_{i-1}(t) - \left(\Gamma_{i,i+1} + \Gamma_{i,i-1}\right) P_i(t) \tag{1}$$

where $P_i(t)$ are the site occupation probabilities, and the transition rates are thermally activated with an activation energy related to the height of the barrier to be overcome, $\Gamma = \Gamma_0 \, e^{-\beta E_i}$, where $\beta = 1/k_B T$. This method deals with noninteracting particles, i.e. neither is Coulomb repulsion taken into account nor is self-exclusion regarded; however, these aspects can be readily incorporated into the formalism.

In order to derive the value of the diffusion coefficient, D, we take Fick's first law, $I = -D\nabla n$, where I is the particle current and n the concentration of particles. Then we calculate the diffusion coefficient from the gradient of particle probability at a given current input. We take this gradient from the master equation (1) linearized for a periodic perturbation ($\mathrm{d} / \mathrm{d}t \to i\omega$, where ω is the angular frequency). So we get the frequency-dependent diffusion coefficient in one dimension:

$$D(\omega) = \mathrm{Re}\left[\frac{IN}{P_1 - P_N}\right] \tag{2}$$

For low frequencies, this method gives the well-known value $D(\omega = 0) = \left\{1/\Gamma_{i,i+1}\right\}^{-1}$ [3].

Results

In Fig. 1 the results obtained from numerical calculation of Eq. (2) for a chain of 100 sites are compared with the analytical equation of the continuous time random walk (CTRW) approximation, which takes the form $D(\omega)/D(0) = [i\omega/\omega_m]/\log(1 + i\omega/\omega_m)$ [4], where $\omega_m = \sigma(0)p(E_c)/\beta$, $p(E_c)$ being the probability of the highest energy barriers in the distribution. Figure 1 shows that our simple method based on an extension of Fick's law is in agreement with the more sophisticated approaches available in the literature [4,5]. Therefore we obtain the main features of frequency-dependent diffusivity in a random energy barrier system [4], which can be summarized as follows.

The diffusivity becomes independent of the energy barrier probability distribution at low temperatures. The diffusivity becomes frequency-dependent when the frequency is of the order of ω_m or, in other words, the probability of the highest barrier to be overcome determines the onset of frequency-dependent diffusivity [5]. Then the most important contribution to the dc conductivity is due to the highest barrier, so that a greater part of the barriers become irrelevant to the low-frequency diffusivity. Since the frequency axes in Fig. 1 have not been rescaled in any way, we note that our method describes quite correctly the onset of frequency-dependent diffusivity.

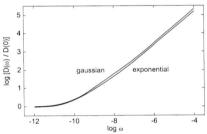

Figure 1. Log-log plot of numerical results of low-temperature normalized diffusivity in 1D system as a function of the frequency for two different energy-barrier distributions. Numerical results based on Eq. (2) (points) are compared with the CTRW approximation (lines).

Conclusions

We suggested a new and simple procedure to calculate the frequency-dependent diffusion coefficient in one dimension, which enables the application of Fick's law in the analysis of electron transport in organic conductors.

[1] G. Garcia-Belmonte, J. Bisquert, E. C. Pereira, F. Fabregat-Santiago, Appl. Phys. Lett. 78, 1185 (2001).
[2] G. Garcia-Belmonte, J. Bisquert, E. C. Pereira, F. Fabregat-Santiago, J. Electroanal. Chem. 508, 48 (2001).
[3] S. Alexander, J. Bernasconi, W. R. Schneider, R. Orbach, Rev. Mod. Phys. 53, 175 (1981).
[4] J. C. Dyre, J. Appl. Phys. 64, 2456 (1988).
[5] J. C. Dyre, Phys. Rev. B 49, 11709 (1994).

Macromol. Symp. **2004**, *212*, 575-580

Quenching of Exciplex and Charge-Transfer Complex Fluorescence of Poly(*N*-vinylcarbazole) - Benzanthrone System

R. M. Siegoczyński, * *W. Ejchart*

Faculty of Physics, Warsaw University of Technology, Koszykowa 75, 00-662
Warsaw, Poland
E-mail: siego@if.pw.edu.pl

Summary: When benzanthrone (Bt), a weak electron acceptor, is doped into poly(*N*-vinylcarbazole) (PVCz) solution or film, an excited carbazole chromophore (D*) interacts with Bt to form a new exciplex state, which gives a broad fluorescence band (λ_{max} = 440 nm) in solution and a new state, which gives broad fluorescence (λ_{max} = 550 nm) in the film. In order to elucidate the origin of these new states, we have studied the results of experiments for absorption, concentration dependence of the excimer and exciplex fluorescence quenching, both in solution and in the film, and electric field-induced fluorescence quenching in the film. Taking into account that (i) the new state formation in the PVCz film containing small amounts of Bt enhances the photocurrent in the absorption region, where the photon energy is insufficient to excite polymer molecules directly into the conduction state, (ii) the 550 nm fluorescence of the PVCz - Bt system in film is only partly quenched by electric field, (iii) the appearance of structureless tail in the fluorescence excitation spectrum, the charge transfer interaction model of the PVCz - Bt system in film is proposed.

Keywords: exciplex; fluorescence; photo-carrier generation; poly(*N*-vinylcarbazole)

Introduction

The problem of charge carrier photogeneration in organic semiconductors has been of great importance and interest in the fields of electrophotographic science and solar energy conversion technology. The fact that poly(*N*-vinylcarbazole) (PVCz) is one of the few photoconducting polymers with a very long carrier range[1], which makes it useful in electrophotography application, has stimulated great interest in its electrical and optical properties[2]. The carbazole side groups generally determine the properties, with the polymer chain configuration simply determining the relative position of the side groups. The emission

 DOI: 10.1002/masy.200450875

properties, energy transfer, and the carrier generation mechanism in this polymer have been studied.

Although the absorption of light in the near UV region of the spectrum is associated with an electronic transition in the monomer chromophore, the fluorescence in thin film samples of PVCz has been associated with emission from the two excimer-like emitting sites[3-6]. No monomer emission can be seen[4]. The broad emission at the longer wavelength ($\lambda_{max} = 415$ nm) is assigned to the sandwich-like excimer fluorescence and the broad emission in the shorter-wavelength region ($\lambda_{max} = 380$ nm) to the second excimer fluorescence[4-9]. The excimers occur in the samples only at special sites, where the side groups are favourably aligned.

When benzanthrone (Bt), a weak electron acceptor, is doped in a poly(N-vinylcarbazole) solution or film, an excited carbazole chromophore (D*) interacts with Bt to form a new exciplex state, which gives a broad fluorescence band ($\lambda_{max} = 440$ nm) in solution and a new state, which gives broad fluorescence ($\lambda_{max} = 550$ nm) in the film.

In order to elucidate the origin of these states, we have studied the results of experiments for absorption, concentration dependence of the excimer and exciplex fluorescence quenching, both in solution and in the film, and electric field-induced fluorescence quenching. Taking into account that (i) the new state formation in the PVCz film containing small amounts of Bt molecules enhances the photocurrent in the absorption region, where the photon energy is insufficient to excite the polymer molecules directly to the conduction state, (ii) the 550 nm fluorescence of the PVCz - Bt system in the film is only partly quenched by electric field, (iii) the appearance of a structureless tail in the fluorescence excitation spectrum, the C-T interaction model of PVCz - Bt system in film and the symmetric structure of this state is proposed.

Experiment

A. Materials

The required PVCz obtained by radiation polymerisation and commercially available Bt were dissolved in benzene and precipitated out with methanol several times. No interfering emission from the solvents (benzene and cyclohexanone) could be detected under the experimental conditions used here. The dilute polymer solutions (typically 10^{-2} - 10^{-3} M in monomer units) were deoxygenated prior to the study by purging with purified argon gas. The solvent-cast PVCz film (≈ 15 μm thick) containing a known amount of Bt on a NESA quartz plate was equipped with evaporated gold electrodes on the surface. All absorption spectra were recorded on a Specord UV.

B. Instrumentation and measurements

UV absorption spectra were recorded with a Unicam SP-500 spectrophotometer.

Photostationary emission and emission excitation spectra were recorded on a SF-1 Cobrabid spectrophotometer and corrected for photoresponse. Emission was viewed at about 45° to the exciting light in front of the samples.

Quenching of fluorescence measurements was carried out as shown in Figure 1. The square-wave voltage (20 – 1000 Hz) up to 220 V was applied to the samples and the modulated fluorescence was detected with a lock-in amplifier. The excitation wavelength, 360 nm was chosen to excite the bulk of the sample.

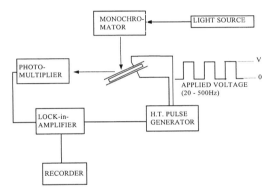

Figure 1. Schematic diagram of the instrumentation for fluorescence measurements in the presence of electric field.

Results and Discussion

In the pioneering work[10] Klöpffer determined the dominant mechanism for energy transfer to be the exciton diffusion. In this case the excitons are electronic excited states of the carbazole chromophores, which hop randomly from one side group to another on either the same or different polymer chains with the excimer sites. As the singlet exciton in PVCz covers many carbazole groups during its lifetime, a fairly high exciplex-forming efficiency can be expected for the π-π* excitation.

In the π-π* absorption region, re-excitation of the trapped carriers by excitons may operate, and, simultaneously, the field-assisted dissociation of the exciplex formed from an excited carbazole group and an electron-accepting compound in the ground state should also be considered.

It is known[8,11] that interaction between PVCz and Bt molecules enhances the photocurrent in the lowest π-π* absorption region of PVCz and we have suggested the following extrinsic carrier photogeneration process, i.e., the field-assisted thermal dissociation of an exciplex into free carriers:

$$D* + A \rightarrow (D^+ .... A^-)* \longrightarrow D^+ + A^- \tag{1}$$

and then

$$A^- + D \xrightarrow{\text{thermalization}} D^- + A \tag{2}$$

$$A^- + D* \xrightarrow{\text{singlet exciton}} D^- + A \tag{3}$$

where D and A represent an electron donor (PVCz) and acceptor (Bt) molecules in the ground state, respectively; $D*$, a singlet exciton in the PVCz film; A^- an immobile anion; D^+ and D^- are ion radicals of PVCz mobile and an immobile charge carrier, respectively. Equation (1) shows the ionisation of an exciplex into a mobile carrier ion in a film. Equations (2) and (3) show the excitation of an electron donor (A^- in this case) thermally or by a singlet exciton.

It is usually difficult for an exciplex to dissociate into free ions in a film with a low permittivity. This dissociation, however, may be possible under conditions, which reduce the energy required for the separation of an ion pair, for example under a strong electric field. Now we show that the 550 nm fluorescence decreased by applying the square-wave voltage. The decrement of this fluorescence $\Delta F = F(0) - F(E)$ dependent on the field strength and the ratio of the decrement to the total fluorescence intensity at 550 nm increases at the applied voltage 220 V.

In order to explain our results, we have attempted to compare these results with Onsager's theory, which predicts the probability, as a function of the field E, that an electron-hole pair produced at distance r_0 will escape geminate recombination.

Next, we consider that the electron-hole pair is assumed to be produced from nonrelaxed encounter exciplex state ($D^{\delta+}...A^{\delta-}$) by thermalization (Yokoyama's model[12]), as shown in Scheme 1:

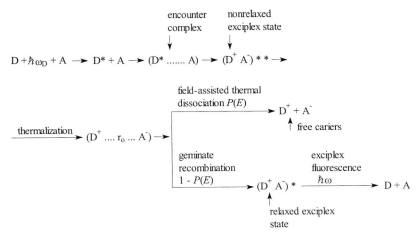

Scheme 1

If we assume Scheme 1 for the present case, the expression for the fluorescence decrement $\Delta F/F$ - using the Onsager formula[13] for the dissociation probability of an ion pair $P(E)$ at the initial separation r_0 is

$$\frac{\Delta F}{F} = \frac{F(0) - F(E)}{F(0)} = \frac{\eta\phi(1 - P(0)) - \eta\phi(1 - P(E))}{\eta\phi(1 - P(0))} = \frac{P(E) - P(0)}{1 - P(0)} \tag{4}$$

where $F(E)$ is the exciplex fluorescence intensity in field E; ϕ the quantum yield of the thermalized electron-hole pair; η the fluorescence quantum efficiency from the relaxed state; $P(E)$ the dissociation probability. The dissociation probability is given by the equation

$$P(E) = \int g(r)f(r,\theta,E)\,d\tau$$
$$g(r) = (4\pi r_0^2)^{-1}\delta(r - r_0) \tag{5}$$
$$f(r,\theta,E) = \exp(-2q/r)\exp[-\beta r(1 + \cos\theta)] \sum_{m,n=0}^{\infty} \frac{(2q)^m \beta^{m+n} r^n (1 + \cos\theta)^{m+n}}{m!(m+n)!}$$

where r_0 is the initial separation of the thermalized electron and hole, θ is the orientation of ion pair relative to the field direction, $q = e^2/2\varepsilon_0 kT$; $\beta = eE/2kT$; e is electron charge, ε_r is relative permittivity; and kT is the thermal energy.

In Figure 2 are shown the calculated curves $[P(E) - P(0)]/[1 - P(0)]$ for various initial separation r_0 with $\varepsilon_r = 3.0$ and $T = 300$ K, together with the experimental data of $\Delta F/F$ (black points).

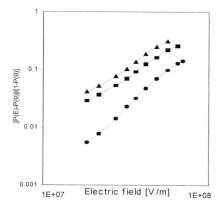

Figure 2. Calculated curves $[P(E) - P(0)]/[1 - P(0)]$ for various initial separation r_0 with $\varepsilon_r = 3.0$ and $T = 300$ K together with the experimental data for $\Delta F/F(0)$ (black points). $\Delta F/F(0)$ - the fluorescence decrement, $r_0 = 22$ Å (PVCz-DMTP)* exciplex, $r_0 = 28$ Å (PVCz-Bt) system, $r_0 = 30$ Å (PVCz$^+$-4CNB$^-$) C-T complex.

The experimental data of $\Delta F/F$ are well reproduced by the theoretical curve for the initial separation of the ion pair, $r_0 = 28$ Å. The r_0 value of 28 Å is between the value 30 Å obtained for (PVCz-4CNB) C-T complex and 22 Å obtained for (PVCz-DMTP) exciplex by Yokoyama et al.[12] (4CNB – tetracyanobenzene, DMTP – dimethyl terephthalate).

These results, showing emission due to the exciton-doped molecule interaction, not only provide a strong support for mechanism of the photocarrier generation via an exciplex and C-T complex in PVCz but may also provide an insight into the detailed processes in the currently accepted mechanism of the extrinsic photocarrier generation in organic molecular crystals.

[1]   R.C. Hughes, *IEEE Trans. Nucl. Sci.* **1971**, 18, 281.
[2]   W. Klöpffer, *Kunststoffe* **1971**, 61, 533.
[3]   G. Giro, G. Orlandi, *Chem. Phys.* **1992**, 160, 145.
[4]   G.E. Johnson, *J. Chem. Phys.* **1975**, 62, 4697.
[5]   W. Klöpffer and D. Fisher, *J. Polym. Sci.* **1973**, 40, 43.
[6]   P.C. Johnson and H.W. Offen, *J. Chem. Phys.* **1971**, 55, 2945.
[7]   A. Itaja et al., *Bull. Chem. Soc. Jpn.* **1976**, 49, 2082.
[8]   R.M. Siegoczyński et al., *Photophysics of Polymers and Electrophotography*, Charles University, Prague 1976, p.178.
[9]   R.M. Siegoczyński, J.Jędrzejewski, *Fifth International Seminar on Polymer Physics*, High Tatra 1987, p.100
[10] W. Klöpffer, *J. Chem. Phys.* **1969**, 50, 2337.
[11] R.M. Siegoczyński et al., *J. Mol. Struct.* **1978**, 45, 445.
[12] M. Yokoyama et al., *J. Chem. Phys.* **1981**, 75, 3006.
[13] L. Onsager, *Phys. Rev.* **1938**, 54, 39.